Continuous Semi-Markov Processes

Continuous Semi-Markov Processes

Boris Harlamov

Series Editor
Nikolaos Limnios

First published in Great Britain and the United States in 2008 by ISTE Ltd and John Wiley & Sons, Inc.

ISTE Ltd
6 Fitzroy Square
London W1T 5DX
UK

John Wiley & Sons, Inc.
111 River Street
Hoboken, NJ 07030
USA

www.iste.co.uk

www.wiley.com

Library of Congress Cataloging-in-Publication Data

Harlamov, Boris.
 Continuous semi-Markov processes/Boris Harlamov.
  p. cm.
 Includes index.
 ISBN 978-1-84821-005-9
 1. Markov processes. 2. Renewal theory. I. Title.
 QA274.7.H35 2007
 519.2'33--dc22

2007009431

British Library Cataloguing-in-Publication Data
A CIP record for this book is available from the British Library
ISBN: 978-1-84821-005-9

Printed and bound in Great Britain by Antony Rowe Ltd, Chippenham, Wiltshire.

Contents

Introduction

Semi-Markov processes are a generalization of Markov processes, having a proper method of investigation and a field of application. In the 1950s, some practical problems, especially problems of the mass service, compelled researchers to look for an adequate mathematical description. Attempts to apply Markov models to these problems were sometimes unsatisfactory due to unjustified conclusions about exponential distribution of corresponding time intervals. In order to reduce this limitation P. Lévy [LEV 54] and B. Smith [SMI 55] almost simultaneously proposed the class of stepped processes, called semi-Markov processes. For these processes the Markov property with respect to a fixed non-random time instant is not fulfilled in general cases, but they remain Markov processes with respect to some random times, for example, jump times of their trajectories. Therefore, such a process is not a Markov process, although it inherits some important properties of Markov processes.

Recently the theory of stepped semi-Markov processes has been developed more intensively. An exhaustive bibliography of corresponding works was compiled by Teugels [TEU 76] (see also Koroluk, Brodi, and Turbin [KOR 74], Koroluk and Turbin [KOR 76], Cheong, de Smit, and Togels [CHE 73], Nollau [NOL 80]). The list of works devoted to practical use of this model is very long (see Kovalenko [KOV 65], Korlat, Kuznecov, Novikov, and Turbin [KOR 91], etc.).

We will consider a more general class of random processes with trajectories in a metric space which are continuous from the right and have limits from the left at any point of the half-line. These processes, called semi-Markov processes (in the general sense), have the Markov property with respect to any intrinsic Markov time such as the first exit time from an open set or any finite iteration of such times. The class of semi-Markov processes includes as a sub-class all stepped semi-Markov processes, and also all strong Markov processes, among them continuous ones.

Apparently for the first time general semi-Markov processes [HAR 74] were defined in [HAR 71b] under the name of *processes with independent sojourn times*.

These processes were called continuous semi-Markov processes in [HAR 72] according to the property of their first exit streams. For these processes it is impossible or not convenient to use specific Markov methods of investigation such as semi-group theory, infinitesimal operators, differential parabolic equations or stochastic differential equations. Thus, it is necessary to develop new methods or to modernize traditional methods of investigation, which do not use the simple Markov property. Some methods are borrowed, firstly, from works on stepped semi-Markov processes (see Lévy [LEV 54], Pyke [PYK 61], Smith [SMI 55, SMI 66], etc.), and secondly, from works on Markov processes (see Dynkin [DYN 59, DYN 63], and also [KEM 70, BLU 68, MEY 67], etc.). Some ideas on non-standard description of processes are taken from works of [COX 67]. [HAR 69, HAR 71a] were attempts of such a description. An investigation of some partial types of continuous non-Markovian semi-Markov processes can be found in [CIN 79]. Problems of ergodicity for processes with embedded stepped semi-Markov processes are investigated by Shurenkov [SHU 77].

The simplest element of description of the semi-Markov process is the pair: the first exit time and the corresponding exit position for the process leaving some open set. However, from the classical point of view, such a pair is not simple because it is not convenient to describe it by finite-dimensional distributions of the process ([KOL 36], [GIH 71], etc.). One of the problems we solve in this book is to eliminate this difficulty by means of changing initial elementary data on the process. In our case distributions of the above first exit pairs or joint distributions for a finite number of such pairs becomes the initial data. From engineering point of view it is not difficult to make a device which fixes first exit times and their compositions, and values of the process at these times. It makes it possible to describe the process in a simple, economic and complete way. Such an approach also has some theoretical advantages over the classic approach. In terms of the first exits the continuity problem of the process has a trivial solution. In addition, the necessary and sufficient conditions for the process to converge weakly are simpler than in classical terms. The same can be said for time change problems. However, when accepting this point of view, one must be ready to meet difficulties of another kind. Efforts in this direction would hardly be justified if there did not exist a class of processes such that their description in terms of first exit times would be natural and convenient. This is the class of semi-Markov processes. For them joint distributions of a finite number of the first exit pairs are determined by repeatedly integrated semi-Markov kernels. The semi-Markov process possesses the Markov property with respect to the first exit time from an open set. It is a simple example of such times. The most general class of such times is said to be the class of intrinsic Markov times, which plays an important role in theory of semi-Markov processes.

One of the main features of the class of semi-Markov processes is its closure with respect to time change transformation of a natural type, which generally does not preserve the usual Markov property. One of the main problems in the theory of semi-Markov processes is the problem of representing such a process in the form of a Markov process, transformed by a time change. At present a simple variant of this problem for stepped semi-Markov process (Lévy hypothesis) under some additional assumption is solved (see [YAC 68]). In general, this problem is not yet solved.

General semi-Markov processes are the natural generalization of general Markov processes (like stepped semi-Markov processes generalize stepped Markov processes). Both of these are mathematical models of real phenomena in the area of natural and humanitarian sciences. They serve to organize human experience and to extend the prognostic possibility of human beings as well as any mathematical model. Evidently semi-Markov models cover a wider circle of phenomena than Markov processes. They are more complex due to the absence of some simplifying assumptions, but they are more suitable for applications for natural phenomena than Markov processes. The use of semi-Markov models makes it possible to give a quantitative description for some new features of the process trajectories, which are not accessible using the Markov model.

The characteristic feature of the semi-Markov process is a set of intervals of constancy in its trajectory. For example, such simple operation as truncation transforms Wiener process W into non-Markov continuous semi-Markov process W_a^b:

$$W_a^b(t) = \begin{cases} a, & W(t) \leq a, \\ W(t), & a < W(t) < b, \quad (t \geq 0) \\ b, & W(t) \geq b, \end{cases}$$

where $a < b$. This process contains intervals of constancy on two fixed levels, but the most typical semi-Markov process possesses intervals of constancy on random spatial positions. In general, such intervals are a consequence of the property of so-called "time run" process with respect to the sequence of states (trace) of the process, which has conditionally independent increments. The famous result of Lévy requires for such a time run process a Poisson field of jumps, which turn into intervals of constancy for an original process itself. This theoretical result finds an unexpected interpretation in chromatography (see [GUT 81]) and in other fields in engineering, where flow of liquid and gas in a porous medium is being used. The continuous semi-Markov model for this movement is reasonable due to the very small inertia of particles of the substance to be filtered. Therefore, the assumption about independence of time intervals, which are taken by non-overlapping parts of the filter while the particle is moving through the filter, is very natural. Moreover, the chromatography curve is a record of the distribution of the first exit time of the particle from the given space interval. Well-known interpretations of intervals of constancy of the particle move trajectory are intervals of delay of particles on the hard phase (convertible adsorption). Such a movement is the experimental fact.

A mathematical theory of reliability is another field of application of continuous semi-Markov processes. For example, the adequate description of an abrasive wear can be given by the inverse process with independent positive increments, i.e. a partial case of a continuous semi-Markov process [VIN 90]. The advantage of such a description appears when analyzing some problems of reliability [HAR 99].

General semi-Markov processes and, in particular, continuous ones can find application in those fields, where the stepped semi-Markov processes are applied, because the former is the limit for a sequence of the latter. Another possible field of application appears when one looks for the adequate model for a Markov-type process, transformed by varying its parameters (see [KRY 77]). The class of semi-Markov processes is more stable relative to such transformations than to Markov ones.

There are many problems of optimal control for random processes, which can be reduced to an optimal choice of a Markov time in order to begin a control action (see [ARK 79, GUB 72, DYN 75, KRY 77, LIP 74, MAI 77, ROB 77, SHI 76]). As a rule, such a Markov time is the first exit time of the process from some set of states. However, the consequence of such a control may be loss of Markovness. An example of such a control is time change depending on a position of the system. Let $(\xi(t))$ be a stationary random process, and $(\psi(t))$ be a time change, i.e. a strictly increasing map $\mathbb{R}_+ \to \mathbb{R}_+$, where $\psi(0) = 0$, $\psi(t) \sim t$ $(t \to \infty)$. Let $(\tilde{\xi}(t))$ be the transformed process. Then

$$\int_0^T \tilde{\xi}(t)\,dt = \int_0^{T'} \xi(t)\psi'(t)\,dt,$$

where $T' = \inf \psi^{-1}[T, \infty) \sim T$ $(T \to \infty)$. The average in time of the transformed process can be more than that of the original process, if $\psi'(t)$ is chosen in a suitable form. For example, it would be the case, if $\psi'(t) = f(\xi(t))$, where f is an increasing function. In this case ψ is uniquely determined if the function ξ is determined. The similar effect of enlargement of the average meaning is possible, when under fixed ξ the time change is some non-degenerated random process. For example, let $\psi'(t)$ be a Poisson process with the intensity $f(\xi(t))$ (Cox process [COX 67]). Then

$$E\int_0^T \tilde{\xi}(t)\,dt = E\int_0^{T'} \xi(t)\,d\psi(t)$$

$$= E\int_0^{T'} \xi(t)f\big(\xi(t)\big)\,dt \sim \int_0^T \xi(t)f\big(\xi(t)\big)\,dt \quad (T \longrightarrow \infty).$$

If the first form determines the random time change, then the second form of time change is twice random. The main distinction of these two kinds of time changes is in their relation to Markovness. The former preserves the Markov property and the latter does not. The process of the second type appears to be semi-Markovian, and the class of such processes is invariant with respect to such a time change. An answer on the practical question about preference of the time change form depends on engineering possibilities.

Let us pay attention to another aspect of the problem. In [HAR 90] the possibility of checking the hypothesis on Markovness of a one-dimensional continuous process is discussed. It was noted that to reveal Markovness of a process with an uncountable set of states is not possible in principle, when probability of the process to be at when any partial one-dimensional distribution of the process is continuous. In addition, it is impossible to check this hypothesis for discrete time and continuous set of states, and also that for continuous time and multi-dimensional $(d \geq 2)$ space of states. It follows from the fact that it is impossible to have a reasonable statistical estimate for conditional probability $P(A|\xi(t) = x)$ for a random process ξ, if $P(\xi(t) = x) = 0$ for any $t \in \mathbb{R}$. The last condition can possibly be true for the one-dimensional process too. However, in this case, in order to check Markovness we can use the infinite sequence of random times, when the trajectory of the process crosses the level x. We can organize a sequence of Markov times containing iterated first exit times from interval (∞, x). For a strictly Markov process any such sequence of marked point processes determines a stepped semi-Markov process. The first test must be checking this property of the point process, although it does not guarantee Markovness of the original process. Therefore, if the first test is positive, the second test must check Markovness with the help of distributions of the first exit times from small neighborhoods of an initial point. For a proper Markov process this distribution has a special limit property while diameters of the neighborhoods tend to zero. Later on we consider two corresponding criteria of Markovness for semi-Markov processes.

It seems to be convincing enough that continuous semi-Markov processes are worth investigating. Their practical usefulness is connected with numerical calculations. Here we will not cover computer problems. The main theme of this book is devoted to theoretical aspects of continuous semi-Markov models. It happens that there are a lot of such aspects.

The book is divided into chapters, sections, subsections, and paragraphs. In every chapter there is its own enumeration of divisions, and as well that of theorems, lemmas, and propositions. All these items, except subdivisions and paragraphs, have double numbers. For example, in 3.14 the number 3 means chapter, and 14 the number of this object in this chapter. Subdivisions have triple numbers. For example, 5.2.3 means the follows: 5 is the chapter number, 2 is that of the section, 3 is the number of subsection in the section. Paragraphs have single numbers. A reference on a paragraph inside a given chapter is given as item 9, where 9 is its number. A reference to a paragraph from another chapter is given as item 2.4, where 2 is the chapter number, and 4 is the paragraph number within this chapter.

It is necessary to include in Chapter 1 some information about stepped semi-Markov processes. It will be assumed that the reader is familiar with this concept. The main focus will be on constructing a measure of the stepped semi-Markov process with the help of the method, which will be further generalized for the general semi-Markov processes. In addition, we give without proof the main result of the theory of stepped semi-Markov processes, namely the ergodic theorem and the corresponding formula of their stationary distribution.

In Chapter 2 a method of investigation of a process with the help of embedded point processes (streams) of the first exits relative to a sequence of subsets of the given metric space of states is discussed. Deducing sequences of subsets, the main tool for analyzing properties of random processes connected with the first exit times, are defined. The structure of the set of regenerative times for a measurable family of probability measures is investigated. Such a name is assigned to a set of Markov times for which the Markov property for the given family of measures is fulfilled for.

In Chapter 3 general semi-Markov processes are defined and investigated. Such a process is characterized by a measurable family of probability measures with the set of regenerative times which includes any first exit time from an open set. Semi-Markov transition functions, transition generating functions, and lambda-characteristic operators are considered. The conditions necessary for a semi-Markov process to be Markov are analyzed.

In Chapter 4 the semi-Markov process is constructed on the base of *a priori* given family of semi-Markov transition functions. It is considered as a special case of more general problem how to construct a process from a consistent system of marked point processes, namely, that of streams which can be interpreted as the streams of the first exit pairs (time, position) from open sets for some process.

In Chapter 5 semi-Markov diffusion processes in a finite-dimensional space are considered. They enable the description in terms of partial differential elliptical equations.

In Chapter 6 properties of trajectories of semi-Markov processes are investigated. For any trajectory, a class of trajectories is defined in such a way that a trajectory from this class differs from the origin one only by some time change. This class can be interpreted as a sequence of states which the system goes through without taking into account the time spent on being in these states. We call this the trace of the trajectory. The individual trajectory from this class is distinguished by time run along the trace. The measure of the semi-Markov process induces a distribution on the set of all traces. This distribution possesses the Markov property of a special kind. In this aspect the projection is related to the Markov process. The special property of the semi-Markov process is exposed by its conditional distribution of time run along the trace. The character of this distribution is reflected in properties of trajectories of the semi-Markov process.

In Chapter 7 we consider conditions for the sequence of stepped semi-Markov processes to converge weak to a general semi-Markov process. Results related to semi-Markov processes are obtained as a consequence of general theorems about weak compactness and weak convergence of probability measures in the space $\mathcal{D}$. These theorems in terms of the first exits times have some comparative advantages with the similar results in terms of the usual finite-dimensional distributions.

In Chapter 8 a problem of representation of the semi-Markov process in the form of a Markov process, transformed by a random time change, is solved. We demonstrate two methods of solving this problem with construction of the corresponding Markov process and time change. The first solution is formulated in terms of the so called lambda-characteristic operator of this process. The corresponding Markov process is represented by its infinitesimal operator. The second solution is obtained in terms of Lévy decomposition of the conditional process with independent positive increments (process of time run). The corresponding Markov process is determined by the distribution of its random trace, and that of the conditional distribution by its time run along the trace. These distributions are used to derive formulae of stationary distributions of corresponding semi-Markov processes.

In Chapter 9 we consider some applications of continuous semi-Markov processes. It seems natural that they can serve as an adequate model for carrying substance through a porous medium. This follows from both some premises about independence, and experimental data of such a phenomenon as chromatography. The latter gives the purest example of a continuous semi-Markov process. We develop semi-Markov models for both the liquid and gas chromatography. The similar application has a geological origin. We consider a continuous semi-Markov model of accumulation of a substance which consists of particles moving up to some non-Markov time and remaining in the stop position forever.

According to accepted convention we use denotations like $P(A, B)$ instead of $P(A \cap B)$. We also will omit braces in expression like $A = \{X_n \in C\}$ as an argument of a measure or expectation: $P(A) = P(X_n \in C)$, $E(f; A) = E(f; X_n \in C)$, and so on.

The author would like to express their deep appreciation for the help given by Yu. V. Linnik, who supported the first works of the author on continuous semi-Markov processes. The author expresses sincere gratitude to all participants of the seminar on probability theory and mathematical statistics at St Petersburg Department of Steklov Mathematical Institute of the Russian Academy of Sciences, and especially to I. A. Ibragimov for their assistance with this work.

Chapter 1

Stepped Semi-Markov Processes

Before investigating general semi-Markov processes, we consider some properties of stepped semi-Markov processes. In this special case we show a structure of the class of the first exit times and corresponding first exit positions of the process. For stepped semi-Markov processes the main element of their constructive description is called a semi-Markov kernel determining the distribution of the pair, consisting of a pair of a time and a position of the process just after the first jump from the initial point. With the help of these kernels it is possible construct the family of probability measures on the set of all stepped right-continuous functions, depending on the initial points of the functions. How are measures of the family coordinated? The answer to this question for the family of semi-Markov distributions (semi-Markov family) follows easily from the construction of an individual measure of the family. In the case of general semi-Markov processes this coordination condition will be accepted as an essential part of the definition.

1.1. Random sequence

1. Random element of a set

According to the fundamental work of Kolmogorov [KOL 36] a probability model begins with a set of elementary events Ω. In this set the system $\mathcal{A}$ of subsets (events) is chosen, representing sigma-algebra, and on which a normalized measure, P, is given. This measure defines probability $P(B)$ of any event $B \in \mathcal{A}$. While defining a random element of some set S, they assume that in S a sigma-algebra of subsets $\mathcal{B}$ is also chosen. Then any measurable function $v : \Omega \to S$ (defined on Ω and having values in S) is said to be a stochastic element of a set S. Take a simple example: a random variable is a measurable map $v : \Omega \to \mathbb{R} \equiv (-\infty, \infty)$ with respect to Borel sigma-algebra of subsets $\mathcal{B}(\mathbb{R})$ on $\mathbb{R}$. The basic characteristic of a random element of a set S (sometimes called a S-valued random variable) is its distribution, namely a new probability measure Q, given on $\mathcal{B}$, connected with the initial measure P by the condition: $(\forall B \in \mathcal{B})\ Q(B) = P(v^{-1}(B))$ where, according to the definition of a measurability of map v, the pre-image $v^{-1}(B)$ of the set B belongs to $\mathcal{A}$, and hence the measure P is defined on it.

In what follows, probability constructions in frames of the given model are connected only with the measure Q or with a set of such measures on the same sigma-algebra $\mathcal{B}$. The connection between them is either not taken into account absolutely or is formulated in terms of these measures. If it is true then it is natural to accept as $(\Omega, \mathcal{A})$ the measurable space $(S, \mathcal{B})$ itself. A stochastic element in this case is each point $s \in S$. To be more precise, in order to remain within the framework of the initial nomenclature, it is the identity mapping $E : S \to S$ $((\forall s \in S)\ E(s) = s)$. It implies the other definition of a stochastic element of a set S: it is a measurable space with a probability measure $(S, \mathcal{B}, Q)$. A point from this space, called a realization (or as a selective value) of the random element, is not a part of the probability theory, which studies only probabilities of various measurable subsets. Any question on an individual realization usually concerns a probability for a measurable subset from S to contain or not contain this realization.

2. Random sequence

According to the definition of a random element of any set, the random sequence of points from a set $\mathbb{X}$ is a triple $(Z, \mathcal{F}, P)$, where Z is a space of all such sequences: $z \in Z \Leftrightarrow z = (z_0, z_1, z_2, \ldots)$ $(z_k \in \mathbb{X})$, $\mathcal{F}$ is a sigma-algebra of subsets of a set Z, and P is a probability measure on $\mathcal{F}$. Let $X_n(z) = z_n$ be a n-th term of a sequence z (n-th coordinate). Thus, X_n is a map $Z \to \mathbb{X}$. The natural requirement to sigma-algebra $\mathcal{F}$ consists of measurability of this map. So we assume the set $\mathbb{X}$ to be measurable, i.e. a sigma-algebra of its subsets is determined. We will assume that $\mathbb{X}$ is a complete, separable, locally compact metric space, with Borel sigma-algebra of subsets $\mathcal{B}(\mathbb{X})$. In particular, if $\mathbb{X}$ is finite or countable (infinite), then $\mathcal{B}(\mathbb{X})$ is generated by all distinct points. Let $\mathbb{N} = \{1, 2, 3, \ldots\}$ and $\mathbb{N}_0 \equiv \mathbb{N} \cup 0$. The fact of a measurability of any map X_n $(n \in \mathbb{N}_0)$ makes σ-algebra $\mathcal{F}$ rich enough, containing any event which can be connected with sequences. Let $\mathcal{F} = \sigma(X_n, n \in \mathbb{N}_0)$, i.e. the least σ-algebra, with respect to which all maps X_n are measurable. The measure P is defined for all events of the form

$$B_n = \{X_0 \in A_0, X_1 \in A_1, \ldots, X_n \in A_n\} \quad (A_k \in \mathcal{B}(\mathbb{X})). \qquad [1.1]$$

On the other hand, the coordinated meanings of P on such events determines this measure on the sigma-algebra $\mathcal{F}$ uniquely. This statement (with a definition of the coordination) is a special case of the Kolmogorov theorem, which in turn is a special example of the theorem about extension of a measure from an algebra on the generated sigma-algebra. It is a basis of the constructive determination of a measure P on $(Z, \mathcal{F})$. Such representation of a measure is natural for a sequence of independent random elements and for a sequence of elements connected in a Markov chain.

3. Sequence of independent random variables

In this case

$$P(B_n) = P(X_0 \in A_0) P(X_1 \in A_n) \cdots P(X_n \in A_n).$$

The constructive representation of P reduces to a sequence of probability distributions (measures) $(F_n)_{n=0}^{\infty}$ on $\mathcal{B}(\mathbb{X})$ where

$$F_k(A_k) = P(X_k \in A_k) = (P \circ X_k^{-1})(A_k) \quad (k \geq 0)$$

and $P(B_n) = F_0(A_0) \cdots F_n(A_n)$. Here and later on we use a denotation $f \circ g$ for a superposition of two functions: $(f \circ g)(x) = f(g(x))$. In particular, all measures F_n can be identical (case independent and identically distributed members of a sequence).

4. Conditional probabilities

According to the definition

$$P(B_1) = \int_{X_0 \in A_0} P(X_1 \in A_1 \mid X_0) P(dz),$$

where $P(C \mid X_0)$ is a conditional probability of an event C with respect to $\mathbb{X}$-valued random variable X_0. This formula can be rewritten with a change of variables. It is known (see Neveu [NEV 69]) that $P(C \mid X_0)$ as a X_0-measurable function can be represented as $g_C(X_0)$, where (with fixed $C \in \mathcal{F}$) g_C is a measurable function on $\mathbb{X}$. Then by a rule of a change of variables

$$\int_{X_0^{-1} A_0} g_C(X_0(z)) P(dz) = \int_{A_0} g_C(x) (P \circ X_0^{-1})(dx) = \int_{A_0} g_C(x) F_0(dx).$$

We will also use more intuitive notation in this and similar cases:

$$g_C(x) = P(C \mid X_0 = x), \quad F_0(dx) = P(X_0 \in dx).$$

Thus, we obtain

$$P(B_1) = \int_{A_0} P(X_1 \in A_1 \mid X_0 = x) P(X_0 \in dx).$$

In a more common case we have

$$P(B_n) = \int_{A_0 \times A_0 \times \cdots \times A_{n-1}} P(X_n \in A_n \mid X_0 = x, \ldots, X_{n-1} = x_{n-1})$$

$$\times P(X_{n-1} \in dx_{n-1} \mid X_0 = x, \ldots, X_{n-2} = x_{n-2}) \cdots$$

$$\times P(X_1 \in dx_1 \mid X_0 = x) P(X_0 \in dx_0).$$

This generalization of the previous formula represents an integrated variant of the formula of total probability in terms of conditional distributions. It can also be used for a constructive representation of a measure P. Namely, we determine F_0, and also for any $n \geq 0$ the kernel $F_n(A \mid x_0, \ldots, x_n)$, where given $(x_0, \ldots, x_n) \in \mathbb{X}^{n+1}$ it is a probability measure on $\mathcal{B}(\mathbb{X})$, and given $A \in \mathcal{B}(\mathbb{X})$ it is a $\mathcal{B}(\mathbb{X}^{n+1})$-measurable function on $\mathbb{X}^{n+1}$ with a value in $[0, 1]$. With the help of these functions the probability is determined

$$P(B_n) = \int_{A_0 \times A_0 \times \cdots \times A_{n-1}} F_n(A_n \mid x_0, \ldots, x_{n-1}) F(dx_{n-1} \mid x_0, \ldots, x_{n-2}) \cdots$$

$$\times F_1(dx_1 \mid x_0) F_0(dx_0).$$

The collection of such probabilities with fixed n determines a probability measure P_n on a set of all finite sequences $(z_0, z_1, \ldots, z_n)$. The set $(P_n)_{n=0}^{\infty}$ is consistent in the sense that $P_n(B'_{n-1}) = P_{n-1}(B_{n-1})$, if $B_{n-1} = \{X_0 \in A_0, \ldots, X_{n-1} \in A_{n-1}\}$ and $B'_{n-1} = \{X_0 \in A_0, \ldots, X_{n-1} \in A_{n-1}, X_n \in \mathbb{X}\}$. Such a set (P_n) determines a unique probability measure P on $\mathcal{F}$ (Ionescu-Tulcea theorem; see Neveu [NEV 69]), for which $P(B_n) = P_n(B_n)$.

1.2. Markov chain

5. Markov sequence

The constructive representation of a measure P in terms of conditional distributions is most natural for a Markov sequence. In this case

$$P(X_{n+1} \in A_{n+1} \mid X_0, \ldots, X_n) = P(X_{n+1} \in A_{n+1} \mid X_n)$$

for any one $n \in \mathbb{N}_0$. The equality is understood almost everywhere on Z with respect to the measure P. It means

$$P(X_{n+1} \in A_{n+1} \mid X_0 = x_0, \ldots, X_n = x_n) = P(X_{n+1} \in A_{n+1} \mid X_n = x_n)$$

almost everywhere on $\mathbb{X}^{n+1}$ concerning a measure $P \circ (X_0, \ldots, X_n)^{-1}$, where

$$\left(P \circ (X_0, \ldots, X_n)^{-1} \right) (A_0 \times \cdots \times A_n) = P(X_0 \in A_0, \ldots, X \in A_n).$$

The theory of Markov sequences (chains) is a detailed branch of the probability theory (see Kemeny and Snell [KEM 70], etc.). We consider these chains only to develop the Markov formalism (regeneration property) for a family of measures which will be convenient later.

6. Shift and Markov property

Besides the map $X_n : Z \to \mathbb{X}$ we will also consider map $\theta_n : Z \to Z$, where $\theta_n(z_0, z_1, z_2, \ldots) = (z_n, z_{n+1}, z_{n+2}, \ldots)$ (shift). Thus $(\forall z \in Z) \, X_k(\theta_n z) = z_{n+k} = X_{n+k}(z)$, which it is possible to note as functional equality $X_k \circ \theta_n = X_{n+k}$ $(k, n \in \mathbb{N}_0)$. Furthermore, the event $\{X_{n+k} \in A\}$ can be noted as $\{X_k \circ \theta_n \in A\} = \theta_n^{-1}\{X_k \in A\}$. And hence

$$\{X_{n+1} \in A_{n+1}, \ldots, X_{n+k} \in A_{n+k}\}$$
$$= \bigcap_{i=1}^{k} \theta_n^{-1}\{X_i \in A_{n+i}\} = \theta_n^{-1} \bigcap_{i=1}^{k} \{X_i \in A_{n+i}\}.$$

On the other hand, by the Markov property for any $k \in \mathbb{N}$ and $A_{n+i} \in \mathcal{B}(\mathbb{X})$

$$P\left(\bigcap_{i=1}^{k} \theta_n^{-1}\{X_i \in A_{n+i}\} \mid X_0, \ldots, X_n\right)$$
$$= P\left(\bigcap_{i=1}^{k} \theta_n^{-1}\{X_i \in A_{n+i}\} \mid X_n\right) = P\left(\theta_n^{-1} \bigcap_{i=1}^{k} \{X_i \in A_{n+i}\} \mid X_n\right).$$

In addition

$$P\left(X_n \in A_n, \theta_n^{-1}B \mid X_0, \ldots, X_n\right) = I_{\{X_n \in A_n\}}P\left(\theta_n^{-1}B \mid X_0, \ldots, X_n\right),$$
$$P\left(X_n \in A_n, \theta_n^{-1}B \mid X_n\right) = I_{\{X_n \in A_n\}}P\left(\theta_n^{-1}B \mid X_n\right) \quad (B \in \mathcal{F}),$$

where

$$I_A(z) = \begin{cases} 1, & z \in A, \\ 0, & z \notin A. \end{cases}$$

From here it is possible to note

$$P\left(\theta_n^{-1} \bigcap_{i=0}^{k} \{X_i \in A_{n+i}\} \mid X_0, \ldots, X_n\right) = P\left(\theta_n^{-1} \bigcap_{i=0}^{k} \{X_i \in A_{n+i}\} \mid X_n\right).$$

Since $\mathcal{F}$ is generated by events of the form $\bigcap_{i=0}^{k}\{X_i \in A_i\}$, for any event $B \in \mathcal{F}$ we have:

$$P\left(\theta_n^{-1}B \mid X_0, \ldots, X_n\right) = P\left(\theta_n^{-1}B \mid X_n\right). \tag{1.2}$$

7. Notation for integrals

The following denotation relates to an integral of a measurable real function by a measure. If the measure is probability, this integral is called an expectation of this function (random variable). Let μ be a measure on measurable space $(S, \mathcal{B})$, f a measurable numerical function on S and $A \in \mathcal{B}$. Designate

$$\mu(f; A) = \int_A f \, d\mu. \qquad [1.3]$$

This denotation, common for the general theory of a measure, is convenient for dealing with many measures. In a special case, when $\mu = P$, a probability measure, this integral refers to an expectation of a random variable $f \cdot I_A$ and is designated as $E(f \cdot I_A) = E(f; A)$.

8. A Markov property in terms of integration

Using the previous denotations according to the definition of conditional probability, we can write

$$P(\theta_n^{-1} B, A) = E(P(\theta_n^{-1} B \mid X_0, \ldots, X_n); A),$$

where $A = \{X_0 \in A_0, \ldots, X_n \in A_n\}$. Under the theorem of extension of a measure (see [KOL 36]) this equality is also true for any $A \in \mathcal{F}_n \equiv \sigma(X_k, k \leq n)$. In a Markov case for these A we have

$$P(\theta_n^{-1} B, A) = E(P(\theta_n^{-1} B \mid X_n); A).$$

Thus, the following equality can be considered as a definition of the Markov sequence:

$$P(\theta_n^{-1} B, A) = E(P(\theta_n^{-1} B \mid X_n); A) \qquad [1.4]$$

for any $n \in \mathbb{N}$, $B \in \mathcal{F}$, $A \in \mathcal{F}_n$.

Furthermore, we shall encounter cases when this equality is fulfilled for n random (i.e. depending on z). We refer to such deterministic or random numbers as regeneration times. Thus, the Markov sequence is a sequence for which any fixed number (the constant instant) is a regeneration time.

9. Markov family of measures

To specify the concept of regeneration, we define a homogenous Markov property and pass from one measure to a coordinated family of measures. Let us assume that for any $n \in \mathbb{N}$ there is a transition function $F_n \colon \mathbb{X} \times \mathcal{B}(\mathbb{X}) \to [0,1]$, i.e. a kernel $F_n(A \mid x) \in [0,1]$ $(x \in \mathbb{X},\ A \in \mathcal{B}(\mathbb{X}))$, such that

$$P\big(X_n \in A_n \mid X_{n-1}\big) = F_n\big(A_n \mid X_{n-1}\big)$$

P-almost surely (a.s.). From here it follows that $(P \circ X_{n-1}^{-1})$-a.s. on $\mathcal{B}(\mathbb{X})$

$$P\big(X_n \in A_n \mid X_{n-1} = x\big) = F_n\big(A_n \mid x\big) \quad (x \in \mathbb{X}).$$

Thus, for a Markov sequence we have the equality

$$P\big(B_n\big) = \int_{A_0} F_{1,\dots,n}\big(A_1 \times \cdots \times A_n \mid x\big) F_0(dx),$$

where $F_0 \equiv P \circ X_0^{-1}$ and F_0-a.s.

$$F_{1,\dots,n}\big(A_1 \times \cdots \times A_n \mid x\big)$$
$$= \int_{A_0 \times \cdots \times A_{n-1}} F_n\big(A_n \mid x_{n-1}\big) F_{n-1}\big(dx_{n-1} \mid x_{n-2}\big) \cdots F_1\big(dx_1 \mid x\big)$$
$$= P\big(X_1 \in A_1, \dots, X_n \in A_n \mid X_0 = x\big).$$

The latter probabilities determine a probability measure $P_x^{(1)}$ on the set of sequences, starting from the point x. In other words, the family of transition functions $(F_n)_1^\infty$ determines a distribution of a Markov sequence within the initial point, i.e. the family of distributions $(P_x^{(1)})_{x \in \mathbb{X}}$. The same method is applicable for determining a family of distributions $(P_x^{(m)})_{x \in \mathbb{X}}$ with the help of the subfamily $(F_n)_m^\infty$, where $m \geq 1$. The connection between these families can be found, considering $P_x^{(1)}$-measure of the event B_{m+n} of the form [1.1]. This event can be represented as $B_m \cap \theta_m^{-1} B_n'$ where $B_n' = \bigcap_{i=1}^n \{X_i \in A_{m+i}\}$. Thus

$$P_x^{(1)}\big(B_{m+n}\big) = E_x^{(1)}\Big(P_{X_m}^{(m+1)}(B_n');\, B_m\Big).$$

From here for any $x \in \mathbb{X},\ B \in \mathcal{F},\ C \in \mathcal{F}_m$ it is true that

$$P_x^{(1)}\big(\theta_m^{-1}B, C\big) = E_x^{(1)}\Big(P_{X_m}^{(m+1)}(B); C\Big).$$

As in the previous case we obtain the equality

$$P_x^{(n+1)}\big(\theta_{n+m}^{-1}B, C\big) = E_x^{(n+1)}\Big(P_{X_{m+n}}^{(m+n+1)}(B); C\Big), \qquad [1.5]$$

where $m, n \in \mathbb{N}$ and $C \in \mathcal{F}_{m+n}$. It is a consistency condition for the family of probability measures $(P_x^{(n)})$ $(n \in \mathbb{N}_0,\ x \in \mathbb{X})$.

10. Condition of regeneration

The family of transition functions (F_n) determines the consistent family of distributions $(P_x^{(n)})_{x \in \mathbb{X}}$, $n \in \mathbb{N}$. We deal generally with a homogenous family of transition functions, when $F_n = F_1 \equiv F$ and with an appropriate homogenous (in time) family of distributions on $\mathcal{F} : P_x^{(n)} = P_x^{(1)} \equiv P_x$. For such a family we have

$$P(\theta_n^{-1} B \mid X_n = x) = P_x(B),$$

$(P \circ X_n^{-1})$-a.s., and the Markov property can be defined as

$$P(\theta_n^{-1} B \mid X_0, \ldots, X_n) = P_{X_n}(B),$$

P-a.s., and also for any $x \in \mathbb{X}$, $B \in \mathcal{F}$, $A \in \mathcal{F}_n$

$$P_x(\theta_n^{-1} B, A) = E_x(P_{X_n}(B); A) \quad (\forall n \in \mathbb{N}_0). \qquad [1.6]$$

It is a condition of coordination inside the family of measures (P_x). We call this property a condition of regeneration.

We came to another definition of a Markov sequence when it is determined within the initial point by a consistent Markov family of probability measures. This point of view on the process will be used to define a general semi-Markov process. Let us emphasize that in the case of Markov sequences, the family of distributions (P_x) and the transition function F determine each other.

11. Distributions of the first exit

Let us consider the first exit time of the sequence z from the set $\Delta \in \mathcal{B}(\mathbb{X})$:

$$\sigma_\Delta(z) = \min \{n \geq 0 : z_n \notin \Delta\},$$

and X_{σ_Δ} is the corresponding first exit position for the case, when $\sigma_\Delta(z) < \infty$: $X_{\sigma_\Delta}(z) = X_{\sigma_\Delta(z)}(z)$. We assume $\sigma_\Delta(z) = 0$ and $X_{\sigma_\Delta}(z) = z_0$, if $z_0 \notin \Delta$. For any $m \in \mathbb{N}$ we have

$$\{\sigma_\Delta \geq m\} = \bigcup_{k=m}^{\infty} \{X_0 \in \Delta, \ldots, X_{k-1} \in \Delta, \ X_k \notin \Delta\} \in \mathcal{F},$$

and also $\{X_{\sigma_\Delta} \in A\} \in \mathcal{F}$, where $A \in \mathcal{B}(\mathbb{X})$. Let

$$F_{\sigma_\Delta}(m, A \mid x) = P_x(\sigma_\Delta \leq m, \ X(\sigma_\Delta) \in A),$$

Then it follows from the definition of the first exit time that for $z_0 \in \Delta$, $\sigma_\Delta(z) = 1 + \sigma_\Delta(\theta_1(z)) \geq 1$ and $X_{\sigma_\Delta} = (X_{\sigma_\Delta} \circ \theta_1)(z)$. From here for $1 \leq m$ we have

$$P_x\big(\sigma_\Delta \leq m,\, X_{\sigma_\Delta} \in A\big)$$
$$= P_x\big(\sigma_\Delta = 1,\, X_1 \in A\big) + P_x\big(1 < \sigma_\Delta,\, 1 + \sigma_\Delta \circ \theta_1 \leq m,\, X_{\sigma_\Delta} \circ \theta_1 \in A\big)$$
$$= F(A\backslash\Delta \mid x) + P_x\big(\theta_1^{-1}\{\sigma_\Delta \leq m - 1,\, X_{\sigma_\Delta} \in A\},\, X_1 \in \Delta\big)$$
$$= F(A\backslash\Delta \mid x) + E_x\big(P_{X_1}(\sigma_\Delta \leq m - 1,\, X_{\sigma_\Delta} \in A);\, X_1 \in \Delta\big).$$

Hence

$$F_{\sigma_\Delta}(m, A \mid x) = F(A\backslash\Delta \mid x) + \int_\Delta F_{\sigma_\Delta}(m - 1, A \mid x_1) F(dx_1 \mid x)$$

is an integral equation with respect to kernel F_{σ_Δ} for the Markov sequence. For $x \notin \Delta$ we assume $F_{\sigma_\Delta}(0, A \mid x) = I_{A\backslash\Delta}(x)$.

1.3. Two-dimensional Markov chain

12. Markov renewal process

We designate $\mathbb{R}_+ = [0, \infty)$, $\overline{\mathbb{R}}_+ = [0, \infty]$. Let W be a set of all sequences of pairs $w = (v_n, z_n)_{n=0}^\infty$, where $v_n \in \overline{\mathbb{R}}_+$ and $z_n \in \overline{\mathbb{X}} \equiv \mathbb{X} \cup \overline{\infty}$, where $\overline{\infty} \notin \mathbb{X}$ (some additional point), and

(1) $v_n = \infty \Leftrightarrow z_n = \overline{\infty}$,

(2) $v_0 = 0$,

(3) $v_n < \infty \Rightarrow 0 < v_1 < \cdots < v_n$,

(4) $v_n \to \infty$ $(n \to \infty)$.

It is a new space of elementary events, on which we consider a special class of Markov sequences. A space of states in this case is $\overline{\mathbb{R}}_+ \times \overline{\mathbb{X}}$, which we consider as a new metric space with some metric generated by the real line metric and that of space $\mathbb{X}$. It would be possible in this case to use all previous notations and conclusions for random sequences with values from the new metric space. However, taking into account some special properties of sequences, we give some new denotations.

Let $\zeta(w)$ be the number of the last finite v_n. If such a number does not exist, let us assume that $\zeta(w) = \infty$. Let $\widetilde{X}_n(w) = v_n$, and $\overline{X}_n(w) = z_n$. Then $\widehat{X}_n(w) = (\widetilde{X}_n(w), \overline{X}_n(w))$. On a set $\{\widetilde{X}_n < \infty\} \equiv \{\zeta \geq n\}$ we define map $\widehat{\theta}_n : W \to W$ by the equality

$$\widehat{\theta}_n(w) = \big((0, z_n), (v_{n+1} - v_n, z_{n+1}), \dots\big),$$

which differs a little from that of θ_n. Let $\mathfrak{B} = \sigma(\widehat{X}_n, n \in \mathbb{N}_0)$; Q be a probability measure on $\mathfrak{B}$. The random sequence $(W, \mathfrak{B}, Q)$ is said to be a two-dimensional Markov sequence with conditionally independent increments of the first component, if $(\forall n \in \mathbb{N}_0)$ $(\forall B \in \mathfrak{B})$ on a set $\{\widetilde{X}_n < \infty\}$ Q-a.s.

$$Q\big(\widehat{\theta}_n^{-1} B \mid \widehat{X}_0, \ldots, \widehat{X}_n\big) = Q\big(\widehat{\theta}_n^{-1} B \mid \overline{X}_n\big).$$

On a set $\{\widetilde{X}_n = \infty\}$ function $\widehat{\theta}_n$ is not defined. On this set the Markov property consists of determined passage $(\infty, \overline{\infty}) \to (\infty, \overline{\infty})$, thus

$$Q\big(\widehat{X}_{n+1} = (\infty, \overline{\infty}) \mid \widehat{X}_0, \ldots, \widehat{X}_n\big) = Q\big(\widehat{X}_{n+1} = (\infty, \overline{\infty}) \mid \widehat{X}_n\big) = 1$$

$(\widehat{X}_n = (\infty, \overline{\infty}))$. This sequence also has the short name: Markov renewal process.

13. Homogenous family of distributions

In order to define homogenous in time (number) sequences we consider transition functions (kernels) for passage from $\mathbb{X}$ to $\mathcal{B}(\mathbb{R} \times \mathbb{X})$. Let $(F_n)_{n=1}^{\infty}$ be a sequence of the kernels $F_n([0, t] \times A \mid x) \in [0, 1]$, where $t \geq 0$, $A \in \mathcal{B}(\mathbb{X})$, $x \in \mathbb{X}$. We define them as

$$Q\big(\widetilde{X}_{n+1} - \widetilde{X}_n \leq t, \overline{X}_{n+1} \in A \mid \overline{X}_n\big) = F_{n+1}\big([0, t] \times A \mid \overline{X}_n\big)$$

Q-a.s. on the set $\{\widetilde{X}_n < \infty\}$. From here

$$Q\big(\widetilde{X}_{n+1} - \widetilde{X}_n \leq t, \overline{X}_{n+1} \in A \mid \overline{X}_n = x\big) = F_{n+1}\big([0, t] \times A \mid x\big)$$

$(Q \circ \overline{X}_n^{-1})$-a.s. on $(\mathbb{X}, \mathcal{B}(\mathbb{X}))$. We call the sequence defined above homogenous in time if $F_n = F_1 \equiv F$. With the help of these kernels as well as in case of simple Markov sequences, we obtain a homogenous in time family of distributions $(Q_x)_{x \in \mathbb{X}}$ on $\mathfrak{B}$, coordinated by the condition of the two-dimensional Markov property with conditionally independent increments of the first component (condition of regeneration). So we have

$$(\forall n \in \mathbb{N}_0) \ (\forall B \in \mathfrak{B}) \ (\forall A \in \mathfrak{B}_n) \ (\forall x \in \mathbb{X})$$

$$Q_x\big(\widehat{\theta}_n^{-1} B, A \cap \{\widetilde{X}_n < \infty\}\big) = E_{Q_x}\big(Q_{\overline{X}_n}(B); A \cap \{\widetilde{X}_n < \infty\}\big),$$

where $\mathfrak{B}_n = \sigma(\widehat{X}_0, \ldots, \widehat{X}_n)$, E_{Q_x} is an integral with respect to the measure Q_x. Thus

$$Q_x\big(\overline{X}_0 \in A_0, (\widetilde{X}_1 - \widetilde{X}_0 \leq t_1, \overline{X}_1 \in A_1), \ldots, (\widetilde{X}_n - \widetilde{X}_{n-1} \leq t_n, \overline{X}_n \in A_n)\big)$$

$$= I_{A_0}(x) \int_{A_1 \times \cdots \times A_n} \prod_{i=1}^{n} F\big([0, t_i], dx_i \mid x_{i-1}\big). \qquad [1.7]$$

$(A_i \in \mathcal{B}(\mathbb{X}), x_0 = x)$. The measure Q_x defined on sets of this form can be uniquely extended to the whole sigma-algebra $\mathfrak{B}$.

We assume that $F([0,\infty) \times \mathbb{X} \mid x) \leq 1$ (sub-probability kernel). A condition $(\exists x \in \mathbb{X}) \, F([0,\infty) \times \mathbb{X} \mid x) < 1$ reflects a possibility of finite $\zeta(w) : Q_x(\zeta < \infty) > 0$. Thus $Q_x(\zeta = 0) = 1 - F([0,\infty) \times \mathbb{X} \mid x)$ and, moreover,

$$Q_x(\zeta = n) = \int_{\mathbb{X}^n} \left(1 - F([0,\infty) \times \mathbb{X} \mid x_n)\right) \prod_{i=1}^{n} F([0,\infty) \times dx_i \mid x_{i-1}) \quad [1.8]$$

If $(\forall x \in \mathbb{X}) \, F([0,\infty) \times \mathbb{X} \mid x) = 1$, then $Q_x(\zeta = \infty) = 1$.

Thus, we again pass from an individual Markov chain with distribution Q to a Markov chain given within the initial point, with a coordinated family of distributions $(Q_x)_{x \in \mathbb{X}}$.

We are interested in conditions on *a priori* given kernel F for the functions Q_x constructed by formula [1.3] which could possibly be extended to a measure on $(W, \mathfrak{B})$. The necessary and sufficient condition for this property requires the kernel F (satisfying usual conditions of a measurability) to be of the form:

$$(\forall t > 0) \, (\forall x \in \mathbb{X}) \quad F^{*n}([0,t] \times \mathbb{X} \mid x) \longrightarrow 0 \, (n \longrightarrow \infty),$$

where $F^{*n}([0,t] \times S \mid x)$ is the convolution:

$$F^{*n}([0,t] \times S \mid x) = \int_0^t F^{*n-1}([0, t - t_1] \times S \mid x_1) F(dt_1, dx_1 \mid x),$$

$(S \in \mathcal{B}(\mathbb{X}))$. This condition formalizes the requirement $v_n \to \infty$ as $n \to \infty$ for all $w \in W$. A sufficient condition for this property to be realized is the existence of such $\varepsilon > 0$ that $(\forall x \in \mathbb{X}) \, F([0, \varepsilon] \times \mathbb{X} \mid x) < 1 - \varepsilon$. Let

$$f(\lambda, A \mid x) = \int_0^\infty e^{-\lambda t} F(dt \times A \mid x).$$

Then $f(\lambda, \mathbb{X} \mid x) < 1 - \varepsilon + \varepsilon e^{-\lambda \varepsilon} \leq \delta < 1$. Furthermore

$$\int_0^\infty e^{-\lambda t} F^{*n}(dt \times \mathbb{X} \mid x) = f^{*n}(\lambda, \mathbb{X} \mid x),$$

where

$$f^{*n}(\lambda, A \mid x) = \int_{\mathbb{X}} f^{*n-1}(\lambda, A \mid x_1) f(\lambda, dx_1 \mid x) \leq \delta^n \longrightarrow 0.$$

We also have

$$f^{*n}(\lambda, \mathbb{X} \mid x) = \int_0^\infty e^{-\lambda t} F^{*n}(dt \times \mathbb{X} \mid x)$$

$$\geq \int_0^t e^{-\lambda t_1} F^{*n}(dt_1 \times \mathbb{X} \mid x)$$

$$\geq e^{-\lambda t} F^{*n}([0,t] \times \mathbb{X} \mid x).$$

14. Independence of increments of the first component

Let us consider on a set $\{\widetilde{X}_n < \infty\}$ conditional probability $Q_x(B_n \mid \overline{X}_0, \ldots, \overline{X}_n)$ of the event

$$B_n = \{\widetilde{X}_1 - \widetilde{X}_0 \leq t_1, \ldots, \widetilde{X}_n - \widetilde{X}_{n-1} \leq t_n\}.$$

We will prove that

$$Q_x(B_n \mid \overline{X}_0, \ldots, \overline{X}_n) = \prod_{k=1}^n Q_x(\widetilde{X}_k - \widetilde{X}_{k-1} \leq t \mid \overline{X}_{k-1}, \overline{X}_k) \qquad [1.9]$$

Q_x-a.s., where

$$Q_x(\widetilde{X}_1 - \widetilde{X}_0 \leq t \mid \overline{X}_0, \overline{X}_1) = Q_x(\widetilde{X}_1 \leq t \mid \overline{X}_1).$$

It is enough for it to prove equality of integrals of these functions by a measure Q_x on any set of the form:

$$B'_n = \{\overline{X}_0 \in A_0, \ldots, \overline{X}_n \in A_n\},$$

where $A_k \in \mathcal{B}(\mathbb{X})$. For $n = 1$ the formula being proved turns to identity. Let it be correct for $n \geq 1$. Then, using the regenerative property of the family (Q_x), we obtain

$$E_{Q_x}(Q_x(B_{n+1} \mid \overline{X}_0, \ldots, \overline{X}_{n+1}); B'_{n+1}) = Q_x(B_{n+1}, B'_{n+1})$$

$$= Q_x(B_n \cap \{\widetilde{X}_{n+1} - \widetilde{X}_n \leq t_{n+1}\}, B'_n \cap \{\overline{X}_{n+1} \in A_{n+1}\})$$

$$= E_{Q_x}(Q_x(\widetilde{X}_{n+1} - \widetilde{X}_n \leq t_{n+1}, \overline{X}_{n+1} \in A_{n+1} \mid \overline{X}_n); B_n \cap B'_n)$$

$$= E_{Q_x}(Q_x(B_n \mid \overline{X}_0, \ldots, \overline{X}_n) Q_x(\widetilde{X}_{n+1} - \widetilde{X}_n \leq t_{n+1}, \overline{X}_{n+1} \in A_{n+1} \mid \overline{X}_n); B'_n).$$

Using the definition of conditional probability with respect to the measure $Q_x(\cdot \mid \overline{X}_n)$ we obtain the Q_x-a.s. equality

$$Q_x(\widetilde{X}_{n+1} - \widetilde{X}_n \leq t_{n+1}, \overline{X}_{n+1} \in A_{n+1} \mid \overline{X}_n)$$

$$= E_{Q_x}(Q_x(\widetilde{X}_{n+1} - \widetilde{X}_n \leq t_{n+1} \mid \overline{X}_n, \overline{X}_{n+1}); \overline{X}_{n+1} \in A_{n+1} \mid \overline{X}_n).$$

This implies that the preceding expression is equal to

$$E_{Q_x}\big(Q_x\big(B_n \mid \overline{X}_0,\ldots,\overline{X}_n\big)Q_x\big(\widetilde{X}_{n+1} - \widetilde{X}_n \le t_{n+1} \mid \overline{X}_n, \overline{X}_{n+1}\big); B'_{n+1}\big).$$

We denote $F(t \mid x, y) = Q_x(\widetilde{X}_1 \le t \mid \overline{X}_1 = y)$. Such a transition function will be covered in Chapter 7.

1.4. Semi-Markov process

15. Stepped random process

Some questions about two-dimensional Markov sequences concern an interpretation of an element of W as a step-function ξ on $[0, \infty)$.

Let $\mathcal{D}_0$ be a set of step-functions $\xi : \mathbb{R} \to \mathbb{X}$ having on each finite interval a finite number of point of discontinuities, constant on each interval between neighboring points of discontinuities, continuous from the right (which defines it in points of discontinuities). The set of measures of Markov renewal processes defined above determines stepped semi-Markov process, that is, the semi-Markov process in such a simple case, when it can be traced step by step on its sequence of states.

Let map $f : W \to \mathcal{D}_0$ put in correspondence to a sequence $w \in W$ a function $\xi = f(w)$ such that $(\forall t \in \mathbb{R}_+)\, \xi(t) = \overline{X}_n(w) \Leftarrow \widetilde{X}_n(w) \le t < \widetilde{X}_{n+1}(w)$. From a condition $\widetilde{X}_n(w) \to \infty\ (n \to \infty)$ it follows that the function ξ is defined everywhere on $\mathbb{R}_+$. Let $\mathcal{F}$ be σ-algebra of subsets of the set $\mathcal{D}_0$, generated by all maps $X_t : \mathcal{D}_0 \to \mathbb{X}\ (t \in \mathbb{R}_+)$, where $X_t \xi = \xi(t)$. Thus, f is a measurable map (with respect to $\mathfrak{B}$ and $\mathcal{F}$), which for any measure Q on $\mathfrak{B}$ induces a measure $P = Q \circ f^{-1}$ on $\mathcal{F}$.

Up to now we did not limit measures Q on $\mathfrak{B}$ by a condition on the neighboring values $\overline{X}_n, \overline{X}_{n+1}$. Now for convenience of use of representation $\xi = f(w)$ we assume that Q is concentrated on $W_0 \subset W$, where W_0 consists of sequences with unequal neighboring values $\overline{X}_n, \overline{X}_{n+1}\ (n \in \mathbb{N}_0)$. More precisely, let $(\forall w \in W_0)$ $\widetilde{X}_n < \infty \Rightarrow \overline{X}_n \ne \overline{X}_{n+1}$. When the original measure Q does not satisfy this condition, we can enlarge $\overline{\mathbb{X}}$, using the space $\overline{\mathbb{X}}' = \overline{\mathbb{X}} \times \mathbb{N}_0$, where the additional coordinate counts numbers of the elements of a sequence. New W' contains the elements $w' = (w'_0, w'_1, w'_2, \ldots)$, where

$$w'_n = \begin{cases} (v_n, z_n, n), & v_n < \infty, \\ (\infty, \overline{\infty}), & v_n = \infty, \end{cases}$$

i.e. realizes a case, when $z'_n \equiv (z_n, n) \ne (z_{n+1}, n+1) = z'_{n+1}$. An example of a Markov renewal process, when it is necessary to accept such enlargement, is a simple renewal process. It corresponds to a case where $\mathbb{X}$ consists of one element. Furthermore, we will consider this example in more detail. However, we start by introducing some denotations connected to functions $\xi \in \mathcal{D}_0$.

16. Shift and stop operators

Let $\theta_t : \mathcal{D}_0 \to \mathcal{D}_0$ be a shift operator, $(\theta_t \xi)(s) = \xi(t + s)$ $(s, t \in \mathbb{R}_+)$; and $\tau_n = \sigma_0^n$ be a time of the n-th discontinuity of function ξ $(n \geq 1)$. This time we also consider for $\xi \in \mathcal{D}_0$ with the number of jumps less than n. In this case we assume $\tau_n(\xi) = \infty$ and $X_{\tau_n}(\xi) = \overline{\infty}$.

Let $\alpha_t : \mathcal{D}_0 \to \mathcal{D}_0$ be an operator of a stopping, $(\alpha_t \xi)(s) = \xi(s \wedge t)$, where $s \wedge t = \min(s, t)$, $s \in \mathbb{R}_+$, $t \in \overline{\mathbb{R}}_+$. This operator is useful for the representation sigma-algebra of subsets of the set $\mathcal{D}_0$, used in the theory of general semi-Markov processes. Let us consider sigma-algebra of subsets (events), preceding time t. From an obvious relation $X_s \circ \alpha_t = X_{s \wedge t}$ it follows that $\mathcal{F}_t = \sigma(X_s \circ \alpha_t, \ s \in \mathbb{R}_+) = \alpha_t^{-1} \sigma(X_1, s \in \mathbb{R}_+) = \alpha_t^{-1} \mathcal{F}$ (see Neveu [NEV 69]).

Now we show another representation for $\mathcal{F}$.

Let $X_\tau(\xi) = X_{\tau(\xi)}(\xi)$, (if $\tau(\xi) = \infty$, we suppose $X_{\tau(\xi)}(\xi) = \overline{\infty}$), and

$$\mathcal{F}' = \sigma\big((\tau_n, X_{\tau_n}), \ n \in \mathbb{N}_0\big).$$

We have $(\forall t \in \mathbb{R}_+)$ $X_t = X_{\tau_n} \Leftarrow \tau_n \leq t < \tau_{n+1}$. Hence X_t is a $\mathcal{F}'$-measurable function, i.e. $\mathcal{F} \subset \mathcal{F}'$. On the other hand

$$\{\tau_1 > s\} = \lim_{n \to \infty} \bigcap_{k=1}^{n} \{X_{sk/n} = X_0\} \in \mathcal{F},$$

$$\{\tau_1 \leq s, X_{\tau_1} \in S\} = \lim_{n \to \infty} \bigcup_{k=1}^{n} \bigcap_{\ell=1}^{k-1} \{X_{s\ell/n} = X_0, \ X_{sk/n} \neq X_0, \ X_{sk/n} \in S\} \in \mathcal{F}.$$

$$[1.10]$$

The inclusions $\{\tau_n \leq s, X_{\tau_n} \in S\} \in \mathcal{F}$ are similarly proved, i.e. $\mathcal{F}' \subset \mathcal{F}$.

We also consider sigma-algebra

$$\mathcal{F}_{\tau_n} = \sigma\big((\tau_k, X_{\tau_k}), \ k \in \{0, 1, \ldots, n\}\big) \quad (n \in \mathbb{N}_0).$$

It is a sigma-algebra of events preceding to a time of the n-th jump (including the n-th jump). It also can be presented with the help of an operator of a stopping such as $\mathcal{F}_{\tau_n} = \alpha_{\tau_n}^{-1} \mathcal{F}$. Essentially, the function $\alpha_{\tau_n} \xi$ coincides with function ξ up to a moment of the jump, and further is equal to a constant value $(\alpha_{\tau_n}\xi)(t) = X_{\tau_n}\xi$ $(t \geq \tau_n)$. Hence $\tau_k(\alpha_{\tau_n}\xi) = \infty$ and $X_{\tau_k}(\alpha_{\tau_n}\xi) = \overline{\infty}$, if $k > n$. Therefore

$$\mathcal{F}_{\tau_n} = \sigma\big((\tau_k \circ \alpha_{\tau_n}, X_{\tau_k} \circ \alpha_{\tau_n}), \ k \in \mathbb{N}_0\big) = \alpha_{\tau_n}^{-1}\mathcal{F}.$$

Note that for $w \in W_0$, $\widetilde{X}_n(w) = \tau_n(f(w))$ and $\overline{X}_n w = X_{\tau_n}(f(w))$. Furthermore, $f(\widehat{\theta}_n w) = \theta_{\tau_n}(f(w))$ on a set $\{\widetilde{X}_n < \infty\} = \{\tau_n \circ f < \infty\}$. From here functional equalities follow: $\widetilde{X}_n = \tau_n \circ f$; $\overline{X}_n = X_{\tau_n} \circ f$ (on a set $\{\tau_n = \infty\}$ assume $X_{\tau_n} = \overline{\infty}$), and $f \circ \widehat{\theta}_n = \theta_{\tau_n} \circ f$ on set $\{\widetilde{X}_n < \infty\}$.

17. Semi-Markov consistency condition

Now we can express a consistency condition for the family of measures (P_x), generated by the family of measures (Q_x), in terms of maps θ_t and α_t. Let us prove that

$$P_x\big(\theta_{\tau_n}^{-1}B, \ A \cap (\tau_n < \infty)\big) = E_x\big(P_{X_{\tau_n}}(B); \ A \cap (\tau_n < \infty)\big) \qquad \text{[1.11]}$$

where $B \in \mathcal{F}$, $A \in \mathcal{F}_{\tau_n}$ (regeneration condition for the family (P_x)). Using representation $\mathcal{F}_{\tau_n} = \alpha_{\tau_n}^{-1}\mathcal{F}$, it is possible to put $A = \alpha_{\tau_n}^{-1}C$, where $C \in \mathcal{F}$. Then

$$P_x\big(\theta_{\tau_n}^{-1}B, \ \alpha_{\tau_n}^{-1}C \cap (\tau_n < \infty)\big) = Q_x\big(f^{-1}\theta_{\tau_n}^{-1}B, B_n\big) = Q_x\big(\widehat{\theta}_n^{-1}f^{-1}B, B_n\big)$$

where $B_n = f^{-1}\alpha_{\tau_n}^{-1}C \cap (\tau_n \circ f < \infty)$. Evidently, it belongs to sigma-algebra $\mathcal{B}_n$. From here, using the condition of regeneration of the family (Q_x), we obtain

$$E_{Q_x}\big(Q_{\overline{X}_n}(f^{-1}B); B_n\big) = E_{Q_x}\big(P_{\overline{X}_n}(B); B_n\big) = E_{Q_x}\big(P_{X_{\tau_n} \circ f}(B); B_n\big)$$
$$= E_x\big(P_{X_{\tau_n}}(B); \ \alpha_{\tau_n}^{-1}C \cap (\tau_n < \infty)\big).$$

The family (P_X), satisfying condition [1.11], is called the semi-Markov (stepped) family of measures. The random process, being determined within the initial point by this family of measures, is said to be a stepped semi-Markov process.

For the semi-Markov family of measures (P_x) the kernel $F([0,t] \times A \mid x)$ means a conditional distribution of a pair (time, point) of the first jump:

$$P_x\big(\tau_1 \leq t, \ X_{\tau_1} \in A\big) = Q_x\big(\widetilde{X}_1 \leq t, \ \overline{X}_1 \in A\big) = F\big([0,t] \times A \mid x\big).$$

We call this kernel a semi-Markov transition function, and the corresponding kernel $f(\lambda, A \mid x)$ a transition generating function of the stepped semi-Markov process.

Furthermore, the regeneration condition of a semi-Markov family of measures (P_x) will be an initial concept. We will not deduce it from properties of more simple objects, as in the case of step-functions. To prepare this passage, we will derive the formula

$$P_x\left(\theta_{\sigma_\Delta}^{-1}B,\ A\cap\{\sigma_\Delta<\infty\}\right)=P_x\left(P_{X_{\sigma_\Delta}}(B),\ A\cap\{\sigma_\Delta<\infty\}\right)$$

where Δ is an open set, $\theta_{\sigma_\Delta}(\xi)=\theta_{\sigma_\Delta(\xi)}(\xi)$ (determined on $\{\sigma_\Delta<\infty\}$), $B\in\mathcal{F}$ and $A\in\mathcal{F}_{\sigma_\Delta}$, $\mathcal{F}_{\sigma_\Delta}$ is a sigma-algebra of events preceding to a time σ_Δ: $\mathcal{F}_{\sigma_\Delta}=\alpha_{\sigma_\Delta}^{-1}\mathcal{F}$. For stepped semi-Markov processes this property can be proved (for an arbitrary measurable Δ):

$$P_x\left(\theta_{\sigma_\Delta}^{-1}B,\ \alpha_{\sigma_\Delta}^{-1}A\cap\{\sigma_\Delta<\infty\}\right)$$

$$=\sum_{k=0}^\infty P_x\left(\theta_{\sigma_\Delta}^{-1}B,\ \alpha_{\sigma_\Delta}^{-1}A\cap\{\sigma_\Delta<\infty\},\ \sigma_\Delta=\tau_k\right)$$

$$=\sum_{k=0}^\infty P_x\left(\theta_{\tau_k}^{-1}B,\ \alpha_{\tau_k}^{-1}A\cap\{\tau_k<\infty\},\ \sigma_\Delta=\tau_k\right).$$

On the other hand $\{\sigma_\Delta=\tau_k\}=\{X_0\in\Delta,\ X_{\tau_1}\in\Delta,\dots,\ X_{\tau_{k-1}}\in\Delta,\ X_{\tau_k}\notin\Delta\}\in\mathcal{F}_{\tau_k}$ and consequently, according to the semi-Markov property of a set (P_x), it is possible to replace $\theta_{\tau_k}^{-1}B$ on $P_{X_{\tau_k}}(B)$ (with corresponding integration). We obtain

$$\sum_{k=0}^\infty E_x\left(P_{X_{\tau_k}}(B);\ \alpha_{\tau_k}^{-1}A\cap\{\tau_k<\infty\},\ \sigma_\Delta=\tau_k\right)$$

$$=E_x\left(P_{X_{\sigma_\Delta}}(B);\ \alpha_{\sigma_\Delta}^{-1}A\cap\{\sigma_\Delta<\infty\}\right).$$

1.5. Stationary distributions for a semi-Markov process

18. Parts of an interval of constancy and their distributions

Let

$$R_t^-(\xi)=t-\tau_{N_t}(\xi),\quad R_t^+(\xi)=\tau_{N_t+1}(\xi)-t,$$

where $N_t(\xi)=\sup\{n:\tau_n(\xi)\le t\}$ is a number of jumps of function ξ up to a time t. These functions give distances from t up to nearest jump time from the left, and that from the right. We have

$$P_x\left(R_t^-\ge r,\ R_t^+\ge s,\ X_t\in A\right)$$

$$=\sum_{k=0}^\infty P_x\left(X_{t_n}\in A,\ \tau_n\le t-r,\ \tau_{n+1}\ge t+s\right).$$

If $\tau_n < \infty$, it is obvious that $\tau_{n+1} = \tau_n + \tau_1 \circ \theta_{\tau_n}$ (the latter is usually designated as $\tau_n \dotplus \tau_1$). Hence, $\{\tau_{n+1} > t + s\} = \{\tau_1 \circ \theta_{\tau_n} > t + s - \tau_n\}$. On a set $\{\tau_n = s_1\}$ the last event is represented as $\theta_{\tau_n}^{-1}\{\tau_1 > t + s - s_1\}$, which makes it possible to use a condition of regeneration for any term of this series:

$$\sum_{k=0}^{\infty} P_x\left(\tau \circ \theta_{\tau_n} X_{\tau_n} \in A,\ \tau_n \leq t - r\right)$$

$$= \sum_{k=0}^{\infty} \int_0^{t-r} E_x\left(P_{X_{\tau_n}}\left(\tau_1 > t + s - s_1\right);\ \tau_n \in ds_1,\ X_{\tau_n} \in A\right)$$

$$= \sum_{k=0}^{\infty} \int_0^{t-r} \int_A P_{x_1}\left(\tau_1 > t + s - s_1\right) P_x\left(\tau_n \in ds_1,\ X_{\tau_n} \in dx_1\right).$$

The strict deduction of this formula can be obtained with the help of passage to the limit with a partition of an interval $[0, t - s]$ on n parts with $n \to \infty$. So we have

$$P_x\left(R_t^- \geq r,\ R_t^+ \geq s,\ X_t \in A\right)$$

$$= \int_0^{t-r} \int_A P_{x_1}\left(\tau_1 > t + s - s_1\right) U\left(x,\ ds_1 \times dx_1\right), \qquad [1.12]$$

where $U(x, [0,t] \times A) = \sum_{k=0}^{\infty} P_x(\tau_n \leq t,\ X_{\tau_n} \in A)$. For fixed x this kernel is a measure of intensity of some random point field. So, let $N(B \times A) = \sum_{n=0}^{\infty} I_{B \times A}(\tau_n, X_n)$ be a number of pairs (time, point) of discontinuity belonging to a set $B \times A$ ($B \in \mathcal{B}(\mathbb{R}_+)$, $A \in \mathcal{B}(\mathbb{X})$). Then

$$E_x\left(N(B \times A)\right) = \sum_{n=0}^{\infty} E_x\left(I_{B \times A}(\tau_n, X_n)\right)$$

$$= \sum_{n=0}^{\infty} P_x\left(\tau_n \in B,\ X_{\tau_n} \in A\right) = U(x,\ B \times A).$$

19. Renewal processes

The equations for distributions of R_t^- and R_t^+ are quite simple in a case $\mathbb{X} = \{x\}$ ($\mathbb{X}$ consists of one point), i.e. for renewal processes. According to the common scheme the renewal process is a Markov renewal process with space of states $\mathbb{X} = \mathbb{N}_0$ and with deterministic passages $\{X_n = n\} \to \{X_{n+1} = n + 1\}$. The formal family of distributions $(P_k)_{k \in \mathbb{N}_0}$ on $(\mathcal{D}_0, \mathcal{F})$ is reduced to one distribution P_0, since this family is homogenous on $k \in \mathbb{N}_0$, namely $P_k(B - k) = P_0(B)$, where for $B \in \mathfrak{B}$, $k \in \mathbb{N}_0$ $B - k = \{\xi \in \mathcal{D}_0 : \xi + k \in B\}$. Moreover

$$P_0\left(\tau_1 \leq t_1,\ \tau_2 - \tau_1 \leq t_2, \ldots, \tau_n - \tau_{n-1} \leq t_n\right)$$

$$= F\left([0, t_1]\right) F\left([0, t_2]\right) \cdots F\left([0, t_n]\right),$$

where $F(A) = F(A, 1 \mid 0) = F(A, k \mid k-1)$ $(\forall k \in \mathbb{N})$. The measure $U(x, B \times A)$ is reduced in this case to a measure $U(B) = \sum_{n=0}^{\infty} P_0(\tau_n \in B)$. The analysis of properties of measure U is the main problem of the renewal theory. In this theory the following property is proved. For any non-lattice distribution F with finite expectation $m > 0$ it is true that

$$U\big([a+t, b+t]\big) \longrightarrow \frac{b-a}{m} \qquad [1.13]$$

for any a, b $(a < b)$ (key renewal theorem – see Feller, II [FEL 67]). For lattice distributions with a pitch h the similar inference is correct for all values from a set $\{kh : k \in \mathbb{N}_0\}$. From here it follows that if $F([0, \infty)) = 1$ and $m < \infty$, then for any bounded function ψ on $\mathbb{R}_+$ there exists a limit

$$\lim_{t \longrightarrow \infty} \int_0^t \psi(t-s) U\big(ds_1\big) = \frac{1}{m} \int_0^{\infty} \psi(t)\, dt. \qquad [1.14]$$

From here

$$P_0\big(R_t^- \geq r, R_t^+ \geq s\big) = \int_0^{t-r} P_0\big(\tau_1 > t+s-s_1\big) U\big(ds_1\big)$$

$$\longrightarrow \frac{1}{m} \int_{r+s}^{\infty} P_0\big(\tau_1 > t\big)\, dt \quad (t \longrightarrow \infty).$$

We have found that distributions of R_t^- and R_t^+ are asymptotically identical with a limit density $p_0(s) = P_0(\tau_1 > s)/m$, where

$$m = E_0\big(\tau_1\big) = \int_0^{\infty} P_0\big(\tau_1 > s\big)\, ds = \int_0^{\infty} x\, F(dx).$$

20. Ergodic theorem

Let us consider the transition function $F(dt, dx_1 \mid x)$ of a stepped semi-Markov process. Let $(\forall x \in \mathbb{X})\ F(\mathbb{R}_+, \mathbb{X} \mid x) = 1$. The kernel $H(dx_1 \mid x) = F(\mathbb{R}_+, dx_1 \mid x)$ is a transition function of some Markov chain, which we call the embedded Markov chain of the given semi-Markov process. A Markov chain is said to be ergodic if there exists a stationary probability distribution $\nu(dx_1)$ such that

$$\nu(A) = \int_{\mathbb{X}} H\big(A \mid x_1\big) \nu\big(dx_1\big) \quad (A \in \mathcal{B}(\mathbb{X}))$$

and for any $x \in \mathbb{X}$ this distribution can be obtain as a weak limit of the sequence of measures $H^{*n}(dx_1 \mid x)$ $(n \to \infty)$, where

$$H^{*n+1}(A \mid x) = \int_{\mathbb{X}} H^{*n}\big(A \mid x_1\big) H\big(dx_1 \mid x\big),$$

and this limit is independent of x. For a stepped semi-Markov process the following generalization of formula [1.14] is true.

THEOREM 1.1. *Let the embedded Markov chain of the given semi-Markov process be ergodic, and let it have as its stationary distribution $\nu(dx)$. In addition, for any $x \in \mathbb{X}$, let the distribution $F(dt, \mathbb{X} \mid x)$ be a non-lattice probability distribution on $\mathbb{R}_+$. Then for any directly integrable by Riemann function $\psi(t, x)$ $(t \geq 0, \ x \in \mathbb{X})$ the following property is true*

$$\lim_{t \longrightarrow \infty} \int_{\mathbb{X}} \int_0^t \psi(t - s_1, \ y) U(x, \ ds_1 \times dy) = \frac{1}{M} \int_{\mathbb{X}} \int_0^\infty \psi(t, y) \, dt \, \nu(dy), \quad [1.15]$$

where

$$M = \int_{\mathbb{X}} \int_0^\infty P_y(\tau_1 > t) \, dt \, \nu(dy).$$

Proof. See Shurenkov [SHU 89, p. 107]. A directly integrable by Riemann function is an absolutely integrable function, which can be approximated uniformly by step-functions, having steps of equal length (see Shurenkov [SHU 89, p. 80]).   □

21. Semi-Markov process

Let us consider probability [1.12] for a stepped semi-Markov process. The right part of this equality can be represented as

$$\int_0^t \int_{\mathbb{X}} \psi(t - s_1, \ y) U(x, \ ds_1 \times dy),$$

where

$$\psi(t, y) = P_y(\tau_1 - s > t) I_A(y) I_{[r, \infty)}(t).$$

The direct integrability by Riemann of the integrand in t follows from its monotonicity as $t > r$. Using theorem 1.15, we obtain for $t \to \infty$

$$P_x(X_t \in A, \ R_t^- \geq r, \ R_t^+ \geq s) \longrightarrow \frac{1}{M} \int_A \int_{r+s}^\infty P_y(\tau_1 > t) \, dt \, \nu(dy). \quad [1.16]$$

In particular,

$$P_x(X_t \in A) \longrightarrow \frac{1}{M} \int_A m(y) \, \nu(dy),$$

where

$$m(y) = \int_0^\infty P_y(\tau_1 > t) \, dt, \qquad M = \int_{\mathbb{X}} m(y) \, \nu(dy).$$

Chapter 2

Sequences of First Exit Times
and Regeneration Times

In this chapter the technique of investigating processes with trajectories continuous from the right and having limits from the left (cádág) is developed. For this purpose we use inserted point processes (streams) of the first exits corresponding to sequences of subsets of metric space of states. These sequences can depend on a trajectory (random) or can be independent (non-random). We will consider either non-random sequences of subsets or the special kind of random sequences. In the latter case each subset is a spherical neighborhood of a point of the first exit from the previous subset. If the radii of all balls are identical, then for any one trajectory a set of time of the first exit is finite on each bounded interval, and, hence, the ordinary stream of the first exits is defined. With a non-random sequence of sets there is a danger of accumulation of infinite number of time of the first exit on a finite interval, even if all these sets are balls of one radius. However, in a sigma-compact phase it is possible to choose such a non-random sequence of balls with identical small radius (deducing sequence) for which such accumulation does not happen for any trajectory. The method of deducing sequences is convenient for an analysis of properties of random processes connected with times of the first exit, and with a proof of a measurability of various maps.

The trajectory space $\mathcal{D}_{[0,1]}(\mathbb{X})$ is usually equipped with the Skorokhod metric (see [GIH 71]). In our case the modification of this metric for space $\mathcal{D} = \mathcal{D}_{[0,\infty)}(\mathbb{X})$, called a Stone-Skorokhod metric, is more natural [STO 63]. The time and points of the first exit, generally speaking, are not continuous maps with respect to this metric. However, they will be continuous on sets of all functions $\xi \in \mathcal{D}$, "correctly going out" from the appropriate regions. At the same time any open set Δ can "be deformed as little as desired" to a set Δ_ε so that a set of functions, "correctly going out" from Δ_ε, is a set of probability 1 [HAR 72]. It enables us to have at our disposal a rather rich class of sets "of correct exit almost surely" in tasks connected with continuity of times and positions of the first exit.

The most fruitful direction for a research of processes with the help of times of the first exit is, of course, the theory of semi-Markov processes, for which the time of the first exit from any open set $\Delta \subset \mathbb{X}$ is a time of regeneration. In other words such a process has a Markov property with respect to the first exit time from any open set. Problems of regeneration times have been investigated by many authors (see Shurenkov [SHU 77], Mainsonneuve [MAI 71, MAI 74], Mürmann [MUR 73], Smith [SMI 55], Taksar [TAK 80]). In this chapter we research a structure of classes of regeneration times connected with families of probability measures depending on a parameter $x \in \mathbb{X}$. The closure of such a class with respect to some natural compositions of Markov times makes it possible to be limited by a rather small set of these times when defining a semi-Markov process for its constructive exposition.

2.1. Basic maps

We will consider a space of cádlág functions (see Gihman and Skorokhod [GIH 71]). In this section elementary properties of four basic maps of this space: $(\tau, X_t, \theta_t, \alpha_t)$ and that of their compositions are proved. Two classes of maps τ are introduced. Their definitions are based on properties of Markov times and times of the first exit from open sets. For completeness we repeat some definitions from Chapter 1.

1. Space of functions and its maps

Let $\mathbb{R}_+ = [0, \infty)$; $\mathbb{X}$ be a full sigma-compact metric space with a metric ρ; let $\mathcal{D} = \mathcal{D}_{\mathbb{R}_+}(\mathbb{X})$ be a set of all continuous from the right and having limits from the left (cádlág) functions $\xi : \mathbb{R}_+ \to \mathbb{X}$; $\overline{\mathbb{R}}_+ = \mathbb{R}_+ \cup \{\infty\}$, where ∞ is a point which is not belonging to $\mathbb{R}_+$ and closing this set from the right; let $\mathcal{T}$ be a set of all maps $\tau : \mathcal{D} \to \overline{\mathbb{R}}_+$. For any $t \in \mathbb{R}_+$ the maps are defined:

$X_t : \mathcal{D} \to \mathbb{X}$, where $X_t(\xi) = \xi(t)$ is a one-coordinate projection,

$\theta_t : \mathcal{D} \to \mathcal{D}$, where $(\forall s \in \mathbb{R}_+)\, (\theta_t \xi)(s) = \xi(s + t)$ is a shift.

The following map is defined for all $t \in \overline{\mathbb{R}}_+$:

$\alpha_t : \mathcal{D} \to \mathcal{D}$, where $(\forall s \in \mathbb{R}_+)\, (\alpha_t \xi)(s) = \xi(s \wedge t)$ is a stopping.

We also consider compositions of the previous maps, namely, for any $\tau \in \mathcal{T}$ the maps are defined:

$X_\tau : \{\tau < \infty\} \to \mathbb{X}$, where $X_\tau(\xi) = X_{\tau(\xi)}(\xi)$,

$\theta_\tau : \{\tau < \infty\} \rightarrow \mathcal{D}$, where $\theta_\tau(\xi) = \theta_{\tau(\xi)}(\xi)$,

$\alpha_\tau : \mathcal{D} \rightarrow \mathcal{D}$, where $\alpha_\tau(\xi) = \alpha_{\tau(\xi)}(\xi)$.

The important role in many consequent constructions is played by an operation of a non-commutative sum $\dotplus$ for functions $\tau \in \mathcal{T}$ (see Itô and McKean [ITO 65], Blumenthal and Getoor [BLU 68]). For $\tau_1 < \infty$ let us assume that $\tau_1 \dotplus \tau_2 = \tau_1 + \tau_2 \circ \theta_{\tau_1}$. The latter sum is defined to be equal to ∞ on the set $\{\tau_1 = \infty\}$ ($\tau_1, \tau_2 \in \mathcal{T}$).

2. Classes of maps

The definition of the following subsets of the set $\mathcal{T}$ reflects basic properties of Markov times and times of the first exit from open sets.

$$\mathcal{T}_a = \{\tau \in \mathcal{T} : (\forall t \in \overline{\mathbb{R}}_+) \ \{\tau \leq t\} = \alpha_t^{-1}\{\tau \leq t\}\}, \qquad [2.1]$$

$$\mathcal{T}_b = \{\tau \in \mathcal{T} : (\forall t \in \mathbb{R}_+) \ \{\tau \leq t\} = \alpha_t^{-1}\{\tau < \infty\}\}. \qquad [2.2]$$

3. Properties of basic maps

PROPOSITION 2.1.

$$(\forall s, t \in \mathbb{R}_+) \qquad X_t \circ \theta_s = X_{s+t} \qquad [2.3]$$

$$(\forall s, t \in \mathbb{R}_+) \qquad \theta_t \circ \theta_s = \theta_{s+t} \qquad [2.4]$$

$$(\forall s, t \in \mathbb{R}_+) \qquad \alpha_t \circ \alpha_s = \alpha_{s \wedge t} \qquad [2.5]$$

$$(\forall s \in \overline{\mathbb{R}}_+, \ t \in \mathbb{R}_+) \quad X_t \circ \alpha_s = X_{s \wedge t} \qquad [2.6]$$

$$(\forall s \in \overline{\mathbb{R}}_+, \ t \in \mathbb{R}_+) \quad \theta_t \circ \alpha_{t+s} = \alpha_s \circ \theta_t \qquad [2.7]$$

Proof. It immediately follows from the definitions, for example, [2.7]:

$$X_u \theta_t \alpha_{t+s}(\xi) = X_{(t+u)\wedge(t+s)}(\xi) = X_{t+(u \wedge s)}(\xi) = X_u \alpha_s \theta_t(\xi). \qquad \square$$

4. Properties of classes of maps

PROPOSITION 2.2. *The following properties of classes $\mathcal{T}_a$ and $\mathcal{T}_b$ are fair:*

$$\mathcal{T}_b \subset \mathcal{T}_a, \qquad [2.8]$$

$$(\forall \tau \in \mathcal{T}_a, \ \forall t \in \mathbb{R}_+) \quad \{\tau \leq t\} \subset \{\tau = \tau \circ \alpha_t\}, \qquad [2.9]$$

$$(\forall \tau \in \mathcal{T}_a) \quad \{\tau = \tau \circ \alpha_\tau\}. \qquad [2.10]$$

Proof. [2.8]. Let $\tau \in T_b$ and $t \in \mathbb{R}_+$. As $\alpha_t \circ \alpha_t = \alpha_t$,

$$\{\tau \leq t\} = \alpha_t^{-1}\{\tau < \infty\} = \alpha_t^{-1}\alpha_t^{-1}\{\tau < \infty\} = \alpha_t^{-1}\{\tau \leq t\}. \qquad [2.11]$$

On the other hand, $(\forall \xi \in \mathcal{D})$ $\alpha_\infty \xi = \xi$ and consequently the equality is true for $t = \infty$.

[2.9]. Let $\tau \in T_a, t \in \mathbb{R}_+, \xi \in \mathcal{D}, \tau(\xi) \leq t$ and $\tau(\xi) < \tau(\alpha_t(\xi))$. Then $(\exists s \leq t)$ $\tau(\xi) \leq s, \tau(\alpha_t(\xi)) > s$. From here $\tau(\alpha_s(\xi)) \leq s$ (since $\alpha_s^{-1}\{\tau \leq s\} = \{\tau \circ \alpha_s \leq s\}$) and $\tau(\alpha_s(\alpha_t(\xi))) = \tau(\alpha_s(\xi)) > s$ (since $\{\tau > s\} = \alpha_s^{-1}\{\tau > s\}$). It is inconsistency. The same with $\tau(\xi) > \tau(\alpha_t(\xi))$.

[2.10]. Let $\tau \in T_a, t \in \mathbb{R}_+$. It is obvious that $\{\tau = t\} = \alpha_t^{-1}\{\tau = t\}$ and $\{\tau = \infty\} = \alpha_\infty^{-1}\{\tau = \infty\}$. Then

$$\mathcal{D} = \bigcup\{\tau = t\} = \bigcup\{\tau \circ \alpha_t = t, \tau = t\}$$

$$= \bigcup\{\tau \circ \alpha_\tau = \tau, \tau = t\} = \{\tau \circ \alpha_\tau = \tau\},$$

where the union is taken by all $t \in \overline{\mathbb{R}}_+$. $\qquad \square$

5. Properties of the combined maps

PROPOSITION 2.3. *Let* $\tau_1, \tau_2 \in T$. *Then*

$$X_{\tau_2} \circ \theta_{\tau_1} = X_{\tau_1 \dotplus \tau_2} \qquad [2.12]$$

$$\theta_{\tau_2} \circ \theta_{\tau_1} = \theta_{\tau_1 \dotplus \tau_2} \qquad [2.13]$$

In addition, let $\tau_2 \in T_a$. *Then*

$$\alpha_{\tau_2} \circ \alpha_{\tau_1} = \alpha_{\tau_1 \wedge \tau_2} \qquad [2.14]$$

$$X_{\tau_2} \circ \alpha_{\tau_1} = X_{\tau_1 \wedge \tau_2} \qquad [2.15]$$

$$\theta_{\tau_2} \circ \alpha_{\tau_2 \dotplus \tau_1} = \alpha_{\tau_1} \circ \theta_{\tau_2} \qquad [2.16]$$

For equalities [2.15] and [2.16] the domain of the left map is contained in the domain of the right map.

Proof. [2.12]. The common domain of both the left and right maps: $\{\tau_2 \circ \theta_{\tau_1} < \infty, \tau_1 < \infty\} = \{\tau_1 \dotplus \tau_2 < \infty\}$. Furthermore

$$X_{\tau_2}\theta_{\tau_1}(\xi) = X_{\tau_2}\theta_{\tau_1(\xi)}(\xi) = X_{(\tau_1 \dotplus \tau_2)(\xi)}(\xi).$$

[2.13] is proved similarly.

[2.14]. The domain of both maps is the whole $\mathcal{D}$. Thus $\alpha_{\tau_2}\alpha_{\tau_1}(\xi) = \alpha_{\tau_2(\alpha_{\tau_1}(\xi))\wedge\tau_1(\xi)}(\xi)$. We have further $\tau_2(\xi) \leq \tau_1(\xi) \Leftrightarrow \tau_2\alpha_{\tau_1(\xi)}(\xi) \leq \tau_1(\xi)\ (\tau_2 \in \mathcal{T}_a)$. From here $\tau_2(\alpha_{\tau_1}(\xi)) \wedge \tau_1(\xi) \leq \tau_2(\xi) \wedge \tau_1(\xi)$.

[2.15]. The domain of the left map $\{\tau_2\alpha_{\tau_1} < \infty\}$ belongs to the domain of the right map $\{\tau_1 \wedge \tau_2 < \infty\}$, because if $\tau_1(\xi) = \tau_2(\xi) = \infty$, then $\tau_2\alpha_{\tau_1}(\xi) = \tau_2(\xi) = \infty$. Thus $X_{\tau_2\alpha_{\tau_1}}(\xi) = X_{\tau_2\alpha_{\tau_1}(\xi)\wedge\tau_1(\xi)}(\xi) = X_{\tau_1\wedge\tau_2}(\xi)$.

[2.16]. The domain of the left map $\{\tau_2 \circ \alpha_{\tau_2\,\dot{+}\,\tau_1} < \infty\}$ belongs to the domain of the right map $\{\tau_2 < \infty\}$, because if $\tau_2(\xi) = \infty$, then $\tau_2\alpha_{\tau_2\,\dot{+}\,\tau_1}(\xi) = \tau_2(\xi) = \infty$. Let $\tau_2\alpha_{\tau_2\,\dot{+}\,\tau_1}(\xi) < \infty$. With $t \in \mathbb{R}_+$ we have $X_t\theta_{\tau_2}\alpha_{\tau_2\,\dot{+}\,\tau_1}(\xi) = X_s(\xi)$, where $s = (t + \tau_2\alpha_{\tau_2\,\dot{+}\,\tau_1}(\xi)) \wedge (\tau_2(\xi) + \tau_1(\theta_{\tau_2}\xi))$, and also $X_t\alpha_{\tau_1}\theta_{\tau_2}(\xi) = X_z(\xi)$, where $z = (t \wedge \tau_1\theta_{\tau_2}(\xi)) + \tau_2(\xi)$. Since $\tau_2 \in \mathcal{T}_a$ and $\alpha_{\tau_2} = \alpha_{\tau_2} \circ \alpha_{\tau_2\,\dot{+}\,\tau_1}$, $\tau_2\alpha_{\tau_2\,\dot{+}\,\tau_1}(\xi) = \tau_2(\xi)$ and $s = z$. $\qquad\square$

6. An associativity of operation $\dot{+}$

PROPOSITION 2.4. *For any* $\tau_i \in \mathcal{T}\ (i = 1, 2, 3)$

$$\tau_1 \dot{+} (\tau_2 \dot{+} \tau_3) = (\tau_1 \dot{+} \tau_2) \dot{+} \tau_3.$$

Proof. It is an immediate corollary of formula [2.13] from proposition 2.3. $\qquad\square$

7. A semigroup property

PROPOSITION 2.5. *The following properties are fair:*

(1) $\tau_1, \tau_2 \in \mathcal{T}_a \Rightarrow \tau_1 \dot{+} \tau_2 \in \mathcal{T}_a$;

(2) $\tau_1, \tau_2 \in \mathcal{T}_b \Rightarrow \tau_1 \dot{+} \tau_2 \in \mathcal{T}_b$;

(3) let A be some set of indexes. Then, if for any $\alpha \in A\ \tau_\alpha \in \mathcal{T}_a$, *then* $\bigvee_{\alpha\in A}\tau_\alpha \in \mathcal{T}_a$, *and if A is finite,*

$$\bigwedge_{\alpha\in A}\tau_\alpha \in \mathcal{T}_a;$$

(4) let A be some set of indexes. Then, if for any $\alpha \in A\ \tau_\alpha \in \mathcal{T}_b$, *then* $\bigvee_{\alpha\in A}\tau_\alpha \in \mathcal{T}_b$, *and if A is finite, then*

$$\bigwedge_{\alpha\in A}\tau_\alpha \in \mathcal{T}_b.$$

Proof. (1) Let $\tau_1, \tau_2 \in \mathcal{T}_a$. Then $(\forall t \in \mathbb{R}_+) \{\tau_1 \dotplus \tau_2 \leq t\} = \bigcup_{a \leq t} \{\tau_1 = a, \tau_2 \circ \theta_a \leq t-a\} = \bigcup_{a \leq t} \{\tau_1 \circ \alpha_t = a, \tau_2 \circ \alpha_{t-a} \circ \theta_a \leq t-a\} = \bigcup_{a \leq t} \{\tau_1 \circ \alpha_t = a, \tau_2 \circ \theta_a \circ \alpha_t \leq t-a\} = \alpha_t^{-1} \bigcup_{a \leq t} \{\tau_1 = a, \tau_2 \circ \theta_a \leq t-a\} = \alpha_t^{-1} \{\tau_1 \dotplus \tau_2 \leq t\}$.

(2) Let $\tau_1, \tau_2 \in \mathcal{T}_b$. Then $(\forall t \in \mathbb{R}_+) \{\tau_1 \dotplus \tau_2 \leq t\} = \bigcup_{a \leq t} \{\tau_1 = a, \tau_2 \circ \theta_a \leq t-a\} = \bigcup_{a \leq t} \{\tau_1 \circ \alpha_t = a, \tau_2 \circ \alpha_{t-a} \circ \theta_a < \infty\} = \bigcup_{a \leq t} \{\tau_1 \circ \alpha_t = a, \tau_2 \circ \theta_a \circ \alpha_t < \infty\} = \alpha_t^{-1} \{\tau_1 \leq t, \tau_2 \circ \theta_{\tau_1} < \infty\} = \alpha_t^{-1} \{\tau_1 < \infty, \tau_2 \circ \theta_{\tau_1} < \infty\} = \alpha_t^{-1} \{\tau_1 \dotplus \tau_2 < \infty\}$.

(3) Let $(\forall \alpha \in A) \tau_\alpha \in \mathcal{T}_a$. Then $\{\bigvee_\alpha \tau_\alpha \leq t\} = \bigcap_\alpha \{\tau_\alpha \leq t\} = \alpha_t^{-1} \bigcap_\alpha \{\tau_\alpha \leq t\} = \alpha_t^{-1} \{\bigvee_\alpha \tau_\alpha \leq t\}$. Furthermore $\{\tau_{\alpha_1} \wedge \tau_{\alpha_2} \leq t\} = \{\tau_{\alpha_1} \leq t\} \cup \{\tau_{\alpha_2} \leq t\} = \alpha_t^{-1} (\{\tau_{\alpha_1} \leq t\} \cup \{\tau_{\alpha_2} \leq t\}) = \alpha_t^{-1} \{\tau_{\alpha_1} \wedge \tau_{\alpha_2} \leq t\}$ $(\alpha_1, \alpha_2 \in A)$.

(4) Let $(\forall \alpha \in A) \tau_\alpha \in \mathcal{T}_b$. Then $\{\bigvee_\alpha \tau_\alpha \leq t\} = \bigcap_\alpha \{\tau_\alpha \leq t\} = \alpha_t^{-1} \bigcap_\alpha \{\tau_\alpha < \infty\} \supset \alpha_t^{-1} \{\bigvee_\alpha \tau_\alpha < \infty\}$. Furthermore $\{\tau_{\alpha_1} \wedge \tau_{\alpha_2} \leq t\} = \{\tau_{\alpha_1} \leq t\} \cup \{\tau_{\alpha_2} \leq t\} = \alpha_t^{-1} (\{\tau_{\alpha_1} < \infty\} \cup \{\tau_{\alpha_2} < \infty\}) = \alpha_t^{-1} \{\tau_{\alpha_1} \wedge \tau_{\alpha_2} < \infty\}$. $\square$

2.2. Markov times

The various denotations concerning Markov times will now be introduced (see Dellacherie [DEL 72], Dynkin [DYN 59, DYN 63], Loève [LOE 62], Neveu [NEV 69], Shiryaev [SHI 80], Protter [PRO 77], etc.). The measurability of basic maps, the representation of sigma-algebra of events preceding a Markov time $\tau \in \mathrm{MT}$, and also the characterization of Markov times (propositions 2.8 and 2.9), known in the literature as the Galmarino theorem, are discussed (see Itô and McKean [ITO 65, p. 113]). The outcomes of this section are not new. A concrete form of probability space and that of basic maps and classes $\mathcal{T}_a$ ($\tilde{\mathcal{T}}_a$) facilitates presentation of results and their proof.

8. Sigma-algebra of subsets

Let $\mathcal{F} = \sigma(X_t, t \in \mathbb{R}_+)$ be the least sigma-algebra of subsets of a set $\mathcal{D}$ such that all maps X_t are measurable with respect to it. Let

$$\mathcal{F}_t = \sigma(X_s, s \leq t), \quad \mathcal{F}_{t+} = \bigcap_{\varepsilon > 0} \mathcal{F}_{t+\varepsilon} \quad (t \in \mathbb{R}_+).$$

Let us designate by

$$\mathrm{MT} = \{\tau \in \mathcal{T} : (\forall t \in \mathbb{R}_+) \{\tau \leq t\} \in \mathcal{F}_t\}$$

a set of Markov times concerning a family of sigma-algebras $(\mathcal{F}_t)_{t \geq 0}$;

$$\mathrm{MT}_+ = \left\{ \tau \in \mathcal{T} : \left(\forall t \in \mathbb{R}_+ \right) \{ \tau \leq t \} \in \mathcal{F}_{t+} \right\}$$

the same definition concerns a family of sigma-algebras $(\mathcal{F}_{t+})_{t \geq 0}$. Let

$$\mathcal{F}_\tau = \left\{ B \in \mathcal{F} : \left(\forall t \in \mathbb{R}_+ \right) B \cap \{ \tau \leq t \} \in \mathcal{F}_t \right\} \quad (\tau \in \mathrm{MT}),$$

$$\mathcal{F}_{\tau+} = \left\{ B \in \mathcal{F} : \left(\forall t \in \mathbb{R}_+ \right) B \cap \{ \tau \leq t \} \in \mathcal{F}_{t+} \right\} \quad (\tau \in \mathrm{MT}_+)$$

be sigma-algebras of events preceding a Markov time τ. Furthermore we will connect to each $\tau \in \mathrm{MT}_+$ one more sigma-algebra contained in $\mathcal{F}_{\tau+}$, more convenient for research of properties of regeneration times.

We will designate $\mathcal{B}(E)$ a Borel sigma-algebra of subsets of a topological space E. If $(A, \mathcal{A})$ and $(E, \mathcal{E})$ are two sets with sigma-algebras of subsets, then $\varphi \in \mathcal{A}/\mathcal{E}$ means that the map $\varphi : A \to E$ is measurable with respect to sigma-algebras $\mathcal{A}$ and $\mathcal{E}$. If $B \subset A$, then $\mathcal{A} \cap B$ means a track of σ-algebra $\mathcal{A}$ on a set B (i.e. all sets of the form $A \cap B$, where $A \in \mathcal{A}$).

9. The Stone metric

Metric We frequently use the map $\beta_\tau : \mathcal{D} \to \mathbb{Y}$, where $\mathbb{Y} = (\mathbb{R}_+ \times \mathbb{X}) \cup \{\overline{\infty}\}$, $\overline{\infty} \notin (\mathbb{R}_+ \times \mathbb{X})$, $\tau \in \mathcal{T}$ and

$$\beta_\tau(\xi) = \begin{cases} \overline{\infty}, & \tau(\xi) = \infty, \\ \left(\tau(\xi), X_{\tau}(\xi) \right), & \tau(\xi) < \infty. \end{cases}$$

Let us consider one-to-one map $B : \mathbb{Y} \to \mathbb{Y}'$, where $\mathbb{Y}' = \left([0,1) \times \mathbb{X} \right) \cup \{\overline{1}\}$ and $\overline{1} \notin [0,1) \times \mathbb{X}$:

$$B(y) = \begin{cases} \overline{1}, & y = \overline{\infty}, \\ \left((2/\pi) \arctan t, x \right), & y = (t, x) \left(t \in \mathbb{R}_+, x \in \mathbb{X} \right). \end{cases}$$

The metric of Stone $\rho_{\mathbb{Y}'}$ on a set $\mathbb{Y}'$ is said to be the following function

$$\rho_{\mathbb{Y}'} \left((t_1, x_1), (t_2, x_2) \right) = \left(1 - t_1 \right) \wedge \left(1 - t_2 \right) \arctan \rho(x_1, x_2) + |t_1 - t_2|,$$

$$\rho_{\mathbb{Y}'} \left(\overline{1}, (t, x) \right) = \rho_{\mathbb{Y}'} \left((t, x), \overline{1} \right) = 1 - t,$$

$$\rho_{\mathbb{Y}'} \left(\overline{1}, \overline{1} \right) = 0,$$

where $t, t_1, t_2 \in [0, 1)$, $x, x_1, x_2 \in \mathbb{X}$. A proof of the metric axioms for function $\rho_{\mathbb{Y}'}$ does not present difficulties (see Stone [STO 63]).

On a set $\mathbb{Y}$ we consider the metric $\rho_{\mathbb{Y}}$ of the form

$$\rho_{\mathbb{Y}}(y_1, y_2) = \rho_{\mathbb{Y}'}\big(B(y_1), B(y_2)\big) \quad (y_1, y_2 \in \mathbb{Y}),$$

converting $\mathbb{Y}$ in separable metric space. Thus, we have

$$\mathcal{B}(\mathbb{Y}) = \sigma\big(\{\overline{\infty}\}, [0, t] \times S \ (t \in \mathbb{R}_+, S \in \mathcal{B}(\mathbb{X}))\big).$$

10. Measurability of combined maps

PROPOSITION 2.6. *Let* $\tau, \tau_1, \tau_2 \in \mathcal{F}/\mathcal{B}(\overline{\mathbb{R}}_+)$. *Then*

(1) $X_\tau \in \mathcal{F} \cap \{\tau < \infty\}/\mathcal{B}(\mathbb{X})$;

(2) $\theta_\tau \in \mathcal{F} \cap \{\tau < \infty\}/\mathcal{F}$;

(3) $\alpha_\tau \in \mathcal{F}/\mathcal{F}$;

(4) $\tau_1 \dotplus \tau_2 \in \mathcal{F}/\mathcal{B}(\overline{\mathbb{R}}_+)$;

(5) $\beta_\tau \in \mathcal{F}/\mathcal{B}(\mathbb{Y})$.

Proof. (1) and (2): see Blumenthal and Getoor [BLU 68, p. 34];

(3), (4), (5): it is obvious. $\qquad\qquad\qquad\qquad\qquad\qquad\qquad\square$

11. Representation of sigma-algebra

PROPOSITION 2.7. *For any* $\tau \in \mathrm{MT}$ *the following representation is fair:*

$$\mathcal{F}_\tau = \alpha_\tau^{-1}\mathcal{F}.$$

Proof. For any $t \geq 0$ let us check the equality $\mathcal{F}_t = \alpha_t^{-1}\mathcal{F}$. For this aim we use representation of an indexed sigma-algebra: for any measurable map M it is true that

$$M^{-1}\sigma(B : B \in \mathcal{A}) = \sigma\big(M^{-1}B : B \in \mathcal{A}\big),$$

where $\mathcal{A}$ is a family of sets [NEV 69]. From here

$$\alpha_t^{-1}\sigma\big(X_s, s < \infty\big) = \sigma\big(X_s \circ \alpha_t, s < \infty\big)$$
$$= \sigma\big(X_{s \wedge t}, s < \infty\big)$$
$$= \sigma\big(X_s, s \leq t\big).$$

The condition $B \in \alpha_t^{-1}\mathcal{F}$ means that $(\exists B' \in \mathcal{F}) \ B = \alpha_t^{-1}B'$. Hence, in particular, $B = \alpha_t^{-1}B$ (to prove it, it is necessary to use the property $\alpha_t \circ \alpha_t = \alpha_t$). Note that with $B \in \mathcal{F}_\tau$ the property is fair $(\forall t \geq 0)$ $B \cap \{\tau = t\} \in \mathcal{F}_t$. Consequently, $B \cap \{\tau = t\} = \alpha_t^{-1} B \cap \{\tau = t\}$. From here for this B the following representation is true:

$$\alpha_\tau^{-1} B = \bigcup_{t \in \overline{\mathbb{R}}_+} \left(\alpha_\tau^{-1} B \cap \{\tau = t\}\right) = \bigcup_{t \in \overline{\mathbb{R}}_+} \left(\alpha_t^{-1} B \cap \{\tau = t\}\right)$$

$$= \bigcup_{t \in \overline{\mathbb{R}}_+} \left(B \cap \{\tau = t\}\right) = B.$$

However, this means that $B \in \alpha_\tau^{-1}\mathcal{F}$. The inverse relation we should show is $(\forall s \geq 0)$ $(\forall S \in \mathcal{B}(\mathbb{X}))$ $\alpha_\tau^{-1}(X_s^{-1} S \cap \{\tau \leq t\}) \in \mathcal{F}_t$. Including this we obtain from the equalities

$$\alpha_t^{-1}\left(X_{s \wedge \tau}^{-1} S \cap \{\tau \leq t\}\right) = X_{t \wedge s \wedge \tau}^{-1} S \cap \left(\alpha_t^{-1}\{\tau \leq t\}\right)$$

$$= X_{s \wedge \tau}^{-1} S \cap \{\tau \leq t\}. \qquad \square$$

A similar theorem is proved in Shiryaev [SHI 76, p. 22].

12. Relation of classes

Let

$$\widetilde{\mathcal{T}}_a = \{\tau \in \mathcal{T} : (\forall t \in \mathbb{R}_+) \{\tau < t\} = \alpha_t^{-1}\{\tau < t\}\}.$$

PROPOSITION 2.8. *The properties are fair:*

(1) $\mathrm{MT} \subset \mathcal{T}_a$;

(2) $\mathrm{MT}_+ \subset \widetilde{\mathcal{T}}_a$.

Proof. It is an obvious corollary of proposition 2.7. $\qquad \square$

13. Measurability of times

PROPOSITION 2.9. *Let* $\tau \in \mathcal{F}/\mathcal{B}(\overline{\mathbb{R}}_+)$. *Then*

(1) $\tau \in \mathcal{T}_a \Rightarrow \tau \in \mathrm{MT}$;

(2) $\tau \in \widetilde{\mathcal{T}}_a \Rightarrow \tau \in \mathrm{MT}_+$.

Proof. It is an obvious corollary of proposition 2.7; see Shiryaev [SHI 76, p. 24] (for denotations see proposition 2.8(2)). $\qquad \square$

2.3. Time of the first exit and deducing sequences

Various denotations connected to streams of the first exit are introduced. Apparently, the most simple stream of the first exit is a sequence of points from space $\mathbb{R}_+ \times \mathbb{X}$, where each next pair is a time and position of the first exit from a spherical neighborhood of a position of the previous pair. If radii of all balls are identical, then for any function $\xi \in \mathcal{D}$ the sequence of time of the first exits for such stream has no points of condensation on a finite part of the half-line. Despite the simplicity and naturalness of such a stream, it is not suitable to research some properties of the process. The fixed sequence of open sets is more convenient for applications of corresponding streams of the first exits. The sequences guaranteeing for all $\xi \in \mathcal{D}$ a lack of points of condensation in a finite part of the half-line are referred to as deducing [HAR 74, HAR 80b]. There are deducing sequences composed of balls as small as may be desired. This fact implies that any function can be approximated uniformly on the whole half-line by step-functions corresponding to streams of the first exit of such a view.

For any $\tau \in \mathrm{MT}_+$ we connect little bit smaller, than $\mathcal{F}_{\tau+}$, sigma-algebra $\mathcal{F}_\tau = \sigma(\alpha_\tau, \tau)$, where only τ itself depends on "an infinitesimally close future". It is measurable with respect to new sigma-algebras by definition. For $\tau \in \mathrm{MT}$ the new definition coincides with old one, see item 8. The method of deducing sequences is applied in order to derive the form of representation of the sigma-algebra $\mathcal{F}_{\tau_1 + \tau_2}$ ($\tau_1, \tau_2 \in \mathrm{MT}_+$), which does not have a place for $\mathcal{F}_{(\tau_1 + \tau_2)+}$. This representation will be used in order to prove "semigroup" property (with respect to $+$) of the set of regeneration times.

14. Time of the first exit

We designate that $\mathfrak{A}$ is a class of all open subsets of a set $\mathbb{X}$, $\widetilde{\mathfrak{A}}$ is a class of all closed subsets of a set $\mathbb{X}$ (see Kelley [KEL 68]), $[\Delta]$ is a closure and $\partial\Delta$ is a boundary of a set $\Delta \subset \mathbb{X}$, $B(x, r)$ an open ball with a center $x \in \mathbb{X}$ and radius $r > 0$.

As a time of the first exit from a set $\Delta \subset \mathbb{X}$ we call the function $\sigma_\Delta \in \mathcal{T}$ of the form

$$\sigma_\Delta = \begin{cases} \inf\left(t \geq 0 : \xi(t) \notin \Delta\right), & (\exists t \geq 0)\ \xi(t) \notin \Delta \\ \infty, & (\forall t \geq 0)\ \xi(t) \in \Delta, \end{cases}$$

(see Dynkin [DYN 59], Blumenthal and Getoor [BLU 68], Keilson [KEI 79], etc.). Thus $\sigma_\mathbb{X} = \infty$. Sometimes it is useful to have the first exit time from an empty set which naturally can be defined as zero: $\sigma_\emptyset = 0$. We consider also the case when choice of Δ depends on ξ, i.e. time of the first exit σ_G, where $G = G(\xi) \in \mathfrak{A}$ and $\sigma_G(\xi) = \sigma_{G(\xi)}(\xi)$. For example, $\sigma_{B(x,r)}(\xi)$, where $x = X_0(\xi)$. We designate the latter Markov time as σ_r, i.e. $\sigma_r \equiv \sigma_{B(X_0, r)}$. The most common time of this aspect, with which we will now deal, is the time $\sigma_{B(X_0,r)\cap\Delta}$ $(\Delta \in \mathfrak{A})$, which, obviously, is equal to $\sigma_r \wedge \sigma_\Delta$.

Let us accept the following denotation for sum of times of the first exit. Let $r > 0$; $\delta = (\Delta_1, \Delta_2, \ldots)$ be a sequence of sets $\Delta_i \subset \mathbb{X}$. Then

$$\sigma_r^n = \sigma_r^{n-1} \dotplus \sigma_r \quad (n \in \mathbb{N}, \ \sigma_r^0 = 0),$$
$$\sigma_\delta^n = \sigma_\delta^{n-1} \dotplus \sigma_{\Delta_n} \quad (n \in \mathbb{N}, \ \sigma_\delta^0 = 0).$$

Let $\mathcal{T}_1 \subset \mathcal{T}$. Designate $T(\mathcal{T}_1)$ "semi-group" with respect to the operation $\dotplus$ generated by a class of functions $\mathcal{T}_1$. It means $\tau_1, \tau_2 \in T(\mathcal{T}_1) \Rightarrow \tau_1 \dotplus \tau_2 \in T(\mathcal{T}_1)$.

15. Properties of the first exit times

Let us designate by

$$\widetilde{\mathcal{T}}_b = \{\tau \in \widetilde{\mathcal{T}}_a : (\forall t \in \mathbb{R}_+) \ \{\sigma_\Delta \leq t\} \supset \alpha_t^{-1}\{\tau < \infty\}\},$$

class $\widetilde{\mathcal{T}}_a$; see proposition 2.8.

PROPOSITION 2.10. *The following properties are fair:*

(1) $\sigma_\Delta, \sigma_r \in \widetilde{\mathcal{T}}_b \cap \mathrm{MT}$ $(r > 0, \ \Delta \in \mathfrak{A})$;

(2) $\sigma_\Delta, \sigma_{[r]} \in \widetilde{\mathcal{T}}_b \cap \mathrm{MT}_+$ $(r > 0, \ \Delta \in \widetilde{\mathfrak{A}})$, *where*

$$\sigma_{[r]}(\xi) = \sigma_{[B(X_0(\xi),r)]}(\xi).$$

Proof. (1) In Gihman and Skorokhod [GIH 73, p. 194] it is proved that for any $\Delta \in \mathfrak{A}$ $\sigma_\Delta \in \mathrm{MT}$. It is similarly proved that $\sigma_r \in \mathrm{MT}$. So we have to prove that $(\forall t \in \mathbb{R}_+)$

$$\{\sigma_\Delta \leq t\} = \alpha_t^{-1}\{\sigma_\Delta < \infty\}, \qquad \{\sigma_r \leq t\} = \alpha_t^{-1}\{\sigma_r < \infty\}.$$

Since $\mathrm{MT} \subset \widetilde{\mathcal{T}}_a$ (proposition 2.8), $(\forall t \in \mathbb{R}_+) \ \{\sigma_\Delta \leq t\} \subset \alpha_t^{-1}\{\sigma_\Delta < \infty\}$. Let $\sigma_\Delta \alpha_t(\xi) < \infty$. Then

$$(\alpha_t \circ \xi)^{-1}(\mathbb{X}\backslash\Delta) \neq \emptyset.$$

If $X_t \alpha_t(\xi) = X_t(\xi) \in \mathbb{X}\backslash\Delta$, then $\sigma_\Delta(\xi) \leq t$. If $X_t \alpha_t(\xi) \in \Delta$, then $(\forall t_1 \geq t)$ $X_{t_1} \alpha_t(\xi) = X_t(\xi) \in \Delta$, and, hence, $(\exists t_2 < t) \ X_{t_2} \alpha_t(\xi) = X_{t_2}(\xi) \in \mathbb{X}\backslash\Delta$ and $\sigma_\Delta(\xi) < t$. From here $\alpha_t^{-1}\{\sigma_\Delta < \infty\} \subset \{\sigma_\Delta \leq t\}$. For σ_r the proof is similar.

(2) Properties $\sigma_\Delta \in \mathrm{MT}_+$ $(\Delta \in \widetilde{\mathfrak{A}})$ follow from a possibility of representation σ_Δ as a limit of decreasing sequence of Markov times $\tau_n \in \mathrm{MT}$ (see Blumenthal and

Getoor [BLU 68, p. 34]). Based on theorem 2.8, $\mathrm{MT}_+ \subset \tilde{\mathcal{T}}_a$. We have to check that $(\forall t \in \mathbb{R}_+)$

$$\{\sigma_\Delta \le t\} \supset \alpha_t^{-1}\{\sigma_\Delta < \infty\},$$

but it is being proved like that of item (1). The time $\sigma_{[r]}$ is investigated in a similar manner.   □

16. Inequalities for the first exit time

PROPOSITION 2.11. *Let* $\Delta \subset \mathbb{X}$ *and* $r > 0$. *Then*

(1) $t_2 \in [t_1, t_1 \dotplus \sigma_\Delta] \Rightarrow t_1 \dotplus \sigma_\Delta = t_2 \dotplus \sigma_\Delta$,

(2) $t_1 \le t_2 \Rightarrow t_1 \dotplus \sigma_\Delta \le t_2 \dotplus \sigma_\Delta$,

(3) $t_1 \le t_2 \Rightarrow t_1 \dotplus \sigma_r \le t_2 \dotplus \sigma_{2r}$.

Proof. (1) It follows from the following representation

$$(t_1 \dotplus \sigma_\Delta)(\xi) = \begin{cases} \inf\left(t \ge t_1 : \xi(t) \notin \Delta\right), & \left(\exists t \ge t_1\right) \xi(t) \notin \Delta, \\ \infty, & \left(\forall t \ge t_1\right) \xi(t) \in \Delta, \end{cases}$$

see Mainsonneuve [MAI 71, MAI 74], Blumenthal and Getoor [BLU 68].

(2) It follows from (1).

(3) It follows from an obvious relation:

$$(\forall \xi \in \mathcal{D}) \quad t_2 \in [t_1, t_1 + \sigma_r \theta_{t_1}(\xi)) \Longrightarrow \left(\forall t \in [t_1, t_1 + \sigma_r \theta_{t_1}(\xi))\right)$$
$$\rho(\xi(t_2), \xi(t)) \le \rho(\xi(t_2), \xi(t_1)) + \rho(\xi(t_1), \xi(t)) < 2r.$$ □

17. Deducing sequence

Let us pay attention to a difference between sequences (σ_r^n) and (σ_δ^n). The former is strictly increasing in n (up to the first n, when it heads towards infinity), while the latter is increasing at the step n only if $X_{\sigma_\delta^{n-1}}$ belongs to the next set Δ_n, otherwise the next term of the sequence is equal to the preceding one. The sequence of point $(X_{\sigma_\delta^i})_1^{n-1}$ "waits" for the next subset, covering the last point. It can be the case that such a covering subset is absent (in spite of the sequence being infinite). However there exist sequences of any small rank that "service" any one function from $\mathcal{D}$.

For a given subset $A \subset \mathbb{X}$ a sequence of subsets of a set $\mathbb{X}$ is referred to as deducing from A if $\delta = (\Delta_1, \Delta_2, \ldots)$ $(\Delta_i \subset \mathbb{X})$ and for any $\xi \in \mathcal{D}$ and $t \in \mathbb{R}_+$ there exists $n \in \mathbb{N}$, such that $X_{\sigma_\delta^n}(\xi) \notin A$ or $\sigma_\delta^n(\xi) > t$.

Let $\mathrm{DS}(\mathfrak{A}_1, A)$ be a set of all sequences deducing from A, composed from the elements of a class $\mathfrak{A}_1$ subsets of a set $\mathbb{X}$. Other abbreviations are:

$$\mathrm{DS}\left(\mathfrak{A}_1\right) = \mathrm{DS}\left(\mathfrak{A}_1, \mathbb{X}\right), \qquad \mathrm{DS}(A) = \mathrm{DS}(\mathfrak{A}, A), \qquad \mathrm{DS} = \mathrm{DS}(\mathfrak{A}, \mathbb{X}),$$

and also

$$\mathrm{DS}(r, A) = \mathrm{DS}\left(\{B(x, r) : x \in \mathbb{X}\}, A\right) \quad (r > 0), \qquad \mathrm{DS}(r) = \mathrm{DS}(r, \mathbb{X}).$$

Deducing from $\mathbb{X}$, a sequence is said to be *deducing* without attributes.

18. Stepped images of functions

For any $\delta \in \mathrm{DS}$ and $r > 0$ the maps L_δ and L_r of type $\mathcal{D} \to \mathcal{D}$ are defined:

$$\left(\forall t \in \mathbb{R}_+\right) \quad X_t\left(L_\delta \xi\right) = X_{\sigma_\delta^n}(\xi), \quad X_t\left(L_r \xi\right) = X_{\sigma_r^n}(\xi),$$

where $\sigma_\delta^n(\xi) \le t < \sigma_\delta^{n+1}(\xi)$ and $\sigma_r^n(\xi) \le t < \sigma_r^{n+1}(\xi)$ respectively.

19. Measurability of maps

PROPOSITION 2.12. *For any $\delta \in \mathrm{DS}(\mathfrak{A} \cup \widetilde{\mathfrak{A}})$ and $r > 0$ the following conditions of measurability are fair*

$$L_\delta \in \mathcal{F}/\mathcal{F}, \qquad L_r \in \mathcal{F}/\mathcal{F}.$$

Proof. It follows from $\mathcal{F}/\mathcal{B}(\mathbb{Y})$-measurability of all $\beta_{\sigma_\delta^n}$ and $\beta_{\sigma_r^n}$ $(n \in \mathbb{N}_0)$ (see proposition 2.6(5)). A measurability of σ_Δ, σ_r can be found in Gihman and Skorokhod [GIH 73, p. 194]. A closure of sets MT and MT_+ concerning operation $\dot{+}$ is proved, for example, by Itô and McKean [ITO 65, p. 114]. $\qquad \square$

20. Existence of deducing sequences

THEOREM 2.1. *The following assertions are fair:*

(1) for any covering $\mathfrak{A}_0$ of the set $\mathbb{X}$, where $\mathfrak{A}_0 \subset \mathfrak{A}$, there exists a deducing sequence composed of elements of this covering;

(2) for any $A \in \mathfrak{A}$ and $r > 0$ there exists $\delta = (\Delta_1, \Delta_2, \ldots) \in \mathrm{DS}(A)$ such that $\Delta_i \subset A$ and $(\forall i \in \mathbb{N})$ diam $\Delta_i \le r$.

Proof. (1) Let $\mathbb{X} = \bigcup_{n=1}^{\infty} \mathbb{X}_n$ and any $\mathbb{X}_n$ belong to $\mathfrak{K}$, where $\mathfrak{K}$ is a set of all compact subsets of a set $\mathbb{X}$. Let $\mathfrak{A}_n$ be finite covering of $\mathbb{X}_n$ and $\mathfrak{A}_n \subset \mathfrak{A}_0$. For any $\xi \in \mathcal{D}$ and $t \in \mathbb{R}_+$ there is $n \in \mathbb{N}$ such that for all $s \leq t\, \xi(s) \in \mathbb{X}_n$. Let for all $s \leq t\, \xi(s) \in \mathbb{X}_n$ and for any finite set $\delta' = (\Delta_1, \ldots, \Delta_m)\, \sigma_{\delta'}(\xi) \leq t$, where $\Delta_i \in \mathfrak{A}_n$ $(1 \leq i \leq m)$, $m \in \mathbb{N}$. Then $t_0 = \sup_{\forall \delta'} \sigma_{\delta'}(\xi) \leq t$ and also there is a consequence $\{\delta'_m\} : t_m \uparrow t_0$, where $t_m = \sigma_{\delta'_m}(\xi)$. Let $x = \lim_{m \to \infty} X_{t_m}(\xi)$ where $x \in \mathbb{X}_n$ since $\mathbb{X}_n$ is closed. Let us consider two possibilities:

(a) $(\exists \Delta' \in \mathfrak{A}_n)\, x, \xi(t_0) \in \Delta'$;

(b) $(\exists \Delta', \Delta'' \in \mathfrak{A}_0)\, x \in \Delta',\, \xi(t_0) \in \Delta'' \backslash \Delta'$.

In case (a) $(\exists m \in \mathbb{N})\, (\forall s \in [\sigma_{\delta'_m}(\xi), t_0))\, \xi(s) \in \Delta'$. From here

$$(\sigma_{\delta'_m} \dotplus \sigma_{\Delta'})(\xi) = (t_0 \dotplus \sigma_{\Delta'})(\xi) > t_0$$

(see proposition 2.11(1)). It is a contradiction.

In case (b) $(\exists m \in \mathbb{N})\, (\forall s \in [\sigma_{\delta'_m}(\xi), t_0))$

$$\xi(s) \in \Delta', \qquad \xi(t_0) \in \Delta'' \backslash \Delta'.$$

From here $(\sigma_{\delta'_m} \dotplus \sigma_{\Delta'})(\xi) = t_0$ and $(\sigma_{\delta'_m} \dotplus \sigma_{\Delta'} \dotplus \sigma_{\Delta''})(\xi) > t_0$. It is a contradiction.

Hence, there is a finite sequence $\delta' = (\Delta_1, \ldots, \Delta'')$ $(\Delta_i \in \mathfrak{A}_n)$ such that $\sigma_{\delta'}(\xi) > t$. A set of all finite sequences composed from sets $\Delta \in \mathfrak{A}_n$ $(n \in \mathbb{N})$ is countable. Let us enumerate all these sequences by numbers of all positive integers and compose from them one infinite sequence δ. Then for any $\xi \in \mathcal{D}$ and $t \in \mathbb{R}_+$ there is a finite subsequence δ' such that $\sigma_{\delta'}(\xi) > t$. Let $\delta' = (\Delta_{k+1}, \ldots, \Delta_{k+m})$ for some k and m. Then $\sigma_{(\Delta_1, \ldots, \Delta_{k+m})} = \sigma_{(\Delta_1, \ldots, \Delta_k)} \dotplus \sigma_{\delta'}$. Under proposition 2.11(2) we have $\sigma_{(\Delta_1, \ldots, \Delta_{k+m})}(\xi) \geq \sigma_{\delta'}(xi) > t$. Hence, $\delta \in \mathrm{DS}$.

(2) From the previous assertion it follows that $\mathrm{DS}(r) = \mathrm{DS}(r, \mathbb{X}) \neq \emptyset$. Let $\delta' = (\Delta_1 \cap A, \Delta_2 \cap A, \ldots)$, where $\delta = (\Delta_1, \Delta_2, \ldots) \in \mathrm{DS}(r)$. We assume that $(\exists \xi \in \mathcal{D})$ $(\exists t \in \mathbb{R}_+)\, (\forall n \in \mathbb{N})$

$$X_{\sigma_{\delta'}^n}(\xi) \in A, \qquad \sigma_{\delta'}^n(\xi) \leq t.$$

Besides there exists $n_0 \in \mathbb{N}$ such that $\sigma_{\delta}^{n_0}(\xi) > t$. As A is open, $X_{\sigma_{\Delta_1 \cap A}}(\xi) \in A \Rightarrow X_{\sigma_{\Delta_1 \cap A}}(\xi) \notin \Delta_1$ and, hence, under proposition 2.11(1)

$$\sigma_{\Delta_1}(\xi) = (\sigma_{\Delta_1 \cap A} \dotplus \sigma_{\Delta_1})(\xi) = \sigma_{\Delta_1 \cap A}(\xi).$$

Let $\sigma_{\delta'}^k(\xi) = \sigma_{\delta}^k(\xi)$ $(k < n)$. Again we have

$$X_{\sigma_{\delta'}^{k+1}}(\xi) = X_{\sigma_{\Delta_{k+1} \cap A} \theta_{\sigma_{\delta'}^k}}(\xi) \in A \Longrightarrow X_{\sigma_{\Delta_{k+1} \cap A} \theta_{\sigma_{\delta'}^k}}(\xi) \notin \Delta_{k+1}$$

and, hence,

$$\sigma_\delta^{k+1}(\xi) = \left(\sigma_\delta^k \dotplus \sigma_{\Delta_{k+1}}\right)(\xi) = \left(\sigma_{\delta'}^k \dotplus \sigma_{\Delta_{k+1} \cap A} \dotplus \sigma_{\Delta_{k+1}}\right)(\xi)$$

$$= -\left(\sigma_{\delta'}^k \dotplus \sigma_{\Delta_{k+1} \cap A}\right)(\xi) = \sigma_{\delta'}^{k+1}(\xi).$$

From here $\sigma_{\delta'}^{n_0}(\xi) = \sigma_\delta^{n_0}(\xi)$. It is a contradiction. $\square$

21. Covering of a compact set

We designate $\mathfrak{K}$ a set of all compact subsets of a set $\mathbb{X}$ (see proof of the previous theorem).

THEOREM 2.2. *Let* $\delta = (\Delta_1, \Delta_2, \ldots) \in$ DS *and* $(\forall n \in \mathbb{N})$ $\Delta_n \neq \mathbb{X}$. *Then* $(\forall K \in \mathfrak{K})$ $(\exists r > 0)$ $(\forall x \in K)$ $(\forall n \in \mathbb{N})$ $(\exists k > n)$

$$B(x, r) \subset \Delta_k.$$

Proof. Let $\delta = (\Delta_1, \Delta_2, \ldots)$, where $(\forall n \in \mathbb{N})$ $\Delta_n \neq \mathbb{X}$ and $(\exists x \in \mathbb{X})$ $(\forall \varepsilon > 0)$ $(\exists n \in \mathbb{N})$ $(\forall k \geq n)$

$$B(x, \varepsilon) \backslash \Delta_k \neq \emptyset.$$

Then, obviously, there exists a sequence $(r_n)_1^\infty$ $(r_n > 0,\ r_n \downarrow 0)$ such that $(\forall n \in \mathbb{N})$ $(\forall k \geq n)$

$$B(x, r_n) \backslash \Delta_k \neq \emptyset.$$

We consider a sequence $(x_n)_0^\infty$, where

$$x_0 \in \mathbb{X},\ x_1 \in B(x, r_{k_1}) \backslash \Delta_{k_1}, \ldots,\ x_n \in B(x, r_{k_n}) \backslash \Delta_{k_n}, \ldots;$$

k_1 is a number of the first set in a sequence δ, covering x_0, and $(\forall n \in \mathbb{N})$ k_{n+1} is a number the first set after Δ_{k_n} covering x_n. Let us construct $\xi \in \mathcal{D}$ as follows: $\xi(t) = x_n$ for $t \in [2 - 2^{1-n}, 2 - 2^{-n})$ $(n \in \mathbb{N}_0)$. If sequence (k_n) is finite and k_M is its last term, we suppose $\xi(t) = x_M$ for $t \geq 2 - 2^{-M}$. If (k_n) is infinite, we suppose $\xi(t) = x$ for $t \geq 2$. Then

$$\left(\sigma_{\Delta_1} \dotplus \cdots \dotplus \sigma_{\Delta_{k_n}}\right)(\xi) = \sum_{i=0}^n 2^{-i} < 2,$$

i.e. $\sigma_\delta^n(\xi) \nrightarrow \infty$, and $\delta \notin$ DS. Hence, if $\delta \in$ DS and $(\forall n \in \mathbb{N})$ $\Delta_n \neq \mathbb{X}$, then $(\forall x \in \mathbb{X})$ $(\exists \varepsilon_x > 0)$ $(\forall n \in \mathbb{N})$ $(\exists k \geq n)$

$$B(x, \varepsilon_x) \subset \Delta_k.$$

Let

$$r(x) = \max\left\{ r : (\forall n \in \mathbb{N}) \ (\exists k \geq n) \ B(x, r) \subset \Delta_k \right\}.$$

Then $r(x)$ is a continuous function. Really, if $\rho(x_1, x_2) < c < r(x_1) \wedge r(x_2)$, then

$$B(x_1, r) \subset \Delta_k \implies B(x_2, r - c) \subset \Delta_k,$$

i.e. $r(x_2) \geq r(x_1) - c$. Similarly $r(x_1) \geq r(x_2) - c$. From here $|r(x_1) - r(x_2)| < c$, and therefore, on any compact set $K \subset \mathbb{X}$ the function $r(x)$ has a positive minimum. $\qquad \square$

22. Representation of sigma-algebra

PROPOSITION 2.13. *Let $r_m \downarrow 0$ and $\delta_m \in \mathrm{DS}(r_m)$. Then*

$$\mathcal{F} = \sigma\left(\beta_{\sigma_{\delta_m}^{k-1}}; k, m \in \mathbb{N} \right) = \sigma\left(\beta_{\sigma_{r_m}^{k-1}; k, m \in \mathbb{N}} \right).$$

The same is fair for $\delta \in \mathrm{DS}(\{[B(x, r_m)]\}_{x \in \mathbb{X}})$ and $\sigma_{[r_m]}$.

Proof. Let $\delta = (\Delta_1, \Delta_2, \ldots) \in \mathrm{DS}(r)$. Then $(\forall t \in \mathbb{R}_+) \ (\forall \xi \in \mathcal{D}) \ (\exists n \in \mathbb{N})$

$$t \in \left[\sigma_\delta^{n-1}(\xi), \sigma_\delta^n(\xi) \right).$$

From here $X_t(\xi), X_t(L_\delta \xi) \in \Delta_n$ and $\rho(X_t(\xi), X_t(L_\delta \xi)) < r$. Hence, $X_t(\xi) = \lim_{m \to \infty} X_t(L_{\delta_m} \xi)$. Therefore $(\forall S \in \mathcal{B}(\mathbb{X}))$

$$\left(X_t \circ L_{\delta_m} \right)^{-1} S \in \sigma\left(\beta_{\sigma_{\delta_m}^{k-1}}; k, m \in \mathbb{N} \right),$$

hence $X_t^{-1} S \in \sigma(\beta_{\sigma_{\delta_m}^{k-1}}; \ k, m \in \mathbb{N})$. The statement for maps σ_r and sequences of closed sets can be similarly proved. $\qquad \square$

23. Note about representation of sigma-algebra

The map L_r corresponds to a so-called "random" deducing sequence of balls depending on a choice of ξ. For non-random deducing sequences a rather stronger statement is fair, namely,

$$\mathcal{F} = \sigma\left(\sigma_{\delta_m}^k; \ k, m \in \mathbb{N} \right).$$

Really, in each $\Delta_m^i \in \delta_m$ let some interior point $x_m^i \in \Delta_m^i$ be chosen. Then for each ξ there is the function $L_{\delta_m}^*(\xi) : (L_{\delta_m}^*(\xi))(t) = x_m^i$, where $\sigma_{\delta_m}^{i-1}(\xi) \leq t < \sigma_{\delta_m}^i(\xi)$ $(i \in \mathbb{N})$, $\sigma_{\delta_m}^0(\xi) = 0$. Thus $\rho(\xi(t), (L_{\delta_m}^*(\xi)))(t)) < r_m$, as $x_m^i, \xi(t) \in \Delta_m^i$.

24. Representation of sigma-algebra in terms of Markov time

PROPOSITION 2.14. *Let* $\tau \in \mathrm{MT}_+$. *Then*

$$\mathcal{F} = \sigma\big(\alpha_\tau, \tau, \theta_\tau\big).$$

Proof. Under proposition 2.13 we have

$$\mathcal{F} = \sigma\big(\beta_\tau, \ \tau \in T\big(\sigma_\Delta, \Delta \in \mathfrak{A}\big)\big)$$

(see item 14). Let $\delta = (\Delta_1, \Delta_2, \ldots) \in \mathrm{DS}$, $t \in \mathbb{R}_+$, $S \in \mathcal{B}(\mathbb{X})$. Then for any $n \geq 0$ we have

$$\big\{\beta_{\sigma_\delta^n} \in [0, t] \times S\big\} = \big\{\sigma_\delta^n \leq t, \ X_{\sigma_\delta^n} \in S\big\} = \big\{\tau = 0, \ \sigma_\delta^n \leq t, \ X_{\sigma_\delta^n} \in S\big\}$$

$$\cup \bigcup_{k=1}^{n} \big\{\tau \in \big(\sigma_\delta^{k-1}, \sigma_\delta^k\big], \ \sigma_\delta^n \leq t, \ X_{\sigma_\delta^n} \in S\big\}$$

$$\cup \big\{\tau > \sigma_\delta^n, \ \sigma_\delta^n \leq t, \ X_{\sigma_\delta^n} \in S\big\}.$$

We have

$$\big\{\tau = 0, \ \sigma_\delta^n \leq t, \ X_{\sigma_\delta^n} \in S\big\} = \big\{\tau = 0, \ \sigma_\delta^n \circ \theta_\tau \leq t, \ X_{\sigma_\delta^n} \circ \theta_\tau \in S\big\},$$

$$\big\{\tau \in \big(\sigma_\delta^{k-1}, \sigma_\delta^k\big], \ \sigma_\delta^n \leq t, \ X_{\sigma_\delta^n} \in S\big\}$$

$$= \big\{\tau \in \big(\sigma_\delta^{k-1}, \sigma_\delta^k\big], \ \tau + \sigma_\delta^{k,n} \circ \theta_\tau \leq t, \ X_{\sigma_\delta^{k,n}} \circ \theta_\tau \in S\big\},$$

where $\sigma_\delta^{k,n} = \sigma_{\Delta_k} \dot{+} \cdots \dot{+} \sigma_{\Delta_n}$ and besides

$$\big\{\tau > \sigma_\delta^n, \ X_{tau_\delta^n} \in S\big\} = \big\{\tau > \sigma_\delta^n, \ \sigma_\delta^n \circ \alpha_\tau \leq t, \ X_{tau_\delta^n} \circ \alpha_\tau \in S\big\}.$$

Finally,

$$\big\{\sigma_\delta^n < \tau\big\} = \bigcup_{t \in \mathbb{R}_+'} \big\{\sigma_\delta^n \circ \alpha_\tau \leq t, \ t < \tau\big\},$$

where $\mathbb{R}_+'$ are all rational $t \in \mathbb{R}_+$. Hence, all components of a set $\{\beta_{\sigma_\delta^n} \in [0, t] \times S\}$ belong to $\sigma(\alpha_\tau, \tau, \theta_\tau)$. It is true as well for the set $\{\beta_{\sigma_\delta^n} = \overline{\infty}\}$. $\quad\square$

25. One more sigma-algebra of preceding events

For $\tau \in \mathrm{MT}_+$ let $\mathcal{F}_\tau = \sigma(\alpha_\tau, \tau)$. Under theorem 2.7 for $\tau \in \mathrm{MT}$

$$\mathcal{F}_\tau = \sigma\big(\alpha_\tau\big) = \sigma\big(\alpha_\tau, \tau\big),$$

i.e. the new definition is coordinated with old one (see item 8).

26. One more representation of a sigma-algebra

Let us denote by $\mathbb{R}'_+$ the all rational set $t \in \mathbb{R}_+$ (this is used in the proof of the previous proposition).

THEOREM 2.3. *For any $\tau_1, \tau_2 \in \mathrm{MT}_+$ we have*

$$\mathcal{F}_{\tau_1 \dotplus \tau_2} = \sigma\left(\mathcal{F}_{\tau_1}, \theta_{\tau_1}^{-1}\mathcal{F}_{\tau_2}\right).$$

Proof. By item 25, $\mathcal{F}_{\tau_1 \dotplus \tau_2} = \sigma(\alpha_{\tau_1 \dotplus \tau_2}, \tau_1 \dotplus \tau_2)$. We have

$$\sigma\left(\mathcal{F}_{\tau_1}, \theta_{\tau_1}^{-1}\mathcal{F}_{\tau_2}\right) = \sigma\left(\alpha_{\tau_1}, \tau_1, \alpha_{\tau_2} \circ \theta_{\tau_1}, \tau_2 \circ \theta_{\tau_1}\right),$$

and

$$\alpha_{\tau_1} = \alpha_{\tau_1} \circ \alpha_{\tau_1 \dotplus \tau_2} \in \sigma\left(\alpha_{\tau_1 \dotplus \tau_2}\right)/\mathcal{F},$$

$$\tau_1 \in \sigma\left(\alpha_{\tau_1 \dotplus \tau_2}, \tau_1 \dotplus \tau_2\right)/\mathcal{B}\left(\overline{\mathbb{R}}_+\right),$$

because

$$\{\tau_1 < t\} = \{\tau_1 < t, \ \tau_1 < \tau_1 \dotplus \tau_2\} \cup \{\tau_1 < t, \ \tau_1 = \tau_1 \dotplus \tau_2\}$$

$$= \{\tau_1 \circ \alpha_{\tau_1 \dotplus \tau_2} < t\} \cup \bigcup_{t \in \mathbb{R}'_+} \{\tau_1 \circ \alpha_{\tau_1 \dotplus \tau_2} \le t, \ t < \tau_1 \dotplus \tau_2\}$$

$$\cup \{\tau_1 \dotplus \tau_2 < t, \ \tau_1 \ge \tau_1 \dotplus \tau_2\} \in \sigma\left(\alpha_{\tau_1 \dotplus \tau_2}, \ \tau_1 \dotplus \tau_2\right)$$

(see proposition 2.14). It is not difficult to show that formula [2.16] is correct with replacement $\mathcal{T}_a$ by $\widetilde{\mathcal{T}}_a$. Hence

$$\alpha_{\tau_2} \circ \theta_{\tau_1} = \theta_{\tau_1} \circ \alpha_{\tau_1 \dotplus \tau_2} \in \sigma\left(\alpha_{\tau_1 \dotplus \tau_2}\right)/\left(\mathcal{F} \cap \{\tau_1 < \infty\}\right),$$

$$\tau_2 \circ \theta_{\tau_1} = \tau_1 \dotplus \tau_2 - \tau_1 \in \sigma\left(\alpha_{\tau_1 \dotplus \tau_2}, \ \tau_1 \dotplus \tau_2\right)/\mathcal{B}\left(\overline{\mathbb{R}}_+\right).$$

From here $\sigma\left(\mathcal{F}_{\tau_1}, \theta_{\tau_1}^{-1}\mathcal{F}_{\tau_2}\right) \subset \mathcal{F}_{\tau_1 \dotplus \tau_2}$. On the other hand, $\tau_1 \dotplus \tau_2 \in \sigma(\tau_1, \tau_2 \circ \theta_{\tau_1})$. Under proposition 2.14 $\mathcal{F} = \sigma(\alpha_{\tau_1}, \tau_1, \theta_{\tau_1})$. For all $S \in \mathcal{F}$ and $t \in \mathbb{R}_+$ we have $\alpha_{\tau_1 \dotplus \tau_2}^{-1} \alpha_{\tau_1}^{-1} S = \alpha_{\tau_1}^{-1} S \in \sigma(\alpha_{\tau_1})$; and also

$$\alpha_{\tau_1 \dotplus \tau_2}^{-1}\{\tau < t\} = \{\tau_1 \circ \alpha_{\tau_1 \dotplus \tau_2} < t, \ \tau_2 \circ \theta_{\tau_1} > 0\} \cup \{\tau_1 \circ \alpha_{\tau_1 \dotplus \tau_2} < t, \ \tau_2 \circ \theta_{\tau_1} = 0\}$$

$$= \{\tau_1 < t, \ \tau_2 \circ \theta_{\tau_1} > 0\} \cup \{\tau_1 \circ \alpha_{\tau_1} < t, \ \tau_2 \circ \theta_{\tau_1} = 0\}$$

$$\in \sigma\left(\tau_1, \alpha_{\tau_1}, \tau_2 \circ \theta_{\tau_1}\right)$$

and

$$\alpha_{\tau_1 \dotplus \tau_2}^{-1} \theta_{\tau_1}^{-1} S = \theta_{\tau_1}^{-1} \alpha_{\tau_2}^{-1} S \in \sigma\left(\alpha_{\tau_2} \circ \theta_{\tau_1}\right).$$

From here $\alpha_{\tau_1 \dotplus \tau_2} \in \sigma(\mathcal{F}_{\tau_1}/\mathcal{F}, \theta_{\tau_1}^{-1}\mathcal{F}_{\tau_2})$ and

$$\mathcal{F}_{\tau_1 \dotplus \tau_2} \subset \sigma\left(\mathcal{F}_{\tau_1}, \theta_{\tau_1}^{-1}\mathcal{F}_{\tau_2}\right). \qquad \square$$

2.4. Correct exit and continuity of points of the first exit

A property of a correct first exit of a trajectory $\xi \in \mathcal{D}$ from a set $\Delta \subset \mathbb{X}$ is defined (see [HAR 77]). It is shown that the class of sets, from which a trajectory has a correct exit with probability one, is rather rich. On the other hand, the pair of the first exit β_{σ_Δ} is continuous on a set of functions correctly going out from Δ, with respect to the metric of Stone-Skorokhod on $\mathcal{D}$ (see Gihman and Skorokhod [GIH 71], Skorokhod [SKO 56, SKO 61], Stone [STO 63]).

27. Correct exit

For any $\Delta \subset \mathbb{X}$ and $r > 0$ let

$$\Delta^{+r} = \left\{ x \in \mathbb{X} : \rho(x, \Delta) < r \right\},$$
$$\Delta^{-r} = \left\{ x \in \mathbb{X} : \rho(x, \mathbb{X}\backslash\Delta) > r \right\}.$$

Let us state that the function $\xi \in \mathcal{D}$ has a correct first exit from Δ, if

(1) $\xi(0) \notin \partial\Delta$;

(2) $\beta_{\sigma_{\Delta^{+r}}}\xi \to \beta_{\sigma_\Delta}\xi \ (r \downarrow 0)$;

(3) $\beta_{\sigma_{\Delta^{-r}}}\xi \to \beta_{\sigma_\Delta}\xi \ (r \downarrow 0)$.

Let $\Pi(\Delta)$ be a set of all $\xi \in \mathcal{D}$, having a correct first exit from a set Δ. In this case, if $\Delta = B(\xi(0), r)$ for given ξ (the choice Δ depends on ξ), we designate the previous set as $\Pi(r) \equiv \Pi_1(r)$. Let

$$\Pi(\Delta_1, \Delta_2) = \left(\Pi(\Delta_1) \cap \{\sigma_{\Delta_1} = \infty\} \right) \cup \left(\Pi(\Delta_1) \cap \theta_{\sigma_{\Delta_1}}^{-1} \Pi(\Delta_2) \right),$$

where $\Delta_1, \Delta_2 \subset \mathbb{X}$ and, according to the definition $\theta_\tau, \theta_\tau^{-1} A \subset \{\tau < \infty\}$ and

$$\Pi(\Delta_1, \dots, \Delta_m) = \left(\bigcup_{k=1}^{m-1} \bigcap_{i=1}^{k} \theta_{\tau_{i-1}}^{-1} \Pi(\Delta_i) \cap \{\tau_k = \infty\} \right) \cup \bigcap_{i=1}^{m} \theta_{\tau_{i-1}}^{-1} \Pi(\Delta_i),$$

where $\tau_0 = 0$, $\tau_i = \tau_{i-1} \dotplus \sigma_{\Delta_i}$, and

$$\Pi(\Delta_1, \Delta_2, \dots) = \bigcap_{m-1}^{\infty} \Pi(\Delta_1, \dots, \Delta_m).$$

Similarly define $\Pi_m(r_1, \dots, r_m)$ $(r_i > 0)$ and $\Pi(r_1, r_2, \dots)$, In particular, all r_i can be equal to each other: $\Pi_m(r) = \Pi_m(r, \dots, r)$.

28. Discontinuity and interval of constancy

Let us designate

$$h(\xi, t) = \bigcup \{\Delta' \in \delta : \sigma_{\Delta'}(\xi) \leq t\} \quad (\xi \in \mathcal{D}, \, t \in \mathbb{R}_+).$$

PROPOSITION 2.15. *The following assertions are fair:*

(1) if $\beta_{\sigma_{\Delta-r}}\xi \nrightarrow \beta_{\sigma_\Delta}$ $(r \downarrow 0, \, \Delta \in \mathfrak{A})$, then there exists a point of discontinuity $t \leq \sigma_\Delta \xi$ of the function ξ, for which $\xi(t-0) \in \partial\Delta$;

(2) let δ be such a family of open sets, that for any $\Delta_1, \Delta_2 \in \delta$ $(\Delta_1 \neq \Delta_2)$ it is either $[\Delta_1] \subset \Delta_2$, or $[\Delta_2] \subset \Delta_1$; besides let $\Delta \in \delta$ be not a maximal element in δ and $\beta_{\sigma_{\Delta+r}}\xi \nrightarrow \beta_{\sigma_\Delta}\xi$ $(r \downarrow 0)$; then

$$\{t \in \mathbb{R}_+ : h(\xi, t) = \Delta\}$$

is an interval of constancy of the function $h(\xi, \cdot)$.

Proof. (1) It is obvious.

(2) For any $\xi \in \mathcal{D}$ we have $\sigma_{\Delta+r}(\xi) \geq \sigma_\Delta(\xi)$ and $\sigma_{\Delta+r}(\xi)$ does not increase with $r \downarrow 0$. If $\sigma_{\Delta+r}(\xi) \downarrow t > \sigma_\Delta(\xi)$ and $t' \in [\sigma_\Delta(\xi), t)$, then $h(\xi, t') \supset \Delta$. On the other hand, let $h(\xi, t') \neq \Delta$ and also there exists $\Delta_1 \supset \Delta$ $(\Delta_1 \in \delta)$ such that $\sigma_{\Delta_1}(\xi) \leq t'$. Since $\sigma_{\Delta+r}(\xi) \downarrow \sigma_{[\Delta]}(\xi)$, we obtain an inconsistency: $[\Delta] \subset \Delta_1$ and $\sigma_{[\Delta]}(\xi) > \sigma_{\Delta_1}(\xi)$. $\qquad \square$

29. Countable subfamily

COROLLARY 2.1. *For any $\xi \in \mathcal{D}$ the family δ from proposition 2.15(2) contains nothing more than countable subfamily δ' such that $\xi \notin \Pi(\Delta)$ with $\Delta \in \delta'$.*

Proof. It follows from denumerability of a set of all point of discontinuities of function ξ and all intervals of a constancy of function $h(\xi, \cdot)$. $\qquad \square$

30. Correct exit with probability 1.

THEOREM 2.4. *Let δ be a family of concentric balls $(B(x_0, r))_{r>0}$ $(x_0 \in \mathbb{X})$. For any probability measure P on $\mathcal{D}$ and any measurable map $\tau : \mathcal{D} \to \overline{\mathbb{R}}_+$ there exists no more than countable subfamily of balls $\delta' \subset \delta$ such that for $\Delta \in \delta'$*

$$P\big(\theta_\tau^{-1}(\mathcal{D} \backslash \Pi(\Delta))\big) > 0.$$

Proof. We consider $\mathcal{B}(\mathcal{D})$-measurable map $f_{k,n} : \mathcal{D} \rightarrow \mathbb{X}$, where $f_{k,n}(\xi) = \xi(t_{k,n} - 0)$ and $t_{k,n}$ is the k-th discontinuity of the function ξ with size of the jump $\geq 1/n$. Then both maps $f_{k,n} \circ \theta_\tau$ and $\rho(f_{k,n} \circ \theta_\tau, x_0)$ on $\{\tau < \infty\}$ are also $\mathcal{B}(\mathcal{D})$-measurable. From here the set of those $r > 0$, for which

$$P(\{\rho(f_{k,n} \circ \theta_\tau, x_0) = r\} \cap \{\tau < \infty\}) > 0,$$

is not more than countable. Let us consider $\mathcal{B}(\mathcal{D})$-measurable map $h(\cdot, t) : \mathcal{D} \rightarrow \mathbb{R}$, where $h(\xi, t) = \sup\{r : \sigma_{B(x_0,r)}(\xi) \leq t\}$. From here $h = h(\cdot, \cdot)$ is $\mathcal{B}(\mathcal{D})$-measurable map $\mathcal{D}$ in $\mathcal{D}_1 = \mathcal{D}_{[0,\infty)}(R^1)$. Let $g_{k,n} : \mathcal{D}_1 \rightarrow \mathbb{R}$ be such a function, where $g_{k,n}(\xi)$ is a value of $\xi \in \mathcal{D}_1$ on the k-th interval of constancy of length not less than $1/n$. Obviously, this map is $\mathcal{B}(\mathcal{D}_1)$-measurable. Then the map $g_{k,n} \circ h \circ \theta_\tau$ is $\mathcal{B}(\mathcal{D}_1)$-measurable on $\{\tau < \infty\}$, and the set of those r for which

$$P(\{g_{k,n} \circ h \circ \theta_\tau = r\} \cap \{\tau < \infty\}) > 0,$$

is not more than countable. Since for $\Delta = B(x_0, r)$

$$\theta_\tau^{-1}(\mathcal{D} \backslash \Pi(\Delta)) \subset \{\tau < \infty\}$$

$$\cap \left(\bigcup_{k,n} \{\rho(f_{k,n} \circ \theta_\tau, x_0) = r\} \cup \bigcup_{k,n} \{g_{k,n} \circ h \circ \theta_\tau = r\} \right),$$

the theorem is proved. □

31. Note about the family of balls

The similar theorem for a family of concentric balls depending on a choice ξ, like $(B(\xi(0), r))_{r>0}$, is fair as well: there exists nothing more than countable set of those r for which $P(\theta_\tau^{-1}(\mathcal{D} \backslash \Pi_1(r))) > 0$.

32. Deducing sequences of correct exit

COROLLARY 2.2. *For any probability measure P on $\mathcal{D}$ and sequence $(r_m)_1^\infty$ $(r_m \downarrow 0)$ there exists a sequence $(\delta_m)_1^\infty$ $(\delta_m \in \text{DS}(r'_m), r'_m \leq r_m)$ such that $P(\bigcap_{m=1}^\infty \Pi(\delta_m)) = 1$.*

Proof. It follows from theorem 2.1 and such an obvious fact that for any $\delta \in \text{DS}$ the replacement of any open set $\Delta \in \delta \in \text{DS}$ by anything larger reduces it to a new sequence, which also is deducing. □

33. Note about a set

According to a note of item 31 and denumerability of a set of those r, for which $P(\Pi(r)) = 1$, there is a sequence $(r_m)_1^\infty$, for which $P(\bigcap_{m=1}^\infty \Pi(r_m)) = 1$, where

$$\Pi(r_m) = \bigcap_{k=1}^\infty \Pi_k(r_m).$$

34. The Skorokhod metric

Let φ be a map of the set $\mathcal{D}$ in a set of all maps $\eta : [0,1] \to \mathbb{Y}'$ (see item 9), defined as follows: $(\forall \xi \in \mathcal{D})$

$$\eta(t) = (\varphi(\xi))(t) = \begin{cases} \bar{1}, & t = 1, \\ \left(t, \xi\left(\tan\frac{\pi}{2}t\right)\right), & t \in [0,1). \end{cases}$$

Let $\overline{\mathcal{D}} = \varphi\mathcal{D}$. Then $\overline{\mathcal{D}}$ is a set of all cádlág maps $\eta : [0,1] \to \mathbb{Y}'$ continuous from the left at point 1 (with respect to metric $\rho_{\mathbb{Y}'}$), and $\eta(1) = \bar{1}$. The Skorokhod metric on a set $\overline{\mathcal{D}}$ is said to be the function $\rho_{\overline{\mathcal{D}}}$ of the following view (see Gihman and Skorokhod [GIH 71, p. 497]):

$$\rho_{\overline{\mathcal{D}}}(\eta_1, \eta_2) = \inf_{\lambda \in \Lambda[0,1]} \left(\sup_{t \leq 1} \rho_{\mathbb{Y}'}\left(\eta_1(t), \eta_2(\lambda(t))\right) + \sup_{t \leq 1} |t - \lambda(t)| \right),$$

where $\eta_1, \eta_2 \in \overline{\mathcal{D}}$, $\Lambda[0,t]$ is the set of all increasing maps of the interval $[0,t]$ on itself $(t > 0)$. For any $\xi_1, \xi_2 \in \mathcal{D}$ we define

$$\rho_{\mathcal{D}}(\xi_1, \xi_2) = \rho_{\overline{\mathcal{D}}}(\varphi(\xi_1), \varphi(\xi_2)).$$

The metric $\rho_{\mathcal{D}}$ is referred to as the Stone-Skorokhod metric. It is known [GIH 71] that this metric transforms space $\mathcal{D}$ into separable metric space.

The metric $\rho_{\mathcal{C}}$ is similarly defined on a set $\mathcal{C}$ of all continuous $\xi \in \mathcal{D}$:

$$\rho(\eta_1, \eta_2) = \sup\left\{\rho_{\mathbb{Y}'}\left(\eta_1(t), \eta_2(t)\right), \ 0 \leq t \leq 1\right\},$$

where $\overline{\mathcal{C}} = \varphi\mathcal{C}$ and $\eta_1, \eta_2 \in \overline{\mathcal{C}}$. It generates a topology of uniform convergence on all finite intervals. Thus, for $\xi_1, \xi_2 \in \mathcal{C}$

$$\rho_{\mathcal{C}}(\xi_1, \xi_2) = \rho_{\overline{\mathcal{C}}}(\varphi(\xi_1), \varphi(\xi_2)).$$

35. Borel sigma-algebra

PROPOSITION 2.16. *The following representations are fair*

$$\mathcal{F} = \mathcal{B}(\mathcal{D}, \rho_{\mathcal{D}}), \qquad \mathcal{F} \cap \mathcal{C} = \mathcal{B}(\mathcal{C}, \rho_{\mathcal{C}}).$$

Proof. For proof of these statements, see Gihman and Skorokhod [GIH 71].   □

36. Distance in the Skorokhod metric

Let $\xi_1, \xi_2 \in \mathcal{D}$ and t be a continuity point of both functions. Let us designate $\rho_{\mathcal{D}}^t(\xi_1, \xi_2)$ a distance in the Skorokhod metric on an interval $[0, t]$:

$$\rho_{\mathcal{D}}^t(\xi_1, \xi_2) = \inf_{\lambda \in \Lambda[0,t]} \left(\sup_{s \leq t} \rho(\xi_1(s), \xi_2(\lambda(s))) + \sup_{s \leq t} |s - \lambda(s)| \right).$$

Similarly for $\xi_1, \xi_2 \in \mathcal{C}$

$$\rho_{\mathcal{C}}^t(\xi_1, \xi_2) = \sup_{s \leq t} \rho(\xi_1(s), \xi_2(s)).$$

37. Estimate from above

THEOREM 2.5. *The following estimates are fair:*

(1) let $\xi_1, \xi_2 \in \mathcal{D}$ and t be a continuity point of both functions; then

$$\rho_{\mathcal{D}}(\xi_1, \xi_2) \leq \frac{4}{\pi} \rho_{\mathcal{D}}^t(\xi_1, \xi_2) + \frac{\pi}{2} - \arctan t;$$

(2) for any $\xi_1, \xi_2 \in \mathcal{C}$ and $t > 0$

$$\rho_{\mathcal{C}}(\xi_1, \xi_2) \leq \rho_{\mathcal{C}}^t(\xi_1, \xi_2) + \frac{\pi}{2} - \arctan t.$$

Proof. (1) Let $a(s) = (2/\pi) \arctan s$ ($s \in \mathbb{R}_+$). We have

$$\rho_{\mathcal{D}}(\xi_1, \xi_2)$$

$$= \inf_{\lambda \in \Lambda[0,1]} \left(\sup_{s < t} \rho_{Y'}((s, \xi_1(a^{-1}(s))), (\lambda(s), \xi_2(a^{-1}\lambda(s)))) + \sup_{s < 1} |s - \lambda(s)| \right)$$

$$\leq \inf_{\lambda \in \Lambda[0,a(t)]} \left(\sup_{s \leq a(t)} \rho_{Y'}((s, \xi_1(a^{-1}(s))), (\lambda(s), \xi_2(a^{-1}\lambda(s)))) \right)$$

$$\vee \left(\sup_{1 > s > a(t)} \rho_{Y'}((s, \xi_1(a^{-1}(s))), (s, \xi_2(a^{-1}(s)))) + \sup_{s \leq a(t)} |s - \lambda(s)| \right).$$

Furthermore

$$\sup_{s\leq a(s)} \rho_{Y'}\big((s,\xi_1(a^{-1}(s))),(\lambda(s),\xi_2(a^{-1}\lambda(s)))\big)$$

$$= \sup_{s\leq a(t)} \big((1-s)\wedge(1-\lambda(s))\arctan\rho(\xi_1(a^{-1}(s)),\xi_2(a^{-1}\lambda(s)))+|s-\lambda(s)|\big)$$

$$\leq \sup_{s\leq a(t)} \arctan\rho(\xi_1(a^{-1}(s)),\xi_2(a^{-1}\lambda(s)))+\sup_{s\leq a(t)}|s-\lambda(s)|$$

$$= \arctan\sup_{s\leq t}\rho(\xi_1(s),\xi_2(\lambda_1(s)))+\sup_{s\leq t}|a(s)-a(\lambda_1(s))|$$

$$\leq \arctan\sup_{s\leq t}\rho(\xi_1(s),\xi_2(\lambda_1(s)))+\frac{2}{\pi}\sup_{s\leq t}|s-\lambda_1(s)|,$$

where $\lambda_1 = a_1^{-1}\circ\lambda\circ a_1$ is an increasing map of $[0,t]$ on itself, a_1 is a contraction of a on $[0,t]$. Besides

$$\sup_{a(t)<s<1} \rho_{Y'}\big((s,\xi_1(a^{-1}(s))),(s,\xi_2(a^{-1}(s)))\big)$$

$$= \sup_{a(t)<s<1} (1-s)\arctan\rho(\xi_1(a^{-1}(s)),\xi_2(a^{-1}(s)))$$

$$\leq (1-a(t))\frac{\pi}{2}=\frac{\pi}{2}-\arctan t.$$

From here

$$\rho_D(\xi_1,\xi_2)\inf_{\lambda_1\in\Lambda[0,t]}\left(\left(\arctan\sup_{s\leq t}\rho(\xi_1(s),\xi_2(\lambda_1(s)))+\frac{2}{\pi}\sup_{s\leq t}|s-\lambda_1(s)|\right)\right.$$

$$\left.\vee\left(\frac{\pi}{2}-\arctan t\right)+\frac{2}{\pi}\sup_{s\leq t}|s-\lambda_1(s)|\right)$$

$$\leq \inf_{\lambda_1\in\Lambda[0,t]}\left(\arctan\sup_{s\leq t}\rho(\xi_1(s),\xi_2(\lambda_1(s)))+\frac{4}{\pi}\sup_{s\leq t}|s-\lambda_1(s)|\right)$$

$$+\frac{\pi}{2}-\arctan t\leq\frac{4}{\pi}\rho_D^t(\xi_1,\xi_2)+\frac{\pi}{2}-\arctan t.$$

(2) It is proved similarly. □

38. *Estimates for metrics*

PROPOSITION 2.17. *The following properties are fair:*

(1) $(\forall t \in \mathbb{R}_+) \, (\forall \varepsilon > 0) \, (\exists \delta > 0)$

$$\rho_D(\xi_1, \xi_2) < \delta \Longrightarrow (\exists \lambda \in \Lambda[0, \infty)) \quad \sup_{s \le t} \rho\big(\xi_1(s), \xi_2(\lambda(s))\big) + \sup_{s \le t} \big|s - \lambda(s)\big| < \varepsilon;$$

(2) $(\forall \xi_1, \xi_2 \in C) \, (\forall t \in \mathbb{R}_+) \, (\forall \varepsilon > 0) \, (\exists \delta > 0)$

$$\rho_C(\xi_1, \xi_2) < \delta \Longrightarrow \rho_D^t(\xi_1, \xi_2) < \varepsilon.$$

Proof. (1) If $\rho_D(\xi_1, \xi_2) < \delta$, then $(\exists \lambda \in \Lambda[0, 1]) \, \sup_{s<1} |s - \lambda(s)| < \delta$ and

$$\sup_{s<1}(1 - s) \wedge \big(1 - \lambda(s)\big) \arctan \rho\big(\xi_1\big(a^{-1}(s)\big), \xi_2\big(a^{-1}(\lambda(s))\big)\big) < \delta,$$

where $a(t) = (2/\pi) \arctan t \ (t \in \mathbb{R}_+)$. Let $t \in \mathbb{R}_+$ and $\delta < (1 - a(t))/(1 + 2/\pi)$. Then $\sup_{s \le t} |a(s) - \lambda(a(s))| < \delta$, and since

$$|\arctan s - \arctan r| \ge \frac{|s - r|}{1 + (s \vee r)^2},$$

$\sup_{s \le t} |s - a^{-1}\lambda(a(s))| < (\pi/2)\delta(1 + \tan^2(\pi/2)(a(t) + \delta))$, where

$$s \vee \big(a^{-1}\lambda\big(a(s)\big)\big) = a^{-1}a(s) \vee \lambda\big(a(s)\big) < a_{-1}\big(a(t) + \delta\big)$$
$$= \tan \frac{\pi}{2}\big(a(t) + \delta\big) < \infty.$$

Furthermore,

$$\sup_{s \le t} \big(1 - a(s)\big) \wedge \big(1 - \lambda\big(a(s)\big)\big) \arctan \rho\big(\xi_1(s), \xi_2\big(a^{-1}\lambda\big(a(s)\big)\big)\big) < \delta,$$

and since $(1 - a(s)) \wedge (1 - \lambda(a(s))) > 1 - a(t) - \delta$,

$$\sup_{s \le t} \rho\big(\xi_1(s), \xi_2\big(a_{-1}\lambda\big(a(s)\big)\big)\big) < \tan \frac{\delta}{1 - a(t) - \delta}.$$

Let us choose δ such that $(\pi/2)\delta(1 + \tan^2(\pi/2)(a(t) + \delta)) \le \varepsilon/2$ and $\tan(\delta/(1 - a(t) - \delta)) \le \varepsilon/2$. Note that $a^{-1} \circ \lambda \circ a \in \Lambda[0, \infty)$. So the first statement is proved.

(2) The second statement is proved similarly.   □

39. Continuity

THEOREM 2.6. *For any $k \in \mathbb{N}$ and $\Delta_1, \ldots, \Delta_k \in \mathfrak{A}$ the map $\beta_{(\Delta_1, \ldots, \Delta_k)}$ is continuous on $\Pi(\Delta_1, \ldots, \Delta_k)$.*

Proof. Let $\Delta \in \mathfrak{A}$, $\sigma_\Delta(\xi) = \infty$, $\xi \in \Pi(\Delta)$ and $\rho_D(\xi_n, \xi) \to 0$ $(n \to \infty)$. Then under proposition 2.17 $(\forall t \in \mathbb{R}_+)$ $(\forall \varepsilon > 0)$ $(\exists \delta_{\varepsilon,t} > 0)$ $(\exists n_0 \in \mathbb{N})$ $(\forall n > n_0)$ the inequality $\rho_D(\xi_n, \xi) < \delta_{\varepsilon,t}$ is sufficient for the conditions

$$\sup_{s \le t} \rho\big(\xi(s), \xi_n(\lambda_n(s))\big) < \varepsilon, \qquad [2.17]$$

$$\sup_{s \le t} |s - \lambda_n(s)| < \varepsilon. \qquad [2.18]$$

to be fulfilled for $\lambda_n \in \Lambda[0, \infty])$. Let $\varepsilon > 0$ such that $\sigma_{\Delta-\varepsilon}(\xi) > t$. Then $\sigma_\Delta(\xi_n \circ \lambda_n) > t$ and, hence, $\lambda_n^{-1}\sigma_\Delta(\xi_n) > t$, $\sigma_\Delta(\xi_n) > \lambda_n(t)$ and $\sigma_\Delta(\xi_n) > t - \varepsilon$. From here, because of arbitrary choice t and ε, we obtain

$$\sigma_\Delta(\xi_n) \longrightarrow \infty, \qquad \beta_{\sigma_\Delta}(\xi_n) \longrightarrow \beta_{\sigma_\Delta}(\xi). \qquad [2.19]$$

Let $\sigma_\Delta(\xi) < \infty$ and $\rho_D(\xi_n, \xi) \to 0$ $(n \to \infty)$. Then the same conditions are sufficient for inequalities [2.17] and [2.18]. Let us take ε and t such that $\sigma_{\Delta+\varepsilon}(\xi) \le t$. Then $\sigma_{\Delta-\varepsilon}(\xi) \le \sigma_\Delta(\xi_n \circ \lambda_n) \le \sigma_{\Delta+\varepsilon}(\xi)$. Evidently, $\sigma_\Delta(\xi_n \circ \lambda_n) = \lambda_n^{-1}\sigma_\Delta(\xi_n)$. Hence, the previous inequality is equivalent to the inequality

$$\lambda_n \sigma_{\Delta-\varepsilon}(\xi) \le \sigma_\Delta(\xi_n) \le \lambda_n \sigma_{\Delta+\varepsilon}(\xi), \qquad [2.20]$$

and consequently

$$\sigma_{\Delta-\varepsilon}(\xi) - \varepsilon \le \sigma_\Delta(\xi_n) \le \sigma_{\Delta+\varepsilon}(\xi) + \varepsilon. \qquad [2.21]$$

Let us consider two possibilities:

(a) $\sigma_\Delta(\xi)$ is a point of continuity of the function ξ; then

$$\rho\big(X_{\sigma_\Delta}(\xi), X_{\sigma_\Delta}(\xi_n \circ \lambda_n)\big)$$
$$\le \rho\big(\xi(\sigma_\Delta(\xi_n \circ \lambda_n)), (\xi_n \circ \lambda_n)(\sigma_\Delta(\xi_n \circ \lambda_n))\big) + \rho\big(\xi(\sigma_\Delta(\xi)), \xi(\sigma_\Delta(\xi_n \circ \lambda_n))\big)$$
$$< \varepsilon + \sup\big\{\rho\big(\xi(\sigma_\Delta(\xi)), \xi(s)\big) : \sigma_{\Delta-\varepsilon}(\xi) \le s \le \sigma_{\Delta+\varepsilon}(\xi)\big\};$$

from here $\beta_{\sigma_\Delta}\xi_n \to \beta_{\sigma_\Delta}\xi$;

(b) $\sigma_\Delta(\xi)$ is a discontinuity point of the function ξ; because $\xi \in \Pi(\Delta)$ we have $\xi(\sigma_\Delta(\xi) - 0) \in \Delta$ and $(\exists r > 0)$

$$X_{\sigma_{\Delta-r}}\xi \notin \Delta, \qquad X_{\sigma_{\Delta-r}}\xi = X_{\sigma_\Delta}\xi;$$

since $\rho(\xi(s), (\xi_n \circ \lambda_n)(s)) < \varepsilon$ $(s \le t)$ we have for $\varepsilon < r$ $\sigma_\Delta(\xi_n \circ \lambda_n) \ge \sigma_\Delta(\xi)$; from here

$$\rho\big(X_{\sigma_\Delta}\xi, X_{\sigma_\Delta}(\xi_n \circ \lambda_n)\big)$$
$$\le \rho\big(\xi(\sigma_\Delta(\xi_n \circ \lambda_n)), (\xi_n \circ \lambda_n)(\sigma_\Delta(\xi_n \circ \lambda_n))\big) + \rho\big(X_{\sigma_\Delta}\xi, \xi(\sigma_\Delta(\xi_n \circ \lambda_n))\big)$$
$$\le \varepsilon + \sup_{\sigma_\Delta(\xi) \le t \le \sigma_{\Delta-r}} \rho\big(X_{\sigma_\Delta}\xi, \xi(t)\big);$$

therefore by right-continuity of functions ξ we have

$$\rho\left(X_{\sigma_\Delta}\xi, X_{\sigma_\Delta}\xi_n\right) \longrightarrow 0, \qquad \beta_{\sigma_\Delta}\xi_n \longrightarrow \beta_{\sigma_\Delta}\xi.$$

Let $\Delta_1,\ldots,\Delta_k \in \mathfrak{A}$, $\tau_0=0$, $\tau_k=\tau_{k-1}\dotplus\sigma_{\Delta_k}$, $\tau_{k-1}(\xi)<\infty$, $\xi\in\Pi(\Delta_1,\ldots,\Delta_k)$, $\rho_D(\xi_n,\xi)\to 0$ and $\beta_{\tau_i}\xi_n \to \beta_{\tau_i}\xi$ ($i \leq k-1$, $n \to \infty$, $k \in \mathbb{N}$), and for all $N > n_0$

$$X_{\tau_i}\xi_n \in \Delta_{i+1} \iff X_{\tau_i}\xi \in \Delta_{i+1}, \qquad X_{\tau_i}\xi_n \notin \left[\Delta_{i+1}\right] \iff X_{\tau_i}\xi \notin \left[\Delta_{i+1}\right]$$

(see item 27 for definition of $\Pi(\Delta)$). There may be three possibilities:

(a) $\tau_k(\xi) = \infty$; then $X_{\tau_{k-1}}\xi \in \Delta_k$ and if $\varepsilon > 0$ is such that $(\tau_{k-1}\dotplus\sigma_{\Delta_k^{-\varepsilon}})(\xi) > t$ then $\tau_k(\xi_n) > t - \varepsilon$; from here $\tau_k(\xi_n) \to \infty$ and $\beta_{\tau_k}(\xi_n) \to \beta_{\tau_k}(\xi)$;

(b) $\tau_k(\xi) < \infty$, $X_{\tau_k}\xi \notin \Delta_k$, and $\tau_k(\xi)$ is a continuity point of the function ξ; evidently in this case $\beta_{\tau_k}(\xi_n) \to \beta_{\tau_k}(\xi)$;

(c) $\tau_k(\xi) < \infty$, $X_{\tau_k}\xi \notin \Delta_k$, and $\tau_k(\xi)$ is a discontinuity point of the function ξ, moreover $\xi(\tau_k - 0) \notin \Delta_k$. In this case we also obtain $\beta_{\tau_k}(\xi_n) \to \beta_{\tau_k}(\xi)$. $\qquad\square$

40. Note about a set

The same technique as in the proof of theorem 2.6 can be applied to prove the statement: $(\forall k \in \mathbb{N})\,(\forall r > 0)$ the function $\beta_{\sigma_r^k}$ is continuous on $\Pi_k(r)$.

2.5. Time of regeneration

Semi-Markov processes, which will be defined in the next chapter, are based on the concept of a regeneration time. It is a Markov time with respect to which the given family of measures possesses the Markov property. Let us recall that Markov property of the process at the time $\tau \in \mathrm{MT}$ is independence future (after τ) from the past (before τ) with the fixed present (just at τ) (see Shurenkov [SHU 77], Blumenthal and Getoor [BLU 68], Keilson [KEI 79], Meyer [MEY 73], etc.). Unfortunately, the most appropriate term for this case, "Markov time", is used in a wide sense without relation to any probability measure (see Gihman and Skorokhod [GIH 73], Dynkin [DYN 63], Dynkin and Yushkevich [DYN 67], Itô and McKean [ITO 65], etc.). The term "time of Markov interference of chance", which is also sometimes used, does not seem to us to be good enough, generally because of its length. The term "time of regeneration" generally speaking is also occupied. They use it usually in a narrower sense, namely, when the independence future from the past fulfils only with respect to the hitting time of one fixed position. The "family" of measures of such a process consists of one measure (see Mainsonneuve [MAI 71, MAI 74], Mürmann [MUR 73], Smith [SMI 55], Taksar [TAK 80], etc.). The most correct term for this class of times would be "Markov regeneration times". However, we will use a shorter name.

In this section some definitions connected with a time of regeneration are discussed. The structure of a set of regeneration times is investigated. Closure of this set with respect to the operation $\dot{+}$ allows us in the following chapter to give a definition of semi-Markov processes, using a rather small set of Markov times.

41. Measures, integrals, space of functions

Let $(\mathbb{E}, \mathcal{E})$ be a set with sigma-algebra of subsets; μ be a measure on $\mathbb{E}$; $A \in \mathcal{E}$; and $f \in \mathcal{E}/\mathcal{B}(\mathbb{R})$ be an integrable function (with respect to μ). In item 1.7 we have introduced designation $\mu(f; A)$ for the integral of the function f by the measure μ on the set A. In a simple case $\mu(f) = \mu(f; \mathbb{E})$. For probability measures we reserve denotation $E_x(f; A)$, $E_{Q_x}(f; A)$ and so on. Let $(B_i, \mathcal{B}_i)$ be sets with sigma-algebras of subsets, $f_i \in \mathcal{E}/\mathcal{B}_i$ be a measurable map $2\,\mathbb{E} \to B_i$ $(i = 1, \ldots, k)$. Let us designate by $\mu \circ (f_1, \ldots, f_k)^{-1}$ a measure induced by a measure μ and maps f_i on measurable space $(B_1 \times \cdots \times B_k, \mathcal{B}_1 \otimes \cdots \otimes \mathcal{B}_k)$. It means that for any $S = S_1 \times \cdots \times S_k$ $(S_i \in \mathcal{B}_i)$

$$\mu \circ (f_1, \ldots, f_k)^{-1}(S) = \mu\big(f_1^{-1} S_1 \cap \cdots \cap f_k^{-1} S_k\big).$$

Let $\mathcal{C}_0(\mathbb{A})$ be a set of continuous and bounded real functions on topological space $\mathbb{A}$.

42. Admissible family of probability measures.

A family of probability measures $(P_x) = (P_x)_{x \in \mathbb{X}}$ on $(\mathcal{D}, \mathcal{F})$ are said to be admissible, if

(a) $(\forall x \in \mathbb{X})\, P_x(X_0 = x) = 1$;

(b) for any $B \in \mathcal{F}$ the map $x \mapsto P_x(B)$ is $\mathcal{B}(\mathbb{X})$-measurable.

43. Time of regeneration

The time $\tau \in \mathrm{MT}_+$ is said to be a time of regeneration of an admissible family of measures (P_x), if for any $x \in \mathbb{X}$ and $B \in \mathcal{F}$ P_x-a.s.

$$P_x\big(\theta_\tau^{-1} B \mid \mathcal{F}_\tau\big) = P_{X_\tau}(B)$$

on the set $\{\tau < \infty\}$.

Let $\mathrm{RT}(P_x)$ be a set of time of regeneration of an admissible family (P_x). If it is clear what family there is a speech about, we write RT.

44. Lambda-continuous family of measures

For any admissible set of measures (P_x) with any $k \in \mathbb{N}$, $\lambda_i > 0$, and with any bounded $\mathcal{B}(\mathbb{X}^k)$-measurable function φ we define a function of $x \in \mathbb{X}$:

$$R(\lambda_1, \ldots, \lambda_k; \varphi \mid x)$$

$$= P_x \left(\int_{0 \le t_1 \le \cdots \le t_k < \infty} \varphi(X_{t_1}, \ldots, X_{t_k}) \prod_{i=1}^{k} \left(\exp\left(-\lambda_i t_i' \right) dt_i \right) \right),$$

where $t_i' = t_i - t_{i-1}$, $t_0 = 0$. Considering this function as an operator on the set of functions-arguments, we can treat it as a multi-dimensional generalization of λ-potential operator $R(\lambda; \varphi | x) \equiv R_\lambda(\varphi)$ (see Blumenthal and Getoor [BLU 68, p. 41]). An admissible set of measures (P_x) are said to be λ-continuous if for any $k \in \mathbb{N}$, and $\lambda_1, \ldots, \lambda_k$, and $\varphi_1, \ldots, \varphi_k \in C_0$ we have $R(\lambda_1, \ldots, \lambda_k; \varphi \mid \cdot) \in C_0$, where $C_0 = C_0(\mathbb{X})$ and

$$\varphi(x_1, \ldots, x_k) = \varphi_1(x_1) \cdots \varphi_k(x_k), \quad (x_1, \ldots, x_k) \in \mathbb{X}^k.$$

45. Other forms of the definition

A class of subsets of a set is said to be a pi-system if this class is closed with respect to all pairwise intersections of its elements (see Dynkin [DYN 59, p. 9]).

PROPOSITION 2.18. *Let $\mathcal{A}, \mathcal{A}_1$ be a pi-system of sets, for which $\mathcal{F} = \sigma(\mathcal{A})$, $\mathcal{F}_\tau = \sigma(\mathcal{A}_1)$, and let V and V_τ be classes of real functions such that $\mathcal{F} = \sigma(V)$, $\mathcal{F}_T = \sigma(V_\tau)$. The following conditions are then equivalent:*

(a) $\tau \in \mathrm{RT}$;

(b) $P_x(\theta_\tau^{-1} B, B_1) = E_x(P_{X_\tau}(B); B_1 \cap \{\tau < \infty\})$;

(c) $E_x(f \circ \theta_\tau; B_1) = E_x(E_{X_\tau}(f); B_1 \cap \{\tau < \infty\})$;

(d) $E_x((f \circ \theta_\tau) \cdot f') = E_x(E_{X_\tau}(f) \cdot f'; \{\tau < \infty\})$;

(e) $E_x(f'; \theta_\tau^{-1} B) = E_x(E_{X_\tau}(B) \cdot f'; \{\tau < \infty\})$,

where each condition (b)–(e) is fulfilled for any $x \in \mathbb{X}$, $B \in \mathcal{A}$, $B_1 \in \mathcal{A}_1$, $f \in V$, $f' \in V'$ correspondingly.

Proof. Proof is implied from sigma-additivity of measures P_x and both definitions of conditional expectation, and time of regeneration (see, e.g., Gihman and Skorokhod [GIH 73, p. 57]).   □

46. Properties of the set of regeneration time

THEOREM 2.7. *The following properties of RT are fair:*

(1) $\tau_1, \tau_2 \in RT \Rightarrow \tau_1 \dotplus \tau_2 \in RT$;

(2) if $(\forall n \in \mathbb{N})$ $\tau_n \in RT$ and $\tau = \tau_n$ on $B_n \in \mathcal{F}_{\tau_n}$, where $B_i \cap B_j = \emptyset$ $(i \neq j)$, $\{\tau < \infty\} \subset \bigcup_{n=1}^{\infty} B_n$, then $\tau \in RT$;

(3) if $(\forall n \in \mathbb{N})$ $\tau_n \in RT$, $\tau_n \geq \tau$, $\tau_n \to \tau$, and (P_x) is a lambda-continuous family, then $\tau \in RT$;

(4) if $(\forall n \in \mathbb{N})$ $\tau_n \in RT$, $\tau_n \leq \tau$ $\tau_n \to \tau$, (P_x) is a lambda-continuous family, and also $(\forall x \in \mathbb{X})$ $P_x(X_{\tau_n} \not\to X_\tau, \tau < \infty) = 0$, then $\tau \in RT$.

Proof. (1) From theorem 2.3 and proposition 2.18(d) it follows that it is enough to prove

$$E_x\left((f \circ \theta_{\tau_1 \dotplus \tau_2}) \cdot (f_2 \circ \theta_{\tau_1}) \cdot f_1\right) = E_x\left(E_{X_{\tau_1 \dotplus \tau_2}}(f) \cdot (f_2 \circ \theta_{\tau_1}) \cdot f_1\right),$$

where $f_1 \in \mathcal{F}_{\tau_1}/\mathcal{B}[0,1]$, $f_2 \in \mathcal{F}_{\tau_2}/\mathcal{B}[0,1]$, $f \in \mathcal{F}/\mathcal{B}[0,1]$, $\{f_1 = 0\} \supset \{\tau_1 = \infty\}$ and $\{f_2 = 0\} \supset \{\tau_2 = \infty\}$. It follows from proposition 2.18(d) and the following equalities

$$E_x\left((f \circ \theta_{\tau_1 \dotplus \tau_2})(f_2 \circ \theta_{\tau_1})f_1\right) = E_x\left((f \circ \theta_{\tau_2} \circ \theta_{\tau_1})(f_2 \circ \theta_{\tau_1})f_1\right)$$

$$= E_x\left((((f \circ \theta_{\tau_2})f_2) \circ \theta_{\tau_1})f_1\right) = E_x\left(E_{X_{\tau_1}}\left((f \circ \theta_{\tau_2})f_2\right)f_1\right)$$

$$= E_x\left(E_{X_{\tau_1}}\left(E_{\tau_2}(f)f_2\right)f_1\right) = E_x\left(((E_{\tau_2}(f)f_2) \circ \theta_{\tau_1})f_1\right)$$

$$= E_x\left(E_{X_{\tau_2} \circ \theta_{\tau_1}}(f) \cdot (f_2 \circ \theta_{\tau_1})f_1\right) = E_x\left(E_{X_{\tau_1 \dotplus \tau_2}}(f) \cdot (f_2 \circ \theta_{\tau_1})f_1\right).$$

(2) Since $\mathcal{F}_{\tau_n} \subset \mathcal{F}_{\tau_n +}$ for $\tau_n \in RT$ we have $(\forall t \in R_+)$

$$\{\tau < \infty\} = \bigcup_{n=1}^{\infty} \{\tau_n < t, B_n\} \in \mathcal{F}_t,$$

From here $\tau \in MT_+$. For such τ according to the definition in item 25 we have $\mathcal{F}_\tau = \sigma(\alpha_\tau, \tau)$, and also

$$\alpha_\tau^{-1} B' \cap B_n = \alpha_{\tau_n}^{-1} B' \cap B_n \in \mathcal{F}_{\tau_n} \quad (B' \in \mathcal{F}),$$

$$\{\tau < t\} \cap B_n = \{\tau_n < t\} \cap B_n \in \mathcal{F}_{\tau_n} \quad (t \in \mathbb{R}_+).$$

Therefore $B \cap B_n \in \mathcal{F}_{\tau_n}$ for any $B \in \mathcal{F}_\tau$, and according to proposition 2.18(c) we have

$$E_x\big(f \circ \theta_\tau; \ B \cap \{\tau < \infty\}\big) = \bigcup_{n=1}^{\infty} E_x\big(f \circ \theta_{\tau_n}; \ B \cap B_n \cap \{\tau < \infty\}\big)$$

$$= \bigcup_{n=1}^{\infty} E_x\big(E_{X_{\tau_n}}(f); \ B \cap B_n \cap \{\tau < \infty\}\big)$$

$$= E_x\big(E_{X_\tau}(f); \ B \cap \{\tau < \infty\}\big),$$

where $f \in \mathcal{F}/\mathcal{B}[0,1]$ and $B \in \mathcal{F}_\tau$.

(3) Obviously, $\tau \in \mathrm{MT}_+$ (see Blumenthal and Getoor [BLU 68, p. 32]). Besides, $\mathcal{F}_\tau \subset \mathcal{F}_{\tau_n}$. Really, $\alpha_\tau = \alpha_\tau \circ \alpha_{\tau_n}$, hence α_τ is $\mathcal{F}_{\tau_n}$-measurable map, and also $\{\tau < t\} = \{\tau < t \leq \tau_n\} \cup \{\tau_n < t\} \in \mathcal{F}_{\tau_n}$, where

$$\{\tau < t \leq \tau_n\} = \alpha_t^{-1}\{\tau < t\} \cap \{t \leq \tau_n\}$$

$$= \alpha_{\tau_n}^{-1}\{\tau < t\} \cap \{t \leq \tau_n\} \in \mathcal{F}_{\tau_n}.$$

Furthermore, $\{\tau_n < \infty\} \subset \{\tau < \infty\}$ and $\{\tau < \infty\} = \bigcup_{n=1}^{\infty}\{\tau_n < \infty\}$. Let

$$f = \int_{0 \leq t_1 \leq \cdots \leq t_k < \infty} \prod_{i=1}^{k} \big(\exp\big(-\lambda_i t_i\big)\varphi_i(X_{t_i})\, dt_i\big),$$

where $\lambda_1, \ldots, \lambda_k > 0$, $\varphi_1, \ldots, \varphi_k \in C$. Then $(\forall \xi \in D)$

$$\big(f \circ \theta_{\tau_n}\big)(\xi) \longrightarrow \big(f \circ \theta_\tau\big)(\xi),$$

and

$$E_{X_{\tau_n}(\xi)}(f) = R\big(\lambda_1, \ldots, \lambda_k; \varphi \mid X_{\tau_n}(\xi)\big)$$

$$\longrightarrow R\big(\lambda_1, \ldots, \lambda_k; \varphi \mid X_\tau(\xi)\big) = E_{X_\tau(\xi)}(f),$$

where $\varphi(x_1, \ldots, x_k) = \varphi_1(x_1) \cdots \varphi_k(x_k)$. From here

$$E_x\big(f \circ \theta_\tau; \ B \cap \{\tau < \infty\}\big) = \lim_{n \to \infty} E_x\big(f \circ \theta_{\tau_n}; \ B \cap \{\tau_n < \infty\}\big)$$

$$= \lim_{n \to \infty} E_x\big(E_{X_{\tau_n}}(f), \ B \cap \{\tau_n < \infty\}\big)$$

$$= E_x\big(E_{X_\tau}(f), \ B \cap \{\tau < \infty\}\big),$$

where $B \in \mathcal{F}_\tau$. Hence, under theorem 2.18(c), $\tau \in \mathrm{RT}$.

(4) Obviously, $\tau \in MT_+$ (see Blumenthal and Getoor [BLU 68, p. 32]). Let $C_\tau = \{\tau = \infty\} \cup \{\tau < \infty, X_{\tau_n} \to X_\tau\}$. Evidently, $C_\tau \in \mathcal{F}_\tau$ and $(\forall x \in \mathbb{X}) \, P_x(C_\tau) = 1$, and besides

$$\mathcal{F}_\tau \cap C_\tau = \sigma\left(\bigcup_{n=1}^\infty \mathcal{F}_{\tau_n}\right) \cap C_\tau.$$

Really, $\{\tau \le t\} = \bigcap_{n=1}^\infty \{\tau_n \le t\} \in \sigma(\bigcup_{n=1}^\infty \mathcal{F}_{\tau_n})$ and $(\forall t \in \mathbb{R}_+)$ on the set C_τ we have $X_t \circ \alpha_{\tau_n} \to X_t \circ \alpha_\tau$. Let f be like that in part 3 of the proof. Then because of lambda-continuity on a set $C_\tau \cap \{\tau < \infty\}$ it is true $E_{X_{\tau_n}}(f) \to E_{X_\tau}(f) \, (n \to \infty)$. Let $\delta \in DS(r)$ and $P_x(\theta_\tau^{-1}\Pi(\delta)) = 1$. Then

$$\left| f \circ \theta_\tau - f \circ \theta_{\tau_n} \right| \le \left| f \circ \theta_\tau - f_r \circ \theta_\tau \right| + \left| f_r \circ \theta_\tau - f_r \circ \theta_{\tau_n} \right|$$
$$+ \left| f_r \circ \theta_{\tau_n} - f \circ \theta_{\tau_n} \right|,$$

where

$$f_r = f \circ L_\delta = \int_{0 \le t_1 \le \cdots \le t_k} \prod_{i=1}^k \left(\exp\left(-\lambda_i t_i \right) \left(\varphi_i \circ X_{t_i} \circ L_\delta \right) dt_i \right).$$

Let us assume without being too specific that all φ_i $(i = 1, \ldots, k)$ are uniformly continuous. Since $(\forall \xi \in \mathcal{D}) \, (\forall t \in \mathbb{R}_+) \, \rho(X_t L_\delta \xi, X_t \xi) < r$, the first and third terms of the right part of the previous inequality tend to zero with $r \to 0$ uniformly on n. On the other hand

$$f_r = \sum_{0 \le l_1 \le \cdots \le l_k} \prod_{i=1}^k \varphi_i\left(X_{\sigma_\delta^{l_i}} \right) \int \prod_{i=1}^k \left(\exp\left(-\lambda_i t_i \right) dt_i \right),$$

where the integral is undertaken on the region

$$\left\{ 0 \le t_i \le \cdots \le t_k; \, \sigma_\delta^{l_i} \le t_1 < \sigma_\delta^{l_1+1}, \ldots, \tau \delta^{l_k} \le t_k < \sigma_\delta^{l_k+1} \right\}$$

and, hence, it is continuous function of $\sigma_\delta^{l_i}, \sigma_\delta^{l_i+1}$, where $i = 1, \ldots, k$. Let $l \ge 0$ and $\delta = (\Delta_1, \Delta_2, \ldots)$. Consider two possibilities on $\{\tau < \infty\}$:

(a) $X_\tau \notin \bigcup_{i=1}^l [\Delta_i]$. Here P_x-a.s. $(\exists n_0 \in \mathbb{N}) \, (\forall n > n_0)$

$$X_{\tau_n} \notin \bigcup_{i=1}^l [\Delta_i].$$

From here $\tau_n \dotplus \sigma_\delta^l = \tau_n \to \tau = \tau \dotplus \sigma_\delta^l$ and on $C_\tau \cap \{\tau < \infty\}$ we have $X_{\sigma_\delta^l} \circ \theta_{\tau_n} = X_{\tau_n} \to X_\tau = X_{\sigma_\delta^l} \circ \theta_\tau$.

(b) $(\exists j \leq l)\ X_\tau \in \Delta_j \cap \bigcap_{i=1}^{j-1} \mathbb{X}\backslash[\Delta_i]$. Here P_x-a.s. $(\exists n_0 \in \mathbb{N})\ (\forall n > n_0)$

$$X_{\tau_n} \in \Delta_j \cap \bigcap_{i=1}^{j-1} \mathbb{X}\backslash[\Delta_i].$$

From here $\tau_n \dot{+} \sigma_\delta^l = \tau \dot{+} \sigma_\delta^l$ (see proposition 2.11(1)), i.e. $\tau_n \dot{+} \sigma_\delta^l \to \tau \dot{+} \sigma_\delta^l$ and on $C_\tau \cap \{\tau < \infty\}$ we have $X_{\sigma_\delta^l} \circ \theta_{\tau_n} \to X_{\sigma_\delta^l} \circ \theta_\tau$. From here it follows that P_x-a.s. $f_r \circ \theta_{\tau_n} \to f \circ \theta_\tau$, where $P_x(\theta_\tau^{-1}\Pi(\delta)) = 1$. Then under corollary 2.2 P_x-a.s. on $\{\tau < \infty\}$ we have $f \circ \theta_{\tau_n} \to f \circ \theta_\tau$. For any $B \in \bigcup_{n=1}^\infty \mathcal{F}_{\tau_n}$ and $A = \bigcap_{n=1}^\infty \{\tau_n < \infty\}$ we have

$$E_x\left(\liminf_{n\to\infty} (f \circ \theta_{\tau_n});\ A \cap B\right) \leq \liminf_{n\to\infty} E_x\left(f \circ \theta_{\tau_n};\ A \cap B\right)$$
$$= \liminf_{n\to\infty} E_x\left(f \circ \theta_{\tau_n};\ \{\tau_n < \infty\} \cap B\right)$$
$$= \liminf_{n\to\infty} E_x\left(E_{X_{\tau_n}};\ \{\tau_n < \infty\} \cap B\right)$$
$$\leq E_x\left(\limsup_{n\to\infty} E_{X_{\tau_n}}(f);\ A \cap B\right).$$

Since $\{\tau < \infty\} \subset \bigcap_{n=1}^\infty \{\tau_n < \infty\}$ and $\{\tau < \infty\} \in \sigma(\bigcup_{n=1}^\infty \mathcal{F}_{\tau_n})$, we obtain

$$E_x\left(f \circ \theta_\tau,\ \{\tau < \infty\} \cap B\right) \leq E_x\left(E_{X_\tau}(f),\ \{\tau < \infty\} \cap B\right).$$

Interchanging the positions of $f \circ \theta_{\tau_n}$ and $E_{X_{\tau_n}}(f)$ in the previous inequalities, we obtain

$$E_x\left(f \circ \theta_\tau,\ \{\tau < \infty\} \cap B\right) \geq E_x\left(E_{X_\tau},\ \{\tau < \infty\} \cap B\right).$$

From here by proposition 2.18(c) $\tau \in RT$. $\qquad\square$

Chapter 3

General Semi-Markov Processes

In this chapter general semi-Markov processes are defined and investigated. In such a process its set of regeneration times contains all first exit times from open subsets of its space of states. Remember that we consider a more general class of regeneration times than that of the renewal theory (Smith [SMI 55]), namely, a regeneration time is a Markov time for which the corresponding admissible family of measures has the Markov property. Obvious examples of semi-Markov processes are a strictly Markov process (Dynkin [DYN 63]) and a stepped semi-Markov process (Korolyuk and Turbin [KOR 76]). As will be shown later on, the class of semi-Markov processes is far from being exhausted. Among them there are continuous non-Markov semi-Markov processes [HAR 71b]. Other possible definitions of semi-Markov processes are considered (see Gihman and Skorokhod [GIH 73], Chacon and Jumison [CHA 79], Çinlar [CIN 79]).

The large role for analysis of semi-Markov processes is played by semi-Markov transition functions. An appropriate countable family of semi-Markov transition functions defines the process uniquely. The problem of existence of a semi-Markov process with the given coordinated family of transition functions is considered in the next chapter. In the second half of this chapter the conditions are analyzed under which the semi-Markov process is Markovian. Properties of semi-Markov transition functions making it possible to judge Markovness of the given semi-Markov process represent a special interest. Defined for this purpose, a so called lambda-characteristic operator (the analog of a characteristic operator of Dynkin) depends linearly on λ (parameter of a Laplace transformation) only in the case when the process is a Markov process.

For some types of semi-Markov processes it is possible to judge the Markovness of the process by a property of its trajectory. One such property, a lack of intervals of constancy, is considered at the end of the chapter.

3.1. Definition of a semi-Markov process

The introduced definition enables us to look at any strictly Markov process as a semi-Markov process of a special kind. Some other definitions are also considered.

1. Semi-Markov process

A semi-Markov (SM) process is said to be the process determined by an admissible family of probability measures (P_x), for which $(\forall \Delta \in \mathfrak{A})\, \sigma_\Delta \in \mathrm{RT}(P_x)$. $(P_x) \in \mathrm{SM}$ means the process with the family of measures (P_x) is semi-Markovian.

2. Strictly Markov process

This is a process for which any time $\tau \in \mathrm{ST}$ is a time of regeneration. Hence, any strictly Markov process is semi-Markovian.

3. The first exit from a spherical neighborhood

PROPOSITION 3.1. *Let* $(P_x) \in \mathrm{SM}$. *Then* $(\forall r > 0)$

$$\sigma_r \in \mathrm{RT}\,(P_x).$$

Proof. We have $(\forall x \in \mathbb{X})\, P_x(X_0 = x) = 1$ and $P_x(\sigma_r = \sigma_\Delta) = 1$, where $\Delta = B(x,r)$. Further $\mathcal{F}_{\sigma_r} \cap \{X_0 = x\} = \mathcal{F}_{\sigma_\Delta} \cap \{X_0 = x\}$ and for any $B \in \mathcal{F}_{\sigma_r}$, $B' = B \cap \{X_0 = x\} \in \mathcal{F}_{\sigma_r} \cap \{X_0 = x\} \subset \mathcal{F}_{\sigma_\Delta}$. From here for any $f \in \mathcal{F}/\mathcal{B}(\mathbb{R})$

$$E_x\left(f \circ \theta_{\sigma_r};\ B \cap \{\sigma_r < \infty\}\right)$$
$$= E_x\left(f \circ \theta_{\sigma_\Delta};\ B' \cap \{\sigma_\Delta < \infty\}\right)$$
$$= E_x\left(E_{X_{\sigma_\Delta}}(f);\ B' \cap \{\sigma_\Delta < \infty\}\right)$$
$$= E_x\left(E_{X_{\sigma_r}}(f);\ B \cap \{\sigma_r < \infty\}\right). \qquad \square$$

4. Jump- and step-functions

The function $\xi \in \mathcal{D}$ is said to be jumped if $(\forall t \in \mathbb{R}_+)\,(\exists \delta > 0)\,(\forall h \in (0,\delta))\, \xi(t) = \xi(t+h)$, i.e. ξ is right-continuous in discrete topology (Kelly [KEL 68]). The function ξ is said to be stepped, if it is jumped and has a finite number of jumps on each limited interval. $\mathcal{D}'$ denotes the class of all jump-functions and $\mathcal{D}_0$ class of all step-functions.

5. Semi-Markov walk

A semi-Markov walk is an admissible set of probability measures (P_x), for which

(1) $(\forall x \in \mathbb{X})\, P_x(\mathcal{D}_0) = 1$;

(2) $\sigma_0 \in \mathrm{RT}(P_x)$, where $\sigma_0(\xi) = \inf(t \geq 0,\ \xi(t) \neq \xi(0))$ the first exit time from an initial point (see Gihman and Skorokhod [GIH 73], Kovalenko [KOV 65], Korolyuk, Brodi and Turbin [KOR 74], Korolyuk and Turbin [KOR 76], Nollau [NOL 80], Pyke [PYK 61], Smith [SMI 55, SMI 66], etc.).

6. Properties of SM walks

PROPOSITION 3.2. *Let (P_x) be an admissible set of probability measures. Then*

(1) if $(\forall x \in \mathbb{X})\, P_x(\mathcal{D}') = 1$ and $(\forall \Delta \in \mathfrak{A})\, \sigma_\Delta \in \mathrm{RT}$, $\sigma_0 \in \mathrm{RT}$;

(2) if $(\forall x \in \mathbb{X})\, P_x(\mathcal{D}_0) = 1$ and $\sigma_0 \in \mathrm{RT}$, $(\forall \Delta \in \mathfrak{A})\, \sigma_\Delta \in \mathrm{RT}$, i.e. the semi-Markov walk is a semi-Markov process.

Proof. (1) We have $(\forall \xi \in \mathcal{D})\, \sigma_0(\xi) = \lim_{r \downarrow 0} \sigma_r(\xi)$. Furthermore $(\forall r > 0)\, \mathcal{F}_{\sigma_0} \subset \mathcal{F}_{\sigma_r}$ and $(\forall t > 0)\, X_{\sigma_r + t} \to X_{\sigma_0 + t}\ (r \to 0)$. From here $(\forall k \in \mathbb{N})\, (\forall t_1, \dots, t_k \in \mathbb{R}_+)$ $(\forall f \in \mathcal{C}(\mathbb{X}^k))\, (\forall B \in \mathcal{F}_{\sigma_0})$

$$E_x\big(f \circ (X_{t_1}, \dots, X_{t_k}) \circ \theta_{\sigma_0};\ B \cap \{\sigma_0 < \infty\}\big)$$

$$= \lim_{r \downarrow 0} E_x\big(f \circ (X_{t_1}, \dots, X_{t_k}) \circ \theta_{\sigma_r};\ B \cap \{\sigma_0 < \infty\}\big)$$

$$= \lim_{r \downarrow 0} E_x\big(E_{X_{\sigma_r}}(f \circ (X_{t_1}, \dots, X_{t_k}));\ B \cap \{\sigma_0 < \infty\}\big)$$

$$= E_x\big(E_{X_{\sigma_0}}(f \circ (X_{t_1}, \dots, X_{t_k}));\ B \cap \{\sigma_0 < \infty\}\big).$$

The latter equality follows from a convergence in discrete topology $X_{\sigma_r} \to X_{\sigma_0}$ $(r \to 0)$ on a set $\mathcal{D}'$ and property $\sigma_r \in \mathrm{RT}$; see proposition 3.1.

(2) We have $(\forall \xi \in \mathcal{D}')\, (\forall \Delta \in \mathfrak{A})\, (\exists k \in \mathbb{N}_0)\, \sigma_\Delta(\xi) = \sigma_0^k(\xi)$, where $\sigma_0^k = \sigma_0^{k-1} + \sigma_0$ $(k \in \mathbb{N},\ \sigma_0^0 = 0)$. Under theorem 2.7(1) $(\forall k \in \mathbb{N})\, \sigma_0^k \in \mathrm{RT}$ and, hence, under theorem 2.7(2) $\sigma_\Delta \in \mathrm{RT}$. $\qquad\square$

7. Quasi-continuity from the left

The admissible family of measures (P_x) is said to be quasi-continuous from the left on the set of its regeneration times if for any non-decreasing sequence (τ_n), where $\tau_n \in \mathrm{RT}$, it is the case that $(\forall x \in \mathbb{X})\, P_x(X_{\tau_n} \not\to X_\tau,\ \tau < \infty) = 0$, where $\tau = \lim \tau_n$ (Gihman and Skorokhod [GIH 73], Dynkin [DYN 59], Blumenthal and Getoor [BLU 68]). Let us denote by QC the class of all such families.

8. Quasi-continuity from the left on RT

The proof of the following theorem is based on an idea borrowed from the work of Dynkin [DYN 63].

THEOREM 3.1. *If for any compact set $K \subset \mathbb{X}$ and $r > 0$ it is the case that*

$$\lim_{c \to 0} \sup_{x \in \mathbb{X}} P_x(\sigma_r \geq c) = 0,$$

then the admissible family of measures (P_x) belongs to QC.

Proof. Let $K_n \uparrow \mathbb{X}$, $K_n \in \mathfrak{K}$, and also $\tau_N \in$ RT, $\tau_N \uparrow \tau$. Then

$$P_x(X_{\tau_N} \not\to X_\tau, \tau < \infty) = P_x\left(\bigcup_n (X_{\tau_N} \not\to X_\tau, \tau < \sigma_{K_n})\right),$$

$$P_x(X_{\tau_N} \not\to X_\tau, \tau < \sigma_{K_n}) = P_x\left(\bigcup_m \bigcap_N \bigcup_{k \geq N} \left(\rho(X_{\tau_k}, X_\tau) \geq \frac{2}{m}, \tau < \sigma_{K_n}\right)\right)$$

$$\leq P_x\left(\bigcup_m \bigcup_M \bigcap_{N \geq M} \left(\sigma_{1/m} \circ \theta_{\tau_N} \leq h, \tau_N < \sigma_{K_n}\right)\right)$$

with any $h > 0$. In addition $(\forall N \geq 1)$

$$P_x(\sigma_{1/m} \circ \theta_{\tau_N} \leq h, \tau_N < \sigma_{K_n}) = E_x(P_{X_{\tau_N}}(\sigma_{1/m} \leq h); \tau_N < \sigma_{K_n})$$

$$\leq \sup_{x \in K_n} P_x(\sigma_{1/m} \leq h).$$

From here

$$P_x\left(\bigcap_N \bigcup_{k \geq N} \left(\rho(X_{\tau_k}, X_\tau) \geq \frac{2}{m}, \tau < \sigma_{K_n}\right)\right)$$

$$\leq P_x\left(\bigcup_M \bigcap_{N \geq M} (\sigma_{1/m} \circ \theta_{\tau_N} \leq h, \tau_N < \sigma_{K_n})\right) \leq \sup_{x \in K_n} P_x(\sigma_{1/m} \leq h) \longrightarrow 0$$

with $h \to 0$. From here $P_x(X_{\tau_N} \not\to X_\tau, \tau < \sigma_{K_n}) = 0$ and $P_x(X_{\tau_N} \not\to X_\tau, \tau < \infty) = 0$. $\qquad \square$

9. Other semi-Markov families of measures

Let $\mathfrak{A}_0 = \bigcup_{m=1}^{\infty} \mathfrak{A}_m$, where $\mathfrak{A}_m = \{\Delta_m^i, i \in \mathbb{N}\}$, $\Delta_m^i \in \mathfrak{A}$, $\bigcup_{i=1}^{\infty} \Delta_m^i = \mathbb{X}$ and diam $\Delta_m^i \leq r_m$, $r_m \to 0$. Let $\mathfrak{A}_0 \subset \mathfrak{A}_1 \subset \mathfrak{A}$. Admissible set of measures (P_x) is called:

(1) semi-Markov with respect to $\mathfrak{A}_0$ or SM($\mathfrak{A}_0$)-process, if $(\forall \Delta \in \mathfrak{A}_0) \, \sigma_\Delta \in RT$,

(2) semi-Markov with respect to $(r_m)_1^\infty$ or SM(r_m)-process, if $(\forall m \in \mathbb{N}) \, \sigma_{r_m} \in$ RT.

10. On correspondence of definitions of SM processes

PROPOSITION 3.3. *Let* $(P_x) \in SM(\mathfrak{A}_0)$, *and* (P_x) *be lambda-continuous. Then*

(1) if $\Delta \in \widetilde{\mathfrak{A}}$ *(closed set),* $\sigma_\Delta \in RT$;

(2) if $(P_x) \in QC$, $(\forall \Delta \in \mathfrak{A}) \, \sigma_\Delta \in RT$ *(i.e.* (P_x) *is a SM process).*

Proof. (1) Let $\Delta \in \widetilde{\mathfrak{A}}$ and $\delta_m \in DS(\mathfrak{A}_m)$. Then $\sigma_\Delta \leq \sigma_\Delta \circ L_{\delta_m} \leq \sigma_{\Delta + r_m}$. Whence $\sigma_\Delta \circ L_{\delta_m} \to \sigma_\Delta \, (m \to \infty)$. From theorem 2.7(2) follows that $\sigma_\Delta \circ L_{\delta_m} \in RT$ and from theorem 2.7(3) follows that $\sigma_\Delta \in RT$.

(2) Let $\Delta \in \mathfrak{A}$ and $\delta_m \in DS(\mathfrak{A}_m)$. Then $\sigma_{\Delta - r_m} \circ L_{\delta_m} \leq \sigma_\Delta$ and $X_{\sigma_{\Delta - r_m}} \circ L_{\delta_m} \notin \Delta^{-r_m}$. By theorem 2.7(2) $\sigma_{\Delta - r_m} \circ L_{\delta_m} \in RT$, $\tau_n = \bigvee_{m=1}^{n}(\sigma_{\Delta - r_m} \circ L_{\delta_m}) \in RT$ and $\tau_n \uparrow \tau \leq \sigma_\Delta$. On the other hand, from the quasi-continuity it follows

$$P_x\left(X_{\tau_n} \not\to X_\tau, \, \tau < \infty\right) = P_x\left(X_\tau \in \Delta, \, \tau < \infty\right)$$

$$= P_x\left(\tau < \sigma_\Delta, \, \tau < \infty\right) = 0.$$

Hence, $P_x(\sigma_\Delta = \tau) = 1$ and $P_x(X_{\tau_m} \not\to X_{\sigma_\Delta}, \, \sigma_\Delta < \infty) = 0$. By theorem 2.7(4) $\sigma_\Delta \in RT$. □

Note that the previous theorem remains fair, if the condition $(P_x) \in SM(\mathfrak{A}_0)$ is replaced by the condition $(P_x) \in SM(r_m)$.

11. Sojourn time in current position

For any function $\xi \in \mathcal{D}$ we define a real non-negative function $J(\xi)$ on positive half-line as follows:

$$J(\xi)(t) = t - 0 \vee \sup\left\{s \leq t : \xi(s) \neq \xi(t)\right\} \quad (t \geq 0).$$

For a step-function this function had been considered in item 1.18: $J(\xi)(t) = R_t^-(\xi)$. It is a piece-wise linear function of a saw-tooth form, determining sojourn time in the current position since the last hitting it before t. On the set of all real functions on the positive half-line we consider a projection X_t and a shift θ_t similar to those on $\mathcal{D}$. Let us designate $(X_\tau J)(\xi) = X_{\tau(\xi)}(J(\xi))$, where $\tau(\xi) < \infty$, $\tau \in MT_+$. Evidently, $(\forall \tau \in MT_+) \, X_\tau J \in \mathcal{F}_\tau \cap \{\tau < \infty\}/\mathcal{B}(\mathbb{R}_+)$ (measurable function).

12. Semi-Markov process by Gihman and Skorokhod

A semi-Markov process by Gihman and Skorokhod (SMGS-process) is said to be an admissible set of probability measures (P_x) on $\mathcal{D}$ with a special property. In order to formulate it let us determine a family of probability measures $(P_{x,s})$ depending on two parameters $x \in \mathbb{X}$ and $s \geq 0$, where

$$P_{x,0}(B) = P_x(B),$$

and for $s > 0$ and x such that $P_x(\sigma_0 > s) > 0$

$$P_{x,s}(B) = P_x\left(\theta_s^{-1}B,\ \sigma_0 > s\right)/P_x\left(\sigma_0 > s\right).$$

The special property of such a family is as follows. For any $x \in \mathbb{X}$ and $\tau \in \mathrm{MT}$ and $B \in \mathcal{F}$

$$P_x\left(\theta_\tau^{-1}B \mid \mathcal{F}_\tau\right) = P_{X_\tau, X_\tau J}(B). \qquad [3.1]$$

In other words, the two-dimensional process $(\xi, J(\xi))$ is strictly Markovian. This definition corresponds to that of a stepped semi-Markov process given in Gihman and Skorokhod [GIH 73, p. 345] (see also Korolyuk and Turbin [KOR 76] and Çinlar [CIN 79]). In [GIH 73] Markovness of the two-dimensional process is proved when the original process is a stepped Markov process. In order to prove the strictly Markov property of this process, let us consider the construction of a Markov time on a set of stepped trajectories.

13. Markov time for step-function

On the set of all step-functions $\mathcal{D}_0$ we consider a sigma-algebra of Borel subsets (generated by the Stoun-Skorokhod metric). In this case it has the form $\mathcal{F} = \sigma(\tau_k, X_{\tau_k};\ k \in \mathbb{N}_0)$, where $\tau_0 = 0$, $\tau_1 = \sigma_0$, $\tau_k = \sigma_0^k$. Let us determine as usual a Markov time with respect to the increasing family of sigma-algebras $\alpha_t^{-1}\mathcal{F}$ $(t \geq 0)$. Let us consider such a time τ. By proposition 2.2 (formula [2.10]), we have $\tau = \tau \circ \alpha_\tau$. On the other hand, for any step-function ξ we have $\alpha_\tau \xi = \alpha_{\tau_{N_\tau}}\xi$, where $N_\tau(\xi)$ is a number of jumps of the function ξ up to the time $\tau(\xi)$: $N_\tau(\xi) = n \Leftrightarrow \tau_n(\xi) \leq \tau(\xi) < \tau_{n+1}(\xi)$. Hence, τ is $\mathcal{F}_{\tau_{N_\tau}}$-measurable random value.

14. Correspondence of classes SM and SMGS

PROPOSITION 3.4. *The following assertions on an admissible family of measures* (P_x) *are true:*

(1) *if* $(P_x) \in \mathrm{SMGS}$, *then* $(P_x) \in \mathrm{SM}$;

(2) *if* (P_x) *determines a SM process with stepped trajectories, then* $(P_x) \in \mathrm{SMGS}$.

Proof. (1) It is obvious that $X_{\sigma_\Delta} J = 0$ for any $\Delta \in \mathfrak{A}$ on $\{\sigma_\Delta < \infty\}$. Then

$$P_x\left(\theta_{\sigma_\Delta}^{-1} B \mid \mathcal{F}_{\sigma_\Delta}\right) = P_{X_{\sigma_\Delta}, X_{\sigma_\Delta} J}(B) = P_{X_{\sigma_\Delta}, 0}(B)$$
$$= P_{X_{\sigma_\Delta}}(B) \quad (B \in \mathcal{F}, \ \sigma_\Delta < \infty).$$

(2) Let (P_x) be a family of measures of a semi-Markov walk (see proposition 3.2). We have $(\forall x \in \mathbb{X}) \, P_x(\sigma_0 > 0) = 1$. Let us consider the (not empty) family $(P_{x,s})$ defined above and check for it the property [3.1], which is equivalent to the following property:

$$P_x\left(\theta_\tau^{-1} B \cap A\right) = E_x\left(P_{X_\tau, X_\tau J}(B); A\right),$$

for any $A \in \mathcal{F}_\tau \cap \{\tau < \infty\}$. It is sufficient to check this property for sets A and B of the form

$$A = A_k \cap \{\tau_k \leq \tau < \tau_{k+1}\}, \qquad B = \{\tau_1 \leq z, \ X_{\tau_1} \in S\} \cap \theta_{\tau_1}^{-1} B'$$

(see proposition 2.18(b)), where $A_k \in \mathcal{F}_{\tau_k}$, $B' \in \mathcal{F}$, $S \in \mathcal{B}(\mathbb{X})$, $z > 0$, $k \in \mathbb{N}_0$. In this case

$$P_x\left(\theta_\tau^{-1} B \cap A\right)$$
$$= P_x\left(\{\tau_1 \theta_\tau \leq z, \ X_{\tau + \tau_1} \in S\} \cap \theta_{\tau + \tau_1}^{-1} B' \cap A_k \cap \{\tau_k \leq \tau < \tau_{k+1}\}\right).$$

By proposition 2.11(1) it follows that $\tau \dotplus \tau_1 = \tau_k \dotplus \tau_1$. Hence

$$P_x\left(\{\tau_{k+1} - \tau \leq z, \ X_{\tau_{k+1}} \in S\} \cap \theta_{\tau_{k+1}}^{-1} B' \cap A_k \cap \{\tau_k \leq \tau < \tau_{k+1}\}\right)$$
$$= P_x\left(\{\tau - \tau_k < \tau_1 \theta_{\tau_k} \leq z + \tau - \tau_k\} \cap \theta_{\tau_k}^{-1}\left(\{X_{\tau_1} \in S\} \cap \theta_{\tau_1}^{-1} B'\right)\right.$$
$$\left. \cap A_k \cap \{\tau - \tau_k \geq 0\}\right).$$

Representing this expression as an integral on the set of all possible values of $\mathcal{F}_{\tau_k}$-measurable random variable $\tau - \tau_k$ with the help of the semi-Markov property, we obtain

$$\int_0^\infty E_x\left(P_{X_{\tau_k}}\left(\{t < \tau_1 \leq z + t\} \cap \{X_{\tau_1} \in S\} \cap \theta_{\tau_1}^{-1} B'\right); \ A_k \cap \{\tau - \tau_k \in dt\}\right)$$
$$= \int_0^\infty E_x\left(P_{X_{\tau_k}}\left(\theta_t^{-1} B \cap \{\tau_1 > t\}\right); \ A_k \cap \{\tau - \tau_k \in dt\}\right).$$

Using definition of the two-parametric family we obtain

$$\int_0^\infty E_x\left(P_{X_{\tau_k}, t}(B) P_{X_{\tau_k}}(\tau_1 > t); \ A_k \cap \{\tau - \tau_k \in dt\}\right).$$

Repeatedly using the semi-Markov property (for inverse conclusion) we have

$$\int_0^\infty E_x\left(P_{X_{\tau_k},t}(B);\ A_k \cap \{\tau - \tau_k \in dt\} \cap \{\tau_1 \theta_{\tau_k} > t\}\right)$$

$$= E_x\left(P_{X_{\tau_k},\tau-\tau_k}(B);\ A_k \cap \{\tau - \tau_k \geq 0\} \cap \{\tau_{k+1} > \tau\}\right). \qquad \square$$

15. Example of a SM but not a SMGS-process

Let us consider a family of degenerate probability measures (P_x) with the phase space $\mathbb{X} = \mathbb{R}$. For any $x \in \mathbb{X}$ the measure P_x is concentrated on a unique trajectory. For $x = 0$ this trajectory represents the function ξ_0 (Figure 3.1), where $\xi_0(t) = 0$ for $t \in [0,1) \cup [2, \infty)$, and $\xi_0(t) = (-1/2)^{n-1}$ for $t \in [2 - 2^{-n+1}, 2 - 2^{-n})$ $(n \geq 0)$. For point $x = (-1/2)^{n-1}$ the measure P_x concentrates on the corresponding shift of the function ξ_0.

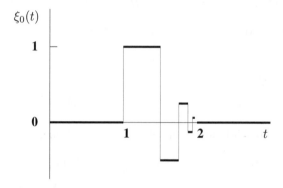

Figure 3.1. *Graphic of the function $\xi_0(t)$*

For other points x trajectories are constant: $P_x(\sigma_0 = \infty) = 1$. These points can be excluded from the space of states without violation the condition for remaining measures to be coordinated. It is a jumped process, but not a stepped process. Its two-dimensional process $(\xi, J(\xi))$ is not strictly Markov, because the first time of accumulation of infinitely many jumps is not a regeneration time of this family of measures. As such, the process is not a SMGS-process. The set of distributions of the process at the first jump times does not determine the process as well. However, for any open set Δ the time σ_Δ is a regeneration time of the family (P_x). Therefore, it is a SM process. For any initial point the corresponding function ξ can be constructed as a limit of a sequence of step-functions $\xi \circ L_r$ $(r \to 0)$, having for every $r > 0$ a finite number of jumps.

3.2. Transition function of a SM process

The possibility of a constructive representation of a SM process is connected with its transition function (see Gihman and Skorokhod [GIH 73], Kuznetsov [KUZ 80]). In contrast to a transition function of Markov process (Dynkin [DYN 63], Blumenthal and Getoor [BLU 68], etc.) a parameter in the transition function of a SM process is a set. For lambda-continuous processes these sets can be taken from some accounting sub-class of all open sets. For lambda-continuous process on the line such a sub-class can be consisted of all intervals with rational ends (two parameters: the extreme points of an interval). A basic condition for a family of SM transition functions to be coordinated is an integral equation for transition functions with set-parameters where one set includes the other. This equation is a consequence of the regeneration property of the first exit time from an open set. In terms of the Laplace transformation of transition functions (by t) these equations accept very simple forms (see Feller [FEL 67]).

16. Transition function

Let (P_x) be an admissible set of probability measures and $\tau \in \mathrm{ST}_+$. Designate

$$F_\tau(B \mid x) = P_x\big((\tau, X_\tau) \in B\big) \quad (B \in \mathcal{B}(\mathbb{R}_+ \times \mathbb{X})),$$

$$f_\tau(\lambda, S \mid x) \equiv \int_0^\infty e^{-\lambda t} F_\tau(dt \times S \mid x) = E_x\big(e^{-\lambda \tau}; \, X_\tau \in S, \, \tau < \infty\big)$$

$(\lambda \geq 0, \, S \in \mathcal{B}(\mathbb{X}))$. The kernel F_τ is said to be a semi-Markov transition function of the family (P_x), corresponding to a Markov time τ; f_τ is its Laplace transformation by time. We call the kernel f_τ a semi-Markov transition generating function.

It is clear that $F_\tau(B \mid \cdot)$, $f_\tau(\lambda, S \mid \cdot)$ are $\mathcal{B}(\mathbb{X})$-measurable functions from x, and $F_\tau(\cdot \mid x)$, $f_\tau(\lambda, \cdot \mid x)$ are sub-probability measures on the appropriate measurable spaces. Many properties of semi-Markov processes (e.g., theorem 3.1) can be formulated in terms of these transition functions.

17. Properties of the transition functions

PROPOSITION 3.5. *Let* $(P_x) \in \mathrm{SM}$, $\tau_1 \in \mathrm{RT}(P_x)$ *and* $\tau_2 \in \mathrm{MT}_+$. *Then*

(1) $F_{\tau_1 \dotplus \tau_2}([0, t) \times S \mid x) = \int_0^t \int_{\mathbb{X}} F_{\tau_1}(dt_1 \times dx_1 \mid x) F_{\tau_2}([0, t - t_1) \times S \mid x)$;

(2) $f_{\tau_1 \dotplus \tau_2}(\lambda, S \mid x) = \int_{\mathbb{X}} f_{\tau_1}(\lambda, dx_1 \mid x) f_{\tau_2}(\lambda, S \mid x_1)$.

Proof. (1) Using representation

$$\{\tau_1 \dotplus \tau_2 < t\} = \bigcup_{n=1}^\infty \bigcup_{k=1}^n \left\{ \tau_1 \in \left[\frac{t(k-1)}{n}, \frac{tk}{n} \right), \, \tau_2 \circ \theta_{\tau_1} < \frac{t(n-k)}{n} \right\},$$

the first formula follows from left-continuity on t of the cumulative distribution function $F_\tau([0, t) \times S \mid x)$, and from the condition of regeneration. Actually,

$$P_x(\tau_1 \in [a, b), \ \tau_2 \circ \theta_{\tau_1} \in [c, d), \ X_{\tau_1 \dotplus \tau_2} \in S)$$
$$= E_x(P_{X_{\tau_1}}(\tau_2 \in [c, d), \ X_{\tau_2} \in S); \ \tau_1 \in [a, b))$$
$$= \int_\mathbb{X} F_{\tau_1}([a, b) \times dx_1 \mid x) F_{\tau_2}([c, d) \times S \mid x_1).$$

(2) The second formula follows from the first one, but it is easier to prove it immediately:

$$f_{\tau_1 \dotplus \tau_2}(\lambda, S \mid x) = E_x\big(\exp(-\lambda(\tau_1 \dotplus \tau_2)); \ \{\tau_1 \dotplus \tau_2 < \infty, X_{\tau_1 \dotplus \tau_2} \in S\}\big)$$
$$= E_x\big(\exp(\lambda\tau_1) E_{X_{\tau_1}}(\exp(-\lambda\tau_2); \ \tau_2 < \infty, \ X_{\tau_2} \in S); \ \tau_1 < \infty\big)$$
$$= \int_\mathbb{X} f_{\tau_1}(\lambda, dx_1 \mid x) f_{\tau_2}(\lambda, S \mid x_1).$$

In this case formula 1 follows from formula 2 as a consequence of the property of the Laplace transformation (see [DIT 65]).   □

18. One sufficient condition of equality of measures

PROPOSITION 3.6. *Let P and P' be two probability measures on $\mathcal{D}$, and $\mathfrak{A}_m$ be a countable open covering of $\mathbb{X}$ with the rank r_m (i.e. $(\forall\Delta \in \mathfrak{A}_m)$ diam $\Delta \leq r_m$), where $r_m \downarrow 0$, and $\delta_m \in DS(\mathfrak{A}_m)$. Let $(\forall k, m \in \mathbb{N})$ $(\forall\lambda_i > 0)$ $(\forall\varphi_i \in C_0)$*

$$E(f_{m,k}; \ \sigma^i_{\delta_m} < \infty) = E'(f_{m,k}; \ \sigma^i_{\delta_m} < \infty),$$

where

$$f_{m,k} = \prod_{i=0}^{k} \exp(-\lambda_i\sigma^i_{\delta_m})(\varphi_i \circ X_{\sigma^i_{\delta_m}}).$$

Then $P = P'$ on $\mathcal{F}$.

Proof. From properties of the Laplace transformation (see Feller [FEL 67]) it follows $(\forall k, m \in \mathbb{N})$

$$P \circ \big(\beta_{\sigma^0_{\delta_m}}, \ldots, \beta_{\sigma^k_{\delta_m}}\big)^{-1} = P' \circ \big(\beta_{\sigma^0_{\delta_m}}, \ldots, \beta_{\sigma^k_{\delta_m}}\big)^{-1}.$$

From here by the Caratheodory extension theorem (see Gihman and Skorokhod [GIH 71], Loève [LOE 62], Shiryaev [SHI 80]) we have: $P \circ L_{\delta_m}^{-1} = P' \circ L_{\delta_m}^{-1}$. Let $f \in C_0(\mathcal{D})$. Since $(\forall \xi \in \mathcal{D})$

$$\sup_{t \in \mathbb{R}_+} \rho\left(L_{\delta_m}\xi(t), \xi(t)\right) \le r_m \longrightarrow 0,$$

we have $\rho_{\mathcal{D}}(L_{\delta_m}\xi, \xi) \to 0$ and

$$\left|E'(f) - E(f)\right| \le \left|E'(f) - E' \circ L_{\delta_m}^{-1}(f)\right| + \left|E \circ L_{\delta_m}^{-1}(f) - E(f)\right|$$
$$\le E'\left(\left|f - f \circ L_{\delta_m}\right|\right) + E\left(\left|f - f \circ L_{\delta_m}\right|\right) \longrightarrow 0. \qquad \square$$

19. Criterion of equality of SM processes

PROPOSITION 3.7. *Let (P_x), (P'_x) be two semi-Markov processes on $\mathcal{D}$; $(\forall m \in \mathbb{N})$ $(\forall \Delta \in \mathfrak{A}_m)$ $f_{\sigma_\Delta} = f'_{\sigma_\Delta}$, i.e. their transition generating functions are identical ($\mathfrak{A}_m$ is defined in proposition 3.6). Then $(P_x) = (P'_x)$.*

Proof. We have $(\forall \delta_m \in \mathrm{DS}(\mathfrak{A}_m))$ $(\forall \lambda_i \ge 0)$ $(\forall \varphi_i \in C_0)$

$$E_x\left(\prod_{i=1}^{k} \exp\left(-\lambda_i(\sigma_{\delta_m}^i - \sigma_{\delta_m}^{i-1})\right)\varphi_i\left(X_{\sigma_{\delta_m}^i}\right); \ \sigma_{\delta_m}^k < \infty\right)$$
$$= \int_{\mathbb{X}^k}\left(\prod_{i=1}^{k} f_{\sigma_{\delta_m}^i}(\lambda_i, dx_i \mid x_{i-1})\varphi_i(x_i)\right),$$

where $\delta_m = (\Delta_m^1, \Delta_m^2, \ldots)$, $\sigma_{\delta_m}^0 = 0$, $x_0 = x$. From here by theorem 3.6 we obtain $(P_x) = (P'_x)$. $\qquad \square$

Let us note that propositions 3.6 and 3.7 would stay valid if in their conditions the times σ_Δ $(\Delta \in \mathfrak{A}_m)$ are replaced by σ_{r_m} $(m \in \mathbb{N})$ and accordingly that of $\sigma_{\delta_m}^k$ by $\sigma_{r_m}^k$.

For semi-Markov functions F_{σ_Δ} and f_{σ_Δ} we will also use simplified labels F_Δ and f_Δ accordingly.

20. Sufficient condition of the semi-Markov property

PROPOSITION 3.8. *Let (P_x) be an admissible family of probability measures. For any $x \in \mathbb{X}$, $\varphi \in C_0$, let $\Delta \in \mathfrak{A}$, $\lambda > 0$ and $\tau \in T(\sigma_\Delta, \Delta \in \mathfrak{A})$ P_x-a.s. on the set $\{\tau < \infty\}$*

$$E_x\left(\left(\exp\left(-\lambda\sigma_\Delta\right)\right) \circ \theta_\tau; \ \tau \dotplus \sigma_\Delta < \infty \mid \mathcal{F}_\tau\right) = f_\Delta(\lambda, \varphi) \circ X_\Delta,$$

where

$$f_\Delta(\lambda, \varphi \mid x) = \int_{\mathbb{X}} \varphi(x_1) f_\Delta(\lambda, dx_1 \mid x), \qquad f_\Delta(\lambda, \varphi) = f_\Delta(\lambda, \varphi \mid \cdot).$$

Then the family of measures (P_x) is semi-Markovian.

Proof. From properties of the Laplace transformation it follows that for any measurable function φ_1 on $\{\tau < \infty\}$ P_x-a.s. $E_x(\varphi_1 \circ \beta_{\sigma_\Delta} \circ \theta_\tau \mid \mathcal{F}_\tau) = E_{X_\tau}(\varphi_1 \circ \beta_{\sigma_\Delta})$. It follows by induction that for any finite set $(\Delta_1, \ldots, \Delta_k)$ $(\Delta_i \in \mathfrak{A})$ and $(\varphi_1, \ldots, \varphi_k)$ $(\varphi_i \in \mathcal{B}(Y)/\mathcal{B}(\mathbb{R}))$ the equality is fulfilled

$$E_x\left(\prod_{i=1}^{k} \varphi_i(\beta_{\sigma_{\Delta_i}}) \circ \theta_\tau \,\Big|\, \mathcal{F}_\tau\right) = E_{X_\tau}\left(\prod_{i=1}^{k} \varphi_i(\beta_{\sigma_{\Delta_i}})\right);$$

and hence for all $B \in \mathcal{F}$ $P_x(\theta_\tau^{-1} B \mid \mathcal{F}_\tau) = P_{X_\tau}(B)$. $\qquad\square$

3.3. Operators and SM walk

Multi-dimensional lambda-potential operators are also used so that Markov processes can be characterized in the class of semi-Markov processes. It is based on the Markov condition expressed in terms of the lambda-potential operator. With the help of transition generating function, a differential (difference) operator similar to the characteristic operator for a Markov process is defined. The duality of the differential and integral operators is proved.

21. Integral operator

An integral operator R on the set of $\mathcal{B}(\mathbb{X}^n)$-measurable functions is defined in item 2.44. For further investigations a summary of this operator will be useful. Let us denote $\mathbb{B}_0$ the class of all bounded measurable functions, For $\tau \in \mathrm{MT}$ we define function

$$R_\tau(\lambda_1, \ldots, \lambda_n; \varphi \mid x) = E_x\left(\int_{s_n < \tau} \varphi(X_{s_1}, \ldots, X_{s_n}) \prod_{i=1}^{n} \left(e^{-\lambda_i t_i} \, dt_i\right)\right),$$

where $s_i = t_1 + \cdots + t_i$, $\lambda_i \geq 0$, $\varphi \in \mathbb{B}_0(\mathbb{X}^n)$. We will use the operator R_τ generally on the class of functions $\varphi = \varphi_1 \otimes \cdots \otimes \varphi_n$ $(\varphi_i \in \mathbb{B}_0)$ (i.e. $\varphi(x_1, \ldots, x_n) = \varphi_1(x_1) \cdots \varphi_n(x_n)$) denoting

$$R_\tau(\lambda_1, \ldots, \lambda_n; \varphi_1, \ldots, \varphi_n \mid x) = E_x\left(\int_{s_n < \tau} \prod_{i=1}^{n} \left(e^{-\lambda_i t_i} \left(\varphi_i \circ X_{s_i}\right) dt_i\right)\right).$$

We call this operator a multi-dimensional lambda-potential operator on the interval $[0, \tau)$ $(\tau \leq \infty)$. If $\tau \equiv \infty$ we omit the subscript τ.

From definition of the operator L_r (see item 2.18) it follows that $(\forall \xi \in \mathcal{D})$ $(\forall t \geq 0)$ $\rho(\xi(t), L_r \xi(t)) < r$. Hence $L_r \xi(s) \to \xi(s)$ for $r \to 0$ uniformly on $s \in \mathbb{R}_+$. Consequently, any semi-Markov process (P_x) can be represented as a limit of the sequence of SM walks $(P_x \circ L_r^{-1})_{x \in \mathbb{X}}$ as $r \to 0$. Chapter 7 of this book is devoted to problems of convergence of sequences of processes. Here we consider a method of evaluating operator $R_{\sigma_\Delta}(\lambda_1, \dots, \lambda_n)$ for SM process when there exists a sequence of Markov times $\tau_r \circ L_r$ $(r \to 0)$, converging on σ_Δ P_x-a.s., where τ_r is a Markov time on the set of step-functions, depending on r (see item 13). If Δ is closed set, then for a general process, such a Markov time can be taken as $\sigma_\Delta \circ L_r$. If Δ is open set and the process belongs to the class QC, then such a Markov time can be taken as $\sigma_{\Delta^{-r}} \circ L_r$ (see item 7 and proposition 3.3). Both these approximations are good enough if for the given x $P_x(\Pi(\Delta)) = 1$, i.e. we can use property of the correct exit (see item 2.27). In this case $P_x(\sigma_\Delta \circ L_r \to \sigma_\Delta) = 1$, and also $(\forall n \in \mathbb{N})$ $(\forall \varphi_i \in \mathcal{C}_0)$ $(\forall \lambda_i > 0)$

$$R_{\sigma_\Delta}(\lambda_1, \dots, \lambda_n; \varphi_1, \dots, \varphi_n \mid x)$$
$$= \lim_{r \to 0} R_{\sigma_\Delta}^r (\lambda_1, \dots, \lambda_n; \varphi_1, \dots, \varphi_n \mid x),$$

where

$$R_{\sigma_\Delta}^r (\lambda_1, \dots, \lambda_n; \varphi_1, \dots, \varphi_n \mid x)$$
$$= E_x \left(\int_{s_n < \sigma_\Delta L_r} \prod_{i=1}^{n} \left(e^{-\lambda_i t_i} \left(\varphi_i \circ X_{s_i} \circ L_r \right) dt_i \right) \right)$$

$(s_n = t_1 + \dots + t_n)$. This stepped approximation can be expressed in terms of transition generating functions. For example,

$$R^r(\lambda; \varphi) = \sum_{k=0}^{\infty} f_{\sigma_r^k} \left(\lambda; \frac{\varphi}{\lambda} \left(1 - f_{\sigma_r}(\lambda; 1) \right) \right),$$

where $f_{\sigma_r}(\lambda; \varphi) = E.(e^{-\lambda \sigma_r}(\varphi \circ X_{\sigma_r}); \sigma_r < \infty)$ is an integral with respect to the transition generating function.

22. Markov property for the operator

In the theory of Markov processes, the one-dimensional operator $R(\lambda; \varphi) = R(\lambda; \varphi \mid \cdot) = R_\lambda \varphi$ corresponds to the definition of a resolvent for a semi-group of operators, called a lambda-potential operator (see item 2.44, Dynkin [DYN 63, p. 44], Gihman and Skorokhod [GIH 73, p. 139], Blumenthal and Getoor [BLU 68, p. 41], Hunt [HAN 62], etc.).

It is not difficult to show that for a Markov process

$$R_{\sigma_\Delta}(\lambda_1,\ldots,\lambda_n;\ \varphi_1,\ldots,\varphi_n)$$
$$= R_{\sigma_\Delta}(\lambda_1;\ \varphi_1 R_{\sigma_\Delta}(\lambda_2,\ldots,\lambda_n;\ \varphi_2,\ldots,\varphi_n)).$$

[3.2]

For an admissible family measures determining these multi-dimensional operators, this condition is a sufficient condition of Markovness (up to the first exit time from the open set Δ). Namely, if for any $n \geq 1$, $\varphi_i \in \mathcal{C}_0$, $\lambda_i > 0$ on the set Δ the condition [3.2] is fulfilled, then for any $x \in \Delta$ and almost all $t \in \mathbb{R}_+$ the Markov property for the measure P_x is fulfilled: $(\forall B \in \mathfrak{F})\ (\forall A \in \mathfrak{F}_t)$

$$P_x(\theta_t^{-1}B,\ A\cap\{t<\sigma_\Delta\})$$
$$= E_x(P_{X_t}(B);\ A\cap\{t<\sigma_\Delta\}).$$

For a semi-Markov family of measures, which is lambda-continuous on the interval $[0,\sigma_\Delta)$ this Markov property is fulfilled for any $t \in \mathbb{R}_+$. A proof of this assertion follows from the Laplace transformation properties (see, e.g., [DIT 66]).

23. Difference operator

For any $\lambda \geq 0$, $\varphi \in \mathbb{B}_0$, $x \in \mathbb{X}$ let

$$f_\lambda(\varphi\mid x) \equiv f_{\sigma_0}(\lambda,\varphi\mid x)$$
$$= E_x\big(e^{-\lambda\sigma_0}(\varphi\circ X_{\sigma_0})\big);\ \sigma_0<\infty).$$

Under the condition $P_x(0<\sigma_0<\infty) > 0$ we define operators

$$A_\lambda\varphi = (f_\lambda\varphi - \varphi)/m_0 \quad (\lambda\geq 0),$$

$$\mathcal{A}_\lambda\varphi = \frac{f_\lambda\varphi - \varphi}{\lambda^{-1}(1 - f_\lambda\mathbb{I})} \quad (\lambda > 0),$$

where $f_\lambda\mathbb{I}(x) = E_x(e^{-\lambda\sigma_0};\ \sigma_0<\infty)$, $f_\lambda\varphi = f_\lambda(\varphi\mid\cdot)$, and $m_0(x) = E_x(\sigma_0;\ \sigma_0<\infty)$ are the expectation of the first time for the initial state to be changed. These operators are said to be lambda-characteristic operators (of the first and second kind correspondingly). In order to formulate properties of SM walk we use the operators of the second kind.

24. Connection of operators

In the following theorem we use denotations:

$$R(k, n) = R_{\sigma_\Delta}(\lambda_k, \ldots, \lambda_n; \varphi_k, \ldots, \varphi_n \mid x) \quad (1 \le k \le n),$$

$$R_\lambda^\Delta \varphi = R_{\sigma_\Delta}(\lambda; \varphi) \quad (\Delta \in \mathfrak{A} \cup \tilde{\mathfrak{A}}, \ \lambda \ge 0, \ \lambda_i > 0, \ \varphi, \varphi_i \in \mathbb{B}_0),$$

$$\psi_k^n = \prod_{i=k}^n \varphi_i, \quad \Pi_i(k, n) = \prod_{k \le j \le n, j \ne i} (\lambda_j - \lambda_i)^{-1} \quad (\lambda_j \ne \lambda_i),$$

$$\Phi_{k,n} = \psi_k^n \Pi_k(k, n) + \sum_{j=k}^{n-1} \psi_k^j \Pi_k(k, j+1) \mathcal{A}_{\lambda_k} R(j+1, n)$$

$(1 \le k \le n)$, N_Δ is a number of jumps of a step-function up to the time σ_Δ:

$$N_\Delta = n \Longleftrightarrow \tau_n < \sigma_\Delta = \tau_{n+1}.$$

Evidently, $\sigma_\Delta < \infty$ implies $N_\Delta < \infty$. These events cannot be identical if the number of jumps is finite on the half-line.

THEOREM 3.2. *Let* $(P_x) \in$ SM *and* $\Delta \in \mathcal{B}(\mathbb{X})$. *Then*
(1) if (P_x) *is a SM walk, then*
    (a) $R_\lambda^\Delta \mathcal{A}_\lambda \varphi = f_{\sigma_\Delta}(\lambda; \varphi) - \varphi\,(\lambda > 0)$,
    (b) if $P_x(\sigma_\Delta < \infty) = 1$, *then* $R_0^\Delta \mathcal{A}_0 \varphi = f_{\sigma_\Delta}(0; \varphi \mid x) - \varphi(x)$;
(2) if $P_x(\sigma_0 > 0) = 1$ *and* $P_x(X_{\sigma_0} \ne x \mid \sigma_0 < \infty) = 1$, *then*

$$\mathcal{A}_{\lambda_1} R(1, n)(x) = -\Phi_{1,n}(x).$$

Proof. (1) Using denotation $\tau_k = \sigma_0^k$, we have for $\lambda > 0$

$$R_\lambda^\Delta \mathcal{A}_\lambda \varphi = E. \left(\int_0^{\sigma_\Delta} e^{-\lambda t} (\mathcal{A}_\lambda \varphi \circ X_t)\, dt; \ N_\Delta < \infty \right)$$

$$+ E. \left(\int_0^{\sigma_\Delta} e^{-\lambda t} (\mathcal{A}_\lambda \varphi \circ X_t)\, dt; \ N_\Delta = \infty \right).$$

The first term of this sum is equal to

$$\sum_{n=0}^\infty E. \left(\sum_{k=0}^n (\mathcal{A}_\lambda \varphi \circ X_{\tau_k}) \lambda^{-1} e^{-\lambda \tau_k} (1 - e^{-\lambda \sigma_0 \circ \theta_{\tau_k}}); \ N_\Delta = n \right)$$

$$= \sum_{k=0}^\infty E. ((\mathcal{A}_\lambda \varphi \circ X_{\tau_k}) \lambda^{-1} e^{-\lambda \tau_k} (1 - e^{-\lambda \sigma_0 \circ \theta_{\tau_k}}); \ k \le N_\Delta < \infty).$$

The second term is equal to

$$\sum_{k=0}^{\infty} E.\left(\left(A_\lambda \varphi \circ X_{\tau_k}\right)\lambda^{-1}e^{-\lambda \tau_k}\left(1 - e^{-\lambda \sigma_0 \theta \tau_k}\right); \ N_\Delta = \infty\right),$$

which when added to the first term gives

$$\sum_{k=0}^{\infty} E.\left(\left(A_\lambda \varphi \circ X_{\tau_k}\right)\lambda^{-1}e^{-\lambda \tau_k}\left(1 - e^{-\lambda \sigma_0 \theta \tau_k}\right); \ k \leq N_\Delta\right).$$

Since $\{k \leq N_\Delta\} \in \mathcal{F}_{\tau_k}$, the semi-Markov property can be used to obtain the following

$$\sum_{k=0}^{\infty} E.\left(\left(A_\lambda \varphi \circ X_{\tau_k}\right)\lambda^{-1}e^{-\lambda \tau_k}\left(1 - \left(f_{\sigma_0}(\lambda; \mathbb{I}) \circ X_{\tau_k}\right)\right); \ k \leq N_\Delta\right)$$

$$= \sum_{k=0}^{\infty} E.\left(\left(\left(f_{\sigma_0}(\lambda; \varphi) - \varphi\right) \circ X_{\tau_k}\right)e^{-\lambda \tau_k}; \ k \leq N_\Delta\right)$$

$$= \sum_{k=0}^{\infty} E.\left(e^{-\lambda(\sigma_0 \theta \tau_k)}\left(\varphi \circ X_{\tau_{k+1}}\right) - e^{-\lambda \tau_k}\left(\varphi \circ X_{\tau_k}\right); \ k \leq N_\Delta\right)$$

$$= \sum_{n=0}^{\infty} E.\left(\sum_{k=0}^{n}\left(e^{-\lambda \tau_{k+1}}\left(\varphi \circ X_{\tau_{k+1}}\right) - e^{-\lambda \tau_k}\left(\varphi \circ X_{\tau_k}\right)\right); \ N_\Delta = n\right)$$

$$+ E.\left(\sum_{k=0}^{\infty}\left(e^{-\lambda \tau_{k+1}}\left(\varphi \circ X_{\tau_{k+1}}\right) - e^{-\lambda \tau_k}\left(\varphi \circ X_{\tau_k}\right)\right); \ N_\Delta = \infty\right)$$

$$= \sum_{n=0}^{\infty} E.\left(e^{-\lambda \tau_{n+1}}\left(\varphi \circ X_{\tau_{n+1}}\right) - \left(\varphi \circ X_0\right); \ N_\Delta = n\right) - E.\left(\varphi \circ X_0; \ N_\Delta = \infty\right)$$

$$= E.\left(e^{-\lambda \sigma_\Delta}\left(\varphi \circ X_{\sigma_\Delta}\right) - \left(\varphi \circ X_0\right); \ N_\Delta < \infty\right) - E.\left(\varphi \circ X_0; \ N_\Delta = \infty\right)$$

$$= f_{\sigma_\Delta}(\lambda; \varphi) - \varphi.$$

So assertion (a) is proved.

Similarly we can prove assertion (b). In this case we do not need to construct $\mathcal{F}_{\tau_k}$-measurable event $\{N_\Delta \geq k\}$ with the help of two non-measurable parts $\{k \leq N_\Delta < \infty\}$ and $\{N_\Delta = \infty\}$. The second distinction is that after replacing $(A_0 \varphi \cdot m_0) \circ X_{\tau_k}$ with $(f_{\sigma_0}(0; \varphi) - \varphi) \circ X_{\tau_k}$ and using the semi-Markov property we have to show explicitly the domain of integration taking into account that $\{N_\Delta \geq k\} \subset \{\tau_k < \infty\}$ and $\{\tau_{k+1} < \infty\} \subset \{\tau_k < \infty\}$. Namely

$$E_x\left(\left(f_{\sigma_0}(0; \varphi) - \varphi\right) \circ X_{\tau_k}; \ N_\Delta \geq k\right)$$

$$= E_x\left(\left(\varphi \circ X_{\tau_{k+1}}\right)I\left(\tau_{k+1} < \infty\right) - \left(\varphi \circ X_{\tau_k}\right)I\left(\tau_k < \infty\right); \ N_\Delta \geq k\right).$$

(2) We have at point x

$$f_{\lambda_1} R(1,n) - R(1,n) = E.\left(- \int_{t_1 < \sigma_0, s_n < \sigma_\Delta} \prod_{i=1}^n \left(e^{-\lambda_i t_i} \left(\varphi_i \circ X_{s_i}\right) dt_i\right) \right)$$

$$= E.\left(- \int_{s_n < \sigma_0} \prod_{i=1}^n \left(e^{-\lambda_i t_i} \left(\varphi_i \circ X_{s_i}\right) dt_i\right) \right)$$

$$+ \sum_{k=1}^{n-1} E.\left(- \int_{S_k} \prod_{i=1}^n \left(e^{-\lambda_i t_i} \left(\varphi_i \circ X_{s_i}\right) dt_i\right) \right),$$

where $S_k = s_k < \sigma_0 \le s_{k+1}, s_n < \sigma_\Delta$. The first term is equal to

$$\psi_1^n E.\left(- \int_{s_n < \sigma_0} \prod_{i=1}^n \left(e^{-\lambda_i t_i} \, dt_i\right) \right).$$

In the second term the summand with a number k is equal to

$$E.\left(- \int_{S_k} \prod_{i=1}^n \left(e^{-\lambda_i t_i} \left(\varphi_i \circ X_{s_i}\right) dt_i\right) \right)$$

$$= E.\left(- \int_{s_k < \sigma_0} \prod_{i=1}^k \left(e^{-\lambda_i t_i} \left(\varphi_i \circ X_{s_i}\right) dt_i\right) \right.$$

$$\left. \times \int_{\sigma_0 \le s_{k+1}, s_n < \sigma_\Delta} \prod_{i=k+1}^n \left(e^{-\lambda_i t_i} \left(\varphi_i \circ X_{s_i}\right) dt_i\right) \right).$$

Assuming $t'_{k+1} = s_{k+1} - \sigma_0$, $t'_i = t_i$ for $i \ge k+2$ and $s'_i = t'_{k+1} + \cdots + t'_i$ for $i \ge k+1$, and using the property $\sigma_\Delta = \sigma_0 + \sigma_\Delta$, we obtain the intrinsic integral to be equal to

$$e^{-\lambda_{k+1}(\sigma_0 - s_k)} \int_{\mathbb{R}_+^{n-k} \cap \{s'_n < \sigma_\Delta\}} \prod_{i=k+1}^n \left(e^{-\lambda_i t'_i} \left(\varphi_i \circ X_{\tau + s'_i}\right) dt'_i\right).$$

Using the semi-Markov property, we obtain the summand with number k:

$$E.\left(- e^{-\lambda_{k+1}\sigma_0} \left(R(k+1,n) \circ X_{\sigma_0}\right) \int_{s_k < \sigma_0} \prod_{i=1}^k \left(e^{-(\lambda_i - \lambda_{k+1})t_i} \left(\varphi_i \circ X_{s_i}\right) dt_i\right) \right)$$

$$= \psi_1^k E.\left(- e^{-\lambda_{k+1}\sigma_0} \left(R(k+1,n) \circ X_{\sigma_0}\right) \int_{s_k < \sigma_0} \prod_{i=1}^k \left(e^{-(\lambda_i - \lambda_{k+1})t_i} \, dt_i\right) \right).$$

Let us suppose temporarily that $\lambda_i \neq \lambda_j$ $(i \neq j)$, and apply the formula

$$e^{-\lambda_{k+1}\tau_1} \int_{s_k < \tau} \prod_{i=1}^{k} \left(e^{-(\lambda_i - \lambda_{k+1})t_i} \, dt_i \right) = \sum_{i=1}^{k+1} e^{-\lambda_i \tau_1} \Pi_i(1, k+1), \qquad [3.3]$$

which can be easily proved by induction using the following identity.

LEMMA 3.1. *For any $n \geq 2$ and $\lambda_i \neq \lambda_j$ $(i \neq j)$*

$$\sum_{i=1}^{n} \Pi_i(1, n) = 0 \quad (n \geq 2, \ \lambda_j \neq \lambda_i).$$

Proof. The given identity is equivalent to the following one: $\forall x \neq \lambda_i$ $(1 \leq i \leq n-1)$

$$\frac{1}{(\lambda_1 - x) \cdots (\lambda_{n-1} - x)} = \sum_{i=1}^{n-1} \Pi_i(1, n-1) \frac{1}{\lambda_i - x}.$$

We prove it by induction. The identity is evident for $n = 2$. Let this identity be true for a given n. Multiplying both parts of the latter equality by $(\lambda_n - x)^{-1}$ we have

$$\frac{1}{(\lambda_1 - x) \cdots (\lambda_n - x)} = \sum_{i=1}^{n-1} \Pi_i(1, n-1) \frac{1}{(\lambda_i - x)(\lambda_n - x)}$$

$$= \sum_{i=1}^{n-1} \Pi_i(1, n-1) \frac{1}{\lambda_n - \lambda_i} \left(\frac{1}{\lambda_i - x} - \frac{1}{\lambda_n - x} \right)$$

$$= \sum_{i=1}^{n-1} \Pi_i(1, n) \frac{1}{\lambda_i - x} + \left(-\sum_{i=1}^{n-1} \Pi_i(1, n) \right) \frac{1}{\lambda_n - x}.$$

However, the coefficient before the fraction in the second term is equal to $\Pi_n(1, n)$ by inductive supposition, which proves the identity for the order $n + 1$. $\qquad \square$

Let us continue the proof of the theorem. For $\lambda_i > 0$ $(i \leq n)$ and $\lambda_{n+1} = 0$ from identity [3.3] we obtain

$$\int_{s_n < \tau_1} \prod_{i=1}^{n} \left(e^{-\lambda_i t_i} \, dt_i \right) = \sum_{i=1}^{n} \left(1 - e^{-\lambda_i \tau_1} \right) \lambda_i^{-1} \Pi_i(1, n). \qquad [3.4]$$

Hence,

$$f_{\lambda_1} R(1,n) - R(1,n)$$

$$= \psi_1^n E. \left(\sum_{i=1}^{n} \left(e^{-\lambda_i \tau_1} - 1 \right) \lambda_i^{-1} \Pi_i(1,n) \right)$$

$$- \sum_{k=1}^{n-1} \psi_1^k E. \left(\sum_{i=1}^{k+1} e^{-\lambda_i \tau_1} \left(R(k+1,n) \circ X_{\tau_1} \right) \Pi_i(1,k+1) \right)$$

$$= \psi_1^n \sum_{i=1}^{n} \lambda_i^{-1} \left(f_{\lambda_i} 1 - 1 \right) \Pi_i(1,n)$$

$$- \sum_{k=1}^{n-1} \psi_1^k \sum_{i=1}^{k+1} f_{\lambda_i} R(k+1,n) \Pi_i(1,k+1)$$

$$= \psi_1^n \sum_{i=1}^{n} \lambda_i^{-1} \left(f_{\lambda_i} 1 - 1 \right) \Pi_i(1,n)$$

$$- \sum_{k=1}^{n-1} \psi_1^k \sum_{i=1}^{k+1} \left(f_{\lambda_i} R(k+1,n) - R(k+1,n) \right) \Pi_i(1,k+1) \equiv \Phi'_{1,n}.$$

From here

$$A_{\lambda_1} R(1,n) = \Phi'_{1,n} \frac{\lambda_1}{1 - f_{\lambda_1} 1}.$$

LEMMA 3.2. *Under the conditions of the theorem the following equality is fair*

$$\Phi'_{1,n} \lambda_1 / \left(f_{\lambda_1} 1 - 1 \right) = \Phi_{1,n}.$$

Proof. In order to be concise, let us designate

$$\mu_i = \left(1 - f_{\lambda_i} 1 \right) / \lambda_1, \qquad d_i z = f_{\lambda_i} z - z \quad \left(z \in \mathbb{B}_0 \right).$$

We have to prove

$$\psi_1^n \Pi_1(1,n) \mu_1 + \sum_{j=1}^{n-1} \psi_1^j \Pi_1(1,j+1) d_1 R(j+1,n)$$

$$= \psi_1^n \sum_{i=1}^{n} \Pi_i(1,n) \mu_i + \sum_{j=1}^{n-1} \psi_1^j \sum_{i=1}^{j+1} \Pi_i(1,j+1) \, d_i R(j+1,n).$$

We prove this equality by induction on n. Assuming $\Pi_1(1,1) = 1$, we see the equality to be fair for $n = 1$. Let it also be fair for all natural numbers up to $n-1$ ($n \geq 2$). Let us find the difference of parts of the equality to be proven:

$$\psi_1^n \sum_{i=2}^{n} \Pi_i(1,n)\mu_i + \sum_{j=1}^{n-1} \psi_1^j \sum_{i=2}^{j+1} \Pi_i(1,j+1)\, d_i R(j+1,n)$$

$$= \psi_1^n \sum_{i=2}^{n} \Pi_i(1,n)\mu_i + \sum_{j=2}^{n-1} \psi_1^j \sum_{i=2}^{j} \Pi_i(1,j+1)\, d_i R(j+1,n) + \Psi,$$

where

$$\Psi = \sum_{k=1}^{n-1} \psi_1^k \Pi_{k+1}(1,k+1)\, d_{k+1} R(k+1,n).$$

Substituting the expression $d_{k+1}R(k+1,n)$ and using the inductive supposition, we discover that it is equal to

$$-\psi_{k+1}^n \Pi_{k+1}(k+1,n)\mu_{k+1} - \sum_{j=k+1}^{n-1} \psi_{k+1}^j \Pi_{k+1}(k+1,j+1)\, d_{k+1} R(j+1,n).$$

Using evident relations

$$\psi_1^k \psi_{k+1}^n = \psi_1^n,$$

$$\Pi_{k+1}(1,k+1)\Pi_{k+1}(k+1,j+1) = \Pi_{k+1}(1,j+1),$$

we obtain

$$\Psi = -\sum_{k=1}^{n-1} \psi_1^n \Pi_{k+1}(1,n)\mu_{k+1}$$

$$-\sum_{k=1}^{n-2}\sum_{j=k+1}^{n-1} \psi_1^j \Pi_{k+1}(1,j+1)\, d_{k+1} R(j+1,n).$$

Changing the order of summands and varying variables, we see the function $-\Psi$ to be equal to the remain part of the difference to be estimated. $\square$

The second assertion of the theorem is proved under condition $\lambda_i \neq \lambda_j$ ($i \neq j$). However, the function $\Phi_{1,n}$ is determined for all $\lambda_i \neq \lambda_1$ (domain of $\Pi_1(1,n)$). The difference of interest is being extended on this region by continuity. In order to extend it on the region of all $\lambda_i > 0$ the function $\Phi_{1,n}$ should be redetermined. At the points of

n-dimensional space of positive vectors $(\lambda_1, \ldots, \lambda_n)$, where some coordinates coincide, let us determine $\Phi_{1,n}$ as a limit of this function values on a sequence of vectors with pair-wise different coordinates. It is not difficult to show that existence of these limits depends on the existence of derivatives of $\mathcal{A}_\lambda g$ by λ, but this function for any $g \in \mathbb{B}_0$ has derivatives of any order at any point. □

3.4. Operators and general SM processes

Both differential and integral operators for a general semi-Markov process are defined and being investigated. In order to evaluate the integral operator on the set of continuous functions a stepped approximation of a semi-Markov process can be used, transforming the process to a semi-Markov walk (item 21). However, if the first exit time from the initial point is equal to zero, it is not convenient to use formulae of SM walks in order to derive conditions of duality of the differential and integral operators for a general semi-Markov process. It is more rational to derive them directly by similar methods.

25. Family of lambda-continuous measures on the interval

Let Δ be a measurable set. An admissible family (P_x) is said to be lambda-continuous on an interval $[0, \sigma_\Delta)$ at point x if $(\forall n \in \mathbb{N})$ $(\forall \lambda_i > 0)$ $(\forall \varphi_i \in \mathcal{C}_0)$ the function $R_{\sigma_\Delta}(\lambda_1, \ldots, \lambda_n; \varphi_1, \ldots, \varphi_n)$ is continuous at point x; the lambda-continuity of a family means the lambda-continuity of it on the whole half-line (see item 2.44). It is not difficult to show that weak continuity of the family (P_x) with respect to the metric ρ_D implies lambda-continuity and lambda-continuity on the interval $[0, \sigma_\Delta)$ for "almost all" $\Delta \in \mathfrak{A} \cup \widetilde{\mathfrak{A}}$ (see theorem 2.4 and item 2.34). In Chapter 7 it will be shown that weak continuity is equivalent to lambda-continuity.

Lambda-continuity plays an important role while proving the regeneration property of some Markov times. Thus, for example, if equation [3.2] is fulfilled for a lambda-continuous SM family (P_x) for all $n \in \mathbb{N}$, $\lambda_i > 0$, and $\varphi_i \in \mathcal{C}_0$, then the family is strictly Markovian. It follows from the property that the class $\mathrm{RT}(P_x)$ is closed with respect to convergence from above (see theorem 2.7, and Blumenthal and Getoor [BLU 68]).

26. Lambda-characteristic operator

We are interested in properties of kernels $f_{\sigma_r}(\lambda, dx_1 \mid x)$ for small $r > 0$ (see denotations of item 16 and 2.14). Let

$$\mathbb{X}_r = \{x \in \mathbb{X} : P_x(\sigma_r < \infty) > 0\}$$

and $\mathbb{X}_0 = \bigcup_{r>0} \mathbb{X}_r$. Evidently, $\mathbb{X}_0 = \{x \in \mathbb{X} : P_x(\sigma_0 < \infty) > 0\}$. In most cases we will assume that $(\forall x \in \mathbb{X}_0)\ P_x(\sigma_0 < \infty) = 1$ and $\mathbb{X}_0$ is an open set. It is not difficult to show that the set $\mathbb{X}_0$ is open if the family (P_x) is lambda-continuous on the whole $\mathbb{X}$. For $x \in \mathbb{X}_r$ let us consider the value

$$A_\lambda^r(\varphi \mid x) = \frac{1}{m_r(x)}\big(f_{\sigma_r}(\lambda;\varphi \mid x) - \varphi(x)\big),$$

where $\lambda \geq 0$, $m_r(x) = P_x(\sigma_r;\ \sigma_r < \infty)$, $\varphi \in \mathbb{B}_0$. For $\lambda > 0$ and $\tau = \infty$ let us determine $e^{-\lambda\tau}(\varphi \circ \pi_\tau) = 0$. In this case

$$f_{\sigma_r}(\lambda, \varphi \mid x) \equiv P_x\big(e^{-\lambda\sigma_r}(\varphi \circ \pi_{\sigma_r});\ \sigma_r < \infty\big) = P_x\big(e^{-\lambda\sigma_r}(\varphi \circ \pi_{\sigma_r})\big).$$

For $x \in \mathbb{X}_0$ let $V_\lambda(x)$ mean the set of all $\varphi \in \mathbb{B}_0$, what the limit

$$A_\lambda(\varphi \mid x) = \lim_{r \to 0} A_\lambda^r(\varphi \mid x)$$

is determined for. If function φ is such that this limit exists for any $x \in S$, then it determines the function $A_\lambda(\varphi \mid x)$ on S. In other words, the operator $A_\lambda : V_\lambda(S) \to \mathbb{B}(S)$ is determined, where $V_\lambda(S) = \bigcap_{x \in S} V_\lambda(x)$, $\mathbb{B}(S)$ is the set of all measurable real functions on S. The operator A_λ, determined above, is said to be a lambda-characteristic operator (of the first kind) (see Dynkin [DYN 63, p. 201]). Evidently, the characteristic operator of Dynkin, if it is determined for given φ and x, coincides with the lambda-characteristic operator for $\lambda = 0$. We distinguish strict (narrow) and wide sense of the definition of the operator A_λ. It depends on whether this operator is a limit for an arbitrary sequence of $\Delta \downarrow x$, or only for decreasing sequence of balls. For the purposes of this chapter it is enough to consider the operators in a wide sense. Furthermore, we will give some sufficient conditions for the existence of lambda-characteristic operators in a strict sense.

Some properties of SM processes are convenient to formulate in terms of lambda-characteristic operators of the second kind: $\mathcal{A}_\lambda\varphi = \lim_{r \to 0} \mathcal{A}_\lambda^r\varphi$, where

$$\mathcal{A}_\lambda^r\varphi = \frac{f_{\sigma_r}(\lambda;\varphi) - \varphi}{\lambda^{-1}\big(1 - f_{\sigma_r}(\lambda;\mathbb{I})\big)},$$

which is determined at point x on the set of functions $V_\lambda(x)$. Using operators $\mathcal{A}_\lambda$ is actual for processes with $m_r = \infty$ $(\forall r > 0)$. For example, it is actual for process of maxima of the standard Wiener process (see Chapter 9). In this case existence of the function $A_{\lambda_i}\mathbb{I}$ is not necessary. Instead the limit

$$a(\lambda_1, \lambda_2 \mid x) = \lim_{r \to 0} a_r(\lambda_1, \lambda_2 \mid x) \quad (\lambda_1 \geq 0,\ \lambda_2 > 0)$$

is used, where

$$a_r(\lambda_1, \lambda_2 \mid x) = \frac{1 - f_{\sigma_r}(\lambda_1;\mathbb{I} \mid x)}{1 - f_{\sigma_r}(\lambda_2;\mathbb{I} \mid x)}.$$

In particular,

$$a_r(0, \lambda \mid x) = \frac{P_x(\sigma_r = \infty)}{1 - f_{\sigma_r}(\lambda; \mathbb{I} \mid x)}.$$

The existence condition for this limit is denoted as $\mathbb{I} \in V_{\lambda_2}^{\lambda_1}(x)$. Evidently, if $\mathbb{I} \in V_{\lambda_i}(x)$ $(i = 1, 2)$, then

$$a(\lambda_1, \lambda_2 \mid x) = A_{\lambda_1}(\mathbb{I} \mid x)/A_{\lambda_2}(\mathbb{I} \mid x),$$

and if operator A_λ is determined, then $\mathcal{A}_\lambda \varphi = -\lambda \, A_\lambda \varphi / A_\lambda \mathbb{I}$. From here $A_\lambda \mathbb{I} = -\lambda$. Obviously, $\mathcal{A}_0 \varphi = A_0 \varphi$ if the last operator is determined. The meaning of the ratio $(1 - f_{\sigma_r}(\lambda; \mathbb{I}))/\lambda|_{\lambda=0}$ is accepted to be a limit of this expression as $\lambda \to 0$.

27. Local and pseudo-local operators

According to definitions item 26 with $P_x(\sigma_0 > 0) > 0$ (i.e. there exists an interval of constancy at the start point of the trajectory) and $\lambda > 0$ we have

$$A_\lambda(\varphi \mid x) = \frac{E_x\left(e^{-\lambda\sigma_0}(\varphi \circ X_{\sigma_0})\right) - \varphi(x)}{m_0(x)}.$$

It corresponds to the definition of the operator of the first kind from item 23. In this sense the operator $\mathcal{A}_\lambda \varphi$, defined in item 26, is a generalization of the operators of the second kind from item 23. In contrast to a Markov process, when the probability $P_x(\sigma_0 > 0)$ can accept only two values: 0 and 1, a semi-Markov process admits this probability equal to any value in the interval $[0, 1]$. At the time σ_0 the trajectory of the SM process can behave differently. It can be a.s. a point of continuity of the process. In this case $\mathcal{A}_\lambda \varphi = \varphi \, A_\lambda \mathbb{I}$. If σ_0 is a point of discontinuity with a positive probability, then the operator A_λ at point x depends, in general, on all the values of the function φ.

The operator A_λ is said to be local at point x if $(\forall \varphi_1, \varphi_2)\ \varphi_1 \in V_\lambda(x)$ and $\varphi_1 = \varphi_2$ in some neighborhood of point x implies $\varphi_2 \in V_\lambda(x)$ and $A_\lambda(\varphi_1 \mid x) = A_\lambda(\varphi_2 \mid x)$. Similarly, a local property of the operator $\mathcal{A}_\lambda$ is defined. For locality of A_0 at point x it is necessary and sufficient that $(\forall R > 0)$

$$\frac{P_x(\sigma_r = \sigma_R < \infty)}{m_r(x)} \longrightarrow 0 \quad (r \longrightarrow 0)$$

(see Gihman and Skorokhod [GIH 73], Dynkin [DYN 63]). Furthermore, the rather weaker property of a process is used, which can be expressed by properties of its lambda-characteristic operator for $\lambda > 0$.

The operator A_λ is said to be pseudo-local at point x if for any $R > 0$

$$\frac{E_x\left(1 - e^{-\lambda\sigma_r};\ \sigma_r = \sigma_R < \infty\right)}{m_r(x)} \longrightarrow 0 \quad (r \longrightarrow 0). \qquad [3.5]$$

Correspondingly, the definition of a pseudo-local operator $\mathcal{A}_\lambda$ uses another normalization:

$$\frac{E_x\left(1 - e^{-\lambda\sigma_r};\ \sigma_r = \sigma_R < \infty\right)}{P_x\left(1 - e^{-\lambda\sigma_r}\right)} \longrightarrow 0 \quad (r \longrightarrow 0). \qquad [3.6]$$

Here we assume $P_x(\sigma_0 < \infty) = 1$. Thus, a local operator of the process having a finite interval of constancy at the beginning of its trajectory and admitting discontinuity at the end of this interval cannot be pseudo-local. Let us note that pseudo-locality of operators A_λ and $\mathcal{A}_\lambda$ does not depend on λ: if an operator is pseudo-local for one $\lambda > 0$, it is pseudo-local for any $\lambda > 0$. In order to prove this assertion the inequality $(\forall \lambda_1, \lambda_2,\ 0 < \lambda_1 < \lambda_2)\ (\forall t > 0)$ can be used:

$$1 < \frac{1 - e^{-\lambda_2 t}}{1 - e^{-\lambda_1 t}} < \frac{\lambda_2}{\lambda_1}, \qquad [3.7]$$

which follows from monotone decreasing of the function $(1 - e^{-x})/x$.

28. Example of a process with discontinuous trajectories

The locality of the operator A_0 and the pseudo-locality of A_λ for any $\lambda > 0$ follows from a.s. continuity of trajectories of the process (P_x).

Let us show an example of a SM process with discontinuous trajectories such that $(\forall x \in \mathbb{X})\ (\forall \lambda > 0)$ its operator A_λ is pseudo-local, but A_0 is not local at point x.

Let $\mathbb{X} = \mathbb{R}_+$ and P_x be probability measures of the process $\xi(t) = x + t + N_t$, where N_t is the Poisson process with intensity c. Moreover

$$P_x\left(\sigma_r \le t\right) = \begin{cases} 1 - e^{-ct}, & t < r, \\ 1, & t \ge r, \end{cases}$$

$$m_r(x) = \frac{1}{c}\left(1 - e^{-cr}\right),$$

$$E_x\left(1 - e^{-\lambda\sigma_r}\right) = \int_0^r \left(1 - e^{-\lambda t}\right)ce^{-ct}dt + \left(1 - e^{-\lambda r}\right)e^{-cr}$$

$$= \left(1 - e^{-(c+\lambda)r}\right)\frac{\lambda}{c+\lambda},$$

$$E_x\left(1 - e^{-\lambda\sigma_r};\ \sigma_r = \sigma_R\right) \le \lambda r P_x\left(\sigma_r = \sigma_R\right).$$

However, for $r < R < 1$ $P_x(\sigma_r = \sigma_R) = P_x(N_r > 0) = cr + o(r)$ $(r \to 0)$.

29. Correspondence of characteristic operators

THEOREM 3.3. *Let (P_x) be an admissible family. The following assertions are fair:*

(1) if $x \in \mathbb{X}_0$, $\mathbb{I}(\cdot) \in V_\lambda(x)$, and operator A_λ is pseudo-local at point x, and the function φ is continuous at point x, then

        (a) $(\varphi \in V_0(x)) \Leftrightarrow (\varphi \in V_\lambda(x))$;

        (b) $A_\lambda(\varphi \mid x) = A_0(\varphi \mid x) + \varphi(x)(A_\lambda(\mathbb{I} \mid x) - A_0(\mathbb{I} \mid x))$;

(2) if $x \in \mathbb{X}_0$, $\mathbb{I} \in V_\lambda^\mu(x)$ $(\lambda, \mu > 0)$, and operator A_λ is pseudo-local, and function φ is continuous at point x, then

        (a) $(\varphi \in V_\lambda(x)) \Leftrightarrow (\varphi \in V_\mu(x))$,

        (b) $\lambda^{-1} A_\lambda(\varphi \mid x) + \varphi(x) = a(\mu, \lambda \mid x)(\mu^{-1} A_\mu(\varphi \mid x) + \varphi(x))$.

Proof. (1) In the case of $P_x(\sigma_r < \infty) = 1$ we have

$$A_\lambda^r(\varphi \mid x) = A_0^r(\varphi \mid x) + \varphi(x)\left(A_\lambda^r(\mathbb{I} \mid x) - A_0^r(\mathbb{I} \mid x)\right)$$

$$+ \frac{1}{m_r(x)} E_x\left((e^{-\lambda\sigma_r} - 1)(\varphi \circ X_{\sigma_r} - \varphi \circ X_0)\right); \quad \sigma_r < \infty.$$

In this case

$$\left| \frac{1}{m_r(x)} E_x\left((e^{-\lambda\sigma_r} - 1)(\varphi \circ X_{\sigma_r} - \varphi \circ X_0)\right) \right|$$

$$\leq - \sup_{x_1 \in B(x,R)} |\varphi(x) - \varphi(x_1)|\lambda$$

$$+ 2\sup_{x \in \mathbb{X}} |\varphi(x)| \frac{1}{m_r(x)} E_x\left(1 - e^{-\lambda\sigma_r}\right); \quad \sigma_r = \sigma_R < \infty.$$

From continuity and boundedness of the function φ it proves the convergence of both the right and left parts of the equality.

(2) We have

$$A_\lambda^r(\varphi \mid x) = \frac{\lambda}{E_x\left(1 - e^{-\lambda\sigma_r}\right)} E_x\left(e^{-\lambda\sigma_r}\varphi(X_{\sigma_r}) - \varphi(X_0)\right)$$

$$= \frac{E_x\left(-(1 - e^{-\lambda\sigma_r})\varphi(X_{\sigma_r}) + (1 - e^{-\mu\sigma_r})\varphi(X_{\sigma_r})\right)}{\lambda^{-1} E_x\left(1 - e^{-\lambda\sigma_r}\right)}$$

$$+ \frac{\lambda}{\mu} a_r(\mu, \lambda \mid x) A_\mu^r(\varphi \mid x)$$

$$= -\lambda\varphi(x) + \lambda\varphi(x)a_r(\mu, \lambda \mid x) + \frac{\lambda}{\mu}a_r(\mu, \lambda \mid x)\mathcal{A}_\mu^r(\varphi \mid x)$$

$$- \lambda\frac{E_x\big((1 - e^{-\lambda\sigma_r})(\varphi(X_{\sigma_r}) - \varphi(X_0))\big)}{E_x(1 - e^{-\lambda\sigma_r})}$$

$$+ \lambda a_r(\mu, \lambda \mid x)\frac{E_x\big((1 - e^{-\mu\sigma_r})(\varphi(X_{\sigma_r}) - \varphi(X_0))\big)}{E_x(1 - e^{-\mu\sigma_r})}.$$

According to pseudo-locality two last terms tend to zero. Convergence of the remaining terms to a limit proves the second assertion. □

30. Difference of operators of the first kind

Let us note as a simple consequence of theorem 3.3(1) the following formula:

$$A_\mu(\varphi \mid x) - A_\lambda(\varphi \mid x) = \big(A_\mu(\mathbb{I} \mid x) - A_\lambda(\mathbb{I} \mid x)\big)\varphi(x),$$

which fulfills for the same x, φ, λ, μ, which is sufficient for theorem 3.3.

31. The meaning of the operator

In the following theorem we use denotations from theorem 3.2.

THEOREM 3.4. *Let* $(P_x) \in SM$; $\Delta \in \mathfrak{A}$; $x \in \mathbb{X}_0 \cap \Delta$; *the family* (P_x) *be lambda-continuous on the interval* $[0, \sigma_\Delta)$ *at point* x; *for any* $\lambda > 0$ *the operator* A_λ *be pseudo-local at point* x; *and also for any* $\lambda_1, \lambda_2 > 0$ $\mathbb{I} \in V_{\lambda_2}^{\lambda_1}$. *Then* $(\forall\mu \geq 0)$ $(\forall n \in \mathbb{N})$ $(\forall\lambda_i > 0, \lambda_i \neq \lambda_1)$ $(\forall\varphi_i \in C_0)$

$$R(1, n) \in V_\mu(x), \quad A_{\lambda_1}\big(R(1, n) \mid x\big) = -\Phi_{1,n}(x).$$

where $\Phi_{1,n}(x)$ *corresponds to denotations of theorem 3.2 while substituting into the formula the meaning of* A_λ, *determined in item 26.*

Proof. According to theorem 3.3(2), and due to continuity of $R(1, n)$ it is sufficient to prove the existence of a limit of function $A_{\lambda_1}^r R(1, n)$ $(r \to 0)$, and to find its meaning. In theorem 3.2(2) it is proved that under $B(x, r) \subset \Delta$

$$f_{\sigma_r}\big(\lambda_1, R(1, n) \mid x\big) - R(1, n)(x)$$

$$= -E_x\left(\int_{s_n < \sigma_r} \prod_{i=1}^{n} (e^{-\lambda_i t_i}(\varphi_i \circ X_{s_i})) \, dt_i\right)$$

$$- \sum_{k=1}^{n-1} E_x\left(e^{-\lambda_{k+1}\sigma_r}\big(R(k+1, n) \circ X_{\sigma_r}\big)\right. \qquad [3.8]$$

$$\left. \times \int_{s_k < \sigma_r} \prod_{i=1}^{k} (e^{-(\lambda_i - \lambda_{k+1})t_i}(\varphi_i \circ X_{s_i})) \, dt_i\right).$$

In this case we cannot take the function φ_i out of the integral as $r > 0$. However, using continuity of φ_i at point x, and the existence of limits $a(\lambda_i, \lambda_j \mid x)$, and formula [3.4], we obtain for $r \to 0$, and $\lambda_i \neq \lambda_j$ $(i \neq j)$

$$\frac{E_x\left(\int_{s_n < \sigma_r} \prod_{i=1}^{n} \left(e^{-\lambda_1 t_i} \left(\varphi_i \circ X_{s_i}\right) dt_i\right)\right)}{\lambda_1^{-1}\left(1 - f_{\lambda_1}^r(\mathbb{I} \mid x)\right)}$$

$$\longrightarrow \psi_1^n(x) \sum_{i=1}^{n} \Pi_i(1,n) a\left(\lambda_i, \lambda_1 \mid x\right),$$

and also

$$E_x\left(e^{-\lambda_{k+1}\sigma_r}\left(R(k+1,n)\circ X_{\sigma_r}\right)\int_{s_k<\sigma_r}\prod_{i=1}^{k}\left(e^{-(\lambda_i-\lambda_{k+1})t_i}\left(\varphi_i\circ X_{s_i}\right)dt_i\right)\right)$$

$$\sim \psi_1^k(x) E_x\left(e^{-\lambda_{k+1}\sigma_r}\left(R(k+1,n)\circ X_{\sigma_r}\right)\int_{s_k<\sigma_r}\prod_{i=1}^{k}\left(e^{-(\lambda_i-\lambda_{k+1})t_i}dt_i\right)\right)$$

$$= \psi_1^k(x) \sum_{i=1}^{k+1} \Pi_i(1,k+1)\left(f_{\lambda_i}^r\left(R(k+1,n)\mid x\right) - R(k+1,n)(x)\right).$$

Divided by $\lambda_1^{-1}(1 - f_{\lambda_1}^r(\mathbb{I} \mid x))$, the last expression is equal to

$$\psi_1^k(x) \sum_{i=1}^{k+1} \Pi_i(1,k+1)\left(\frac{\lambda_1}{\lambda_{k+1}} a_r\left(\lambda_{k+1}, \lambda_1\right) \mathcal{A}_{\lambda_{k+1}}^r\left(R(k+1,n)\mid x\right)\right.$$

$$\left. + \lambda_1\left(a_r\left(\lambda_{k+1}, \lambda_1 \mid x\right) - a_r\left(\lambda_i, \lambda_1 \mid x\right)\right)R(k+1,n \mid x) + \varepsilon_r\right),$$

where ε_r is equal to

$$\frac{E_x\left(\left(e^{-\lambda_1\sigma_r} - e^{-\lambda_{k+1}\sigma_r}\right)\left(R(k+1,n)\circ X_{\sigma_r} - R(k+1,n)\circ X_0\right)\right)}{\lambda_1^{-1}\left(1 - f_{\lambda_1}^r(\mathbb{I} \mid x)\right)},$$

and tends to zero, according to the pseudo-locality of the operator $\mathcal{A}_\lambda$ and the continuity of $R(k+1,n)$. Now convergence to a limit of the function $\mathcal{A}_{\lambda_1}^r R(1,n \mid x)$ can be proved by induction on $n \geq 1$ based for $n = 1$:

$$\mathcal{A}_{\lambda_1}^r R(1,1 \mid x) = \frac{E_x\left(\int_0^{\sigma_r} e^{-\lambda_1 t}\left(\varphi_1 \circ X_t\right) dt\right)}{\lambda_1^{-1}\left(1 - f_{\lambda_1}^r(\mathbb{I} \mid x)\right)}.$$

It converges to the limit $\varphi_1(x)$. Taking into account theorem 3.3, and the evident relation $a(\lambda_3, \lambda_2) \times a(\lambda_2, \lambda_1) = a(\lambda_3, \lambda_1)$ we can write the obtained formula as follows $A_{\lambda_1} R(1, n \mid x) = -\widetilde{\Phi}_{1,n}$, where

$$\widetilde{\Phi}_{1,n} = \psi_1^n(x) \sum_{i=1}^n \Pi_i(1, n) \frac{\lambda_1}{\lambda_i} a(\lambda_i, \lambda_1 \mid x)$$

$$+ \sum_{k=1}^{n-1} \psi_1^k(x) \sum_{i=1}^{k+1} \Pi_i(1, k+1) \frac{\lambda_1}{\lambda_i} a(\lambda_i, \lambda_1 \mid x) A_{\lambda_i}(R(k+1, n) \mid x).$$

It remains to prove that $\Phi_{1,n} = \widetilde{\Phi}_{1,n}$, but it can be shown in the same way as lemma 3.2 is proved. □

32. Another representation of operators

Under conditions of theorem 3.4 the formula $A_{\lambda_1} R(1, n) = -\overline{\Phi}_{1,n}$ is fair, where

$$\overline{\Phi}_{1,n} = \psi_1^n \sum_{i=1}^n \Pi_i(1, n) \frac{\lambda_1}{\lambda_i} a(\lambda_i, \lambda_1)$$

$$- \lambda_1 \sum_{k=1}^{n-1} \psi_1^k R(k+1, n) \sum_{i=1}^{k+1} \Pi_i(1, k+1) a(\lambda_i, \lambda_1).$$

Using the identity from lemma 3.1 this formula follows from $A_{\lambda_1} R(1, n) = -\widetilde{\Phi}_{1,n}$ and the substitution

$$\frac{\lambda_1}{\lambda_i} a(\lambda_i, \lambda_1) A_{\lambda_i} R(k+1, n) = A_{\lambda_1} R(k+1, n) + \lambda_1 R(k+1, n)$$

$$- \lambda_1 R(k+1, n) a(\lambda_i, \lambda_1).$$

It is true due to theorem 3.3(2).

33. Inverse operator

Theorem 3.2(1) makes plausible suppositions that $R_{\sigma_\Delta}(\lambda; A_\lambda \varphi) = f_{\sigma_\Delta}(\lambda; \varphi) - \varphi$ ($\lambda \geq 0$) for a class of function wide enough. Let us denote

$$W_\lambda(\Delta) = \{\varphi \in V_\lambda(\Delta) : R_{\sigma_\Delta}(\lambda; A_\lambda \varphi) = f_{\sigma_\Delta}(\lambda; \varphi) - \varphi\} \quad (\lambda \geq 0),$$

$$\mathcal{W}_\lambda(\Delta) = \{\varphi \in \mathcal{V}_\lambda(\Delta) : R_{\sigma_\Delta}(\lambda; A_\lambda \varphi) = f_{\sigma_\Delta}(\lambda; \varphi) - \varphi\} \quad (\lambda > 0).$$

34. Domain of the inverse operator

PROPOSITION 3.9. *Let* $(P_x) \in$ SM *and* $\Delta \subset \mathbb{X}_0$ *be such that at least one of the two following conditions is fulfilled (for denotations, see item 7):*

(a) Δ *is a closed set;*

(b) Δ *is an open set, and* $(P_x) \in QC$.

(1) If $(\forall x \in \Delta)\, P_x(\sigma_\Delta) < \infty$, $\varphi \in \mathcal{C}_0$ *and the following conditions are fulfilled:*

   (A) $(\exists M > 0)\, (\forall r > 0)\, \sup_{x \in \Delta} |A_0^r(\varphi \mid x)| \le M$;

   (B) $(\forall x \in \Delta)\, A_0^r(\varphi \mid x) \to A_0(\varphi \mid x)\, (r \to 0)$;

   (C) the function $A_0\varphi$ *is continuous on* Δ;

then $\varphi \in \mathcal{W}_0(\Delta)$.

(2) If $\lambda > 0$, $\varphi \in \mathcal{C}_0$ *and the following conditions are fulfilled:*

   (D) $(\exists M > 0)\, (\forall r > 0)\, \sup_{x \in \Delta} |A_\lambda^r(\varphi \mid x)| \le M$;

   (E) $(\forall x \in \Delta)\, A_\lambda^r(\varphi \mid x) \to A_\lambda(\varphi \mid x)\, (r \to 0)$;

   (F) the function $A_\lambda\varphi$ *is continuous on* Δ;

then $\varphi \in \mathcal{W}_\lambda(\Delta)$.

Proof. In fact, the following formulae are proved in theorem 3.2(1):

$$R_{\sigma_{\Delta(r)}}^r\left(0; A_0^r\varphi\right) = f_{\sigma_{\Delta(r)}L_r}(0; \varphi) - \varphi,$$

$$R_{\sigma_{\Delta(r)}}^r\left(\lambda; A_\lambda^r\varphi\right) = f_{\sigma_{\Delta(r)}L_r}(\lambda; \varphi) - \varphi \quad (\lambda > 0),$$

where $\varphi \in \mathcal{C}_0$, $\Delta(r)$ is a measurable set depending on r. If condition (a) is fulfilled we assume $\Delta(r) = \Delta$. If condition (b) is fulfilled we assume $\Delta(r) = \Delta^{-r}$ (denotation, see items 10, 21 and 26). In both cases for $r \to 0$ $f_{\sigma_{\Delta(r)}\circ L_r}(0; \varphi) \to f_{\sigma_\Delta}(0; \varphi)$ and $f_{\sigma_{\Delta(r)}\circ L_r}(\lambda; \varphi) \to f_{\sigma_\Delta}(\lambda; \varphi)$ on the set Δ (see proposition 3.3), and also for any $x \in \Delta$ $|\sigma_{\Delta(r)} \circ L_r - \sigma_\Delta| \to 0$ P_x-a.s. According to the Egorov theorem [KOL 72, p. 269], for any $x \in \Delta$ and $\varepsilon > 0$ there exists a set $E_\varepsilon \subset \Delta$ with the measure $\mu_0(\Delta \backslash E_\varepsilon) \equiv R_{\sigma_\Delta}(0; I_{\Delta \backslash E_\varepsilon} \mid x) < \varepsilon$, on which the convergence $A_0^r\varphi \to A_0\varphi$ is uniform. In this case

$$\left| E_x\left(\int_0^{\sigma_{\Delta(r)}\circ L_r} \left(A_0^r\varphi \circ X_t \circ L_r\right) dt \right) - E_x\left(\int_0^{\sigma_\Delta} \left(A_0\varphi \circ X_t\right) dt \right) \right|$$

$$= \varepsilon_1 + \varepsilon_2 + \varepsilon_3 + \varepsilon_4,$$

where

$$\varepsilon_1 \le M\, E_x\left(\left|\sigma_{\Delta(r)} \circ L_r - \sigma_\Delta\right|\right),$$

$$\varepsilon_2 \le E_x\left(\int_0^{\sigma_\Delta} \left|A_0\varphi \circ X_t - A_0\varphi \circ X_t \circ L_r\right| dt \right),$$

$$\varepsilon_3 \le E_x\left(\int_0^{\sigma_\Delta} \left|A_0^r\varphi \circ X_t \circ L_r - A_0\varphi \circ X_t \circ L_r\right| \left(I_{E_\varepsilon} \circ X_t\right) dt \right),$$

$$\varepsilon_4 \le 2M E_x\left(\int_0^{\sigma_\Delta} \left(I_{\Delta \backslash E_\varepsilon} \circ X_t\right) dt \right) \le 2M\varepsilon.$$

The first of these members tends to zero under both conditions (a) and (b); the second one tends to zero because of boundedness condition (A) and continuity condition (B); the third one tends to zero because of uniform convergence on the chosen set; the fourth one can be made as small as is desired at the expense of choice of ε. The first assertion is proved. Similarly, the second assertion can also be proved.   □

Note that in the second assertion a size of the set Δ is not specified at all. In particular, it can be $\Delta = \mathbb{X}$, which corresponds to integration on the whole half-line. In this case the second assertion establishes sufficient conditions for the equality

$$R(\lambda; A_\lambda \varphi) = -\varphi.$$

35. Dynkin's formulae

Let us note that formulae

$$R_\tau(\lambda; A_\lambda \varphi) = f_\tau(\lambda; \varphi) - \varphi \quad (\lambda > 0), \tag{3.9}$$

$$R_\tau(0; A_0 \varphi) = f_\tau(0; \varphi) - \varphi \quad (\tau < \infty), \tag{3.10}$$

justified in proposition 3.9 for $\tau = \sigma_\Delta$, can be considered as variants of the Dynkin's formulae (see [DYN 63, p. 190]), which sets the connection between the infinitesimal operator of a Markov process, the resolvent of a semi-group of transition operators, and the semi-Markov transition generating function of the Markov process. While deriving them, the Markov properties of the process were essentially being used. Therefore our formulae, derived for a general semi-Markov process, can be considered as some generalizations of the Dynkin's formulae. Dynkin applied these formulae in order to prove the existence of the characteristic operator on the same class of functions that the infinitesimal operator is determined on. In the case of a general semi-Markov process the Dynkin's formulae in terms of infinitesimal operator are not in general true at least because for a non-Markov semi-Markov process the infinitesimal operator does not coincide with the characteristic one. We will apply these formulae in order to prove the existence of the lambda-characteristic operator in strict sense, if this operator is determined in wide sense, i.e. in such a view as it is included in the formulae, proved in proposition 3.9.

36. Operators in a strict sense

Up to now operators A_λ and $\mathcal{A}_\lambda$ have been considered as limits of fractions determined by a sequence of balls $(B(x, r))_{r>0}$ for $r \to 0$. Proposition 3.9 opens possibility for the limits

$$\widetilde{A}_\lambda(\varphi \mid x) = \lim_{\Delta \downarrow x} \widetilde{A}_\lambda^\Delta(\varphi \mid x),$$

$$\widetilde{\mathcal{A}}_\lambda(\varphi \mid x) = \lim_{\Delta \downarrow x} \widetilde{\mathcal{A}}_\lambda^\Delta(\varphi \mid x)$$

to exist for any admissible sequence of neighborhoods of point x. Here

$$\widetilde{A}_\lambda^\Delta(\varphi \mid x) = \frac{f_{\sigma_\Delta}(\lambda; \varphi \mid x) - \varphi(x)}{m_\Delta(x)} \quad (\lambda \geq 0),$$

$$\widetilde{\mathcal{A}}_\lambda^\Delta(\varphi \mid x) = \frac{f_{\sigma_\Delta}(\lambda; \varphi \mid x) - \varphi(x)}{\lambda^{-1}\left(1 - f_{\sigma_\Delta}(\lambda; \mathbb{I} \mid x)\right)} \quad (\lambda > 0),$$

$m_\Delta(x) = E_x(\sigma_\Delta; \sigma_\Delta < \infty)$. The convergence $\Delta_n \downarrow x$ of a sequence of neighborhoods $(\Delta_n)_1^\infty$ of point x means that $(\forall \varepsilon > 0)$ $(\exists m \geq 1)$ $(\forall n > m)$ $\Delta_n \subset B(x, \varepsilon)$. The operators $\widetilde{A}_\lambda \varphi$ and $\widetilde{\mathcal{A}}_\lambda \varphi$, determined on corresponding classes of functions (evidently, more narrow than that defined in item 26), are said to be lambda-characteristic operators in strict sense (of the first and second kind, correspondingly). Let $\widetilde{V}_\lambda(x)$ and $\widetilde{\mathcal{V}}_\lambda(x)$ be domains of the operators in strict sense at point x.

For the difference operators, determined in item 23, evidently there is no distinction between strict and narrow sense. In general it is not an evident fact (maybe it is not true). In any case the existence of operators in strict sense requires a special analysis. For operators in strict sense it is natural to re-formulate the definition of pseudo-locality in item 27. The limit in this definition with respect to a sequence of balls must be replaced by a limit with respect to a sequence of neighborhoods of more general forms. So, operators A_λ and $\mathcal{A}_\lambda$ are said to be pseudo-local in strict sense at point x if for any $R > 0$ the following ratios

$$\frac{E_x\left(1 - e^{-\lambda \sigma_{\Delta_n}}; \sigma_{\Delta_n} = \sigma_R\right)}{m_{\Delta_n}(x)}, \quad \frac{E_x\left(1 - e^{-\lambda \sigma_{\Delta_n}}; \sigma_{\Delta_n} = \sigma_R\right)}{1 - f_{\sigma_{\Delta_n}}(\lambda; \mathbb{I} \mid x)} \quad (\lambda > 0)$$

tend to zero for any admissible sequence $\Delta_n \downarrow x$. This property takes place, for example, if $P_x(\mathcal{C}) = 1$. Admissibility has to be corrected in every partial case. Since we admit sequences of closed sets, the first condition of admissibility of the sequence (Δ_n), converging to $\{x\}$, we assume the condition $(\forall n \geq 1)$ $(\exists r > 0)$ $B(x, r) \subset \Delta_n$.

From pseudo-locality in strict sense it follows the theorem about representation of operators $\widetilde{A}_\lambda \varphi$ and $\widetilde{\mathcal{A}}_\lambda \varphi$, similar to theorem 3.3 and formulae item 30, is fair. On the other hand, from representation

$$A_\lambda^\Delta(\varphi \mid x) = A_0^\Delta(\varphi \mid x) + \varphi(x)\left(A_\lambda^\Delta(\mathbb{I} \mid x) - A_0^\Delta(\mathbb{I} \mid x)\right)$$

$$+ \frac{1}{m_\Delta(x)} E_x\left((e^{-\lambda \sigma_\Delta} - 1)(\varphi \circ X_{\sigma_\Delta} - \varphi \circ X_0)\right)$$

the problem of pseudo-locality in strict sense reduces to the existence problem for operators in strict sense.

37. Existence of operators in strict sense

THEOREM 3.5. *Let* $(P_x) \in SM$, $x \in \mathbb{X}_0$, $\Delta_n \downarrow x$. *Then*

(1) if $(\exists n \geq 1)$ $\varphi \in W_0(\Delta_n)$ *and the function* $A_0\varphi$ *is continuous at point* x, *then* $\varphi \in \tilde{V}_0(x)$ *and* $\tilde{A}_0(\varphi \mid x) = A_0(\varphi \mid x)$;

(2) if $(\exists n \geq 1)$ $\varphi \in \mathcal{W}_\lambda(\Delta_n)$ *and the function* $A_\lambda\varphi$ *is continuous at point* x, *then* $\varphi \in \tilde{V}_\lambda(x)$ *and* $\tilde{A}_\lambda(\varphi \mid x) = A_\lambda(\varphi \mid x)$ $(\lambda > 0)$.

Proof. According to the definition of operators R_{σ_Δ} from item 21, the following estimates are fair

$$\inf_{x_1 \in \Delta} A_0(\varphi \mid x_1) \leq \frac{R_{\sigma_\Delta}(0; A_0\varphi \mid x)}{m_\Delta(x)} \leq \sup_{x_1 \in \Delta} A_0(\varphi \mid x_1),$$

$$\inf_{x_1 \in \Delta} A_\lambda(\varphi \mid x_1) \leq \frac{R_{\sigma_\Delta}(\lambda; A_\lambda\varphi \mid x)}{\lambda^{-1}(1 - f_{\sigma_\Delta}(\lambda; \mathbb{I} \mid x))} \leq \sup_{x_1 \in \Delta} A_\lambda(\varphi \mid x_1).$$

According to the definition of regions W and $\mathcal{W}$ from item 33, it means that

$$\inf_{x_1 \in \Delta} A_0(\varphi \mid x_1) \leq \tilde{A}_0^\Delta(\varphi \mid x) \leq \sup_{x_1 \in \Delta} A_0(\varphi \mid x_1),$$

$$\inf_{x_1 \in \Delta} A_\lambda(\varphi \mid x_1) \leq \tilde{A}_\lambda^\Delta(\varphi \mid x) \leq \sup_{x_1 \in \Delta} A_\lambda(\varphi \mid x_1).$$

Existence of the limits follows from continuity of operators in a neighborhood of point x. $\qquad \square$

38. Arbitrary Markov time

The Dynkin's formulae [3.9] and [3.10] (see item 35), have proven under known suppositions for the time $\tau = \sigma_\Delta$, can be justified for τ from a more wide sub-class of Markov times.

Let us denote by T_λ $(\lambda > 0)$ a class of all Markov times that the Dynkin's formula [3.9] is fulfilled for. Let class of $\varphi \in C_0$ be such that the operator A_λ is determined and bounded on it.

PROPOSITION 3.10. *Let* (P_x) *be an admissible family of probability measures on* $\mathcal{D}$. *Then*

(1) if $\tau_1, \tau_2 \in T_\lambda$ *and* $\tau_1 \in RT(P_x)$, *then* $\tau_1 \dotplus \tau_2 \in T_\lambda$;

(2) if $\tau_n \in T_\lambda$ $(n \geq 1)$ and $\tau_n \downarrow \tau$, then $\tau \in T_\lambda$.

Proof. (1) Using the regeneration property we have for any integrable function ψ

$$R_{\tau_1 \dotplus \tau_2}(\lambda; \psi)$$

$$= P\!\left(\int_0^{\tau_1 \dotplus \tau_2} e^{-\lambda t} \left(\psi \circ \pi_t \right) dt \right)$$

$$= P\!\left(\int_0^{\tau_1} e^{-\lambda t} \left(\psi \circ \pi_t \right) dt \right) + P\!\left(e^{-\lambda \tau_1} \int_0^{\tau_2 \circ \theta_{\tau_1}} e^{-\lambda t} \left(\psi \circ \pi_t \circ \theta_{\tau_1} \right) dt \right)$$

$$= R_{\tau_1}(\lambda; \psi) + f_{\tau_1}\!\left(\lambda; R_{\tau_2}(\lambda; \psi) \right).$$

Assuming $\psi = \mathcal{A}_\lambda \varphi$, we obtain the required equality

$$R_{\tau_1 \dotplus \tau_2}(\lambda; \mathcal{A}_\lambda \varphi) = f_{\tau_1 \dotplus \tau_2}(\lambda; \varphi) - \varphi.$$

(2) The second assertion follows from boundedness of $\mathcal{A}_\lambda \varphi$ and continuity from the right of the process trajectories. It implies that $f_{\tau_n}(\lambda; \varphi) \to f_\tau(\lambda; \varphi)$ as $n \to \infty$. $\qquad \square$

The application of the Dynkin's formulae for more general Markov times is connected, firstly, with a generalization of theorem 3.2(1) for semi-Markov walks, and secondly, with a convergence of a sequence of embedded semi-Markov walks to the semi-Markov process, and with corresponding limit formulae.

3.5. Criterion of Markov property for SM processes

In a class of SM walks the sub-class of Markov processes is characterized by an exponential distribution of a sojourn time in each state from the sequence of states of the process. In addition, if $F(dt, dx_1 \mid x)$ is a transition function of a semi-Markov walk, in a Markov case it is representable as

$$H\!\left(dx_1 \mid x \right) \exp\left(-a(x)t \right) dt,$$

where $H(dx_1 \mid x)$ is a Markov kernel on a phase space $\mathbb{X}$ and $a(x)$ is some positive function. The sub-class of Markov processes in a class of general semi-Markov processes is also characterized by some property of a semi-Markov transition function. Such a characteristic property is exhibited in an asymptotic behavior of a semi-Markov transition function $F_{\sigma_r}(dt, dx_1 \mid x)$ with $r \to 0$. This asymptotic behavior is reflected in properties of the lambda-characteristic operator A_λ. With some regularity conditions the semi-Markov process is Markovian only in the case when a family of operators (A_λ) is a linear function on a parameter λ.

39. Necessary condition of the Markov property

THEOREM 3.6. *Let $(P_x) \in$ SM; let the multi-dimensional potential operators of the process satisfy equation [3.2] (Markov condition); (P_x) be a lambda-continuous on the interval $[0, \sigma_\Delta)$ family in point x, and $\mathbb{I} \in \bigcap_{\lambda \geq 0} V_\lambda(x)$. Then*

(1) if $P_x(\sigma_0 < \infty) = 1$, and $P_x(X_{\sigma_0} \neq x, \, 0 < \sigma_0 < \infty) = 1$, then

$$f_{\sigma_0}(\lambda; \varphi \mid x) = f_{\sigma_0}(0; \varphi \mid x) \frac{1/m_0(x)}{1/m_0(x) + \lambda};$$

(2) if A_λ is pseudo-local in point x, then $A_\lambda(\mathbb{I} \mid x) = -\lambda$.

Proof. (1) Using theorem 3.2 we obtain the formula

$$A_{\lambda_1} R(1,2) = \frac{\varphi_1 \varphi_2}{\lambda_2 - \lambda_1} + \frac{\varphi_1 A_{\lambda_1} R(2,2)}{\lambda_2 - \lambda_1}.$$

On the other hand, from equation [3.2] it follows that $A_{\lambda_1} R(1,2) = \varphi_1 R(2,2)$. Hence we obtain an equation

$$A_{\lambda_1} R(2,2) = -\varphi_2 + (\lambda_2 - \lambda_1) R(2,2). \qquad [3.11]$$

Hence the relation is fair

$$A_{\lambda_1} R(2,2) - A_{\lambda_3} R(2,2) = (\lambda_3 - \lambda_1) R(2,2) \quad (\lambda_3 > 0, \, \lambda_3 \neq \lambda_1).$$

In this relation, valid for any $\varphi_2 \in C_0$ and $\lambda_2 > 0$, we can use the limit

$$\lim_{\lambda_2 \to \infty} \lambda_2 R(2,2) = \varphi_2$$

(see, e.g., [DIT 65]), and replace $R(2,2)$ by an arbitrary function $\varphi_2 \in C_0$:

$$A_{\lambda_1} \varphi_2 - A_{\lambda_3} \varphi_2 = (\lambda_3 - \lambda_1) \varphi_2.$$

Consequently, for any $\lambda_3 > 0$ the relation is fair

$$\frac{f_{\lambda_1} \varphi_2 - \varphi_2}{\lambda_1^{-1}(1 - f_{\lambda_1} \mathbb{I})} + \lambda_1 \varphi_2 = \frac{f_{\lambda_3} \varphi_2 - \varphi_2}{\lambda_3^{-1}(1 - f_{\lambda_3} \mathbb{I})} + \lambda_3 \varphi_2.$$

On the right part, letting $\lambda_3 \to 0$, we obtain

$$\frac{f_{\lambda_1} \varphi_2 - \varphi_2}{\lambda_1^{-1}(1 - f_{\lambda_1} \mathbb{I})} + \lambda_1 \varphi_2 = \frac{f_0 \varphi_2 - \varphi_2}{m_0}.$$

From here we obtain

$$f_{\lambda_1} \varphi_2 = \varphi_2 f_{\lambda_1} \mathbb{I} + \frac{(f_0 \varphi_2 - \varphi_2)(1 - f_{\lambda_1} \mathbb{I})}{m_0 \lambda_1}.$$

For the given point $x \in \Delta$ we consider a sequence of continuous functions φ_n, equal to one at point x, and equal to zero outside of the ball $B(x, r_n)$ $(r_n \to 0)$. Substituting φ_n instead of φ_2, and passing to a limit, we get an equation with respect to $f_{\lambda_1} \mathbb{I}$, and its solution

$$f_{\lambda_1} \mathbb{I} = \frac{1/m_0}{1/m_0 + \lambda_1}.$$

Substituting the obtained meaning in the previous formula, we obtain

$$f_{\lambda_1} \varphi_2 = f_0 \varphi_2 \frac{1/m_0}{1/m_0 + \lambda_1},$$

which correspond to conditional mutual independence of the first exit time and that of position from the position x with an exponent distribution of the first exit time. It is a well-known Markov property in its traditional view.

(2) Formula [3.11] with corresponding definition of the operator $\mathcal{A}_\lambda$ stays true in the case when operator A_λ is determined and pseudo-local at point x. Therefore

$$A_{\lambda_1} R(2,2) = \varphi_2 \frac{A_{\lambda_1} \mathbb{I}}{\lambda_1} - (\lambda_2 - \lambda_1) R(2,2) \frac{A_{\lambda_1} \mathbb{I}}{\lambda_1}.$$

Applying theorem 3.3 to the operator of a continuous function $R(2,2)$, we obtain

$$A_0 R(2,2) = \varphi_2 \frac{A_{\lambda_1} \mathbb{I}}{\lambda_1} - R(2,2) \left(\frac{\lambda_2 A_{\lambda_1} \mathbb{I}}{\lambda_1} - A_0 \mathbb{I} \right).$$

From here for any $\lambda_3 > 0$ $(\lambda_3 \neq \lambda_1)$

$$\left(R(2,2) - \frac{\varphi_2}{\lambda_2} \right) \left(\frac{A_{\lambda_1} \mathbb{I}}{\lambda_1} - \frac{A_{\lambda_3} \mathbb{I}}{\lambda_3} \right) = 0.$$

For the given point x let $\varphi_2(x) = 1$ and $\varphi_2(x_1) < 1$ for $x_1 \neq x$. Since $x \in \mathbb{X}_0$ $R(2,2)(x) < \lambda_2^{-1}$ for any $\lambda_2 > 0$. Therefore, it is necessary that $\lambda_3^{-1} A_{\lambda_3}(\mathbb{I} \mid x) = \lambda_1^{-1} A_{\lambda_1}(\mathbb{I} \mid x)$. Hence, there exists $b(x)$ such that $(\forall \lambda \geq 0)$ $A_\lambda(\mathbb{I} \mid x) = -\lambda b(x)$, where $0 \leq b(x) \leq 1$. Let $b(x) < 1$. In this case we have two limits:

$$\lim_{r \to 0} \frac{-A_\lambda^r(\mathbb{I} \mid x)}{\lambda} \longrightarrow b(x) \quad (r \longrightarrow 0),$$

$$\lim_{\lambda \to 0} \frac{-A_\lambda^r(\mathbb{I} \mid x)}{\lambda} \longrightarrow 1 \quad (\lambda \longrightarrow 0).$$

From here a partial derivative of the function $-A_\lambda^r(\mathbb{I} \mid x)/\lambda$ on λ for $\lambda = 0$ must not be bounded (tends to $-\infty$ as $r \to 0$). In this case

$$\frac{-A_\lambda^r(\mathbb{I} \mid x)}{\lambda} = \frac{1 - f_{\sigma_r}(\lambda; \mathbb{I} \mid x)}{\lambda m_r(x)} = \frac{1}{\lambda m_r(x)} E_x\left(1 - e^{-\lambda \sigma_r}\right)$$

$$= \frac{1}{m_r(x)} E_x\left(\int_0^{\sigma_r} e^{-\lambda t}\, dt\right) = \frac{1}{m_r(x)} E_x\left(\sigma_r \int_0^1 e^{-\lambda \sigma_r t}\, dt\right),$$

$$\left|\frac{\partial}{\partial \lambda}\left(\frac{A_\lambda^r(\mathbb{I} \mid x)}{\lambda}\right)\right| = \frac{1}{m_r(x)} E_x\left((\sigma_r)^2 \int_0^1 t e^{-\lambda \sigma_r t}\, dt\right) \le \frac{E_x\left((\sigma_r)^2\right)}{2 m_r(x)}.$$

Consequently, under this supposition there must be a non-bounded ratio of the second moment to the first one as $r \to 0$. Let us show that it is not true.

LEMMA 3.3. *Let P be a probability measure on $(\mathcal{D}, \mathcal{F})$; $(\exists r_0 > 0)$ $(\exists n \ge 1)$, $E(\sigma_{r_0}^n) < \infty$; and for any r $(0 < r \le r_0)$ the moment $E(\sigma_{r_0})$ be positive. Then the ratio $E((\sigma_r)^n)/E(\sigma_r)$ is a non-decreasing function of r on the interval $(0, r_0)$.*

Proof. Let $0 < r_1 < r_2 \le r_0$. Let us represent the product of integrals as an integral on a product of spaces. We have

$$E\left((\sigma_{r_2})^n\right) E(\sigma_{r_1}) - E\left((\sigma_{r_1})^n\right) E(\sigma_{r_2})$$

$$= \int_{\mathcal{D}^2} \left((\sigma_{r_2}(\xi_1))^n \sigma_{r_1}(\xi_2) - (\sigma_{r_1}(\xi_1))^n \sigma_{r_2}(\xi_2)\right) P(d\xi_1) P(d\xi_2)$$

$$= \int_{\mathcal{D}^2} \sigma_{r_2}(\xi_1) \sigma_{r_1}(\xi_2) \left((\sigma_{r_2}(\xi_1))^{n-1} - (\sigma_{r_1}(\xi_1))^{n-1}\right) P(d\xi_1) P(d\xi_2) \ge 0.$$

$\square$

Since for a Markov process under conditions of theorem 3.6 the ratio $E_x((\sigma_{r_0})^2)/E_x(\sigma_{r_0})$ is finite (see [GIH 73, p. 195]), the supposition that $b(x) < 1$ leads to contradiction. $\square$

Note that for a Markov process the result of lemma 3.3 can be strengthened as follows.

PROPOSITION 3.11. *If conditions of theorem 3.6(2) are fulfilled, then for any $n \ge 2$ and $x \in \mathbb{X}$*

$$\lim_{r \to 0} E\left((\sigma_r)^n\right)/E(\sigma_r) = 0.$$

Proof. It follows from the Taylor formula for function e^{-x} in a neighborhood of zero and that of the proven theorem. $\qquad\qquad\qquad\qquad\qquad\qquad\qquad\qquad\qquad\qquad\quad$ □

40. Correlation with infinitesimal operator

The Markov property of a stochastically continuous process (P_x) makes it possible to use some results of the theory of contractive operator semi-groups. In particular, if for any $x \in \Delta$ $P_x(\sigma_\Delta > t) < 1$, then the operator $T_t^\Delta(\varphi \mid x) = E_x(\varphi \circ X_t;\ \sigma_\Delta > t)$ is contractive. In addition, let us suppose that the semi-group of operators $(T_t^\Delta)_{t \geq 0}$ is continuous from the right on t and is a Feller semi-group, i.e. every operator of the semi-group maps the set of all continuous bounded functions C_0 in itself. Let A^Δ be an infinitesimal operator of the semi-group of such operators. Let the domain of the operators be a set of such functions $\varphi \in C_0$ that for any $x \in \Delta$ there exists the limit

$$A^\Delta(\varphi \mid x) = \lim_{t \to 0} \frac{1}{t} \left(T_t^\Delta(\varphi \mid x) - \varphi(x) \right).$$

Then for any function $g \in C_0$ and $\lambda > 0$ there exists a solution of the operator equation

$$\lambda\varphi - A^\Delta\varphi = g$$

and this solution is represented by the formula $\varphi = R_\lambda^\Delta g \in C_0$, where, according to our denotations, $R_\lambda^\Delta g = R_{\sigma_\Delta}(\lambda;\ g)$ (see Dynkin [DYN 63, p. 43], Venzel [VEN 75], Gihman and Skorokhod [GIH 73], Ito [ITO 63]). It is well-known that such a Markov process is strictly Markovian [DYN 63, p. 144] and consequently, it is semi-Markovian. From here, according to formula [3.8], for $n = 1$ we have

$$f_\lambda^r(\varphi \mid x) - \varphi(x) = f_\lambda^r\left(R_\lambda^\Delta g \mid x\right) - R_\lambda^\Delta(g \mid x) = -E_x\left(\int_0^{\sigma_r} e^{-\lambda t} g(X_t)\, dt \right).$$

Hence we obtain the first Dynkin's formula:

$$f_\lambda^r(\varphi \mid x) - \varphi(x) = -E_x\left(\int_0^{\sigma_r} e^{-\lambda t}\left((\lambda\varphi - A^\Delta\varphi) \circ X_t \right) dt \right). \qquad [3.12]$$

Passing to a limit as $\lambda \to 0$, we obtain the second Dynkin's formula:

$$f_0^r(\varphi \mid x) - \varphi(x) = E_x\left(\int_0^{\sigma_r} \left(A^\Delta\varphi \circ X_t \right) dt \right). \qquad [3.13]$$

From the first formula we obtain $\mathcal{A}_\lambda^\Delta \varphi = A^\Delta \varphi - \lambda\varphi$. It is the lambda-characteristic operator of the second kind for a semi-Markov process, stopped at the first exit time from Δ. From the second formula it follows that $\mathcal{A}_0^\Delta \varphi = A^\Delta \varphi$. It is the lambda-characteristic operator of the first kind for this process with $\lambda = 0$. If the condition of pseudo-locality is fulfilled, then $\mathcal{A}_\lambda^\Delta$ and $\mathcal{A}_0^\Delta$ does not depend on Δ. Hence A^Δ also does not depend on Δ. In particular, from the first formula it follows that for any $\mu > 0$ $A_\lambda \varphi + \lambda \varphi = A_\mu \varphi + \mu \varphi$. Comparing this formula with that of theorem 3.3(2), we obtain in this case $a(\mu, \lambda) = \mu/\lambda$. This means that $A_\lambda^r \mathbb{I}/\lambda \sim A_\mu^r \mathbb{I}/\mu$ as $r \to 0$. Using boundedness of these ratios, and passing to converging subsequences, by lemma 3.3, we convince ourself that $A_\lambda \mathbb{I} = -\lambda$.

41. When R belongs to W

THEOREM 3.7. *Let* $(P_x) \in$ SM, *and* $\Delta \subset \mathbb{X}_0$ *be such that at least one of two following conditions is fulfilled (see item 7):*

(a) Δ *is a closed set;*

(b) Δ *is an open set, and* $(P_x) \in$ QC.

Besides, let (P_x) *be lambda-continuous family on interval* $[0, \sigma_\Delta)$, *and also for any* $\lambda_1, \lambda_2 > 0$ $\mathbb{I} \in \mathcal{V}_{\lambda_2}^{\lambda_1}(\Delta)$, *and function* $a(\lambda_1, \lambda_2)$ *is continuous on* Δ. *Then* $(\forall n \in \mathbb{N})$ $(\forall \lambda_i > 0, \ \lambda_i \neq \lambda_j)$ $(\forall \varphi_i \in \mathcal{C}_0)$

$$R(1, n) \in \mathcal{W}_{\lambda_1}(\Delta), \qquad R_{\sigma_\Delta}\big(\lambda_1; A_{\lambda_1} R(1, n)\big) = -R(1, n),$$

where $R(1, n) = R_{\sigma_\Delta}(\lambda_1, \ldots, \lambda_n; \ \varphi_1, \ldots, \varphi_n)$ *(see item 24).*

Proof. It is sufficient to check that for function $R(1, n)$ (which is continuous, under the theorem condition) all the conditions of proposition 3.9(2) are fulfilled for positive functions φ_i. Convergence to a limit follows from theorem 3.4. Continuity of the limit follows from formula of item 32, and that of the theorem conditions. Let us show boundedness of a pre-limit function. In order to do it we use representation [3.8] from the proof of theorem 3.4:

$$\big| f_{\sigma_r}(\lambda_1; R(1, n)) - R(1, n) \big|$$

$$\leq \Psi_1^n E. \left(\int_{s_n < \sigma_r} \prod_{i=1}^{n} \big(e^{-\lambda_i t_i} \, dt_i \big) \right)$$

$$+ \sum_{k=1}^{n-1} \Psi_1^k \overline{R}(k+1, n) E. \left(e^{-\lambda_{k+1} \sigma_r} \int_{s_k < \sigma_r} \prod_{i=1}^{k} \big(e^{-(\lambda_i - \lambda_{k+1}) t_i} \, dt_i \big) \right),$$

where $B(x, r) \subset \Delta$,

$$\Psi_k^n = \prod_{i=k}^{n} \sup \big\{ \varphi_i(x) : x \in \Delta \big\},$$

$$\overline{R}(k+1, n) = \sup \big\{ R(k+1, n)(x) : x \in \Delta \big\}.$$

Let $0 < \lambda_0 \leq \lambda_i$ $(i = 1, \ldots, n)$. In this case $\overline{R}(k+1, n) \leq \Psi_{k+1}^n \lambda_0^{n-k}$. Furthermore, it is not difficult to show that

$$\int_{s_n < \sigma_r} \prod_{i=1}^{n} \left(e^{-\lambda_i t_i} \, dt_i\right) \leq \lambda_0^{-n}\left(1 - e^{-\lambda_0 \sigma_r}\right).$$

Similarly the other integral can be estimated. Thus we obtain

$$\left|A_{\lambda_1}^r\left(R(1,n) \mid x\right)\right| \leq n\lambda_0^{-n} \lambda_1 \Psi_1^n.$$

Hence, $R(1,n) \in \mathcal{W}_{\lambda_1}(\Delta)$. The final formula we obtain from the evident equality $f_{\sigma_\Delta}(\lambda_1, R(1,n)) = 0$, which follows from the relation $\sigma_\Delta \circ \theta_{\sigma_\Delta} = 0$. $\qquad\square$

42. Sufficient condition of the Markov property

THEOREM 3.8. *Let the conditions of theorem 3.7 be fulfilled and $(\forall \lambda, \mu > 0)$ $a(\lambda, \mu) = \lambda/\mu$ on Δ. Then for the semi-Markov family (P_x) the Markov property [3.2] is fulfilled.*

Proof. By theorem 3.7 the following equality is fulfilled:

$$R_{\sigma_\Delta}\left(\lambda_1; A_{\lambda_1} R(1, n)\right) = -R(1, n).$$

Besides, according to formula of item 32, and that of lemma 3.1, we have

$$A_{\lambda_1} R(1, n) = -\sum_{k=1}^{n-1} \psi_1^k R(k+1, n) \sum_{i=1}^{k+1} \Pi_i(1, k+1)\lambda_i.$$

LEMMA 3.4. *For any positive $\lambda_i \neq \lambda_j$ $(i \neq j)$ for $n \geq 3$ it is true that*

$$\sum_{i=1}^{n} \Pi_i(1, n)\lambda_i = 0;$$

for $n = 2$ the left part is equal to -1.

Proof. We have

$$\sum_{i=1}^{n} \Pi_i(1, n)\lambda_i = \sum_{i=1}^{n-1} \Pi_i(1, n)\lambda_i + \lambda_n \Pi_n(1, n)$$

$$= \sum_{i=1}^{n-1} \Pi_i(1, n-1)\left(\frac{\lambda_n}{\lambda_n - \lambda_i} - 1\right) + \lambda_n \Pi_n(1, n)$$

$$= \sum_{i=1}^{n} \Pi_i(1, n)\,\lambda_n - \sum_{i=1}^{n-1} \Pi_i(1, n-1).$$

By lemma 3.1, the latter expression is equal to zero as $n \geq 3$. The case $n = 2$ is evident. It also follows from this expression, if we define $\Pi_1(1,1) = 1$. □

Applying this lemma we obtain $\mathcal{A}_{\lambda_1} R(1, n) = -\varphi_1 R(2, n)$. Thus the equation [3.2] is satisfied. □

3.6. Intervals of constancy and SM processes

Intervals of constancy play an important role in the theory of semi-Markov processes. Every lambda-continuous semi-Markov process with trajectories without intervals of constancy is a Markov process. Some inversion of this result is fair as well. Thus, the absence of intervals of constancy in trajectories has to be reflected in properties of the lambda-characteristic operator of the process. This properties are corrected for some intervals of constancy of special view. Absence of intervals of constancy for a Markov process can be used for a new approach to a correlation of a lambda-characteristic operator and an infinitesimal one. The whole solution of the problem of intervals of constancy will be considered in Chapter 6 in terms of the Lévy measure.

43. Intervals of constancy

The function $\xi \in \mathcal{D}$ is said to have no interval of constancy on the right of point $t \geq 0$, if for any $\varepsilon > 0$ the function ξ is not constant on the interval $[t, t + \varepsilon)$. Let $\mathcal{D}_t^*$ be a set of functions not having intervals of constancy on the right of point t. Then $\mathcal{D}^* = \bigcap_{t \in \mathbb{R}_+} \mathcal{D}_t^*$ is the set of functions not having intervals of constancy on the whole half-line. Evidently, $\mathcal{D}^* = \bigcap_{t \in \mathbb{R}'_+} \mathcal{D}_t^*$, where $\mathbb{R}'_+$ is any set which is dense everywhere in the half-line, for example, the set of all rational positive numbers. From here

$$P(\mathcal{D}^*) = 1 \iff (\forall t \in \mathbb{R}'_+) P(\mathcal{D}_t^*) = 1.$$

44. Another sufficient condition of the Markov property

THEOREM 3.9. *Let (P_x) be a lambda-continuous SM family of measures. If $(\forall x \in \mathbb{X})$ $P_x(\mathcal{D}^*) = 1$, then (P_x) is a Markov family (Markov process).*

Proof. Let $t \in \mathbb{R}_+$ and $\sigma_{r,t} = (t \dot{+} \sigma_0) \circ L_r = \sigma_r^{k+1}$, where $\sigma_r^k \leq t < \sigma_r^{k+1}$. In this case $\sigma_{r,t} \in \mathrm{RT}$, and for any $\xi \in \mathcal{D}_t^*$ $\sigma_{r,t}(\xi) \to t$ $(r \to 0)$. For $B \in \mathcal{F}_t \subset \mathcal{F}_{\sigma_{r,t}}$, and for f of the form

$$f = \int_{\mathbb{R}_+^n} \prod_{i=1}^n \left(e^{-\lambda_i t_i} \left(\varphi_i \circ X_{s_i} \right) dt_t \right)$$

$(s_i = t_1 + \cdots + t_i, \ \varphi_i \in \mathcal{C}_0, \ \lambda_i > 0)$ we have equalities

$$E_x\big(f \circ \theta_t; B\big) = \lim_{r \to 0} E_x\big(f \circ \theta_{\sigma_{r,t}}; \ B \cap \{\sigma_{r,t} < \infty\}\big)$$

$$= \lim_{r \to 0} E_x\big(E_{X_{\sigma_{r,t}}}(f); \ B \cap \{\sigma_{r,t} < \infty\}\big) = E_x\big(E_{X_t}(f); B\big),$$

since $(\forall t_1 \in \mathbb{R}_+) \ X_{\sigma_{r,t}+t_1} \to X_{t+t_1}$, $P_x(\sigma_{r,t} < \infty) \to 1$ and the function $E_x(f)$ is continuous on x (follows from lambda-continuity). Therefore $t \in \mathrm{RT}$. $\qquad\square$

45. Another necessary condition of the Markov property

THEOREM 3.10. *Let (P_x) be a Markov family, and $(\forall x \in X) \ P_x(\mathcal{D}_0^*) = 1$ (there is no interval of constancy at the beginning of a trajectory; see item 43). Then $(\forall x \in X) \ P_x(\mathcal{D}^*) = 1$.*

Proof. Let $(\exists x \in X) \ (\exists t \in \mathbb{R}_+) \ P_x(\mathcal{D}_t^*) < 1$. We have $\xi \in \mathcal{D}_t^* \Leftrightarrow \theta_t \xi \in \mathcal{D}_0^*$. From here $P_x(\mathcal{D}_t^*) = P_x(\theta_t^{-1}\mathcal{D}_0^*) = E_x(P_{X_t}(\mathcal{D}_0^*)) = 1$. This is a contradiction. $\qquad\square$

46. Infinite interval of constancy

THEOREM 3.11. *Let $(P_x) \in \mathrm{SM}$; $\Delta \subset \mathbb{X}_0$ $(\Delta \in \mathfrak{A})$; $\mathbb{I} \in V_\lambda^0(\Delta)$; and $(\forall x \in \Delta)$ $a(0, \lambda \mid x) = 0$. Then $(\forall x \in \Delta) \ P_x$-a.s. an infinite interval of constancy is absent until the first exit time from Δ.*

Proof. Let $\tau(\xi) = \inf\{t \geq 0 : \xi$ be constant on $[t, \infty)\}$. Evidently $\tau \in \mathcal{F}/\mathcal{B}(\bar{\mathbb{R}}_+)$, but $\tau \notin \mathrm{MT}_+$ (if $\tau \not\equiv \infty$). From the definition of τ it follows that $\tau = \lim_{r \to 0} \tau \circ L_r$, where $\tau \circ L_r \leq \tau$, and therefore $(\forall \lambda > 0)$ as $r \to 0$ we have $E_x(e^{-\lambda\tau} \circ L_r) \to E_x(e^{-\lambda\tau})$. On the other hand,

$$E_x\big(e^{-\lambda\tau} \circ L_r\big) = \sum_{k=0}^{\infty} E_x\left(e^{-\lambda\sigma_r^k}; \ \sigma_r^k < \infty, \ \sigma_r^{k+1} = \infty\right)$$

$$= \sum_{k=0}^{\infty} E_x\left(e^{-\lambda\sigma_r^k} P X_{\sigma_r^k}(\sigma_r = \infty); \ \sigma_r^k < \infty\right)$$

$$= R_\lambda^r\left(\frac{\lambda P.(\sigma_r = \infty)}{1 - f_{\sigma_r}(\lambda, \mathbb{I})}\ \bigg|\ x\right) = \lambda R_\lambda^r\big(a_r(0, \lambda) \mid x\big)$$

(see theorem 3.2 and item 21). However, by the definition of $a_r(0, \lambda)$ $(\lambda > 0)$ (see item 26), the latter expression tends to zero (passing to a limit can be justified in the same way as that of proposition 3.9). Consequently, $E_x(e^{-\lambda\tau}) = 0$. Hence $(\forall x \in X)$ $P_x(\tau = \infty) = 1$. $\qquad\square$

47. Intervals of constancy before jumps

THEOREM 3.12. *Let* $(P_x) \in SM$; $\Delta \subset \mathbb{X}_0$ $(\Delta \in \mathfrak{A})$; *the operator* A_λ *is pseudo-local on* Δ. *Then* $(\forall x \in \mathbb{X})$ *until the first exit time from* Δ *in a trajectory of the process intervals of constancy before any point of discontinuity* P_x-a.s. *are absent (it means a point of discontinuity is not the right end of an interval of constancy).*

Proof. Let $\tau_\varepsilon(\xi) < \infty$ be the first jump time of the trajectory ξ with a jump value of more than ε, i.e. if $\tau_\varepsilon(\xi) = t$, then $\rho(\xi(t-0), \xi(t)) > \varepsilon$. Let

$$\tau'_\varepsilon(\xi) = \inf\left\{t < \tau_\varepsilon(\xi) : \text{on the interval } [t, \tau_\varepsilon(\xi)), \; \xi \text{ is constant}\right\},$$

if such t does not exist, we assume $\tau'_\varepsilon(\xi) = \tau_\varepsilon(\xi)$. Without being too specific it can be assumed that P_x-a.s. there are no jumps with value equal to ε. Then P_x-a.s. $\tau_\varepsilon \circ L_r \to \tau_\varepsilon$ and $\tau'_\varepsilon \circ L_r \to \tau'_\varepsilon$. From here

$$E_x\left(\left(e^{-\lambda\tau'_\varepsilon} - e^{-\lambda\tau_\varepsilon}\right) \circ L_r\right) \longrightarrow E_x\left(e^{-\lambda\tau'_\varepsilon} - e^{-\lambda\tau_\varepsilon}\right).$$

On the other hand, for any sufficiently small $r > 0$

$$E_x\left(\left(e^{-\lambda\tau'_\varepsilon} - e^{-\lambda\tau_\varepsilon}\right) \circ L_r\right)$$

$$= \sum_{k=0}^{\infty} E_x\left(e^{-\lambda\sigma_r^k}\left(1 - e^{-\lambda\sigma_r}\right) \circ \theta_{\sigma_r^k}; \; \sigma_r^k < \tau_\varepsilon, \; \sigma_r^{k+1} = \tau_\varepsilon < \infty\right).$$

Since from $t < \tau_\varepsilon$ it follows that $\tau_\varepsilon = t \dotplus \tau_\varepsilon$ (τ_ε is the so called terminal Markov time; see [BLU 68, p. 78]), the latter expression is equal to

$$\sum_{k=0}^{\infty} E_x\left(e^{-\lambda\sigma_r^k} E_{X_{\sigma_r^k}}\left(1 - e^{-\lambda\sigma_r}; \; \sigma_r = \tau_\varepsilon < \infty\right); \; \sigma_r^k < \tau_\varepsilon < \infty\right)$$

$$\leq \sum_{k=0}^{\infty} E_x\left(e^{-\lambda\sigma_r^k} E_{X_{\sigma_r^k}}\left(1 - e^{-\lambda\sigma_r}; \; \sigma_r = \tau_\varepsilon < \infty\right); \; \sigma_r^k < \infty\right)$$

$$= R_\lambda^r\left(\frac{\lambda E.\left(1 - e^{-\lambda\sigma_r}; \; \sigma_r = \tau_\varepsilon < \infty\right)}{1 - f_{\sigma_r}(\lambda; \mathbb{I})} \,\middle|\, x\right).$$

Evidently, $\tau_\varepsilon \geq \sigma_{\varepsilon/2}$, and hence for $r < \varepsilon/2$

$$E_x\left(1 - e^{-\lambda\sigma_r}; \; \sigma_r = \tau_\varepsilon < \infty\right) \leq E_x\left(1 - e^{-\lambda\sigma_r}; \; \sigma_r = \sigma_{\varepsilon/2} < \infty\right).$$

Then the integral is not more than

$$\lambda R_\lambda^r\left(\frac{E.\left(1 - e^{-\lambda\sigma_r}; \; \sigma_r = \sigma_{\varepsilon/2} < \infty\right)}{1 - f_{\sigma_r}(\lambda; \mathbb{I})} \,\middle|\, x\right).$$

According to the pseudo-locality of $\mathcal{A}_\lambda$ and boundedness of the integrand for any $r > 0$, this value tends to zero as $r \to 0$. Therefore $E_x(e^{-\lambda \tau'_\varepsilon} - e^{-\lambda \tau_\varepsilon}) = 0$. Hence, $(\forall x \in \mathbb{X})$ for all $\varepsilon > 0$ $P_x(\tau'_\varepsilon = \tau_\varepsilon) = 1$. Let τ^k_ε be the k-th jump time with the jump value of more than ε, and $[(\tau^k_\varepsilon)', \tau^k_\varepsilon)$ be the maximal interval of constancy on the left of τ^k_ε. Then for all ε (excluding at most countable number of them)

$$P_x\big((\tau^k_\varepsilon)' < \tau^k_\varepsilon\big) = P_x\big(\tau^{k-1}_\varepsilon \dotplus (\tau_\varepsilon)' < \tau^{k-1}_\varepsilon \dotplus (\tau_\varepsilon)\big)$$

$$= P_x\Big(\theta^{-1}_{\tau^{k-1}_\varepsilon}(\tau'_\varepsilon < \tau_\varepsilon),\ \tau^{k-1}_\varepsilon < \infty\Big)$$

$$= E_x\Big(P_{X(\tau^{k-1}_\varepsilon)}(\tau'_\varepsilon < \tau_\varepsilon);\ \tau^{k-1}_\varepsilon < \infty\Big) = 0$$

(evidently $\tau^k_\varepsilon \in \mathrm{RT}$). Any jump time of ξ belongs to the set $\{(\tau^k_{\varepsilon_n}),\ k, n \in \mathbb{N}\}$ $(\varepsilon_n \downarrow 0)$.

□

48. Another correlation with an infinitesimal operator

We show once more a way to prove linearity of the operator A_λ on λ. Let $\sigma_{r,t} = (t \dotplus \sigma_0) \circ L_r$ (see theorem 3.9). In this case $\sigma_{r,t} = \sigma^{k+1}_r \Leftrightarrow \sigma^k_r \le t < \sigma^{k+1}_r$. It is not difficult to show theorem 3.2(1) to be true for such a compound Markov time as $\tau \equiv \sigma_{c,t}$ $(c > 0)$. Let us assume conditions ensuring convergence

$$R^r_\tau(\lambda; \mathcal{A}^r_\lambda \varphi) \longrightarrow R_\tau(\lambda; \mathcal{A}_\lambda \varphi), \quad f^r_\tau(\lambda; \varphi) \longrightarrow f_\tau(\lambda; \varphi) \quad (r \longrightarrow 0).$$

In this case we would obtain the formula

$$R_\tau(\lambda; \mathcal{A}_\lambda \varphi) = f_\tau(\lambda; \varphi) - \varphi.$$

If a trajectory of the original process does not contain any interval of constancy, then, evidently, $\tau \downarrow t$ as $c \downarrow 0$. Applying proposition 3.10(2), we would receive the formula

$$E.\left(\int_0^t e^{-\lambda s}(\mathcal{A}_\lambda \varphi \circ X_s)\, ds\right) = e^{-\lambda t} E.\big(\varphi \circ X_t\big) - \varphi.$$

Dividing terms of this equality by t and tending t to zero, we obtain the formula as a limit

$$\mathcal{A}_\lambda \varphi = A\varphi - \lambda\varphi,$$

which is justified for a Markov process in item 40, and follows $A_\lambda \mathbb{I} = -\lambda$.

Chapter 4

Construction of Semi-Markov Processes using Semi-Markov Transition Functions

For a semi-Markov process an exposition of its distribution with the help of all finite-dimensional distributions $P(X_{t_1} \in S_1, \ldots, X_{t_k} \in S_k)$ ($k \in \mathbb{N}$, $t_i \in \mathbb{R}_+$, $S_i \in \mathcal{B}(\mathbb{X})$) [KOL 36] is not natural. It would be more convenient to describe them in terms of semi-Markov transition functions $F_\tau(dt, dx_1 \mid x)$, because for them an analogy of the Kolmogorov-Chapman equation holds [GIH 73]. However, the exposition with the help of functions F_τ means it is necessary to analyze a set of new finite-dimensional distributions, joint distributions of pairs of the first exit. The first problem is to construct such a distribution for an arbitrary family of open subsets of the phase space, and then to compose these distributions into a projective system of finite-dimensional measures. We solve this problem for an arbitrary process with trajectories from $\mathcal{D}$. The second problem is to extend a projective limit of this system to a probability measure on $(\mathcal{D}, \mathcal{F})$. It can be shown that an *a priori* given projective system of such distributions cannot always be extended. Conditions of such extensions are given in this chapter.

The presentation is adapted to semi-Markov processes [HAR 74]. Having the theory of extension of a projective system of measures up to a probability measure on $(\mathcal{D}, \mathcal{F})$, it is not difficult to give conditions under which an *a priori* given set of semi-Markov transition functions determines a semi-Markov process. Construction of finite-dimensional distributions in this case has some peculiarities on a comparison with a Markov case. It is connected with a lack of a linear ordering property in the set of first exit times. Besides, semi-Markov transition functions should have properties which are exhibited in an infinite sequence of iterations of such functions [RUS 75].

In contrast to methods of the theory of Markov processes [GIH 71], the conditions for trajectories of a semi-Markov process to be continuous in terms of semi-Markov transition functions look very simple. The necessary and sufficient condition for continuity is concentration of distributions of the first exit points on the boundaries of the corresponding sets. The consistency conditions for distributions of the first exit pairs, which are only sufficient in a common case, are necessary and sufficient for continuous processes.

Note that it is possible to construct a theory of semi-Markov processes for which the set of regeneration times consists of the first exit times from closed subsets. The investigation shows that such theory even has some advantages connected with continuity of these times with respect to any decreasing sequence of closed sets. However, at the same time in this case a construction of joint distributions of several pairs of the first exit requires more complicated conditions.

4.1. Realization of an infinite system of pairs of the first exit

For each function $\xi \in \mathcal{D}$ and sequence of open sets $(\Delta_1, \ldots, \Delta_k)$ there exists "a pair" $\beta_\tau \xi$, where $\tau = \sigma_{\Delta_1} \dot{+} \cdots \dot{+} \sigma_{\Delta_k}$. We will deal with finite sets of such pairs, where each pair is a point from space $Y = (\mathbb{R}_+ \times \mathbb{X}) \cup \{\overline{\infty}\}$, corresponding to sequences $(\Delta_1, \ldots, \Delta_k)$ ($k \geq 1$, $\Delta_i \in \mathfrak{A}$). Before proving a theorem about extension of a measure it is important to find conditions when the *a priori* given infinite sequence of such finite sets of pairs can be realized, i.e. when a function $\xi \in \mathcal{D}$ exists such that each pair from this sequence is a pair of the first exit for this function from the corresponding open sets.

1. Full collection, chains

Let $\mathcal{K}(\mathfrak{A}_1)$ be a set of all finite sequences $\mathfrak{z} = (\Delta_1, \ldots, \Delta_k)$ ($k \in \mathbb{N}_0$, $\Delta_i \in \mathfrak{A}_1$), where $\mathfrak{A}_1 \subset \mathfrak{A}$ is some system of open subsets of a set $\mathbb{X}$; for $k = 0$ we set $\mathfrak{z} = (\emptyset)$ (empty sequence).

We say that $\mathfrak{z}_1$ precedes $\mathfrak{z}_2$, or $\mathfrak{z}_1$ is less than $\mathfrak{z}_2$, if $\mathfrak{z}_1$ is an initial piece of a sequence $\mathfrak{z}_2$; and designate $\mathfrak{z}_1 \prec \mathfrak{z}_2$ ($\mathfrak{z}_1, \mathfrak{z}_2 \in \mathcal{K}(\mathfrak{A}_1)$). For example, $\mathfrak{z}' \prec \mathfrak{z}$, where $\mathfrak{z} = (\Delta_1, \ldots, \Delta_k)$, $\mathfrak{z}' = (\Delta_1, \ldots, \Delta_{k-1})$; and also $(\emptyset) \prec \mathfrak{z}$ for all non-empty $\mathfrak{z} \in \mathcal{K}(\mathfrak{A}_1)$.

Let us denote by $\mathcal{Z}(\mathfrak{A}_1)$ a set of all non-empty finite subsets of a set $\mathcal{K}(\mathfrak{A}_1)$.

We call $z \in \mathcal{Z}(\mathfrak{A}_1)$ as full collection if $(\emptyset) \in z$ and for any non-empty $\mathfrak{z} \in z$: $\mathfrak{z}_1 \prec \mathfrak{z} \Rightarrow \mathfrak{z}_1 \in z$. Let $\mathcal{Z}_0(\mathfrak{A}_1)$ be a set of all full collections $z \in \mathcal{Z}(\mathfrak{A}_1)$. A full collection contains a unique minimal element (empty set), but can contain more than one maximal element.

A linearly ordered full collection is said to be a chain. A chain z contains a unique maximal element, $\max z$. Let $\mathcal{Z}'_0(\mathfrak{A}_1)$ be a set of all chains $z \in \mathcal{Z}_0(\mathfrak{A}_1)$.

Each $z \in \mathcal{Z}_0(\mathfrak{A}_1)$ is representable as $z = \bigcup_{i=1}^k z^i$, where $z^i \in \mathcal{Z}'_0(\mathfrak{A}_1)$. Let us designate by $c(z)$ the least number k in such representation. It is a number of maximal chains in z; $|z|$ is a number of non-empty $\mathfrak{z} \in z$.

Let us designate by $\mathrm{rank}(z)$ the maximal diameter of an open set what the definition of z is formulated from $\mathrm{rank}(z) = \max(\mathrm{rank}(z^i) : 1 \le i \le c(z))$, where $\mathrm{rank}(z^i) = \mathrm{rank}(\max(z^i)) = \max(\mathrm{diam}\,\Delta_{ij}) : 1 \le j \le m$, if $\max(z^i) = (\Delta_{i1}, \dots, \Delta_{im})$.

Let $(\mathfrak{z}_n)_1^\infty$ $(\mathfrak{z}_n \in \mathcal{K}(\mathfrak{A}_1))$ be a non-decreasing sequence; denote by $\lim \mathfrak{z}_n$ either the maximal term of this sequence, or the unique infinite extension of all these finite sequences.

In case when $\mathfrak{A}_1 = \mathfrak{A}$ (all open sets), we will write $\mathcal{K}$, $\mathcal{Z}$, $\mathcal{Z}_0$, $\mathcal{Z}'_0$.

2. Correct map and realizing function

For $y \in Y$ the following denotations are introduced

$$k_1(y) = \begin{cases} t, & y = (t, x), \\ \infty, & y = \overline{\infty}, \end{cases} \qquad k_2(y) = x,$$

where $y = (t, x)$, $t \in \mathbb{R}_+$, $x \in \mathbb{X}$. Map $\zeta : z \to Y$ $(z \in \mathcal{Z})$ is said to be correct, if the following conditions are fulfilled:

(1) $(\emptyset) \in z \Rightarrow k_1(\zeta(\emptyset)) = 0$;

(2) $\mathfrak{z}_1, \mathfrak{z}_2 \in z$, $k_1(\zeta(\mathfrak{z}_1)) = k_1(\zeta(\mathfrak{z}_2)) < \infty \Rightarrow k_2(\zeta(\mathfrak{z}_1)) = k_2(\zeta(\mathfrak{z}_2))$;

(3) $\mathfrak{z}_1, \mathfrak{z}_2 \in z$, $\mathfrak{z}_1 \prec \mathfrak{z}_2 \Rightarrow k_1(\zeta(\mathfrak{z}_1)) \le k_1(\zeta(\mathfrak{z}_2))$;

(4) $\mathfrak{z} = (\Delta_1, \dots, \Delta_n) \in z \Rightarrow \zeta(\mathfrak{z}) = \overline{\infty}$ or $k_2(\zeta(\mathfrak{z})) \in \mathbb{X} \setminus \Delta_n$, and if $\Delta_n = \mathbb{X}$, then $\zeta(\mathfrak{z}) = \overline{\infty}$;

(5) if $\mathfrak{z}_1, \mathfrak{z}_2, \mathfrak{z}_3 \in z$, $\mathfrak{z}_1 = (\Delta_1, \dots, \Delta_n)$, and $\mathfrak{z}_2 = (\Delta_1, \dots, \Delta_{n+1})$, and $k_1(\zeta(\mathfrak{z}_1)) \le k_1(\zeta(\mathfrak{z}_3)) < k_1(\zeta(\mathfrak{z}_2))1$, that $k_2(\zeta(\mathfrak{z}_3)) \in \Delta_{n+1}$.

Let $B(z)$ be a set of all such correct maps.

For any $z \in \mathcal{Z}$ and $\zeta : z \to Y$, for which conditions (1) and (2) of previous definitions are fulfilled and $(\emptyset) \in z$, the function $L\zeta \in \mathcal{D}$ is defined, where $(L\zeta)(t) = k_2(\zeta(\mathfrak{z}_0))2$, if $(\exists \mathfrak{z}_0 \in z)$ such that $k_1(\zeta(\mathfrak{z}_0)) \le t$ and for any $\mathfrak{z} \in z$ either $k_1(\zeta(\mathfrak{z})) \le k_1(\zeta(\mathfrak{z}_0))$, or $k_1(\zeta(\mathfrak{z})) > t$.

3. Realization of correct map

PROPOSITION 4.1. *If $z \in \mathcal{Z}_0$ and $\zeta \in B(z)$, then $(\forall \mathfrak{z} \in z)\ \zeta(\mathfrak{z}) = \beta_{\tau_\mathfrak{z}}(L\zeta)$, where $\tau_\mathfrak{z} = 0$, if $\mathfrak{z} = (\varnothing)$, and $\tau_\mathfrak{z} = \sigma_{\Delta_1} \dotplus \cdots \dotplus \sigma_{\Delta_k}$ if $\mathfrak{z} = (\Delta_1, \ldots, \Delta_k)$.*

Proof. Let $z = \bigcup_{i=1}^{k} z^i$, $c(z) = k \geq 1$, and also

$$z^i = \left\{ (\varnothing), (\Delta_{i1}), \ldots, (\Delta_{i1}, \ldots, \Delta_{in_i}) \right\}$$

$(n_i \in \mathbb{N}_0,\ \Delta_{ij} \in \mathfrak{A})$. Let us assume $\mathfrak{z}_{ij} = (\Delta_{i1}, \ldots, \Delta_{ij})$, $y_{ij} = \zeta(\mathfrak{z}_{ij})$, $y_0 = \zeta(\varnothing)$ and $\xi = L\zeta$. Then $\zeta(\varnothing) = \beta_{\tau_\varnothing}\xi$, and $k_1(y_0) = 0$, $k_2(y_0) = \xi(0)$. Let $\beta_{\tau_{\mathfrak{z}_{ij}}}\xi = \zeta(\mathfrak{z}_{ij})$ for some i and $j < n_i$ and $k_1(\zeta(\mathfrak{z}_{ij})) < \infty$. Then $\tau_{\mathfrak{z}_{ij+1}}\xi = (\tau_{\mathfrak{z}_{ij}} \dotplus \sigma_{\Delta_{ij+1}})\xi \leq k_1(\zeta(\mathfrak{z}_{ij+1}))$, because either $k_2(\zeta(\mathfrak{z}_{ij+1})) = \xi(k_1(\zeta(\mathfrak{z}_{ij+1}))) \notin \Delta_{ij+1}$, or $k_1(\zeta(\mathfrak{z}_{ij+1})) = \infty$. On the other hand, for any $\mathfrak{z}$ $k_1(\zeta(\mathfrak{z}_{ij})) \leq k_1(\zeta(\mathfrak{z})) < k_1(\zeta(\mathfrak{z}_{ij+1})) \Rightarrow k_2(\zeta(\mathfrak{z})) \in \Delta_{ij+1}$. Hence, for any t

$$k_1\big(\zeta(\mathfrak{z}_{ij})\big) \leq t < k_1\big(\zeta(\mathfrak{z}_{ij+1})\big) \implies \xi(t) \in \Delta_{ij+1},$$

i.e. $\tau_{\mathfrak{z}_{ij+1}}\xi \geq k_1(\zeta(\mathfrak{z}_{ij+1}))$. From here $\beta_{\tau_{\mathfrak{z}_{ij+1}}}\xi = \zeta(\mathfrak{z}_{ij+1})$.   □

4. Admissible sequence

The sequence $(z_n)_1^\infty$ $(z_n \in \mathcal{Z}_0(\mathfrak{A}_1))$ is referred to as admissible if the following conditions are fulfilled:

(1) $z_n \subset z_{n+1}$ $(n \in \mathbb{N})$;

(2) for any $r > 0$ there exists an increasing sequence of maximal chains $(z_n^{in})_{n=1}^\infty$ $(z_n^{in} \subset z_n)$ such that $\lim \max z_n^{in} \in \mathrm{DS}(r)$ (deducing sequence of rank r).

Let $\Sigma_0(\mathfrak{A}_1)$ be a set of all admissible sequences, constructed for a class of open sets $\mathfrak{A}_1$, and $\Sigma_0 = \Sigma_0(\mathfrak{A})$.

5. Realization of a sequence

The sequence of maps $(\zeta_n)_1 : \infty$ $(\zeta_n \in B(z_n))$ is said to be consistent on the sequence $(z_n)_1^\infty$ of elements from $\mathcal{Z}$ if $(\forall n_1, n_2 \in \mathbb{N})\ \zeta_{n_1} = \zeta_{n_2}$ on $z_{n_1} \cap z_{n_2}$. A consistent sequence of maps can be extended uniquely on the union of all z_n. It means that there exists a map $\zeta : \bigcup_n z_n \mapsto \mathbb{Y}$ such that $(\forall n)\ \zeta_n = \zeta$ on z_n (projective limit). It is interesting to clarify conditions for a map ζ to be realized with the help of some function $\xi \in \mathcal{D}$ just like it is possible for a finite map (see proposition 4.1): $(\forall \mathfrak{z} \in \bigcup_n z_n)\ \beta_{\tau_\mathfrak{z}}\xi = \zeta(\mathfrak{z})$. This function (if any) is said to be a realizing function.

In the following theorem we use denotation

$$\mathfrak{z}^{-\varepsilon} = (\Delta_1, \dots, \Delta_{k-1}, \Delta_k^{-\varepsilon}),$$

where $\mathfrak{z} = (\Delta_1, \dots, \Delta_{k-1}, \Delta_k)$ and $\Delta^{-\varepsilon} = \{x \in \Delta : \rho(x, \mathbb{X} \setminus \Delta) > \varepsilon\}$. For a full collection z, consisting of $c(z)$ chains z^i, and for a correct map ζ on it we denote

$$r_t(z, \zeta) = \min \left(\operatorname{rank}(z^i) : 1 \le i \le c(z), \ k_1(\zeta(\mathfrak{z}^i)) > t \right) \quad (t > 0).$$

THEOREM 4.1. *Let* $(\zeta_n)_1^\infty$ $(\zeta_n \in B(z_n))$ *be a consistent sequence of correct maps with a projective limit* ζ, *where* $(z_n)_1^\infty \in \Sigma_0$, *and some of the following conditions be fulfilled:*

(1) if $\delta^{(i_n)} \equiv \lim \max z_n^{i_n} \in DS$, *then* $k_1(\zeta_n(\mathfrak{z}_n^{i_n})) \to \infty$, *where* $\mathfrak{z}_n^{i_n} = \max z_n^{i_n}$;

(2) for any $t > 0$ *and* $\mathfrak{z} = (\Delta_1, \dots, \Delta_k) \in \bigcup_{n=1}^\infty z_n$ *such that* $k_1(\zeta_n(\mathfrak{z})) \le t$, *it is fair that* $\tau_{\mathfrak{z}^{-r_t(n)}} \xi_n \to k_1(\zeta_n(\mathfrak{z}))$ *as* $n \to \infty$, *where* ξ_n *is a realizing function for* (z_n, ζ_n) *(constructed in proposition 4.1) and* $r_t(n) = r_t(z_n, \zeta_n)$;

(3) for any $\mathfrak{z} = (\Delta_1, \dots, \Delta_k) \in \bigcup_{n=1}^\infty z_n$ *if* $k_2(\zeta(\mathfrak{z}')) \in \Delta_k$ *and* $k_1(\zeta(\mathfrak{z})) < \infty$, *then* $k_2(\zeta(\mathfrak{z})) \in \partial \Delta_k$, *where it is assumed* $\mathfrak{z}' = (\Delta_1, \dots, \Delta_{k-1})$.

Then

(A) condition (1) together with (2) is sufficient for a realizing function to exist;

(B) condition (1) together with (2) is necessary for existence of a realizing function, which exits correctly from any set of the sequence $\mathfrak{z} \in \bigcup_{n=1}^\infty z_n$ *(see item 2.27 for definition of correct exit);*

(C) condition (1) together with (2) and (3) is necessary and sufficient for a continuous realizing function ξ *to exist.*

Proof. (A) Let us prove the existence of a limit $(\forall t \in \mathbb{R}_+) \lim \xi_n(t)$ for $n \to \infty$. In order to do this we use a deducing sequence of a rank as small as desired, which is constructed according to the growth of sequence (z_n). We also use condition (1) which implies that the first coordinate of ζ_n on elements of this sequence exceeds t for sufficiently large n. From here for any $r > 0$ and $t > 0$ $(\exists n_1 \in \mathbb{N})$ $(\forall n \ge n_1)$ $r_t(z_n, \zeta_n) < r$ and $k_1(\zeta_n(\mathfrak{z}_n^i)) > t$ for some chain z_n^i with a rank not more than r. In this case $(\forall m \ge n, \ s < t) \ \rho(\xi_n(s), \xi_m(s)) < r$. Let us show the latter assertion is true. Let $\max z_n^i = (\Delta_1, \dots, \Delta_N)$ and $s \in [k_1(\zeta_n(\mathfrak{z}')), \ k_1(\zeta_n(\mathfrak{z})))$, where $\mathfrak{z}' = (\Delta_1, \dots, \Delta_{k-1})$ and $\mathfrak{z} = (\Delta_1, \dots, \Delta_k)$ $(1 \le k \le N)$. In this case the values $\xi_n(s), \xi_m(s)$ of the function ξ_n, ξ_m, constructed in proposition 4.1 are determined by nearest from the left points $k_1(\zeta_n(\mathfrak{z}(n))), k_1(\zeta_n(\mathfrak{z}(m)))$ for some $\mathfrak{z}(n) \in z_n$, $\mathfrak{z}(m) \in z_m$, which belongs to the interval $[k_1(\zeta_n(\mathfrak{z}')), s]$. These values are $k_2(\zeta_n(\mathfrak{z}(n)))$ and $k_2(\zeta_n(\mathfrak{z}(m)))$ correspondingly. According to the definition of a correct exit both of them belong to an open set Δ_k, the diameter of which is not more than r. Therefore, for any $t > 0$ the sequence $(\xi_n(t))$ converges in itself, and due to completeness of the space $\mathbb{X}$ there exists a limit of this sequence, which we denote as $\xi(t)$. Evidently, the function ξ_n converges on ξ uniformly on each bounded interval. It follows that the limit function (like all pre-limit functions) belongs to space $\mathcal{D}$. In addition, it is obvious that for all $\mathfrak{z} \in \bigcup_{n=1}^{\infty} z_n$ $\xi(k_1(\zeta_n(\mathfrak{z}))) = k_2(\zeta_n(\mathfrak{z}))$.

Now we will prove that $(\forall \mathfrak{z} \in \bigcup_{n=1}^{\infty} z_n)$ $\beta_{\tau_\mathfrak{z}}\xi = \zeta_n(\mathfrak{z})$. Let us prove this property by induction for all chains. For $\mathfrak{z} = (\emptyset)$ it is obviously true. Let us consider the chain $z_k^i = \{\mathfrak{z}_0^i, \mathfrak{z}_1^i, \ldots, \mathfrak{z}_k^i\}$, where $\mathfrak{z}_0^i = (\emptyset)$, $\mathfrak{z}_1^i = (\Delta_1)$, $\mathfrak{z}_k^i = (\Delta_1, \ldots, \Delta_k)$ $(k \geq 0)$. We assume it to belong to the union of all full collections z_n. Let $\beta_{\tau(\mathfrak{z}_k^i)}\xi = \zeta_n(\mathfrak{z}_k^i)$. With respect to $\zeta_n(\mathfrak{z}_{k+1}^i)$ there may be four possibilities.

(a) If $\zeta_n(\mathfrak{z}_k^i) = \infty$, then $\zeta_n(\mathfrak{z}_{k+1}^i) = \infty$, but if $\tau_{\mathfrak{z}_k^i}\xi = \infty$, then $\tau_{\mathfrak{z}_{k+1}^i}\xi = \infty$. Therefore $\beta_{\tau(\mathfrak{z}_{k+1}^i)}\xi = \zeta_n(\mathfrak{z}_{k+1}^i)$.

(b) Let $\zeta_n(\mathfrak{z}_k^i) \neq \infty$ and $k_2(\zeta_n(\mathfrak{z}_k^i)) \notin \Delta_{k+1}$. Then $\beta_{\tau(\mathfrak{z}_{k+1}^i)}\xi = \beta_{\tau(\mathfrak{z}_k^i)}\xi$. In this case for any big enough n and $\tau_{\mathfrak{z}_k^i}\xi < t$ (when $\rho(\xi_n(s), \xi(s)) < r_t(n)$ for all $s < t$) $\tau_{(\mathfrak{z}_{k+1}^i)-r_t(n)}\xi_n = \tau_{\mathfrak{z}_k^i}\xi$. Therefore, by condition (2), $k_1(\zeta_n(\mathfrak{z}_{k+1}^i)) = \tau_{\mathfrak{z}_k^i}\xi$ and, consequently, $\beta_{\tau(\mathfrak{z}_{k+1}^i)}\xi = \zeta_n(\mathfrak{z}_{k+1}^i)$.

(c) Let $\zeta_n(\mathfrak{z}_k^i) \neq \infty$, $k_2(\zeta_n(\mathfrak{z}_k^i)) \in \Delta_{k+1}$, $k_1(\zeta_n(\mathfrak{z}_{k+1}^i)) = t < \infty$. Since $k_2(\zeta_n(\mathfrak{z}_{k+1}^i)) \notin \Delta_{k+1}$, we have $\tau_{\mathfrak{z}_{k+1}^i}\xi \leq t$. On the other hand, for big enough n $s < t$ it is true that $\rho(\xi(s), \xi_n(s)) \leq r_t(n)$. Therefore $\tau_{(\mathfrak{z}_{k+1}^i)-r_t(n)}\xi_n \leq \tau_{\mathfrak{z}_{k+1}^i}\xi$, and from condition (2) it follows that $\tau_{\mathfrak{z}_{k+1}^i}\xi \geq t$. Hence $\beta_{\tau(\mathfrak{z}_{k+1}^i)}\xi = \zeta_n(\mathfrak{z}_{k+1}^i)$.

(d) Let $\zeta_n(\mathfrak{z}_k^i) \neq \infty$, $k_2(\zeta_n(\mathfrak{z}_k^i)) \in \Delta_{k+1}$, $\zeta_n(\mathfrak{z}_{k+1}^i) = \infty$. For any $t \in \mathbb{R}_+$ and big enough n it follows that $\rho(\xi(s), \xi_n(s)) \leq r_t(n)$ for all $s < t$. From here $\tau_{\mathfrak{z}_{k+1}^i}\xi \geq \tau_{(\mathfrak{z}_{k+1}^i)-r_t(n)}\xi_n > t$. Hence $\tau_{\mathfrak{z}_{k+1}^i}\xi = \infty$ and $\zeta_n(\mathfrak{z}_{k+1}^i) = \beta_{\tau_{\mathfrak{z}_{k+1}^i}}\xi$.

(B) For the given sequence $(z_n) \in \Sigma_0$ let us consider function $\xi \in \Pi(\bigcup_{n=1}^{\infty} z_n)$ (correct exit from every set of a countable system of open sets; see item 2.27). For any n let us determine a map $\zeta_n : z_n \to \mathbb{Y}$ of the form $\zeta_n(\mathfrak{z}) = \beta_\mathfrak{z}\xi$. Evidently, this map is correct, and the family of maps is consistent. The first condition of the theorem is fulfilled for any function $\xi \in \mathcal{D}$ due to the definition of a deducing sequence. Let us check the second condition to be fulfilled. If $k_1(\zeta_n(\mathfrak{z}_k^i)) = \tau_{\mathfrak{z}_k^i}\xi \leq t < \infty$ and n is big enough, then $\rho(\xi(s), \xi_n(s)) < r_t(n)$ $(s < t)$, where ξ_n is step-function constructed by values of ξ at the points $\tau_\mathfrak{z}\xi$ $(\mathfrak{z} \in z_n)$ according to the method of proposition 4.1. Therefore $\tau_{(\mathfrak{z}_k^i)-2r_t(n)}\xi \leq \tau_{(\mathfrak{z}_k^i)-r_t(n)}\xi_n \leq \tau_{\mathfrak{z}_k^i}\xi$. According to the correct exit property $\tau_{(\mathfrak{z}_k^i)-2r_t(n)}\xi \to \tau_{\mathfrak{z}_k^i}\xi$. Hence, $\tau_{(\mathfrak{z}_k^i)-r_t(n)}\xi_n \to \tau_{\mathfrak{z}_k^i}\xi = k_1(\zeta_n(\mathfrak{z}_k^i))$. If $\tau_{\mathfrak{z}_k^i} = \infty$ and t is any value as big as desired, then for n big enough $\rho(\xi(s), \xi_n(s)) < r_t(n)$ and $\tau_{(\mathfrak{z})-2r_t(n)}\xi > t/2$. From here $\tau_{(\mathfrak{z}_k^i)-r_t(n)}\xi_n > t/2$ and $\tau_{(\mathfrak{z}_k^i)-r_t(n)}\xi_n \to \infty = k_1(\zeta(\mathfrak{z}_k^i))$.

(C) If ξ is a continuous function, it exits correctly from any system of open sets. As such, it satisfies conditions (1) and (2). Condition (3) follows immediately from the continuity of ξ. It remains to prove that under condition (3) the realizing function can be chosen continuous, but for any $t > 0$ and big enough n on the interval $[0, t]$ the function ξ_n has jumps with values not more than $r_t(n)$. It means that on this interval the limit function ξ has jumps with values not more than $2r_t(n)$. Since $r_t(n) \downarrow 0$ as $n \to \infty$ ξ is continuous. $\square$

4.2. Extension of a measure

One sufficient condition on the existence of a projective limit of a family of probability measures is proved. Each element of the family is a probability measure on some partition of the given set. Such a partition of space $\mathcal{D}$ gives, for example, map $\xi \to (\beta_{\tau_1}\xi, \ldots, \beta_{\tau_k}\xi)$, where $\tau_i \in T(\sigma_\Delta, \ \Delta \in \mathfrak{A})$ (see item 2.14). This example will be used later.

6. Maps of partitioning and enlargement

Let Ω be a fixed set; $\mathcal{U}$ a directed set (set of indexes) with an order relation $\leq$; $B(u)$ a partition of a set Ω (a class of non-intersecting subsets, consisting the whole set Ω in their union) corresponding to an index $u \in \mathcal{U}$. It is assumed that B is an isotone map: $u_1 \leq u_2 \Rightarrow B(u_1) \prec B(u_2)$, i.e. partition $B(u_1)$ is more rough, than $B(u_2)$. This means that for any $a_2 \in B(u_2)$ exists $a_1 \in B(u_1)$ such that $a_2 \subset a_1$. Let $K_u : \Omega \to B(u)$ be a map of a partition, where $K_u \omega = a \in B(u)$, if $\omega \in a \subset \Omega$; $K_{u_1}^{u_2} : B(u_2) \to B(u_1)$ $(u_1 \leq z_u)$ is a map of enlargement, where $K_{u_1}^{u_2} a_2 = a_1$, if $a_2 \subset a_1$, where $a_1 \in B(u_1)$, $a_2 \in B(u_2)$. In this case, if $u_1 \leq u_2 \leq u_3$, then $K_{u_1}^{u_3} = K_{u_1}^{u_2} \circ K_{u_2}^{u_3}$, and $K_{u_1} = K_{u_1}^{u_2} \circ K_{u_2}$. We call this property a consistency of projection operators (partition and enlargement).

7. Determining sequence of functionals

Let Σ be a class of sequences $(u_n)_1^\infty$ $(u_n \in \mathcal{U})$, possessing the following properties:

(1) if $(u_n)_1^\infty \in \Sigma$, then $u_n \leq u_{n+1}$ $(n \in \mathbb{N})$;

(2) for any sequences $(u'_n)_1^\infty$ $(u'_n \in \mathcal{U})$ there exists a sequence $(u_n)_1^\infty \in \Sigma$ such that $(\forall n \in \mathbb{N})$ $u'_n \leq u_n$.

A sequence of non-negative functionals $(h_n)_1^\infty$ is referred to as determining for a sequence of partitions $(B(u_n))_1^\infty$, if from a condition $a_n \in B(u_n)$, $a_n \supset a_{n+1}$ $(n \in \mathbb{N})$ and $\liminf h_n(a_n) = 0$ follows $\bigcap_{n=1}^\infty a_n \neq \emptyset$.

8. Theorem about extension of a measure

A family of probability measures $(P_u)_{u \in \mathcal{U}}$ on $\mathcal{B}(B(u))$ is said to be consistent, if from relation $u_1 \leq u_2$ it follows that $P_{u_1} = P_{u_2} \circ (K_{u_1}^{u_2})^{-1}$. A unique function P determined on the algebra of sets

$$\bigcup_{u \in \mathcal{U}} K_u^{-1} \mathcal{B}(B(u))$$

by the equality $P \circ K_u^{-1} = P_u$ is an additive function of sets (projective system and limit; see [BLU 68]).

THEOREM 4.2. *Let the following conditions be fulfilled:*

(1) each partition $B(z)$ is a Hausdorff topological space;

(2) $(\forall u_1, u_2 \in \mathcal{U}, u_1 \leq u_2)$ $K_{u_1}^{u_2} \in \mathcal{B}(B(u_2))/\mathcal{B}(B(u_1))$ (measurability of enlargement maps);

(3) for any finite measure μ on $\mathcal{B}(B(u_2))$, any $u_1 \leq u_2$, $\varepsilon > 0$ and $B \in \mathcal{B}(B(u_2))$, where $\mu(B) > 0$, there exists compact $B' \subset B$, on which $K_{u_1}^{u_2}$ is continuous and $\mu(B \setminus B') < \varepsilon$;

(4) there exists a class of majorizing sequences Σ, and for any sequence $(u_n)_1^{\infty} \in \Sigma$ there is a sequence of $\mathcal{B}(B(u_n))$-measurable determining functionals $(h_n)_1^{\infty}$;

(5) a consistent family of probability measures $(P_u)_{u \in \mathcal{U}}$ on $\mathcal{B}(B(u))$ $(u \in \mathcal{U})$ is determined;

(6) for any $\varepsilon > 0$ and $(u_n)_1^{\infty} \in \Sigma$ $P_{u_n}(h_n \geq \varepsilon) \to 0$ as $n \to \infty$.

Then there exists probability measure P on $\sigma(K_u^{-1}(\mathcal{B}(B(u))), u \in \mathcal{Z})$, such that $P_u = P \circ K_u^{-1}$.

Proof. It is sufficient to prove continuity of P. Let (M_n) be a sequence of sets such that $(\forall n \in \mathbb{N})$ $M_n \supset M_{n+1}$; $M_n \in K_{u_n}^{-1} \mathcal{B}(B(u_n))$; $M_n = K_{u_n}^{-1} S_n$, $S_n \in \mathcal{B}(B(u_n))$ and $P(M_n) \geq p > 0$. By the property of class Σ it is enough to assume that $(u_n)_1^{\infty} \in \Sigma$. A proof of the continuity is here divided into four stages.

(1) Let $(h_n)_1^{\infty}$ be a determining sequence for $(B(u_n))_1^{\infty}$ and $A_{n_k} = \{a \in B(u_{n_k}) : h_{n_k}(a) \leq 1/k\}$, where n_k is chosen such that $P_{u_{n_k}}(A_{n_k}) \geq 1 - p \cdot 2^{-k-1}$. Let $S'_{n_1} = A_{n_1} \cap S_{n_1}$ and $S'_{n_k} = (K_{u_{n_{k-1}}}^{u_{n_k}})^{-1} S'_{n_{k-1}} \cap A_{n_k} \cap S_{n_k}$ $(k \geq 2)$. Then $S'_{n_k} \in \mathcal{B}(B(u_{n_k}))$ and has properties $(\forall k \in \mathbb{N})$:

(a) $S'_{n_k} \supset K_{u_{n_k}}^{u_{n_{k+1}}} S'_{n_{k+1}}$;

(b) $P_{u_{n_k}}(S'_{n_k}) \geq p/2$;

(c) $a \in S'_{n_k} \Rightarrow h_{n_k}(a) \leq 1/k$.

(2) Let $S''_{n_k} \subset S'_{n_k}$, $S''_{n_k} \in \mathfrak{K}$ (compact), $K^{u_{n_k}}_{u_{n_{k-1}}}$ is continuous on S''_{n_k}, and $P_{u_{n_k}}(S'_{n_k} \setminus S''_{n_k}) < 2^{-k-2}p$. Let $S^{\circ}_{n_1} = S''_{n_1}$, and $S^{\circ}_{n_k} = (K^{u_{n_k}}_{u_{n_{k-1}}})^{-1} S^{\circ}_{n_{k-1}} \cap S''_{n_k}$ $(k \geq 2)$. Then $S^{\circ}_{n_k} \in \mathcal{B}(B(u_{n_k}))$ and has properties:

(a) $S^{\circ}_{n_k} \supset K^{u_{n_{k+1}}}_{u_{n_k}} S^{\circ}_{n_{k+1}}$;

(b) $P_{u_{n_k}}(S^{\circ}_{n_k}) \geq p/4$;

(c) $S^{\circ}_{n_k} \in \mathfrak{K}$ and $K^{u_{n_k}}_{u_{n_{k-1}}}$ is continuous on it;

(d) $a \in S^{\circ}_{n_k} \Rightarrow h_{n_k}(a) \leq 1/k$.

(3) Let $S^{*}_{n_k} = \bigcap_{\ell \geq k} K^{u_{n_\ell}}_{u_{n_k}} S^{\circ}_{n_\ell}$. Then $S^{*}_{n_k} \in \mathcal{B}(B(u_{n_k}))$, and has properties:

(a) $S^{*}_{n_k} = K^{u_{n_{k+1}}}_{u_{n_k}} S^{*}_{n_{k+1}}$,

(b) $P_{u_{n_k}}(S^{*}_{n_k}) \geq p/4$,

(c) $a \in S^{*}_{n_k} \Rightarrow h_{n_k}(a) \leq 1/k$.

In order to prove (a) it is necessary to establish that

$$\left(\forall a \in S^{*}_{n_k}\right) \quad \left(K^{u_{n_{k+1}}}_{u_{n_k}}\right)^{-1} a \cap S^{*}_{n_{k+1}} \neq \varnothing.$$

We have

$$\left(K^{u_{n_{k+1}}}_{u_{n_k}}\right)^{-1} a \cap S^{*}_{n_{k+1}} = \left(K^{u_{n_{k+1}}}_{u_{n_k}}\right)^{-1} a \cap \bigcap_{\ell \geq k+1} K^{u_{n_\ell}}_{u_{n_{k+1}}} S^{\circ}_{n_\ell},$$

and

$$\left(K^{u_{n_{k+1}}}_{u_{n_k}}\right)^{-1} a \cap \bigcap_{\ell \geq k+1}^{N} K^{u_{n_\ell}}_{u_{n_{k+1}}} S^{\circ}_{n_\ell} \neq \varnothing \quad (\forall N \geq k+1),$$

since

$$a \in \bigcap_{\ell \geq k+1}^{N} K^{u_{n_\ell}}_{u_{n_k}} S^{\circ}_{n_\ell} = K^{u_{n_{k+1}}}_{u_{n_k}} \bigcap_{\ell \geq k+1}^{N} K^{u_{n_\ell}}_{u_{n_{k+1}}} S^{\circ}_{n_\ell}$$

$$= K^{u_{n_{k+1}}}_{u_{n_k}} K^{u_{n_N}}_{u_{n_{k+1}}} S^{\circ}_{n_N} = K^{u_{n_N}}_{u_{n_k}} S^{\circ}_{n_N}.$$

From here, since $S^{\circ}_{n_{k+1}}$ is compact and its intersection with $(K^{u_{n_{k+1}}}_{u_{n_k}})^{-1} a$ and all $K^{u_{n_\ell}}_{u_{n_{k+1}}} B^{\circ}_{n_\ell}$ are closed, property (a) follows.

(4) For any $a_{n_1} \in S^{*}_{n_1}$ there exists $a_{n_2} \in S^{*}_{n_2}$ such that $a_{n_1} = K^{u_{n_2}}_{u_{n_1}} a_{n_2}$, and if $\{a_{n_1}, \ldots, a_{n_k}\}$ is a sequence such that $a_{n_i} \in S^{*}_{n_i}$ and $a_{n_i} = K^{u_{n_{i+1}}}_{u_{n_i}} a_{n_{i+1}}$ $(i = 1, \ldots, k-1)$, then it can be continued with the preservation of these properties.

By the Hausdorff theorem on a maximal inserted circuit (see [BIR 84, p. 252], [KEL 68, p. 352], [KOL 72, p. 36], [SKO 70, p. 22]) there exists sequence $\{a_{n_k}\}_1^\infty$ such that $a_{n_k} \in S^*_{n_k} \subset B(u_{n_k})$, $a_{n_k} = K^{u_{n_{k+1}}}_{u_{n_k}} a_{n_{k+1}}$ $(k=1,2,\ldots)$ and $h_{n_k}(a_{n_k}) \to 0$. This sequence can be uniquely determined on all skipped numbers n: if $n_{k-1} < n \le n_k$, then $a_n = K^{u_{n_k}}_{u_n} a_{n_k}$ and since $K^{u_3}_{u_1} = K^{u_2}_{u_1} K^{u_3}_{u_2}$ $(u_1 \le u_2 \le u_3)$, we have $a_n = K^{u_{n+1}}_{u_n} a_{n+1}$ $(n \in \mathbb{N})$. Then according to definition of a determining sequence $\bigcap_{n=1}^\infty K^{-1}_{u_n} a_n = \bigcap_{n=1}^\infty a_n \ne \emptyset$. Hence, there is $\omega \in \Omega$ such that $(\forall n \in \mathbb{N})$ $\omega \in M_n$. $\square$

4.3. Construction of a measure with the given system of distributions of pairs of the first exit

The theorem on extension of a measure proved in the previous section is applied to a construction of a measure with a given family of distributions of pairs of the first exit.

9. Partitions of the set of functions

For any $z \in \mathcal{Z}_0$ the set of correct maps, $B(z)$, can be interpreted as a partition of a set $\mathcal{D}$. The element of this partition, corresponding to map $\zeta \in B(z)$, is a set of all $\xi \in \mathcal{D}$ such that $\zeta(\mathfrak{z}) = \beta_{\tau_{\mathfrak{z}}} \xi$ for all $\mathfrak{z} \in z$. Obviously, $\mathcal{Z}_0$ is an ordered set with respect to the relation of inclusion, and B is an isotone map. Let $K_z : \mathcal{D} \mapsto B(z)$, $(K_z \xi)(\mathfrak{z}) = \beta_{\tau_{\mathfrak{z}}} \xi$ $(\mathfrak{z} \in z)$ be an operation of projection, and $K^{z_2}_{z_1} : B(z_2) \mapsto B(z_1)$ $(z_1 \subset z_2)$; $(\forall \zeta \in B(z_2))$ $K^{z_2}_{z_1} \zeta$ be a narrowing (restriction) of ζ from z_2 to z_1. It is clear that the narrowing again leads to a correct maps ζ, and the class of all maps of projection and narrowing is consistent in the sense of item 6. In this case $z_1 \subset z_2 \subset z_3 \Rightarrow K^{z_3}_{z_1} = K^{z_2}_{z_1} \circ K^{z_3}_{z_2}$ and $K_{z_1} = K^{z_2}_{z_1} \circ K_{z_2}$ (consistency of maps).

10. Theorem on construction of a measure

In the following theorem and up to the end of this chapter we assume that $\mathfrak{A}_0 = \bigcup_{n=1}^\infty \mathfrak{A}(r_n)$, where $\mathfrak{A}(r_n)$ is some covering of set $\mathbb{X}$ by open sets of rank r_n, $r_n \downarrow 0$, and $\mathfrak{A}_1 \subset \mathfrak{A}$ is a pi-system of open sets, containing $\mathfrak{A}_0$ (see item 3.9).

THEOREM 4.3. *Let (P_z) $(z \in \mathcal{Z}(\mathfrak{A}_1))$ be a consistent family of probability measures on measurable topological spaces $(B(z), \mathcal{B}(B(z)))$ with a projective limit P such that:*

(a) $(\forall (z_n) \in \Sigma_0(\mathfrak{A}_1))$ $(\forall t > 0)$ $P_{z_n}(k_1(\zeta(\mathfrak{z}^{in}_n)) \le t) \to 0$ $(n \to \infty)$, where $\mathfrak{z}^{in}_n = \max z^{in}_n$, z^{in}_n is a maximal chain from z_n such that $\lim \mathfrak{z}^{in}_n \equiv \delta^{in} \in \mathrm{DS}$;

(b) $(\forall (z_n) \in \Sigma_0(\mathfrak{A}_1))$ $(\forall t > 0)$ $(\forall \varepsilon > 0)$ $(\forall \mathfrak{z} \in \bigcup_n z_n)$

$$P_{z_n}\left(k_1\left(\zeta_n(\mathfrak{z})\right) \le t, \ k_1\left(\zeta_n(\mathfrak{z})\right) - \tau_{\mathfrak{z}-r_t(n)} \xi_n \ge \varepsilon\right) \longrightarrow 0 \quad (n \longrightarrow \infty);$$

(c) for any $\mathfrak{z} = (\Delta_1, \ldots, \Delta_k) \in z$ $(k \geq 1)$

$$P_z\big(k_1\big(\zeta(\mathfrak{z}')\big)\big) < k_1\big(\zeta(\mathfrak{z})\big) < \infty, \ k_2\big(\zeta(\mathfrak{z})\big) \notin \partial\Delta_k\big) = 0,$$

where $\mathfrak{z}' = (\Delta_1, \ldots, \Delta_{k-1})$.

Then:

(1) condition (a) together with (b) is sufficient for a probability measure P on $(\mathcal{D}, \mathcal{F})$ *to exist, where* $P_z = P \circ K_z^{-1}$ $(\forall z \in \mathcal{Z}(\mathfrak{A}_1))$;

(2) conditions (a) and (b) are necessary for a process with the measure P to exist, where either this process possesses the property of correct exit with respect to any $z \in \mathcal{Z}(\mathfrak{A}_1)$ (i.e. $P(\Pi(z)) = 1$), or it is quasi-continuous from the left (see [DYN 59, p. 150], [BLU 68, p. 45], and also item 3.7);

(3) condition (a) together with (b) and (c) is sufficient and necessary for measure P to exist, there it is an extension of all measures P_z such that $P(\mathcal{C}) = 1$ ($\mathcal{C}$ is the set of all continuous $\xi \in \mathcal{D}$).

Proof. (1) It is sufficient to check for the given family the conditions (1)–(4) and (6) of theorem 4.2.

The first condition (topology). Every $\zeta : z \mapsto \mathbb{Y}$ is determined by a collection of number $|z|$ of points $\{\zeta(\mathfrak{z}_1), \ldots, \zeta(\mathfrak{z}_{|z|})\}$ $(\mathfrak{z}_i \in z)$. Hence, any $B(z)$ is a Hausdorff topological space and is a subset of the set $\mathbb{Y}^{|z|}$.

The second condition (measurability and continuity). Map $K_{z_1}^{z_2} : B(z_2) \to B(z_1)$ $(z_1 \subset z_2)$ is continuous and, consequently, it is measurable on the whole $B(z_2)$, since a convergence in $\mathbb{Y}^{|z_2|}$ is equivalent to the coordinate-wise convergence.

The third condition (regularity). Every $B(z)$ $(z \in \mathcal{Z})$ is a locally-compact Hausdorff topological space with a countable basis. For such a space for any finite measure μ on $\mathcal{B}(B(z))$, any $S \in \mathcal{B}(B(z))$ $(\mu(S) > 0)$, and any $\varepsilon > 0$ there exists a compact $S' \subset S$, for which $\mu(S \setminus S') < \varepsilon$ (see [BLU 68, p. 8]).

The fourth condition (majorizing and determining sequences). Class $\Sigma_0(\mathfrak{A}_1)$ of sequences $(z_n)_1^\infty$ $(z_n \in \mathcal{Z}_0(\mathfrak{A}_1))$ is a Σ-class by item 7. In order to use theorem 4.2, let us construct for every $(z_n)_1^\infty \in \Sigma_0(\mathfrak{A}_1)$ a sequence of determining functionals $(h_n)_1^\infty$. Sequence $(\zeta_n)_1^\infty$ can be interpreted as a decreasing sequence of subsets of the set $\mathcal{D}$, where the n-th term of the sequence is an element of partition $B(z_n)$. By theorem 4.1, a convergence of sequences:

$$\overline{\rho}\big(\infty, k_1\big(\zeta(\mathfrak{z}_n^{in})\big)\big) \longrightarrow 0 \quad \big(\lim \max z_n^{in} \in \mathrm{DS}\big),$$

$$\overline{\rho}\big(k_1\big(\zeta(\mathfrak{z})\big), \tau_{\mathfrak{z}-r_t(n)}\xi_n\big) \longrightarrow 0 \quad \bigg(t \in \mathbb{N}, \ \mathfrak{z} \in \bigcup_n z_n, \ k_1\big(\zeta(\mathfrak{z})\big) \leq t\bigg)$$

as $n \to \infty$ is sufficient for the existence of a function $\xi \in \mathcal{D}$, realizing all this ζ_n (in other words, so that a non-empty intersection of the decreasing sequence of subsets can exist). Hence, we have to construct a sequence of functionals $(h_n)_1^\infty$ such that $h_n(\zeta_n) \to 0$ if and only if all the previous sequences tend to zero. In order to obtain such a sequence it is sufficient to enumerate the original sequences by natural numbers and to compose a new sequence of weighted sums as follows:

$$h_n(\zeta_n) = \sum_{i=1}^{\infty} 2^{-i} \left(1 \wedge \bar{p}\left(\infty, k_1\left(\zeta_n\left(\mathfrak{z}_n^i \right) \right) \right) \right)$$

$$+ \sum_{i=1}^{\infty} \sum_{j=1}^{\infty} 2^{-i-j} \left(1 \wedge \bar{p}\left(k_1\left(\zeta_n(\mathfrak{z}(i)) \right), \tau_{\mathfrak{z}^{-r_j(n)}} \xi_n \right) \right),$$

where $\mathfrak{z}_n^i$ is the n-th term of the i-th deducing sequence; $\mathfrak{z}(i)$ is the i-th element of the union $\bigcup_n z_n$, $r_j(n) = r_j(z_n, \zeta_n)$.

The sixth condition (convergence to zero). For the above defined determining sequence, condition $P_{z_n}(h_n \geq \varepsilon) \to 0$ is, evidently, equivalent to conditions (b) and (c) combined.

(2) The necessity of conditions (a) and (b) for the extension of all P_z to the measure P of a quasi-left-continuous process follows from the property of a deducing sequence (a), and also from the property of quasi-left-continuity (b): $P(\tau_{\mathfrak{z}^{-r_t(n)}} \to \tau_{\mathfrak{z}}) = 1$. Actually, let $\tau = \lim_{n \to \infty} \tau_{\mathfrak{z}^{-r_t(n)}}$. Then

$$P\left(\tau_{\mathfrak{z}^{-r_t(n)}} \not\to \tau_{\mathfrak{z}} \right) = P\left(\tau < \tau_{\mathfrak{z}} \right) \leq P\left(X_{\tau_{\mathfrak{z}^{-r_t(n)}}} \not\to X_\tau, \tau < \infty \right) = 0,$$

Since $\lim X_{\tau_{\mathfrak{z}^{-r_t(n)}}} \notin \Delta_k$, where $\mathfrak{z} = (\Delta_1, \ldots, \Delta_k)$ and

$$\mathfrak{z}^{-r_t(n)} = \left(\Delta_1, \ldots, \Delta_{k-1}, \Delta_k^{-r_t(n)} \right),$$

and $X_\tau \in \Delta_k$, if $\tau < \tau_{\mathfrak{z}}$.

(3) From theorem 4.1 (for continuous $\xi \in \mathcal{D}$) it follows that the set of continuous functions $\mathcal{C}$ can be represented as follows

$$\{ \xi \in \mathcal{D} : (\forall n, m \in \mathbb{N}) \left(X_{\sigma_{\delta_n}^m} \xi \notin \Delta_{n,m+1} \right) \vee \left(\sigma_{\delta_n}^{m+1} \xi = \infty \right) \vee \left(X_{\sigma_\delta^{m+1}} \in \partial \Delta_{n,m+1} \right) \},$$

where $\delta_n = (\Delta_{n1}, \Delta_{n2}, \ldots) \in DS(r_n)$. From here it follows that the union of conditions (a), (b), and (c) is sufficient for $P(\mathcal{C}) = 1$. The necessity of them is evident. $\square$

4.4. Construction of a projective system of measures by semi-Markov transition functions

A construction of a joint distribution of pairs of the first exit on the basis of a given set of semi-Markov transition functions is somewhat more complicated than similar constructions with the help of Markov transition functions. It is connected with a lack of linear ordering in a set of all first exit times and their iterations. The exception is made for a chain $z \in \mathcal{Z}_0'$, where $z = \{(\emptyset), \mathfrak{z}_1, \ldots, \mathfrak{z}_n\}$ and $\mathfrak{z}_1 \prec \cdots \prec \mathfrak{z}_n$ $(n \geq 1)$. In this case $0 \leq \tau_{\mathfrak{z}_1} \leq \tau_{\mathfrak{z}_2} \leq \cdots \leq \tau_{\mathfrak{z}_n}$. In general the construction is connected by a passage from a full collection of a general view to a chain. Thus, for each $z \in \mathcal{Z}_0$, there exists a one-to-one correspondence between the set of all correct maps $\zeta : z \to Y$ and all chains z' equipped with correct maps $\zeta' : z' \to Y$. This correspondence is formulated in terms of indexes of intersection connected with z.

11. Indexes of intersection

Let

$$z = \{(\emptyset); \mathfrak{z}_{11}, \ldots, \mathfrak{z}_{1n_1}; \mathfrak{z}_{k1}, \ldots, \mathfrak{z}_{kn_k}\} \quad (k \geq 0, \ n_i \geq 1),$$

where $\mathfrak{z}_{ij} = (\Delta_{i1}, \ldots, \Delta_{ij})$, $\Delta_{ij} \in \mathfrak{A}_0 \subset \mathfrak{A}$ $(\Delta_{ij} \neq \mathbb{X})$, i.e. $z \in \mathcal{Z}_0(\mathfrak{A}_0)$. Let $\mathfrak{A}_1$ be a pi-system generated by a class $\mathfrak{A}_0$. For $k \geq 1$ we name as an index of intersection connected with z, vector $\alpha = (j_1, \ldots, j_k)$, where $(\forall s \in \{1, \ldots, k\})$ $1 \leq j_s \leq n_s + 1$. With the help of index α we will determine intersections of sets $\bigcap_{i=1}^{k} \Delta_{ij_i}$. In this denotation we assume $\Delta_{i,n_i+1} = \mathbb{X}$. Let $I(z)$ be collection of all indexes of intersection for given z. In the case of $\alpha_1, \alpha_2 \in I(z)$, $\alpha_1 = (j_{11}, \ldots, j_{1k})$, $\alpha_2 = (j_{21}, \ldots, j_{2k})$ we denote: $\alpha_1 < \alpha_2 \Leftrightarrow (\forall s) j_{1s} \leq j_{2s}$ and $(\exists s) \ j_{1s} < j_{2s}$.

12. Indexes and generated correct maps

For any $z \in \mathcal{Z}_0(\mathfrak{A}_0)$ there is a finite increasing sequence of indexes of intersection $(\alpha_1, \ldots, \alpha_N)$ $(N \geq 0, \ \alpha_i < \alpha_{i+1})$, and also a chain $z' \in \mathcal{Z}_0'(\mathfrak{A}_1)$ $(z' = \{(\emptyset), \mathfrak{z}_1, \ldots, \mathfrak{z}_N\})$ and its correct map $\zeta' \in B(z')$ connected with $\zeta \in B(z)$. We denote $k_1(\zeta(\mathfrak{z}_{ij})) = t_{ij}$, and if $t_{ij} < \infty$, then $k_2(\zeta(\mathfrak{z}_{ij})) = x_{ij}$; $k_1(\zeta'(\mathfrak{z}_i)) = t_i$, and if $t_i < \infty$, then $k_2(\zeta'(\mathfrak{z}_i)) = x_i$. The chain z' and the map $\zeta' \in B(z')$ are determined by the following rules:

(1) The chain z' begins with element $(\emptyset)$. In this case $\zeta'(\emptyset) = \zeta(\emptyset)$.

(2) Let $(\forall i : 1 \leq i \leq k) \ j_{i1} = \min\{j : 1 \leq j \leq n_i + 1, \ k_2(\zeta(\emptyset)) \in \Delta_{ij}\}$. Because of $\Delta_{i,n_i+1} = \mathbb{X}$, the set in braces is not empty. If $(\forall i) j_{i1} = n_i + 1$, then $N = 0$ and the construction is complete. Otherwise $\alpha_1 = (j_{11}, \ldots, j_{k1})$, $\Delta_1 = \bigcap_{i=1}^{k} \Delta_{i,j_{i1}}, \mathfrak{z}_1 = (\Delta_1), t_1 = \bigwedge_{i=1}^{k} t_{ij_{i1}}$. If $t_1 = \infty$, then $N = 1$, and the construction is complete. If $t_1 < \infty$, then $(\exists i) \ t_1 = t_{ij_{i1}}, x_1 = x_{ij_{i1}}$ and $N \geq 1$.

(3) Let the sequence of indexes of intersection $(\alpha_1, \ldots, \alpha_m)$, the chain $\{(\emptyset),$ $\mathfrak{z}_1, \ldots, \mathfrak{z}_m\}$, and meanings of the correct map ζ' on this chain be constructed; let $\alpha_m = (j_{1m}, \ldots, j_{km})$, $\mathfrak{z}_m = (\Delta_1, \ldots, \Delta_m)$, $t_m < \infty$, $x_m \notin \Delta_m$, and also $N \geq m$. Let us construct the index $\alpha_{m+1} = (j_{1,m+1}, \ldots, j_{k,m+1})$, where $(\forall i : 1 \leq i \leq k)$

$$j_{i,m+1} = \begin{cases} j_{im}, & t_m < t_{ij_{im}}, \\ \min\{j : j_{im} < j \leq n_i + 1, \ x_m \in \Delta_{ij}\}, & t_m = t_{ij_{im}}. \end{cases}$$

If $(\forall i \geq 1)\ j_{i,m+1} = n_i + 1$, then $N = m$, and the construction is complete. Otherwise $\Delta_{m+1} = \bigcap_{i=1}^{k} \Delta_{ij_{i,m+1}}$, $\mathfrak{z}_{m+1} = (\Delta_1, \ldots, \Delta_{m+1})$, $t_{m+1} = \bigwedge_{i=1}^{k} t_{ij_{i,m+1}}$. If $t_{m+1} = \infty$, then $N = m + 1$, and the construction is complete. If $t_{m+1} < \infty$, then $(\exists i \geq 1)\ t_{m+1} = t_{ij_{i,m+1}}$, $x_{m+1} = x_{ij_{i,m+1}}$ and $N \geq m + 1$.

The construction is made after a finite number of steps: $N \leq n_1 + \cdots + n_k$. According to definition of a correct map ζ (see item 2), the sequence of indexes is determined by a sequence of the first exit points, because $t_m < t_{sj_{sm}} \Leftrightarrow x_m \in \Delta_{sj_{sm}}$. For any z there exists a finite set $J(z)$ of sequences of indexes of intersections $A = (\alpha_m)_1^N$, which generates a finite partition $(B(z, A))_{A \in J(z)}$ of the set $B(z)$ of all correct maps ζ. Evidently, $L\zeta = L\zeta'$ (see item 2), there ζ' is the correct map of the chain z', constructed above for ζ. From here it follows that for a fixed z the correspondence $\zeta \leftrightarrow (A, z', \zeta')$ is one-to-one. Evidently, the partition $B(z, A)$ is measurable: $(\forall z \in \mathcal{Z}_0(\mathfrak{A}))\ (\forall A \in J(z))\ B(z, A) \in \mathcal{B}(B(z))$.

13. Admissible family of semi-Markov kernels

Let $\mathfrak{A}_1 \subset \mathfrak{A}$. We will consider families of kernels $(Y_\Delta(B \mid x))_{\Delta \in \mathfrak{A}_1}$ $(x \in \mathbb{X},\ B \in \mathcal{B}(\mathbb{R}_+ \times \mathbb{X}))$, which satisfy the following conditions:

(a) $Y_\Delta(\cdot \mid x)\ (x \in \mathbb{X})$ is a sub-probability measure on $\mathcal{B}(\mathbb{R}_+ \times \mathbb{X})$, where

$$Y_\Delta(B \mid x) = \begin{cases} I_B((0, x)), & x \notin \Delta, \\ Y_\Delta(B \cap (\mathbb{R}'_+ \times (\mathbb{X} \setminus \Delta)) \mid x), & x \in \Delta, \end{cases}$$

$\mathbb{R}'_+ = (0, \infty)$;

(b) $Y_\Delta(B \mid \cdot)\ (B \in \mathcal{B}(\mathbb{R}_+ \times \mathbb{X}))$ is a $\mathcal{B}(\mathbb{X})$-measurable function;

(c) $(\forall \Delta_1, \Delta_2 \in \mathfrak{A}_1,\ \Delta_1 \subset \Delta_2)$

$$Y_{\Delta_2}([0, t) \times S \mid x) = \int_0^t \int_{\mathbb{X}} Y_{\Delta_1}(dt_1 \times dx_1 \mid x) Y_{\Delta_2}([0, t - t_1) \times S \mid x_1).$$

We call the family of kernels (Y_Δ) satisfying conditions (a), (b) and (c) an *admissible* family of semi-Markov transition functions. These conditions are necessary for $Y_\Delta = F_{\sigma_\Delta}$ to be semi-Markov transition functions for some semi-Markov process (P_x) (see item 3.17).

Consider also a family of Laplace transformations of semi-Markov transition functions:

$$y_\Delta(\lambda, S \mid x) \equiv \int_0^\infty e^{-\lambda t} Y_\Delta(dt \times S \mid x),$$

where $\lambda \geq 0$. Evidently, $y_\Delta(0, S \mid x) = Y_\Delta(\mathbb{R}_+ \times S \mid x)$.

The following properties correspond to properties (a), (b) and (c):

(d) $(\forall \lambda \geq 0) (\forall x \in \mathbb{X})$ $y_\Delta(\lambda, S \mid x)$ is a sub-probability measure on $\mathcal{B}(\mathbb{X})$, with

$$y_\Delta(\lambda, S \mid x) = \begin{cases} I_S(x), & x \notin \Delta, \\ y_\Delta(S \setminus \Delta \mid x), & x \in \Delta; \end{cases}$$

(e) $(\forall \lambda \geq 0) (\forall S \in \mathcal{B}(\mathbb{X})$ $y_\Delta(\lambda, S \mid \cdot))$ is a $\mathcal{B}(\mathbb{X})$-measurable function;

(f) $(\forall \lambda \geq 0) (\forall \Delta_1, \Delta_2 \in \mathfrak{A}_1, \Delta_1 \subset \Delta_2)$

$$y_{\Delta_2}(\lambda, S \mid x) = \int_{\mathbb{X}} y_{\Delta_1}(\lambda, dx_1 \mid x) y_{\Delta_2}(\lambda, S \mid x_1).$$

Besides, each function $y_\Delta(\lambda, S \mid x)$ is a completely monotone function of λ. Remember that function $f(\lambda)$ $(\lambda \geq 0)$ is said to be *completely monotone* [FEL 67], if it is non-negative, infinitely differentiable and

$$(\forall k \in \mathbb{N}) \quad (-1)^k \frac{\partial^k}{\partial \lambda^k} f(\lambda) \geq 0.$$

We call the family of kernels (y_Δ) satisfying conditions (d), (e) and (f) an *admissible* family of semi-Markov transition generating functions. Conditions (d), (e) and (f) are necessary for $y_\Delta = f_{\sigma_\Delta}$ to be the semi-Markov transition generating functions of some semi-Markov process (P_x) (see item 3.17).

The admissible families (Y_Δ) and (y_Δ) mutually determine each other. Furthermore, we will establish duality of their properties providing existence of corresponding semi-Markov process.

14. Projective system of measures

Let $\mathfrak{A}_1$ be a pi-system of open sets and (Y_Δ) $(\Delta \in \mathfrak{A}_1)$ be a family of kernels satisfying the conditions of item 13. Let us assume z, ζ, A, z', ζ', as those in items 11 and 12. For any $x \in \mathbb{X}$ we determine distribution $P_{x,z}$ on $\mathcal{B}(B(z, A))$ by its values on sets $G \in \mathcal{B}(B(z, A))$ of the form

$$G = \bigcap_{i=1}^N \{\zeta' : t_i - t_{i-1} \in T_i, \ x_i \in S_i \cap \Delta_{i+1}\},$$

where $T_i \in \mathcal{B}(\mathbb{R}_+)$, $S_i \in \mathcal{B}(\mathbb{X})$, $t_0 = 0$ and $\Delta_{N+1} = \mathbb{X}$:

$$P_{x,z}(G) = \int \prod_{i=1}^{N} Y_{\Delta_i}(T_i \times dx_i \mid x_{i-1}),$$

where $x_0 = x$ and integration is made by all $(x_1, \ldots, x_N)$, from the set $(S_1 \cap \Delta_2) \times \cdots \times (S_N \cap \Delta_{N+1})$. The probability of the event with respect to ζ', where $\zeta'(\mathfrak{z}_N) = \overline{\infty}$, is equal to the difference of probabilities of the considered view. Without being too specific we can accept an additional condition that all sets T_i represent intervals $[0, \tau_i)$ $(\tau_i > 0)$. The function of sets $P_{x,z}$ constructed on the pi-system of sets G can be extended uniquely up to a measure on the whole sigma-algebra $\mathcal{B}(B(z, A))$ (see, e.g., [GIH 73, DYN 63]).

THEOREM 4.4. *For any $x \in \mathbb{X}$ the family of measures $(P_{x,z})_{z \in \mathcal{Z}_0(\mathfrak{A}_1)}$ constructed above is a projective system of measures i.e.*

$$\left(\forall z_1, z_2 \in \mathcal{Z}_0(\mathfrak{A}_1),\ z_1 \subset z_2\right) \quad P_{x,z_1} = P_{x,z_2} \circ \left(K_{z_1}^{z_2}\right)^{-1}.$$

Proof. Because of the consistency of projective operators (see item 6 and 9) it is sufficient to prove that $P_{x,z_1} \circ (K_z^{z_1})^{-1}(G_1) = P_{x,z}(G_1)$ $(G_1 \in \mathcal{B}(B(z)))$, where z is obtained from z_1 by deleting the last element in some one chain of those, consisting z_1. For the given z it is sufficient to consider two cases:

$$z_1 = \left\{(\emptyset); \mathfrak{z}_{11}, \ldots, \mathfrak{z}_{1n_1}; \ldots; \mathfrak{z}_{k1}, \ldots, \mathfrak{z}_{kn_k}, \mathfrak{z}_{kn_{k+1}}\right\},$$

$\mathfrak{z}_{kn_{k+1}} = (\mathfrak{z}_{kn_k}, \Delta)$ $(\Delta \neq \mathbb{X})$ (one chain is enlarged), and

$$z_1 = \left\{(\emptyset); \mathfrak{z}_{11}, \ldots, \mathfrak{z}_{1n_1}; \ldots; \mathfrak{z}_{k1}, \ldots, \mathfrak{z}_{kn_k}; \mathfrak{z}_{k+1,1}\right\},$$

where $\mathfrak{z}_{k+1,1} = (\Delta)$ (number of chains is enlarged). We consider only the first case, because the second is reduced to the first one. It is sufficient to prove equality of sets $G_1 = G$, mentioned in the preface of the theorem. For any $\zeta \in G$ the fixed sequence $A \in J(z)$ determines the same order statistics of the first coordinates of pairs $\zeta(\mathfrak{z}_{ij})$ on axis $\mathbb{R}_+$. We consider variants of location of a new point on the time axis corresponding to the additional parameter, the sequence $(\mathfrak{z}_{kn_k}, \Delta) = (\Delta_{k1}, \ldots, \Delta_{kn_k}, \Delta)$. All these variants are contained in the set $(K_z^{z_1})^{-1}B(z, A)$. For each of these variants there exists a corresponding specific condition on a sequence of indexes. Let $s = \min\{m : 0 \leq m \leq N,\ j_{k,m+1} = n_k + 1\}$. This s shows the number of such an index that $t_s = t_{kn_k}$. At this instant the first exit of the function $L\zeta$ from the sequence of sets $\mathfrak{z}_{kn_k}$ happens. Hence, up to this number the sequences of indexes z and z_1 coincide. Indexes with larger numbers can be determined considering the full collection $z_1' = z' \cup z''$, composed of two chains: $z' = \{(\emptyset), \mathfrak{z}_1, \ldots, \mathfrak{z}_N\}$ (old chain) and $z'' = \{(\emptyset), \mathfrak{z}_1, \ldots, \mathfrak{z}_s, (\mathfrak{z}_s, \Delta)\}$ (new chain). The sequence of two-dimensional indexes A'_m for z'_1 is

$$\big((1,1), \ldots, (s,s), (s+1, s+1), \ldots, (s+m, s+1), (s+m, s+2), \ldots, (N, s+2)\big),$$

where m corresponds to including a value of $k_1(\zeta((\mathfrak{z}_s, \Delta)))$ into interval (t_{s+m-1}, t_{s+m}) (where $1 \le m \le N - s + 1$, $t_0 = 0$, $t_{N+1} = \infty$). The appearance of a pair with meaning of the second coordinate $s+2$ in this sequence means the first exit from the additional set Δ in the chain z''. The sequence

$$A'_{N-s+2} = \big((1,1), \ldots, (s,s), (s+1, s+1), \ldots, (N+1, s+1)\big),$$

with $k_1(\zeta((\mathfrak{z}_s, \Delta))) = \infty$ corresponds to the case when the sequence (process) does not exit from the set Δ. According to the rule of construction of the measure on a set G by the family of kernels (Y_Δ), the measure P_{x, z_1} of the set $(K_z^{z_1})^{-1}G$ determined for $N - s + 2$ possible chains with corresponding sequences of indexes $A'_1, \ldots, A'_{N-s+2}$, is equal to

$$\int \prod_{i=1}^{s} Y_{\Delta_i}(T_i \times dx_i \mid x_{i-1}) H_s(x_s),$$

where the region of integration is $(S_1 \cap \Delta_2) \times \cdots \times (S_s \cap \Delta_{s+1})$. In this integral

$$H_s(x_s) = \sum_{m=1}^{N-s} p_m(x_s) + p(x_s) + q(x_s),$$

where

$$p_m(x_s) = \int \prod_{i=s+1}^{s+m-1} Y_{\Delta_i \cap \Delta}(T_i \times dx_i \mid x_{i-1}) g_m(x_{s+m}),$$

and the region of integration is

$$\big(S_{s+1} \cap \Delta_{s+2} \setminus \Delta_{s+1}\big) \times \cdots \times \big(S_{s+m-1} \cap \Delta_{s+m} \setminus \Delta_{s+m-1}\big).$$

Besides,

$$g_m(x_{s+m-1}) = \int_0^{\tau_{s+m}} \int_{\Delta_{s+m}} Y_{\Delta_{s+m} \cap \Delta}(dt \times dy \mid x_{s+m-1})$$

$$\times \int_{S_{s+m} \cap \Delta_{s+m+1}} Y_{\Delta_{s+m}}\big([0, \tau_{s+m} - t) \times dx_{s+m} \mid y\big) h_m(x_{s+m}),$$

where

$$h_m(x_{s+m}) = \int \prod_{i=s+m+1}^{N} Y_{\Delta_i}(T_i \times dx_i \mid x_{i-1}),$$

and the region of integration is

$$\left(S_{s+m+1} \cap \Delta_{s+m+2}\right) \times \cdots \times \left(S_N \cap \Delta_{N+1}\right) \quad \left(\Delta_{N+1} = \mathbb{X}\right).$$

While interpreting these formulae it should be taken into account that the sum of the empty set of summands is equal to zero, and the product of the empty set of terms is equal to 1. In addition, we have

$$p(x_s) = \int \prod_{i=s}^{N} Y_{\Delta_i \cap \Delta}\left(T_i \times dx_i \mid x_{i-1}\right) Y_{\Delta}\left(\mathbb{R}_+ \times \mathbb{X} \mid x_N\right),$$

$$q(x_s) = \int \prod_{i=s}^{N} Y_{\Delta_i \cap \Delta}\left(T_i \times dx_i \mid x_{i-1}\right)\left(1 - Y_{\Delta}\left(\mathbb{R}_+ \times \mathbb{X} \mid x_N\right)\right),$$

where the region of integration is

$$\left(S_{s+1} \cap \Delta_{s+2} \setminus \Delta_{s+1}\right) \times \cdots \times \left(S_N \cap \Delta_{N+1} \setminus \Delta_N\right) \quad \left(\Delta_{N+1} = \mathbb{X}\right).$$

On the other hand, according to the definition of an admissible family of kernels (item 13), for any $f \in \mathbb{B}$ and $1 \leq s \leq N$

$$\int_S Y_{\Delta_s}\left(T_s \times dx_1 \mid x_0\right) f(x_1)$$

$$= \int_0^{\tau_s} \int_{\mathbb{X}} Y_{\Delta_s \cap \Delta}\left(dt \times dy \mid x_0\right) \int_S Y_{\Delta_s}\left([0, \tau_s - t) \times dx_1 \mid y\right) f(x_1)$$

$$= \int_0^{\tau_s} \int_{\Delta_s} Y_{\Delta_s \cap \Delta}\left(dt \times dy \mid x_0\right) \int_S Y_{\Delta_s}\left([0, \tau_s - t) \times dx_1 \mid y\right) f(x_1)$$

$$+ \int_{S \setminus \Delta_s} Y_{\Delta_s \cap \Delta}\left(T_s \times dx_s \mid x_{s-1}\right) f(x_s).$$

Applying this decomposition repeatedly we obtain

$$H_s(x_s) = \int \prod_{i=s+1}^{N} Y_{\Delta_i}\left(T_N \times dx_i \mid x_{i-1}\right),$$

where the region of integration is

$$\left(S_{s+1} \cap \Delta_{s+2}\right) \times \cdots \times \left(S_N \cap \Delta_{N+1}\right) \quad \left(\Delta_{N+1} = \mathbb{X}\right),$$

and, consequently, $P_{x,z}(G) = P_{x,z_1}\left(\left(K_z^{z_1}\right)^{-1} G\right).$ □

4.5. Semi-Markov processes with given transition functions

The theorem about a limit of a projective system of consistent distributions of pairs of the first exit proved in section 4.3 is applied to a projective system constructed with the help of semi-Markov transition functions in section 4.4. The conditions are given for the projective limit to be a semi-Markov process.

15. Existence of an admissible set of measures

THEOREM 4.5. *Let $\mathfrak{A}_1$ be a pi-system of open sets, containing $\mathfrak{A}_0 = \bigcup_{n=1}^{\infty} \mathfrak{A}_{r_n}$ ($r_n \downarrow 0$, $\mathfrak{A}_{r_n}$ is a covering of rank r_n); $(Y_\Delta)_{\Delta \in \mathfrak{A}_1}$ be an admissible family of kernels (see item 13) and besides some of the following conditions are fulfilled:*

(a) for any $t \in \mathbb{R}_+$ and $\delta = (\Delta_1, \Delta_2, \ldots) \in DS(\mathfrak{A}_1)$ as $n \to \infty$

$$\int \prod_{i=1}^{n} Y_{\Delta_i} (dt_i \times dx_i \mid x_{i-1}) \longrightarrow 0,$$

where the region of integration is $\mathbb{X}^n \times \{t_1 + \cdots + t_n \le t\}$;
(b) $(\forall \Delta \in \mathfrak{A}_1)(\forall \varepsilon > 0)(\forall x \in \Delta)$

$$\int_{\mathbb{X}} Y_\Delta ([\varepsilon, \infty) \times \mathbb{X} \mid x_1) Y_{\Delta-r} (\mathbb{R}_+ \times dx_1 \mid x) \longrightarrow 0 \quad (r \longrightarrow 0);$$

(c) $(\forall \Delta \in \mathfrak{A}_1)(\forall x \in \Delta)(\forall S \in \mathcal{B}(\mathbb{X}))$

$$Y_\Delta (\mathbb{R}_+ \times S \mid x) = Y_\Delta (\mathbb{R}_+ \times (\partial \Delta \cap S) \mid x).$$

Then

(1) conditions (a) and (b) combined are sufficient for $(\forall x \in \mathbb{X})$ the finitely additive function of sets P_x, corresponding to the family of probability measures $(P_{x,z})$ ($z \in \mathcal{Z}_0(\mathfrak{A}_1)$), constructed in theorem 4.4, to be a probability measure on $(\mathcal{D}, \mathcal{F})$, and $(P_x)_{x \in \mathbb{X}}$ to be an admissible family of probability measures (see item 2.42);

(2) conditions (a) and (b) are necessary for the process to exist, which is determined by an admissible family of probability measures $(P_x)_{x \in \mathbb{X}}$, which either possesses the property of the correct exit with respect to any $z \in \mathcal{Z}(\mathfrak{A}_1)$, or is quasi-continuous from the left (see item 3.7);

(3) conditions (a), (b) and (c) combined are sufficient and necessary for the process to exist, which is determined by an admissible family of probability measures $(P_x)_{x \in \mathbb{X}}$, and $(\forall x \in \mathbb{X})$ is P_x-a.s. continuous.

Proof. (1) Condition (a) of theorem 4.3 follows immediately from condition (a) of this assertion, because the latter is a reformulation of the previous one in terms of kernels Y_Δ. Let us show fulfilment of condition (b) of theorem 4.3 for fixed $x \in \mathbb{X}$. Let $\mathfrak{z} = (\Delta_1, \ldots, \Delta_k)$ and $x \equiv x_0 \in \Delta_1$. Evidently, $(\forall r > 0)$ for big enough n

$$P_{x,z_n}\left(r_t(n) \geq r\right) \leq P_{x,z_n}\left(k_1\left(\zeta_n\left(\mathfrak{z}_n^{in}\right)\right) \leq t\right),$$

where $\lim \mathfrak{z}_n^{in} \in DS(r)$ and $\lim \mathfrak{z}_n^{in} \subset \bigcup_n z_n$, according to the definition of a correct sequence (z_n). On the other hand

$$P_{x,z_n}\left(k_1\left(\zeta_n(\mathfrak{z})\right) \leq t, \; k_1\left(\zeta(\mathfrak{z})\right) - \tau_{\mathfrak{z} - r_t(n)}\xi_n \geq \varepsilon\right)$$

$$\leq P_{x,z_n}\left(r_t(n) \geq r\right) + P_{x,z_n}\left(k_1\left(\zeta_n(\mathfrak{z})\right) < \infty, \; k_1\left(\zeta(\mathfrak{z})\right) - \tau_{\mathfrak{z} - r}\xi_n \geq \varepsilon\right).$$

We consider the case when $t_1 < \cdots < t_k$ (t_i is the first exit time from the i-th subsequence of $\mathfrak{z}$). Variants can be considered similarly when at least one of the inequalities (but not all of them) turns to an equality. In this case according to construction the second summand is equal to

$$\int \prod_{i=1}^{k-1} Y_{\Delta_i}\left(\mathbb{R}_+ \times dx_i \mid x_{i-1}\right) \int_{\Delta_k} Y_{\Delta_k}^{-r}\left(\mathbb{R}_+ \times dy \mid x_{k-1}\right) Y_{\Delta_k}\left([\varepsilon, \infty) \times \mathbb{X} \mid y\right),$$

where the external integral is determined on the set $\Delta_1 \times \cdots \times \Delta_{k-1}$. The internal integral is bounded and tends to zero under condition (b) of the assertion. From here fulfilment of condition (b) of theorem 4.3 follows. Thus, the family of probability measures (P_x) is constructed. The admissibility of this family follows from condition $P_{x,z}(k_2(\zeta(\emptyset)) = x) = 1$ and measurability of all functions $Y_\Delta(B \mid \cdot)$.

(2) The necessity of condition (a) is evident. A proof of necessity of condition (b) repeats the proof of assertion (2) of theorem 4.3 for $\mathfrak{z} = (\Delta)$.

(3) Evidently, condition (c) of theorem 4.3 follows from condition (c) of the present assertion. From here follows the third assertion.   □

16. Semi-Markovness of the constructed family

THEOREM 4.6. *Let (P_x) be an admissible family of probability measures, constructed in theorem 4.5. Then $(\forall \Delta \in \mathfrak{A}_1)$ $\sigma_\Delta \in RT(P_x)$.*

Proof. Let $\Delta \in \mathfrak{A}_1$ and $\delta \equiv (\Delta_1, \Delta_2, \ldots) \in DS(\mathfrak{A}_1)$. The sequence $\delta \cap \Delta \equiv (\Delta_1 \cap \Delta, \Delta_2 \cap \Delta, \ldots)$ is then deduced from Δ (see items 2.17 and 2.20). Let $Q_1 = \bigcap_{i=1}^{k} \beta_{\sigma_{\delta \cap \Delta}^{-i}}^{-1}(T_i \times S_i)$ and $Q_2 = \bigcap_{j=1}^{n} \beta_{\sigma_\delta^j}^{-1}(T_j' \times S_j')$, where $T_i, T_j' \in \mathcal{B}(\mathbb{R}_+)$, $S_i, S_j' \in \mathcal{B}(\mathbb{X})$, $k, n \in \mathbb{N}$ (denotation σ_δ^n; see item 2.14). Then

$$P_x\left(\theta_{\sigma_\Delta}^{-1} Q_2 \cap Q_1 \cap \{\sigma_\Delta < \infty\}\right)$$

$$= \sum_{m=0}^{\infty} P_x\left(\theta_{\sigma_\Delta}^{-1} Q_2 \cap Q_1 \cap \{\sigma_\Delta = \sigma_{\delta \cap \Delta}^m\} \cap \{\sigma_\Delta < \infty\}\right).$$

In this case

$$\theta_{\sigma_{\delta\cap\Delta}^m}^{-1} Q_2 \cap Q_1 \cap \{\sigma_\Delta = \sigma_{\delta\cap\Delta}^m\} \cap \{\sigma_\Delta < \infty\} \in K_z^{-1} \mathcal{B}(B(z)),$$

where $z = \{(\varnothing), \mathfrak{z}_1, \ldots, \mathfrak{z}_n\}$ is a chain with the common term of the form $\mathfrak{z}_m = (\Delta_1 \cap \Delta, \ldots, \Delta_m \cap \Delta)$ (see item 9). According to the construction of such a probability, the sequence of meanings $\zeta(\mathfrak{z}_i)$ determines a semi-Markov renewal process, i.e. a two-dimensional Markov chain of special view, considered in Chapter 1. The probability $P_x(\theta_{\sigma_\Delta}^{-1} Q_2 \cap Q_1 \cap \{\sigma_\Delta = \sigma_{\delta\cap\Delta}^m\})$ is determined by iterative method on the basis of the family of kernels $Y_{\Delta_i\cap\Delta}$ and Y_{Δ_j} $(i = 1, \ldots, m; \; j = 1, \ldots, n)$. In this case according to the construction of measures $P_{x,z}$ we have

$$P_x\left(\theta_{\sigma_\Delta}^{-1} Q_2 \cap Q_1 \cap \{\sigma_\Delta = \sigma_{\delta\cap\Delta}^m\} \cap \{\sigma_\Delta < \infty\}\right)$$
$$= E_x\left(P_{\pi_{\sigma_\Delta}}(Q_2); \; Q_1 \cap \{\sigma_\Delta = \sigma_{\delta\cap\Delta}^m\} \cap \{\sigma_\Delta < \infty\}\right)$$

(see items 1.10 and 1.17). Therefore

$$P_x\left(\theta_{\sigma_\Delta}^{-1} Q_2, \; Q_1 \cap \{\sigma_\Delta < \infty\}\right) = E_x\left(P_{\pi_{\sigma_\Delta}}(Q_2); \; Q_1 \cap \{\sigma_\Delta < \infty\}\right).$$

Since the rank of the sequence δ can be as small as desired (see item 2.20), the sigma-algebras $\mathcal{F}_{\sigma_\Delta}(\mathcal{F})$ are generated by sets Q_1 (Q_2) (see item 2.22). From here the previous formula is true for all $Q_1 \in \mathcal{F}_{\sigma_\Delta}$ and $Q_2 \in \mathcal{F}$.   □

17. A system of transition generating functions

Conditions of theorem 4.5 can be reformulated in terms of semi-Markov transition generating functions.

PROPOSITION 4.2. *Let* (y_Δ) $(\Delta \in \mathfrak{A}_1)$ *be an admissible family of semi-Markov transition generating functions, corresponding to the family of semi-Markov transition functions (kernels)* (Y_Δ) $(\Delta \in \mathfrak{A}_1)$ *(see item 13). Then conditions (a), (b) and (c) of theorem 4.5 are equivalent to corresponding conditions (d), (e) and (f):*

(d) for any $\lambda > 0$ *and* $\delta = (\Delta_1, \Delta_2, \ldots) \in \mathrm{DS}(\mathfrak{A}_1)$ *for* $n \to \infty$ *it is fair*

$$\int_{\mathbb{X}^n} \prod_{i=1}^n y_{\Delta_i}(\lambda, dx_i \mid x_{i-1}) \longrightarrow 0;$$

(e) $(\forall \lambda > 0)$ $(\forall \Delta \in \mathfrak{A}_1)$ $(\forall x \in \Delta)$

$$\int_{\mathbb{X}} y_\Delta(\lambda, \mathbb{X} \mid x_1) y_{\Delta-r}(0, dx_1 \mid x) \longrightarrow y_\Delta(0, \mathbb{X} \mid x) \quad (r \longrightarrow 0);$$

(f) $(\forall \lambda > 0)$ $(\forall \Delta \in \mathfrak{A}_1)$ $(\forall x \in \Delta)$ $(\forall S \in \mathcal{B}(\mathbb{X}))$

$$y_\Delta(\lambda, S \mid x) = y_\Delta(\lambda, \partial\Delta \cap S \mid x).$$

Proof. (1) Let us designate

$$F_n(t) = \int \prod_{i=1}^{n} Y_{\Delta_i}(dt_i \times dx_i \mid x_{i-1}).$$

Applying the Laplace transformation to this function we obtain

$$\int_0^\infty e^{-\lambda t} F_n(t)\, dt = \frac{1}{\lambda} \int_{\mathbb{X}^n} \prod_{i=1}^{n} y_{\Delta_i}(\lambda, dx_i \mid x_{i-1}).$$

Hence, properties (a) and (d) are equivalent.

(2) Let us designate

$$G_r(t) = \int_{\mathbb{X}} Y_\Delta\big([t, \infty) \times \mathbb{X} \mid x_1\big) Y_{\Delta-r}\big(\mathbb{R}_+ \times dx_1 \mid x\big).$$

Applying the Laplace transformation to this function and using equality $Y_\Delta(\mathbb{R}_+ \times S \mid x) = y_\Delta(0, S \mid x)$, we obtain

$$\int_0^\infty e^{-\lambda t} G_r(t)\, dt = \frac{1}{\lambda} \int_{\mathbb{X}} \big(y_\Delta(0, \mathbb{X} \mid x_1) - y_\Delta(\lambda, \mathbb{X} \mid x_1)\big) y_{\Delta-r}(0, dx_1 \mid x).$$

From here by boundedness of $G_r(t)$ implication (b)$\Rightarrow$(e) follows. For an arbitrary function $G_r(t)$ the inverse implication is true for almost all t. However, for monotone in t functions (e.g., $G_r(t)$) from convergence almost for all t, the convergence for all t follows. Hence the implication (e)$\Rightarrow$(b) is true.

(3) The equivalence of conditions (c) and (f) is evident.   □

18. Note on construction of the process

For the given family of semi-Markov kernels the constructed process is proved to be semi-Markov with respect to $\mathfrak{A}_1$ and, in particular, to be proper semi-Markov, if $\mathfrak{A}_1 = \mathfrak{A}$. For most applications a set of regeneration times of rather simple view can be used, for example, first exit times from open sets, their finite iterations with respect to operation $\dotplus$, and some limits. From a theoretical point of view it is preferable to enlarge this class, extracting only some main properties of the first exit times. A class of such Markov times, called intrinsic Markov times, is mentioned in Chapter 6. Including these times into the set of regeneration times determines the class of semi-Markov processes in strict sense. It gains logical completeness, although it loses the possibility of a constructive description of the process with the help of semi-Markov kernels.

Chapter 5

Semi-Markov Processes
of Diffusion Type

A semi-Markov processes of diffusion type is described by a differential second-order equation. As for a Markov process it is possible to define a class of such processes on the basis of local behavior of their transition functions.

The family of Markov transition functions determines a semi-group of operators with an infinitesimal operator of differential type. Coefficients of this differential operator are Kolmogorov's coefficients, a local drift and a local variance, determined by a limit behavior of the process in a fixed time t as $t \to 0$.

For semi-Markov processes the family of Markov transition functions generally speaking does not form a semi-group, so it is useless for them to be investigated with the help of an infinitesimal operator. However, for these processes there is another aspect of local characterization, with the help of distributions of the first exit time and the corresponding first exit position (called the first exit pair) of the process, leaving a small neighborhood of the initial point. It is shown that in the case of regular asymptotics of the distribution, when a diameter of a neighborhood tends to zero, the transition generating function of the process satisfies a differential second-order elliptic equation with coefficients determined by the first terms of this asymptotics.

It is interesting to reverse this outcome. The inverse problem consists of (1) construction of a semi-Markov process with transition generating functions, satisfying the *a priori* given differential equation, and (2) investigation asymptotic properties of these functions as solutions of a Dirichlet problem with various regions of determination of this equation.

5.1. One-dimensional semi-Markov processes of diffusion type

The continuous semi-Markov processes on the line represent the most natural class of processes, which are easy to study with the help of the first exit time. In this case $\mathbb{X} = \mathbb{R} \equiv (-\infty, \infty)$. The integral equation to which the transitional generating functions submit turns into a difference equation, and under some regularity conditions it turns into a differential equation. The lambda-characteristic operator of such a processes is in the form of a differential operator of the second order. Such a process can be obtained from a Markov process with the help of special transformation, which, as will be seen in the following chapter, is equivalent to a time change.

5.1.1. *Differential equation for transition functions*

A transition generating function $f_G(\lambda, dx_1 \mid x)$ of a continuous semi-Markov process on the line in the case when G is a finite interval and $x \in G$ breaks up to two functions $g_G(\lambda, x)$ and $h_G(\lambda, x)$ corresponding to the first exit from the interval through the left or right end [HAR 80b, HAR 83]. If these functions tend to non-degenerate limits when the ends of the interval tend to some internal point x, and first three coefficients of the asymptotics are continuous, and functions g_G and h_G are twice differentiable on x, these functions submit to differential second-order equations. In the following section the condition about a differentiability of the functions will be removed.

1. Functions g and h

Let $\mathbb{X} = \mathbb{R}$; $(P_x) \in SM$; $(\forall x \in \mathbb{X})\ P_x(\mathcal{C}) = 1$. If $G = (a, b)\ (a < b)$ and $x \in [G]$, then

$$f_G(\lambda, \mathbb{X} \mid x) = f_G(\lambda, \{a\} \mid x) + f_G(\lambda, \{b\} \mid x).$$

Let us designate $f_G(\lambda, \{a\} \mid x) = g_G(\lambda, x)$ and $f_G(\lambda, \{b\} \mid x) = h_G(\lambda, x)$. In addition $h_G(\lambda, a) = g_G(\lambda, b) = 0$ and $g_G(\lambda, a) = h_G(\lambda, b) = 1$. Functions $g_G(\lambda, x)$, and $h_G(\lambda, x)$ $(G \in \mathfrak{A}_0, \lambda > 0, x \in \mathbb{X})$ completely determine a semi-Markov process, where $\mathfrak{A}_0$ is the set of all finite open intervals on the line.

If $G_1 \subset G_2$, then $\sigma_{G_2} = \sigma_{G_1} \dotplus \sigma_{G_2}$, and the formula of theorem 3.5(2) can be rewritten as

$$f_{G_2}(\lambda, S \mid x) = \int_{\mathbb{X}} f_{G_1}(\lambda, dx_1 \mid x) f_{G_2}(\lambda, S \mid x_1).$$

In the case of $G_2 = (a, b)$, $G_1 = (c, d)$ $(a \leq c < d \leq b)$ and $x \in [G_1]$ this equation passes in a system of two equations:

$$g_{G_2}(\lambda, x) = g_{G_1}(\lambda, x)g_{G_2}(\lambda, c) + h_{G_1}(\lambda, x)g_{G_2}(\lambda, d), \qquad [5.1]$$

$$h_{G_2}(\lambda, x) = g_{G_1}(\lambda, x)h_{G_2}(\lambda, c) + h_{G_1}(\lambda, x)h_{G_2}(\lambda, d). \qquad [5.2]$$

Assuming $d = b$, we discover that $g_{G_2}(\lambda, x)$ does not increase on (a, b). Similarly, $h_{G_2}(\lambda, x)$ does not decrease.

2. Symmetric neighborhood of an initial point

If $G_1 = B(x, r) \equiv (x - r, x + r)$, we write $h_{G_1}(\lambda, x) = h_r(\lambda, x)$ and $g_{G_1}(\lambda, x) = g_r(\lambda, x)$ $(r > 0)$. Thus, the previous system of equations passes in the following system of difference equations:

$$g_G(\lambda, x) = g_r(\lambda, x)g_G(\lambda, x - r) + h_r(\lambda, x)g_G(\lambda, x + r),$$

$$h_G(\lambda, x) = g_r(\lambda, x)h_G(\lambda, x - r) + h_r(\lambda, x)h_G(\lambda, x + r),$$

where $B(x, r) \subset G$. Let us assume that for any $\lambda \geq 0$ and $x \in G_0$ (G_0 is an open interval) the following expansions are fair:

$$g_r(\lambda, x) = C_1(\lambda, x) + A_1(\lambda, x)r + B_1(\lambda, x)r^2/2 + o(r^2), \qquad [5.3]$$

$$h_r(\lambda, x) = C_2(\lambda, x) + A_2(\lambda, x)r + B_2(\lambda, x)r^2/2 + o(r^2) \qquad [5.4]$$

uniformly on x in each finite interval. Let all coefficients of this expansion be continuous functions of x, and both C_1 and C_2 is positive. Under these conditions we will prove that g_G and h_G are twice continuously differentiable and satisfy the differential equation of the second order, which will be shown below.

3. Corollaries from conditions on C factors

LEMMA 5.1. *Let* $(\forall \lambda \geq 0)$ $(\forall x \in \mathbb{X})$

$$g_r(\lambda, x) \longrightarrow C_1(\lambda, x), \quad h_r(\lambda, x) \longrightarrow C_2(\lambda, x) \quad (r \longrightarrow 0)$$

uniformly on x in each finite interval $G \subset G_0$; C_1 and C_2 are continuous on x and positive. Then $(\forall \lambda \geq 0)$

(1) $(\forall x \in G)$ $C_1(\lambda, x) + C_2(\lambda, x) = 1$ *and C_i do not depend on λ;*

(2) functions $h_G(\lambda, x)$ is strictly increasing, and $g_G(\lambda, x)$ is strictly decreasing in $x \in G$:

(3) if $(\exists x \in G)$ $h_G(\lambda, x) + g_G(\lambda, x) = 1$ *then* $(\forall x \in G)$ $h_G(\lambda, x) + g_G(\lambda, x) = 1$.

Proof. (1) Since

$$g_G(\lambda, x) = C_1(\lambda, x)g_G(\lambda, x - 0) + C_2(\lambda, x)g_G(\lambda, x + 0)$$

and $g_G(\lambda, x)$ almost everywhere on G is continuous, then from continuity and positiveness of C_1 and C_2 it follows $C_1(\lambda, x) + C_2(\lambda, x) = 1$. However, as C_1 and C_2 are principal terms of expansion of functions g_r and h_r correspondingly, which are non-increasing on λ, they also do not depend on λ.

(2) Let $G = (a, b)$ and $2r < b - a$. We have

$$h_G(\lambda, b - r) = h_r(\lambda, b - r) + g_r(\lambda, b - r)h_G(\lambda, b - 2r)$$
$$= C(b - r) + \big(1 - C(b - r)\big)h_G(\lambda, b - 2r) + \varepsilon_r,$$

where $\varepsilon_r \to 0$ $(r \to 0)$. From here

$$h_G(\lambda, b - 2r) = 1 - \big(h_G(\lambda, b - 2r) - h_G(\lambda, b - r) + \varepsilon_r\big)\big(C(b - r)\big)^{-1},$$

and consequently $h_G(\lambda, b-0)=1$. In addition $g_G(\lambda, a+0)=1$. Hence, $h_G(\lambda, a+0) = g_G(\lambda, b - 0) = 0$. From here

$$h_G(\lambda, x - r) = h_{(a,x)}(\lambda, x - r)h_G(\lambda, x) \longrightarrow h_G(\lambda, x),$$
$$h_G(\lambda, x + r) = h_{(x,b)}(\lambda, x + r) + g_{(x,b)}(\lambda, x + r)h_G(\lambda, x)$$

$h_G(\lambda, x + r) \to h_G(\lambda, x)$. The same is true for $g_G(\lambda, x)$, i.e. $h_G(\lambda, x)$ and $g_G(\lambda, x)$ are continuous on $x \in G$. Furthermore, from the positiveness of C_1, C_2 it follows that $(\exists r_0 > 0)$ $(\exists \varepsilon > 0)$ $(\forall x \in G)$ $(\forall r < r_0)$ $h_r(\lambda, x) \geq \varepsilon$. Then $h_G(\lambda, x) \geq \prod_{k=0}^{n} h_r(\lambda, x+kr) > 0$, where $n = (b-x)/r$, an integer, and $r \leq x-a$, i.e. $h_G(\lambda, x)$ (and also $g_G(\lambda, x)$) is positive on G. From here $g_G(\lambda, x)$, $h_G(\lambda, x)$ are less than 1 for $x \in G$, and since $h_G(\lambda, x_1) = h_{(a,x_2)}(\lambda, x_1)h_G(\lambda, x_2)$, $h_G(\lambda, x_1) < h_G(\lambda, x_2)$ $(x_1 < x_2)$. Similarly, $g_G(\lambda, x_1) > g_G(\lambda, x_2)$.

(3) Let $x_0 \in G$ and $h_G(\lambda, x_0) + g_G(\lambda, x_0) = 1$. Then for $x \in G$ and $x < x_0$ we have

$$h_G(\lambda, x_0) = h_{(x,b)}(\lambda, x_0) + g_{(x,b)}(\lambda, x_0)h_G(\lambda, x),$$
$$g_G(\lambda, x_0) = g_{(x,b)}(\lambda, x_0)g_G(\lambda, x),$$

hence

$$h_G(\lambda, x) + g_G(\lambda, x) = \big(1 - h_{(x,b)}(\lambda, x_0)\big)/g_{(x,b)}(\lambda, x_0) \geq 1.$$

The same for $x > x_0$. $\qquad\square$

4. Corollaries from conditions on A and C factors

LEMMA 5.2. *Let* $(\forall \lambda \geq 0)$ $(\forall x \in G_0)$

$$h_r(\lambda, x) = C(x) + A_2(\lambda, x)r + o(r),$$
$$g_r(\lambda, x) = 1 - C(x) + A_1(\lambda, x)r + o(r)$$

uniformly on x on each finite interval, coefficients of expansion are continuous on x, $C(x)$ and $1 - C(x)$ are positive. Then $(\forall x \in G_0)$

(1) $C(x) = 1/2$;

(2) $A_2(\lambda, x) = -A_1(\lambda, x)$, *and* $A_2(\lambda, x)$, $A_1(\lambda, x)$ *do not depend on λ.*

Proof. For $x \in G \subset G$ we have

$$g_G(\lambda, x) = C(x)g_G(\lambda, x + r) + \big(1 - C(x)\big)g_G(\lambda, x - r)$$
$$+ A_2(\lambda, x)rg_G(\lambda, x + r) + A_1(\lambda, x)rg_G(\lambda, x - r) + o(r),$$

hence

$$\big(g_G(\lambda, x) - g_G(\lambda, x - r)\big)\big(1 - C(x)\big) - \big(g_G(\lambda, x + r) - g_G(\lambda, x)\big)C(x)$$
$$= A_2(\lambda, x)rg_G(\lambda, x + r) + A_1(\lambda, x)rg_G(\lambda, x - r) + o(r).$$

Similarly we have

$$\big(h_G(\lambda, x) - h_G(\lambda, x - r)\big)\big(1 - C(x)\big) - \big(h_G(\lambda, x + r) - h_G(\lambda, x)\big)C(x)$$
$$= A_2(\lambda, x)rh_G(\lambda, x + r) + A_1(\lambda, x)rh_G(\lambda, x - r) + o(r).$$

It is known that derivative of a non-decreasing (non-increasing) function exists almost everywhere on the interval (see, e.g., [NAT 74, p. 199]). From here it follows that derivative of the strictly increasing (strictly decreasing) function is negative (positive) almost everywhere. Let us divide both parts of the previous expressions by r and direct it to zero. Let $G' \subset G$ be the set on which there are negative derivatives $g'_G(\lambda, x)$, and positive derivative $h'_G(\lambda, x)$. For $x \in G'$ we have

$$\big(1 - 2C(x)\big)g'_G(\lambda, x) = \big(A_2(\lambda, x) + A_1(\lambda, x)\big)g_G(\lambda, x),$$
$$\big(1 - 2C(x)\big)h'_G(\lambda, x) = \big(A_2(\lambda, x) + A_1(\lambda, x)\big)h_G(\lambda, x).$$

Since $g_G(\lambda, x)$ and $h_G(\lambda, x)$ are positive, the supposition that at least one of the coefficients, $1 - 2C(x)$ or $A_2(\lambda, x) + A_1(\lambda, x)$, differs from zero brings a contradiction. From the continuity of functions C and A_i it follows $1 - 2C(x) = 0$ and $A_2(\lambda, x) + A_1(\lambda, x) = 0$ for all $x \in G$, and since both functions $A_2(\lambda, x)$ and $A_1(\lambda, x)$ do not increase on λ, they do not depend on λ. $\qquad\square$

Under conditions of lemma 5.2 let us designate

$$\frac{1}{2}A(x) \equiv A_2(\lambda, x) = -A_1(\lambda, x).$$

5. Corollaries from conditions on B factors

LEMMA 5.3. *Let* $(\forall \lambda \geq 0)$

$$g_r(\lambda, x) = 1/2 - A(x)r/2 + B_1(\lambda, x)r^2/2 + o(r^2),$$
$$h_r(\lambda, x) = 1/2 + A(x)r/2 + B_2(\lambda, x)r^2/2 + o(r^2),$$

[5.5]

and $(\exists G_1 \in \mathfrak{A}_0)$ $(\exists x \in G_1)$ $g_{G_1}(\lambda, x)$ *and* $h_{G_1}(\lambda, x)$ *are twice differentiable at point* x. *Then:*

(1) $(\forall G \in \mathfrak{A}_0, \ G_1 \subset G)$ $g_G(\lambda, x)$, $h_G(\lambda, x)$ *are twice differentiable in a point* x *and their derivatives satisfy the equations:*

$$g_G''(\lambda, x) + 2A(x)g_G'(\lambda, x) + \big(B_1(\lambda, x) + B_2(\lambda, x)\big)g_G(\lambda, x) = 0,$$
$$h_G''(\lambda, x) + 2A(x)h_G'(\lambda, x) + \big(B_1(\lambda, x) + B_2(\lambda, x)\big)h_G(\lambda, x) = 0;$$

(2) $(\forall \lambda > 0)$ $B_1(\lambda, x) + B_2(\lambda, x) < 0$; $-\partial B_i(\lambda, x)/\partial \lambda$ *is a completely monotone function of* λ *($i = 1, 2$); if* $g_{G_1}(0, x) + h_{G_1}(0, x) = 1$ *then* $B_1(0, x) + B_2(0, x) = 0$.

Proof. (1) We have

$$g_G(\lambda, x) = \frac{1}{2}\big(g_G(\lambda, x - r) + g_G(\lambda, x + r)\big)\frac{1}{2}A(x)r\big(g_G(\lambda, x + r) - g_G(\lambda, x - r)\big)$$
$$+ \frac{1}{2}B_1(\lambda, x)r^2 g_G(\lambda, x - r) + \frac{1}{2}B_2(\lambda, x)r^2 g_G(\lambda, x + r) + o(r^2),$$

from which the first equation for $G = G_1$ follows. Similarly the second equation can be derived. From equations [5.1] and [5.2] it follows that these formulae are true for any interval including G_1.

(2) We have $(\forall \lambda > 0)$

$$g_{G_1}(\lambda, x) + h_{G_1}(\lambda, x) = 1 + \big(B_1(\lambda, x) + B_2(\lambda, x)\big)\frac{r^2}{2} + o(r^2) < 1.$$

Hence, $B_1(\lambda, x) + B_2(\lambda, x) < 0$. Furthermore, according to the definition, $g_r(\lambda, x)$ and $h_r(\lambda, x)$ are completely monotone functions of λ; and since $B_1(\lambda, x)$ and $B_2(\lambda, x)$

are limits of the sequences $(2(g_r(\lambda, x) - 1/2 + A(x)r/2)/r^2)$ and $(2(g_r(\lambda, x) - 1/2 - A(x)r/2)/r^2)$ $(r > 0)$ for $r \to 0$ correspondingly, the property of the derivative of a completely monotone function of $\lambda > 0$ is fulfilled (except perhaps the non-negativeness of the function itself), i.e. $(\forall \lambda > 0)$ $(\forall k \in \mathbb{N})$ $(-1)^k \partial^k B_i(\lambda, x)/\partial \lambda^k \geq 0$.

If $g_{G_1}(0, x) + h_{G_1}(0, x) = 1$, then, evidently, for $(x - r, x + r) \subset G$ $g_r(0, x) + h_r(0, x) = 1$, hence $B_1(0, x) + B_2(0, x) = 0$. □

5.1.2. Construction of semi-Markov process with given factors of the differential equations

We have shown that for some class of transition generating functions of a one-dimensional SM process, when a parameter of such a function is a neighborhood of the initial point and the diameter of this neighborhood (length of an interval) tends to zero, it satisfies a second-order differential equation depending on two coefficients. In the present section the problem of construction of a semi-Markov process by a second-order differential equation with the given coefficients is decided [HAR 80b]. The properties of transition generating functions of the constructed process make it possible to specify factors of the asymptotic expansion and to omit the conditions on differentiability in lemma 5.3.

6. Consistency of solutions

Let us consider the differential equation

$$f'' + 2A(x)f' + 2B(\lambda, x)f = 0, \tag{5.6}$$

where A, B are measurable functions. Moreover there is a twice differentiable solution of this equation on an interval $G_0 \in \mathfrak{A}_0$, and $(\forall \lambda > 0)$ $(\forall x \in G_0)$ $B(\lambda, x) < 0$.

THEOREM 5.1. *Let $(\forall G \in \mathfrak{A}_0 \cap G_0)$ $\overline{g}_G$ and $\overline{h}_G$ be solutions of differential equation [5.6] on an interval G where for $G = (a, b)$ the solutions accept values $\overline{g}_G(\lambda, a) = \overline{h}_G(\lambda, b) = 1$ and $\overline{g}_G(\lambda, b) = \overline{h}_G(\lambda, a) = 0$. Then the family of such solutions $(\overline{g}_G, \overline{h}_G)_{G \in \mathfrak{A}_0 \cap G_0}$ satisfies a system of difference equations: $(\forall \lambda > 0)$ $(\forall G_1, G_2 \in \mathfrak{A}_0 \cap G_0, G_1 \subset G_2)$ $(\forall x \in G_1)$*

$$\overline{g}_{G_2}(\lambda, x) = \overline{g}_{G_1}(\lambda, x)\overline{g}_{G_2}(\lambda, c) + \overline{h}_{G_1}(\lambda, x)\overline{g}_{G_2}(\lambda, d), \tag{5.7}$$

$$\overline{h}_{G_2}(\lambda, x) = \overline{g}_{G_1}(\lambda, x)\overline{h}_{G_2}(\lambda, c) + \overline{h}_{G_1}(\lambda, x)\overline{h}_{G_2}(\lambda, d), \tag{5.8}$$

where $G_1 = (c, d)$.

Proof. We designate by $\widehat{g}_{G_2}(\lambda, x)$ and $\widehat{h}_{G_2}(\lambda, x)$ the right parts of the equations [5.7] and [5.8]. On interval $G_1 = (c, d) \subset G_2$ $\widehat{g}_{G_2}, \widehat{h}_{G_2}$ satisfy equation [5.6] as linear combinations of solutions $\overline{g}_{G_1}$ and $\overline{h}_{G_1}$ of this equation on G_1. On the other hand, $\overline{g}_{G_2}(\lambda, c) = \widehat{g}_{G_2}(\lambda, c)$, $\overline{g}_{G_2}(\lambda, d) = \widehat{g}_{G_2}(\lambda, d)$, and under the uniqueness theorem, which is ensured by the negative value of factor B [SAN 54], we have $\overline{g}_{G_2}(\lambda, x) = \widehat{g}_{G_2}(\lambda, x)$ for all $x \in G_1$. The same is true for the second pair of functions: $\overline{h}_{G_2}(\lambda, x) = \widehat{h}_{G_2}(\lambda, x)$ on G_1. □

7. Minimum principle

Let f be the solution of the equation

$$f'' + 2A(x)f' + 2B(\lambda, x)f = F(x)$$

with non-negative boundary conditions where $(\forall x \in G)$ $F(x) \leq 0$ and $B(\lambda, x) < 0$, and let $x_0 \in G$ be a point of its negative minimum. Then at this point we have $f'(x_0) = 0$ and $f''(x_0) = -2B(\lambda, x_0)f(x_0) + F(x_0) < 0$. It contradicts a necessary condition of a smooth function f to have a local minimum at the point x_0, which is: $f''(x_0) \geq 0$. Hence, $f \geq 0$. This conclusion is an example of the so called minimum principle (see, e.g., [GIL 89]). A dual maximum principle is fair as well. In this case, $F(x) \geq 0$ and boundary values of the solution are non-positive. Thus $f \leq 0$.

From the minimum principle it follows that solutions $\overline{g}_G$ and $\overline{h}_G$ of equation [5.6] are non-negative and do not exceed 1. On the other hand, from equations [5.1], [5.2] it follows that the first of them does not increase, and the second one does not decrease on interval $G = (a, b)$. From the negative value of coefficient B a rather more strong assertion follows: these solutions are positive inside the interval where they are determined and are strictly monotone (they decrease and increase correspondingly).

8. Asymptotics of solution with a small parameter

THEOREM 5.2. *Let $\overline{g}_G(\lambda, \cdot)$, $\overline{h}_G(\lambda, \cdot)$ be solutions of the equation [5.6] with boundary conditions specified in theorem 5.1, where $G = (a, b)$ and $a < x_0 < b$. Let A be differentiable, and A', $B(\lambda, \cdot)$ be continuous on x in some neighborhood of a point x_0. Then with $a \uparrow x_0$, $b \downarrow x_0$ the expansions are fair*

$$\overline{g}_G(\lambda, x_0) = \frac{b - x_0}{b - a} - A(x_0)\frac{(b - x_0)(x_0 - a)}{b - a}$$

$$+ \frac{1}{3}B(\lambda, x_0)\frac{(b - x_0)(x_0 - a)(2b - a - x_0)}{b - a} \quad\quad [5.9]$$

$$- \frac{1}{3}(A^2(x_0) + A'(x_0))\frac{(b - x_0)(x_0 - a)(a + b - 2x_0)}{b - a}$$

$$+ (b - x_0)(x_0 - a) \cdot O(b - a),$$

$$\overline{h}_G(\lambda, x_0) = \frac{x_0 - a}{b - a} + A(x_0)\frac{(b - x_0)(x_0 - a)}{b - a}$$

$$+ \frac{1}{3}B(\lambda, x_0)\frac{(b - x_0)(x_0 - a)(b - 2a + x_0)}{b - a}$$

$$+ \frac{1}{3}\left(A^2(x_0) + A'(x_0)\right)\frac{(b - x_0)(x_0 - a)(a + b - 2x_0)}{b - a} \qquad [5.10]$$

$$+ (b - x_0)(x_0 - a) \cdot O(b - a),$$

Proof. Let $G = (a, b)$. It is known (see Sobolev [SOB 54, p. 296]) that a solution of equation $y'' = g(x)$ on an interval (a, b) under conditions $y(a) = y_1$, $y(b) = y_2$ has the form

$$y(x) = y_1\frac{b - x}{b - a} + y_2\frac{x - a}{b - a} - \frac{b - x}{b - a}\int_a^x g(t)(t - a)\, dt$$

$$- \frac{x - a}{b - a}\int_x^b g(t)(b - t)\, dt. \qquad [5.11]$$

Let us represent equation [5.6] as $f'' = q(x)$, where $q = -2Af' - 2Bf$. Substituting q in the previous expression, we obtain for equation [5.6] a formal solution, which reduces to an integral equation with respect to f:

$$f(x) = f(a)\frac{b - x}{b - a} + f(b)\frac{x - a}{b - a} + \int_a^b K_G(\lambda, x, t)f(t)\, dt, \qquad [5.12]$$

where the kernel $K_G(\lambda, x, t)$ has a view

$$K_G(\lambda, x, t) = \begin{cases} \dfrac{b - x}{b - a}\left(-2A(t) + \left(2B(\lambda, t) - 2A'(t)\right)(t - a)\right), & a < t < x, \\[2mm] \dfrac{x - a}{b - a}\left(2A(t) + \left(2B(\lambda, t) - 2A'(t)\right)(b - t)\right), & x < t < b. \end{cases}$$

From here

$$\overline{h}_G(\lambda, x) = \frac{x - a}{b - a} + \int_a^b K_G(\lambda, x, t)\overline{h}_G(\lambda, t)\, dt.$$

Furthermore

$$K_G^{(2)}(x, t) \equiv \int_a^b K_G(x, s)K_G(s, t)\, ds$$

$$= \begin{cases} -\dfrac{(b - x)(2t - a - x)}{b - a}2A^2(x) + (b - x)(x - a)Y_1(x, t), & a < t < x, \\[2mm] \dfrac{(x - a)(2t - b - x)}{b - a}2A^2(x) + (b - x)(x - a)Y_2(x, t), & x < t < b \end{cases}$$

functions Y_1 and Y_2 are limited in $G_0 \times G_0$. We also have

$$K_G^{(3)}(x,t) \equiv \int_a^b K_G^{(2)}(x,s)K_G(s,t)\,ds = (b-x)(x-a)Z(x,t),$$

where function Z is limited in $G_0 \times G_0$. The integral equation for $\overline{h}_G(\cdot)$ can be written as

$$\overline{h}_G(x) = \frac{x-a}{b-a} + \sum_{i=1}^3 \int_a^b K_G^{(i)}(x,t)\frac{t-a}{b-a}\,dt + \int_a^b K_G^{(4)}(x,t)\overline{h}_G(t)\,dt.$$

From here we obtain

$$\overline{h}_G(x) = \frac{x-a}{b-a} + \int_a^x \left(\frac{b-x}{b-a}\big(-2A(x) + (2B(\lambda,x) - 2A'(x))(t-a)\big)\right.$$
$$\left. - \frac{(b-x)(2t-a-x)}{b-a}2A^2(x)\right)\frac{t-a}{b-a}\,dt$$
$$+ \int_x^b \left(\frac{x-a}{b-a}\big(2A(x) + (2B(\lambda,x) - 2A'(x))(b-t)\big)\right.$$
$$\left. + \frac{(x-a)(2t-b-x)}{b-a}2A^2(x)\right)\frac{t-a}{b-a}\,dt + (b-x)(x-a)O(b-a)$$
$$= \frac{x-a}{b-a} + A(x)\frac{(x-a)(b-x)}{b-a} + \frac{1}{3}B(\lambda,x)\frac{(x-a)(b-x)(b-2a+x)}{b-a}$$
$$+ \frac{1}{3}\big(A^2(x) + A'(x)\big)\frac{(x-a)(b-x)(b+a-2x)}{b-a} + (b-x)(x-a)O(b-a)$$

uniformly on $x \in G$. The asymptotic expansion for function $\overline{h}_G$ is proved. Similarly the expansion for $\overline{g}_G$ can be obtained, but it is easier to deduce it from the proved expansion by a replacement of variable $\overline{x} = -x$. Let us designate $\overline{f}(\overline{x}) = f(-\overline{x}) = f(x)$. If on G the initial function f satisfies the original equation $f'' + 2A(x)f' + 2B(\lambda,x)f = 0$ with boundary conditions $f(a) = 1$, $f(b) = 0$ (i.e. $f = \overline{g}_G$ in our denotation), then $\overline{f}$ satisfies the equation

$$\overline{f}'' - 4\overline{A}(\overline{x})\overline{f}' + 2\overline{B}(\lambda,\overline{x})\overline{f} = 0$$

with boundary conditions $\overline{f}(-a) = 1$ and $\overline{f}(-b) = 0$, i.e. it plays the role of $\overline{h}_{(-G)}$ for the new equation. Hence, in order to derive the expansion of function $\overline{g}_G(x)$ it is sufficient to replace $\{a, b, x, A(x)\}$ on $\{-b, -a, -x, -A(-x)\}$ in obtained expansion for $\overline{h}_G$ (we need not change a sign before $A'(-x)$, since the sign of this factor varies twice; the second time it happens after differentiation by the new variable $\overline{x}$). In an outcome we obtain the required expression. $\qquad\square$

9. Note on the theorem

In the previous theorem the existence of limits of ratios $(b - x_0)/(b - a)$ and $(x_0 - a)/(b - a)$ is not supposed. If these limits exist, for example, $x_0 - a = cr$, and $b - x_0 = dr$ $(r \to 0)$, the obtained formulae reduce to expansions on degrees $b - a$ or r.

COROLLARY 5.1. *Let $c = d = 1$ (sequence of symmetric neighborhoods). Then under conditions of theorem 5.2 the expansions are fair*

$$\overline{g}_{(x_0-r,x_0+r)}(\lambda, x_0) = \frac{1}{2}\left(1 - A(x_0)r + B(\lambda, x_0)r^2 + o(r^2)\right) \qquad [5.13]$$

$$\overline{h}_{(x_0-r,x_0+r)}(\lambda, x_0) = \frac{1}{2}\left(1 + A(x_0)r + B(\lambda, x_0)r^2 + o(r^2)\right). \qquad [5.14]$$

Proof. It immediately follows from the previous theorem. □

10. Pseudo-maximum of a function

A point x from the domain (region of determination) of a function f is referred to as a point of pseudo-maximum of this function with a step $r > 0$, if two conditions are fulfilled:

(1) $f(x - r) = f(x + r)$;

(2) $f(x + r) - 2f(x) + f(x - r) \le 0$, where points $x - r$, $x + r$ also belong to the domain.

LEMMA 5.4. *Let a continuous function f have a point of a maximum x_M inside some interval from its domain. Then there are sequences $(r_n)_1^\infty$ $(r_n > 0, r_n \to 0)$ and $(x_n)_1^\infty$ $(x_n \to x_M)$ such that for any $n \ge 1$ the function f has in a point x_n a pseudo-maximum with a step r_n.*

Proof. If point x_M belongs to a closure of some interval of constancy of the function f then, obviously, the lemma is proved. Let the point x_M not belong to a closure of any interval of constancy of the function f. Then for big enough n the following points also belong to the same interval as x_M:

$$a_n = \sup\left\{x < x_M : f(x) \le f(x_M) - 1/n\right\},$$

$$b_n = \inf\left\{x > x_M : f(x) \le f(x_M) - 1/n\right\}.$$

Let us assume $r_n = (b_n - a_n)/2$ and $x_n = (b_n + a_n)/2$. Then we have $f(x_n) > f(x_M) - 1/n$, and also $f(x_n - r_n) = f(a_n) = f(x_M) - 1/n$ and $f(x_n + r_n) = f(b_n) = f(x_M) - 1/n$. Due to the continuity of f the points a_n, b_n tend to x_M as $n \to \infty$. □

11. Comparison with solutions of the equation

LEMMA 5.5. *Let $\lambda > 0$ and expansion [5.5] of lemma 5.3 is fulfilled uniformly on $x \in G$, where A, A', $B_1(\lambda, \cdot)$, $B_2(\lambda, \cdot)$ are continuous on x, and in addition, the last two functions are negative. Then transition generating functions g_G and h_G of the corresponding semi-Markov process are twice differentiable on x and coincide on the interval G with solutions $\overline{g}_G$ and $\overline{h}_G$ of differential equation [5.6] with $B_1(\lambda, \cdot) = B_2(\lambda, \cdot) = B(\lambda, \cdot)$.*

Proof. We use the expansion of the solutions of differential equation [5.6] obtained in corollary 5.13 for central points of an interval. Consider a difference $z(x) = h_G(x) - \overline{h}_G(x)$. It is a continuous function, accepting zero values on the ends of an interval G. Moreover, the following representation is fair (for brevity, we omit arguments λ and x of the coefficients):

$$z(x) = z(x - r)\left(1/2 - Ar/2 + B_1 r^2/2\right) + z(x + r)\left(1/2 + Ar/2 + B_2 r^2/2\right)$$
$$+ \overline{h}_G(x - r)(B_1 - B) + \overline{h}_G(x + r)(B_2 - B) + o(r^2),$$

or, by taking into account the differentiability of $\overline{h}_G$,

$$-z(x) + (1/2)\left(z(x - r) + z(x + r)\right)$$
$$+ Ar\left(z(x + r) - z(x - r)\right) + z(x - r)B_1 r^2 + z(x + r)B_2 r^2\right)$$
$$+ o(r^2) = 0.$$

Let us assume that this function has a positive maximum inside interval G. Then according to lemma 5.4 for this function there exists a sequence (r_n), converging to zero, and a sequence of points of pseudo-maxima (x_n) converging on the point of maximum with the appropriate steps such that $z(x_n - r_n) = z(x_n + r_n)$ and $-z(x_n) + (1/2)(z(x_n - r_n) + z(x_n + r_n)) \leq 0$. Substituting these sequences in the previous expression, dividing by r_n and letting n tend to infinity, we obtain a contradiction, because the factors B_1, B_1 are negative. Applying the lemma about pseudo-maximum to function $-z(x)$, we prove that the original function has no points of a negative minimum inside an interval as well. Hence, $h_G = \overline{h}_G$ on the given interval. It is obvious that the same is fair for the second pair of functions: $g_G = \overline{g}_G$. From lemma 5.3 by uniqueness of asymptotic expansions of a solution of the differential equation (corollary 5.13) we find that condition $B_1 = B_2 = B$ is necessary.   □

12. Derivation of differential equation from local properties

THEOREM 5.3. *Let (g_G, h_G) $G \in \mathfrak{A}_0$ be a family of semi-Markov transition generating functions of a continuous SM process. Let expansions [5.3], [5.4] from item 2 be* *fulfilled uniformly on x in every finite interval $G \subset G_0$, moreover $(\forall \lambda > 0)$ $C_i(\lambda, \cdot)$, $A_i(\lambda, \cdot)$, $\partial A_i(\lambda, \cdot)/\partial x$, $B_i(\lambda, \cdot)$ be continuous on x and $C_i(\lambda, \cdot) > 0$ $(i = 1, 2)$. Then*

(1) $C_1(\lambda, \cdot) = C_2(\lambda, \cdot) = 1/2$;

(2) $A_1(\lambda, \cdot) = -A_2(\lambda, \cdot) \equiv (1/2)A(\cdot)$ do not depend on λ;

(3) $B_1(\lambda, \cdot) = B_2(\lambda, \cdot) \equiv B(\lambda, \cdot) \leq 0$;

(4) $(\forall \lambda > 0)$ $B(\lambda, \cdot) < 0$, $(\forall x \in G)$ the function $-(\partial/\partial \lambda)B(\lambda, x)$ is completely monotone on λ;

(5) g_G and h_G are twice differentiable and satisfy differential equation [5.6].

Proof. It follows from lemmas 5.1, 5.2, 5.3 and 5.5. □

DEFINITION 5.1. *A semi-Markov process with transition generating functions satisfying equation [5.6] with some coefficients A and B, or, which amounts to the same thing, with transition generating functions admitting expansions [5.3] and [5.4] with some coefficients A_i, B_i, C_i, is said to be a semi-Markov process of diffusion type.*

13. Probabilistic solution

Let us denote

$$n_G(\lambda) = -\sup \{ B(\lambda, x) : x \in G \}.$$

THEOREM 5.4. *Let $\overline{g}_G(\lambda, x)$ and $\overline{h}_G(\lambda, x)$ be solutions of equation [5.6] with boundary conditions specified in theorem 5.1. Let function A be limited on G, $(\forall x \in G)$ function $-\partial B(\lambda, x)/\partial \lambda$ be completely monotone on λ, $(\forall \lambda > 0)$ $B(\lambda, x) < 0$ and $n_G(\lambda) \to \infty$ as $\lambda \to \infty$. Then $(\forall x \in G)$ the functions $\overline{g}_G(\lambda, x)$ and $\overline{h}_G(\lambda, x)$ are Laplace transformed sub-probability measures on $[0, \infty)$, not containing an atom in zero*

Proof. From the minimum principle (see item 7) it follows that $\overline{g}_G(\lambda, x)$ and $\overline{h}_G(\lambda, x)$ are not negative. Let

$$f(x) = 1 - \overline{g}_G(\lambda, x) - \overline{h}_G(\lambda, x).$$

Then $f(a) = f(b) = 0$ and

$$f''(x) + 2A(x)f'(x) + 2B(\lambda, x)f(x) = 2B(\lambda, x).$$

Hence, by minimum principle $f(x) \geq 0$, i.e.

$$\overline{g}_G(\lambda, x) + \overline{h}_G(\lambda, x) \leq 1.$$

Let us prove the complete monotonicity of functions $\bar{g}_G(\lambda, x)$ and $\bar{h}_G(\lambda, x)$ on λ. In this case, according to the Bernstein theorem [FEL 67, p. 505], they are the Laplace images of some sub-probability measures on $[0, \infty)$. A proof of complete monotonicity of a solution of equation [5.6] would be significantly simpler if we assume in advance that it is infinite-differentiable on parameter λ. We do not know it and therefore we will use the following criterion of complete monotonicity.

Denote $\Delta^h f(s) = f(s+h) - f(s)$ and for any $n \geq 1$

$$\Delta^{h_1,\ldots,h_{n+1}} f(s) = \Delta^{h_{n+1}} \left(\Delta^{h_1,\ldots,h_n} f \right)(s).$$

LEMMA 5.6. *A function f is completely monotone on interval $(0, b)$ if and only if* $(\forall n \in N) \, (\forall s > 0) \, (\forall h_i > 0) \, (\forall s > 0, \, s < b - \sum_{i=1}^{n} h_i)$

$$(-1)^n \Delta^{h_1,\ldots,h_n} f(s) \geq 0.$$

Proof. Representing a finite difference as an integral of a derivative, we obtain proof of necessity of the proposed condition. Representing a derivative as a limit of relative increments of the function, and noting that any jump implies non-monotonicity of the finite difference of the second order, we obtain proof of sufficiency of the proposed condition. □

Let us continue to prove the theorem. Denote for brevity $\Delta^{h_1,\ldots,h_n} = \Delta_n$. Condition $f \geq 0$ means fulfillment of the condition of lemma 5.6 for $n = 0$, if to define $\Delta_0 f = f$. Let already it be proved that for all $k \in \{0, 1, \ldots, n\} \, (-1)^k \Delta_k f \geq 0$. We have

$$\Delta_{n+1} \left(f'' + 2A(x) f' + 2B f \right) = 0$$

$(B = B(\lambda, x))$. Evidently, the operator of differentiation on x can be transposed with the operator of finite difference on λ. Hence, it is true that

$$\left(\Delta_{n+1} f \right)'' + 2A(x) \left(\Delta_{n+1} f \right)' + 2\Delta_{n+1}(Bf) = 0.$$

Consider the n-th difference of the product of two functions: $\Delta_n(Bf)$. Let θ_h be the shift operator on class of functions, determined on $\mathbb{R}_+$ (see item 2.16). In this case we can write $\Delta^h f = \theta_h f - f$. Denote $H_n = (h_1, \ldots, h_n) \, (n \in \mathbb{N}, \, h_i > 0)$ (elements of this sequence can be repeated); let C_k be a combination of k elements from the sequence H_n (i.e. sub-sequence); $|C_k|$ be the sum of all elements composing the combination C_k; let the set $H_n \setminus C_k$ mean the residual sub-sequence (if it is not empty); $\Delta^\emptyset$ and θ_0 mean operators of identity.

LEMMA 5.7. *For the n-th difference of a product of two functions the following formula is true*

$$\Delta^{H_n}(B\,f) = \sum_{k=0}^{n} \sum_{C_k} \left(\Delta^{C_k} B\right)\left(\theta_{|C_k|}\Delta^{H_n \backslash C_k} f\right),$$

where the intrinsic sum is on all combinations of C_k.

Proof. It is passed for all variants by induction on the basis of induction for $n = 1$: $\Delta^h(Bf) = B\Delta^h f + \Delta^h B\theta_h f$. While proving we take into account that operator Δ^H does not vary with permutation of members of sequence H. We also take into account permutability of operators Δ and θ and properties of these operators: $\Delta^{S_1}\Delta^{S_2} = \Delta^{(S_1,S_2)}$ where (S_1, S_2) is a sequence composed with two sub-sequences S_1 and S_2, and $\theta_{h_1}\theta_{h_2} = \theta_{h_1+h_2}$ $(h_i > 0)$. Let us prove the formula in its simple variant when all h_i are equal to h. In this case $\Delta^{C_k} = \Delta_k$, $\Delta^{H_n \backslash C_k} = \Delta_{n-k}$, $\theta_{|C_k|} = \theta_{kh}$ and the formula we have to prove turns to a simple form

$$\Delta_n(B\,f) = \sum_{k=0}^{n} \binom{n}{k}\left(\Delta_k B\right)\left(\theta_{kh}\Delta_{n-k}f\right).$$

Let us apply operator Δ_1 to this formula and use the formula of the first difference of a product:

$$\Delta_1(Bf) = B\left(\Delta_1 f\right) + \left(\Delta_1 B\right)\left(\theta_h f\right).$$

We obtain

$$\Delta_{n+1}(Bf) = \sum_{k=0}^{n} \binom{n}{k}\left(\Delta_k B\right)\left(\theta_{kh}\Delta_{n+1-k}f\right)$$

$$+ \sum_{k=0}^{n} \binom{n}{k}\left(\Delta_{k+1} B\right)\left(\theta_{(k+1)h}\Delta_{n-k}f\right)$$

$$= \sum_{k=0}^{n} \binom{n}{k}\left(\Delta_k B\right)\left(\theta_{kh}\Delta_{n+1-k}f\right)$$

$$+ \sum_{k=1}^{n} \binom{n}{k-1}\left(\Delta_k B\right)\left(\theta_{kh}\Delta_{n+1-k}f\right)$$

$$= \sum_{k=0}^{n+1} \binom{n+1}{k}\left(\Delta_k B\right)\left(\theta_{kh}\Delta_{n+1-k}f\right).$$

So, the formula is proved by induction. $\qquad\square$

Let us continue to prove complete monotonicity of function f. Applying the formula of lemma 5.7 we can write the obtained equation as

$$\left(\Delta_{n+1}f\right)'' + 2A(x)\left(\Delta_{n+1}f\right)' + 2B(\lambda, x)\Delta_{n+1}f$$

$$= -\sum_{k=0}^{n}\sum_{C_k}\left(\Delta^{C_k}B\right)\left(\theta_{|C_k|}\Delta^{H_n \setminus C_k}f\right),$$

Since function $-\partial B/\partial\lambda$ is completely monotone, and under supposition $(\forall k \leq n)$ $(-1)^k\Delta_k f \geq 0$, for all even n the summands on the right are positive (because in each summand superscripts of two of its factors have different evenness). From here the right part is not negative. Moreover, evidently, function $\Delta_{n+1}f$ $(n \geq 0)$ has zero meanings at the ends of the interval G. Hence according to the maximum principle $\Delta_{n+1}f \leq 0$. For odd n the right part of the equation is not positive and, hence, $\Delta_{n+1}f \geq 0$. From here it follows that $(\forall n \in N)$ $(-1)^n\Delta_n f \geq 0$, i.e., according to lemma 5.6, f is a completely monotone function on $\lambda \in (0, \infty)$. Hence, $\overline{g}_G(\lambda, x)$ and $\overline{h}_G(\lambda, x)$ are completely monotone functions on $\lambda \in (0, \infty)$, and there exist measures $\overline{F}_G(dt \times \{a\} \mid x)$ and $\overline{F}_G(dt \times \{b\} \mid x)$ on $\mathbb{R}_+$ $(G = (a, b))$ which determine the following representations

$$\overline{g}_G(\lambda, x) = \int_0^\infty e^{-\lambda t}\overline{F}_G \mid \left(dt \times \{a\}|x|\right),$$

$$\overline{h}_G(\lambda, x) = \int_0^\infty e^{-\lambda t}\overline{F}_G \mid \left(dt \times \{b\}|x|\right).$$

Now let us prove that measure $\overline{F}_G$ does not contain an atom at zero. In order to prove this property it is sufficient to prove that $\overline{g}_G(\lambda, x)$ and $\overline{h}_G(\lambda, x)$ tend to zero as $\lambda \to \infty$. To do this we compare these solutions with solutions of the similar equation, but with constant coefficients. Let

$$A_3 = \sup\left\{A(x) : x \in G\right\}, \qquad B_0(\lambda) = \sup\left\{B(\lambda, x) : x \in G\right\}.$$

Then equation $f'' + 2A_3f' + 2B_0(\lambda)f = 0$ with boundary conditions $f(a) = 0$, $f(b) = 1$ has a solution

$$\widehat{h}_G(\lambda, x) = e^{A_3(b-x)}\frac{\text{sh}(x - a)\sqrt{A_3^2 - 2B_0(\lambda)}}{\text{sh}(b - a)\sqrt{A_3^2 - 2B_0(\lambda)}}.$$

Let $f(\lambda, x) = \widehat{h}_G(\lambda, x) - \overline{h}_G(\lambda, x)$. Then $f(\lambda, a) = f(\lambda, b) = 0$ and besides

$$f''(\lambda, x) + 2A_3 f'(\lambda, x) + 2B_0(\lambda)f(\lambda, x)$$

$$= 2(A(x) - A_3)\overline{h}'_G(\lambda, x) + 2\left(B(\lambda, x) - B_0(\lambda)\right)\overline{h}_G(\lambda, x).$$

However, from the consistency of solutions of equation [5.6] and non-negative value of $\overline{h}_G$ it follows that $\overline{h}_G(\lambda, x)$ does not increase on G (see item 1). Hence the right part of the previous equation is not positive, and as such $f(\lambda, x) \geq 0$. Yet $(\forall x \in G)$

$$\widehat{h}_G(\lambda, x) \sim e^{2A_3(b-x)} e^{-(b-x)2\sqrt{A_3^2 - B_0(\lambda)/2}} \longrightarrow 0$$

as $\lambda \to \infty$, since $B_0(\lambda) \to -\infty$. Hence, $(\forall x \in G)$ $\overline{h}_G(\lambda, x) \to 0$ $(\lambda \to \infty)$. Similarly it can be proved that $\overline{g}_G(\lambda, x) \to 0$ $(\lambda \to \infty)$. In this case $\overline{g}_G(\lambda, x)$ is compared with a solution of differential equation $f'' + 4A_4 f' + 2B_0(\lambda)f = 0$ with boundary conditions $f(a) = 1$, $f(b) = 0$, where $A_4 = \inf\{A(x) : x \in G\}$.   □

14. Sub-probability kernel connected with solution

Let us determine a sub-probability kernel $y_G(\lambda, dx_1 \mid x)$ connected to a solution of equation [5.6] on interval $G \in \mathfrak{A}_0$ as follows: $(\forall x \in \mathbb{X})$ $(\forall S \in \mathcal{B}(\mathbb{X}))$

$$y_G(\lambda, S \mid x) = \begin{cases} \overline{g}_G(\lambda, x)I_S(a) + \overline{h}_G(\lambda, x)I_S(b), & x \in G \\ I_S(x), & x \notin G, \end{cases}$$

where $G = (a, b)$. By theorem 5.1, the kernels (y_G) $(G \in \mathfrak{A}_0)$ are connected by the condition $(\forall G_1 \subset G_2)$

$$y_{G_2}(\lambda, \varphi) = y_{G_1}\big(\lambda, y_{G_2}(\lambda, \varphi)\big),$$

where, as above,

$$y_G(\lambda, \varphi \mid x) = \int_{\mathbb{X}} \varphi(x_1) y_G(\lambda, dx_1 \mid x) \quad (\varphi \in \mathbb{B}).$$

Let $\delta_n = (G_1, \ldots, G_n)$ $(G_i \in \mathfrak{A}_0)$. Let $y_{\delta_n}(\lambda, \varphi)$ denote an iterated kernel, determined by induction:

$$y_{\delta_n}(\lambda, \varphi) = y_{G_1}\big(\lambda, y_{\delta_n^2}(\lambda, \varphi)\big),$$

where $\delta_n^i = (G_i, \ldots, G_n)$ $(i \leq n)$ and $y_{\delta_n}(\lambda, dx_1 \mid x)$ is a corresponding iterated kernel for the sequence δ_n. We assume $y_{\delta_n}(\lambda, \varphi) = \varphi$ when $n = 0$ $(\delta_0 = \emptyset)$.

15. Behavior of iterated kernels

In order to construct a semi-Markov process with transition generating functions $y_G(\lambda, S \mid x)$ it remains for us to check that the necessary conditions for a cádlág process exist. Namely, the trajectory of the process has to leave any finite time interval with the help of any deducing sequence of open sets (in this case, open intervals). In other words, the following condition must be fulfilled: $(\forall \lambda > 0)\,(\forall x \in \mathbb{X})\,(\forall \delta \in \mathrm{DS})$

$$y_{\delta_n}(\lambda, \mathbb{I} \mid x) \longrightarrow 0 \quad (n \longrightarrow \infty).$$

Using sigma-compactness of space $\mathbb{X}$ (in this case, the real line) it is convenient to divide the proof into two parts. Firstly, we find a sufficiently large compact such that a trajectory exits from it not before a fixed time. Furthermore it is being proved that a step-process constructed on the deducing sequence either exits from the compact or remains in it. In this case the sequence of jump times tends to infinity (see item 2.17).

Let $K \subset \mathbb{X}$ and $\overline{K} = \mathbb{X} \setminus K$. We denote

$$y_G^K(\lambda, \varphi \mid x) = \int_K y_G(\lambda, dx_1 \mid x)\varphi(x_1),$$

$$y_{\delta_n}^K(\lambda, \varphi) = y_{G_1}^K\big(\lambda, y_{\delta_n^2}^K(\lambda, \varphi)\big),$$

$$M_G = \sup\big\{|A(x)| : x \in G\big\}.$$

In order to prove the next theorem we use the existence of a finite epsilon-net for a pre-compact subset of a metric space [KOL 72, p. 100]. Remember that an epsilon-net of the set K with parameter $\varepsilon > 0$ is said to be such a set $T \subset \mathbb{X}$ that $(\forall x \in K)$ $(\exists x_1 \in T)\,\rho(x, x_1) < \varepsilon$; pre-compact is called the set with a compact closure.

THEOREM 5.5. *Let coefficients A and $B(\lambda, \cdot)$ in equation [5.6] be continuous on $x \in \mathbb{X}$, moreover $(\forall G \in \mathfrak{A}_0)\,(\forall \lambda > 0)\,n_G(\lambda) > 0$ and $r^2 n_{(-r,r)} \to \infty$; $M_{(-r,r)}/ (r\,n_{(-r,r)}(\lambda)) \to 0$ as $r \to \infty$. Then $(\forall \lambda > 0)\,(\forall x \in \mathbb{X})\,(\forall \delta \in \mathrm{DS})\,y_{\delta_n}(\lambda, \mathbb{I} \mid x) \to 0\ (n \to \infty)$ where $\delta_n = (G_1, \ldots, G_n)$ is an initial piece of the sequence $\delta = (G_1, G_2, \ldots)$.*

Proof. We divide the proof into four parts.

(1) From the Markov property of the family of kernels (y_G), proved in theorem 5.1, we obtain

$$y_{\delta_n}(\lambda, \mathbb{I} \mid x_0) = \int_{\mathbb{X}^n} \prod_{i=1}^{n} y_{G_i}(\lambda, dx_i \mid x_{i-1}) = y_{\delta_n}^K(\lambda, \mathbb{I} \mid x_0)$$

$$+ \sum_{i=0}^{n-1} \int_K y_{\delta_n^i}^K(\lambda, dx_1 \mid x) \int_{\overline{K}} y_{G_{i+1}}(\lambda, dx_2 \mid x_1) y_{\delta_n^{i+2}}(\lambda, \mathbb{I} \mid x_2).$$

On the other hand,

$$y_K\left(\lambda, \overline{K} \mid x_0\right) \geq \sum_{i=0}^{n-1} \int_K y_{\delta_i}^K\left(\lambda, dx_1 \mid x_0\right) y_{G_{i+1}}\left(\lambda, \overline{K} \mid x_1\right).$$

This inequality for SM processes is a consequence of evident inequality $\sigma_K \leq \sigma_K \circ L_{\delta_n}$. For an arbitrary family of kernels (y_G) it follows from the representation

$$
\begin{aligned}
y_{G_1}\left(\lambda, \overline{K} \mid x_0\right) &= \int_{\mathbb{X}} y_{G_1 \cap K}\left(\lambda, dx_1 \mid x_0\right) y_{G_1}\left(\lambda, \overline{K} \mid x_1\right) \\
&= \int_{\overline{K}} y_{G_1 \cap K}\left(\lambda, dx_1 \mid x_0\right) y_{G_1}\left(\lambda, \overline{K} \mid x_1\right) \\
&\quad + \int_{K \backslash G_1} y_{G_1 \cap K}\left(\lambda, dx_1 \mid x_0\right) y_{G_1}\left(\lambda, \overline{K} \mid x_1\right) \\
&= \int_{\overline{K}} y_{G_1 \cap K}\left(\lambda, dx_1 \mid x_0\right) y_{G_1}\left(\lambda, \overline{K} \mid x_1\right) \\
&\leq y_{G_1 \cap K}\left(\lambda, \overline{K} \mid x_0\right) \leq y_K\left(\lambda, \overline{K} \mid x_0\right) \\
&= y_{G_1 \cap K}\left(\lambda, \overline{K} \mid x_0\right) + \int_K y_{G_1 \cap K}\left(\lambda, dx_1 \mid x_0\right) y_K\left(\lambda, \overline{K} \mid x_1\right).
\end{aligned}
$$

Let the inequality be true for all sequences of length n. Then

$$
\sum_{i=0}^{n} \int_K y_{\delta_i}^K\left(\lambda, dx_1 \mid x_0\right) y_{G_{i+1}}\left(\lambda, \overline{K} \mid x_1\right)
$$

$$
= y_{G_1}\left(\lambda, \overline{K} \mid x_0\right)
$$

$$
+ \int_K y_{G_1}\left(\lambda, dx_1 \mid x_0\right) \sum_{i=1}^{n} \int_K y_{\delta_i^2}^K\left(\lambda, dx_2 \mid x_1\right) y_{G_{i+1}}\left(\lambda, \overline{K} \mid x_2\right)
$$

$$
= \int_{\overline{K}} y_{G_1 \cap K}\left(\lambda, dx_1 \mid x_0\right) \sum_{i=0}^{n} \int_K y_{\delta_i}^K\left(\lambda, dx_2 \mid x_1\right) y_{G_{i+1}}\left(\lambda, \overline{K} \mid x_2\right)
$$

$$
+ \int_{K \backslash G_1} y_{G_1 \cap K}\left(\lambda, dx_1 \mid x_0\right) \sum_{i=1}^{n} \int_K y_{\delta_i^2}^K\left(\lambda, dx_2 \mid x_1\right) y_{G_{i+1}}\left(\lambda, \overline{K} \mid x_2\right).
$$

Obviously, in the first term the sum not more than 1 for any n. In the second term, according to the assumption, the sum is not more than $y_K\left(\lambda, \overline{K} \mid x_1\right)$. From here the latter expression is not more than

$$y_{G_1 \cap K}\left(\lambda, \overline{K} \mid x_1\right) + \int_{K \backslash G_1} y_{G_1 \cap K}\left(\lambda, dx_1 \mid x\right) y_K\left(\lambda, \overline{K} \mid x_1\right)$$

$$\leq y_{G_1 \cap K}\left(\lambda, \overline{K} \mid x_0\right) + \int_K y_{G_1 \cap K}\left(\lambda, dx_1 \mid x\right) y_K\left(\lambda, \overline{K} \mid x_1\right) = y_K\left(\lambda, \overline{K} \mid x_0\right).$$

The inequality is proved. From here it follows that

$$y_{\delta_n}\left(\lambda, \mathbb{I} \mid x\right) \leq y_{\delta_n}^K\left(\lambda, \mathbb{I} \mid x\right) + y_K(\lambda, \overline{K} \mid x).$$

(2) Estimate from above the function $y_G(\lambda, \mathbb{I} \mid x)$. Let $G = (a, b)$. Consider the function

$$y(x) = c_G \left(\frac{a + b - 2x}{b - a}\right)^2 + 1 - c_G,$$

where

$$c_G = n_G(\lambda) \left(\frac{4}{(b-a)^2} + \frac{8M_G}{b-a} + n_G(\lambda)\right)^{-1}.$$

Let $u(x) = y(x) - y_G(\lambda, \mathbb{I} \mid x)$. We have $u(a) = u(b) = 0$ and in addition $u'' + 4A(x)u' + 2B(\lambda, x)u = F(x)$, where

$$F(x) = 2B(\lambda, x)\left(c_G\left(\frac{a + b - 2x}{b - a}\right)^2 + 1 - c_G\right)$$

$$+ \frac{8c_G}{(b-a)^2} + A(x)\frac{16c_G(2x - a - b)}{(b-a)^2}$$

$$\leq -2n_G(\lambda)(1 - c_G) + \frac{8c_G}{(b-a)^2} + \frac{16M_G c_G}{b-a} = 0.$$

Hence, $(\forall x \in G)\ y(x) \geq y_G(\lambda, \mathbb{I} \mid x)$. For x, situated in the middle of the interval, we obtain

$$y_r(\lambda, \mathbb{I} \mid x) \equiv y_{(x-r, x+r)}(\lambda, \mathbb{I} \mid x) \leq 1 - c_{(x-r, x+r)}.$$

From here, in particular, it follows that $(\forall x \in \mathbb{R})\ y_r(\lambda, \mathbb{I} \mid x) \to 0\ (r \to \infty)$, because in this case $r^2 n_{(-r,r)} \to \infty$,

$$M_{(-r,r)}/\left(rn_{(-r,r)}(\lambda)\right) \longrightarrow 0$$

and consequently $c_{(-r,r)} \to 1$.

(3) Prove the formula

$$y_{\delta_n}(\lambda, \mathbb{I} \mid x) \leq \min \left\{ y_{G_i}(\lambda, \mathbb{I} \mid x) : 1 \leq i \leq n \right\}.$$

Evidently, $y_{G_1}(\lambda, y_{G_2}(\lambda, \mathbb{I}) \mid x) \leq y_{G_1}(\lambda, \mathbb{I} \mid x)$. From here $(\forall i : 1 \leq i \leq n)$

$$y_{\delta_n}(\lambda, \mathbb{I} \mid x) \leq y_{\delta_i}(\lambda, \mathbb{I} \mid x) \leq y_{G_1}(\lambda, \mathbb{I} \mid x).$$

If $x \notin G_1$, then $y_{G_1}(\lambda, y_{G_2}(\lambda, \mathbb{I}) \mid x) = y_{G_2}(\lambda, \mathbb{I} \mid x)$. If $x \notin G_2$, then $y_{G_1}(\lambda, y_{G_2}(\lambda, \mathbb{I}) \mid x) \leq 1 = y_{G_2}(\lambda, \mathbb{I} \mid x)$. If $x \in G_1 \cap G_2$, then

$$y_{G_1}\left(\lambda, y_{G_2}(\lambda, \mathbb{I}) \mid x\right)$$

$$= \int_{\mathbb{X} \setminus (G_1 \cap G_2)} y_{(G_1 \cap G_2)}(\lambda, dx_1 \mid x) y_{G_1}\left(\lambda, y_{G_2}(\lambda, \mathbb{I}) \mid x\right)$$

$$\leq \int_{\mathbb{X} \setminus (G_1 \cap G_2)} y_{(G_1 \cap G_2)}(\lambda, dx_1 \mid x) y_{G_2}\left(\lambda, \mathbb{I} \mid x_1\right) = y_{G_2}(\lambda, \mathbb{I} \mid x).$$

Therefore $(\forall i : 2 \leq i \leq n)$

$$y_{\delta_i}(\lambda, \mathbb{I} \mid x) = y_{\delta_{i-2}}\left(\lambda, y_{G_{i-1}}\left(\lambda, y_{G_i}(\lambda, \mathbb{I})\right) \mid x\right)$$

$$\leq y_{\delta_{i-2}}\left(\lambda, y_{G_i}(\lambda, \mathbb{I}) \mid x\right) \leq y_{G_i}(\lambda, \mathbb{I} \mid x).$$

(4) Let $K = (x_0 - r, x_0 + r)$. Then $(\forall \varepsilon > 0)$ $(\exists r > 0)$ $y_K(\lambda, \overline{K} \mid x_0) < \varepsilon$. In order to prove the theorem it is sufficient to know that $y_{\delta_n}^K(\lambda, \mathbb{I} \mid x_0)$ tends to zero, where $\delta \in \mathrm{DS}$. According to theorem 2.2, $(\forall \delta \in \mathrm{DS})$ $(\exists s > 0)$ $(\forall x \in K)$ the interval $(x - s, x + s)$ is covered an infinite number of times by elements of the sequence $\delta = (G_1, G_2, \ldots)$, i.e. $(\forall k \in \mathbb{N})$ $(\exists n \in \mathbb{N}, n > k)$ $(x - s, x + s) \subset G_n$. Let T be a finite $s/2$-net of set K. Then there exists such a sequence of integers $0 = n(0) < n(1) < n(2) < \cdots$ that $(\forall i \in \mathbb{N})$ $(\forall x \in T)$ $(\exists m : n(i) < m \leq n(i+1))$

$$(x - s, x + s) \subset G_m.$$

In this case

$$y_{\delta_{n(k+1)}}^K(\lambda, \mathbb{I} \mid x) = y_{\delta_{n(k)}}^K\left(\lambda, y_{\delta_{n(k+1)}^{n(k)+1}}^K(\lambda, \mathbb{I}) \mid x\right) \leq y_{\delta_{n(k)}}^K\left(\lambda, y_{\delta_{n(k+1)}^{n(k)+1}}(\lambda, \mathbb{I}) \mid x\right)$$

$$\leq y_{\delta_{n(k)}}^K\left(\lambda, \min\left\{ y_{G_i}(\lambda, \mathbb{I}) : n(k) < i \leq n(k+1) \right\} \mid x\right).$$

Consider the function

$$z(x) = \min\left\{ y_{G_i}(\lambda, \mathbb{I} \mid x) : n(k) < i \leq n(k+1) \right\}$$

on set K. For any point $x_1 \in K$ there exists $x_2 \in T$ such that $|x_1 - x_2| < s/2$. Let m_2 be a number of a member of the sequence δ, which covers s-neighborhood of the point x_2: $(x_2 - s, x_2 + s) \subset G_{m_2}$ $(n(k) < m_2 \le n(k+1))$. Then $G(s) \equiv (x_1 - s/2, x_1 + s/2) \subset G_{m_2}$ and

$$z(x_1) \le y_{G_{m_2}}(\lambda, \mathbb{I} \mid x_1) \le y_{G(s)}(\lambda, \mathbb{I} \mid x_1) \le 1 - c,$$

where, according to the second part of the proof

$$c \ge n_K(\lambda) \left(\frac{4}{s^2} + \frac{8 M_K}{s} + n_K(\lambda) \right)^{-1} > 0.$$

Hence, $y^K_{\delta_{n(k+1)}}(\lambda, \mathbb{I} \mid x) \le (1 - c) y^K_{\delta_{n(k)}}(\lambda, \mathbb{I} \mid x)$ and $y^K_{\delta_n}(\lambda, \mathbb{I} \mid x) \to 0$ as $n \to \infty$. $\qquad\square$

16. Construction of a semi-Markov process

In order to construct a SM process on the basis of coefficients of differential equation [5.6] we are almost ready since:

(1) a consistent system of kernels (y_G) is obtained (item 6);

(2) a probabilistic character of these kernels is established (item 13);

(3) a response of the family of these kernels on a deducing sequence is checked (item 19);

(4) a condition of correct exit is checked (item 9 and 3);

(5) the construction of the process is based on the condition of its continuity (item 8).

From these items only the fourth one requires some explanation. It concerns fulfillment of condition (b) of theorem 4.5 or condition (e) of corollary 4.2. In this case taking into account denotations from item 14 the last condition is formulated as follows: $(\forall \lambda > 0) \, (\forall G \in \mathfrak{A}_0) \, (\forall x \in G)$

$$\left(\bar{g}_G(\lambda, a + r) + \bar{h}_G(\lambda, a + r) \right) \bar{g}_{G-r}(0, x)$$
$$+ \left(\bar{g}_G(\lambda, b - r) + \bar{h}_G(\lambda, b - r) \right) \bar{h}_{G-r}(0, x) \longrightarrow 1 \quad (r \longrightarrow 0),$$

where $G = (a, b)$, which is trivially fulfilled, if

$$\bar{g}_G(\lambda, a + r) \longrightarrow 1, \quad \bar{h}_G(\lambda, b - r) \longrightarrow 1.$$

It means continuity of solutions of the equation on the boundary. Actually it is an initial demand to a solution of the equation with given boundary conditions. Formally this property can be derived from the asymptotic representation of item 8 and lemma 5.1.

THEOREM 5.6. *Let coefficients A and B of differential equation [5.6] satisfy the following conditions:*

(1) function $A(x)$ is continuously differentiable on $x \in \mathbb{X}$;

(2) $(\forall \lambda \geq 0)$ function $B(\lambda, x)$ is continuous on $x \in \mathbb{X}$;

(3) $(\forall x \in \mathbb{X})$ $B(0, x) \leq 0$, function $B(\lambda, x)$ infinitely differentiable and $-\partial B(\lambda, x)/\partial \lambda$ is a completely monotone function on $\lambda \geq 0$;

(4) $(\forall G \in \mathfrak{A}_0)$ $n_G(\lambda) > 0$ for $\lambda > 0$ and $n_G(\lambda) \to \infty$ as $\lambda \to \infty$;

(5) $(\forall \lambda \geq 0)$ $r^2 n_{(-r,r)}(\lambda) \to \infty$ and $M_{(-r,r)}/(r\, n_{(-r,r)}(\lambda)) \to 0$ as $r \to \infty$.

Then there exists a continuous semi-Markov process with transition generating functions (f_G) which are solutions of this equation such that for any interval $G = (a, b)$, $x \in G$ and $\lambda > 0$

$$f_G\big(\lambda, \{a\} \mid x\big) = \bar{g}_G(\lambda, x), \qquad f_G\big(\lambda, \{b\} \mid x\big) = \bar{h}_G(\lambda, x).$$

Proof. Proof can be found in theorems 5.1, 5.2, 5.4 and 5.5. $\qquad\qquad\qquad$ $\square$

Solutions of equation [5.6] for $\lambda = 0$ can be obtained as limits from solutions for positive meanings of λ. Let us note that in equation [5.6] negative meanings of the coefficient $B(0, x)$ are admitted. In this case we obtain $y_G(0, \mathbb{I} \mid x) < 1$ $(x \in G)$. This solutions is interpreted as an absence of exit of a trajectory of the semi-Markov process from set G. It can be the case that the process need not stop at a particular point of the interval. On the contrary, for a process of diffusion type, as a rule, the time and place of stopping are random. It is possible for the process to tend to some limit which is not attainable in a finite time. We will return to this question in Chapters 6 and 9.

5.1.3. *Some properties of the process*

In this subsection some properties of one-dimensional SM processes of diffusion type are considered. In particular, we investigate the existence and a view of lambda-characteristic operators of the first and second kind (strict definition), conditions of markovness, and the representation of the semi-Markov process in the form of a transformed Markov process.

17. *Operator of the second kind*

We consider a semi-Markov process of diffusion type corresponding to equation [5.6] with coefficients satisfying conditions of theorem 5.6. From theorem 5.2 it follows that for any bounded function φ and $x \in G = (a, b)$

$$f_G(\lambda, \varphi \mid x) \equiv g_G(\lambda, x)\varphi(a) + h_G(\lambda, x)\varphi(b)$$

$$= \varphi(x) + \frac{(b - x)(x - a)}{b - a} \left(\frac{\varphi(b) - \varphi(x)}{b - x} - \frac{\varphi(x) - \varphi(a)}{x - a} \right)$$

$$+ A(x) \frac{(b - x)(x - a)}{b - a} (\varphi(b) - \varphi(a)) + B(\lambda, x)(b - x)(x - a)\varphi(x)$$

$$+ (b - x)(x - a)O(b - a) + F,$$

where

$$F = \frac{1}{3} B(\lambda, x) \frac{(b - x)(x - a)}{b - a} ((b - 2a + x)(\varphi(b) - \varphi(x))$$

$$+ (2b - a - x)(\varphi(a) - \varphi(x)))$$

$$+ \frac{1}{3} (A^2(x) + A'(x)) \frac{(b - x)(x - a)(a + b - 2x)}{b - a} (\varphi(b) - \varphi(a)).$$

Besides, this asymptotics with $b \downarrow x$ and $a \uparrow x$ is uniform on x at each bounded interval. In addition, it shows as well that for φ with a bounded derivative the term F is of the order $(b - x)(x - a) O(b - a)$ as $b - a \to 0$ uniformly on each bounded interval. For constant φ this term is equal to zero. In particular,

$$f_G(\lambda, \mathbb{I} \mid x) = 1 + B(\lambda, x)(b - x)(x - a) + (b - x)(x - a)O(b - a).$$

Hence, for the given semi-Markov process the following lambda-characteristic operator of the second kind in the strict sense is determined on the class of all twice differentiable functions φ

$$A_\lambda(\varphi \mid x) \equiv \lim_{G \downarrow x} \frac{\lambda(f_G(\lambda, \varphi \mid x) - \varphi(x))}{1 - f_G(\lambda, \mathbb{I} \mid x)}$$

$$= (\varphi''(x)/2 + A(x)\varphi'(x) + B(\lambda, x)\varphi(x)) \frac{-\lambda}{B(\lambda, x)}.$$

In order to find this limit it is important to note that the denumenator has an order $(b - x)(x - a)$. As will be shown below, an expectation of the first exit time from G is of the same order. This makes it possible to obtain a formula for the lambda-characteristic operator of the first kind in strict sense.

18. Asymptotics of the expectation for a small parameter

THEOREM 5.7. *Let (P_x) be a continuous SM process on the line with transition generating functions h_G and g_G satisfying differential equation [5.6], where $G = (a, b)$,*

$a < x_0 < b$. *Let A be differentiable, and A' and $B(\lambda, \cdot)$ be functions continuous on x, $B(0, x) = 0$, and also let the following partial derivative from the right*

$$-\partial B(\lambda, x)/\partial \lambda|_{\lambda=0} = \gamma(x) \quad (0 < \gamma(x) < \infty)$$

exist and be continuous on x in some neighborhood of the point x_0. Then there exists $m_G(x_0) \equiv E_{x_0}(\sigma_G) \; (0 < m_G(x_0) < \infty)$ and

$$m_G(x_0) = \gamma(x_0)(b - x_0)(x_0 - a) + (b - x_0)(x_0 - a)O(b - a),$$

where $a \uparrow x_0$ and $b \downarrow x_0$.

Proof. Note that under conditions of the theorem $y_G(0, \mathbb{I} \mid x) = 1$. We have to analyze the function

$$m_G(x) = -\frac{\partial}{\partial \lambda} y_G(\lambda, \mathbb{I} \mid x)\big|_{\lambda=0}.$$

Formally the asymptotics of $m_G(x)$ we could receive by differentiation on λ every term of the expansion $y_G(\lambda, \mathbb{I} \mid x) = 1 + B(\lambda, x)(b-x)(x-a) + (b-x)(x-a)O(b-a)$. In order to justify possibility of such differentiation we use the integral representation [5.12] of solutions of the Dirichlet problem from the proof of theorem 5.2. For any $\lambda \geq 0$ the function $y_G(\lambda, \mathbb{I} \mid x)$ satisfies the equation

$$y_G(\lambda, \mathbb{I} \mid x) = 1 + \int_a^b K_G(\lambda, x, t) y_G(\lambda, \mathbb{I} \mid t) \, dt, \tag{5.15}$$

In this case the differentiation on λ for $\lambda = 0$ is justified, for example, by the Arcell theorem [KOL 72, p. 104]. In order to apply it, it is sufficient to have ratios $B(\lambda, x)/\lambda$ and $(1 - y_G(\lambda, \mathbb{I} \mid x))/\lambda$ tending to their limits monotonically. It follows from non-negativeness of the second derivatives of the functions $B(\lambda, x)$ and $y_G(\lambda, \mathbb{I} \mid x)$ on λ. In the outcome of this operation we obtain the equation

$$m_G(x) = M_G(x) + \int_a^b K_G(0, x, t) m_G(t) \, dt, \tag{5.16}$$

where

$$K_G(0, x, t) = \begin{cases} 2\dfrac{b - x}{b - a}\big(-A(t) - A'(t)(t - a)\big) & (t < x), \\[2mm] 2\dfrac{x - a}{b - a}\big(A(t) - A'(t)(b - t)\big) & (t > x), \end{cases}$$

$$M_G(x) = -\int_a^b K'_G(0, x, t) \, dt = 2\frac{b - x}{b - a} \int_a^x \gamma(t)(t - a) \, dt$$

$$+ 2\frac{x - a}{b - a} \int_x^b \gamma(t)(b - t) \, dt.$$

From continuity of function γ in a neighborhood of point x_0 we obtain

$$M_G(x) = \gamma(x)(b-x)(x-a) + (b-x)(x-a)O(b-a)$$

uniformly on x in some neighborhood of point x_0. Since the kernel $K_G(0, x, t)$ is bounded in square $G \times G$, for sufficiently small $b - a$ the solution of equation [5.16] is representable in the form of the series

$$m_G(x) = M_G(x) + \sum_{k=1}^{\infty} \int_a^b K_G^{(k)}(0, x, t) M_G(t)\, dt,$$

uniformly converging in some neighborhood of the point x_0. From here follows the boundedness of function $m_G(x)$. It is enough to know only the first term of its asymptotics in a neighborhood of the point x_0. For this aim we consider three terms of the series and the remainder

$$m_G(x) = M_G(x) + \sum_{k=1}^{3} \int_a^b K_G^{(k)}(0, x, t) M_G(t)\, dt + \int_a^b K_G^{(4)}(0, x, t) m_G(t)\, dt.$$

Furthermore, using positiveness of the kernel $K_G(0, x, t)$ and an asymptotics of the integral

$$\int_a^b K_G(0, x, t)(b-t)(t-a)\, dt = (b-x)(x-a)O(b-a),$$

we discover that three terms of the sum are of the same order. In the last term we have

$$K_G^{(3)}(0, x, t) = (b-x)(x-a)Z(x, t),$$

where function Z is bounded in the square (see theorem 5.2), and therefore this term is of the order $(b-x)(x-a)O(b-a)$. $\square$

From the proved theorem it follows that for a given process there exists a lambda-characteristic operator of the first kind in strict sense on a set of twice differentiable functions. Moreover

$$A_\lambda(\varphi \mid x_0) = (\varphi''(x_0)/2 + A(x_0)\varphi'(x_0) + B(\lambda, x_0)\varphi(x_0))/\gamma(x_0).$$

Note that $\gamma(x) = \lim(-B(\lambda)/\lambda)$ as $\lambda \to 0$. This means that operators of the first and second kind for $\lambda = 0$ coincide.

Both normings used for definition of the lambda-characteristic operators are of the order $(b-x)(x-a)$, depending on the length of an interval and on the position of the initial point inside the interval. It it resonable to consider this expression as a standard norming in the one-dimensional case. In the case of a symmetric situation for the interval of the length $2r$ with respect to the initial point, such a standard norming is r^2. It is interesting to note that the standard norming in a symmetric case does not depend on the dimension of the state space. This fact will be proved in section 5.2 devoted to multi-dimensional processes of diffusion type. An advantage of operators of the second kind (as well that of operators with the standard norming) appears in the case when the limit $\lim(-B(\lambda)/\lambda)$ as $\lambda \to 0$ is equal to infinity. In this case the operator of the first kind does not exist.

A construction of a SM process on the basis of an *a priori* given lambda-characteristic operator uses the identity $(\forall G \in \mathfrak{A}) \, (\forall \varphi \in V_\lambda) \, A_\lambda f_G(\lambda, \varphi) = 0$. The same is true for operators of the second kind both in a strict and in a wide sense. For a diffusion process this identity passes to a differential equation considered here. However, the question about existence and uniqueness of a general semi-Markov process corresponding to the equation $A_\lambda \varphi = 0$ remains open.

19. Criterion of Markovness for SM processes

THEOREM 5.8. *Let (P_x) be a continuous semi-Markov process on the line with transition generating functions satisfying differential equation [5.6] with boundary conditions specified in theorem 5.1. Let coefficient A be differentiable, and functions A', $B(\lambda, \cdot)$ and γ be continuous on x. This process is Markovian if and only if $(\forall \lambda \geq 0)$ $B(\lambda, x) = -\lambda \gamma(x)$.*

Proof. Necessity. Let us check the fulfillment of the conditions of theorem 3.6(2). The continuity of transition generating functions provides (λ, Δ)-continuity of the family of measures (P_x). The pseudo-locality of the operator A_λ follows from continuity of trajectories of the process. According to item 18, $A_\lambda(\mathbb{I} \mid x) = B(\lambda, x)/\gamma(x)$. Hence, a necessary condition of Markovness is the condition $B(\lambda, x)/\gamma(x) = -\lambda$.

Sufficiency. Let us check the fulfillment of the conditions of theorems 3.7 and 3.8. From the continuity of trajectories of the process the quasi-left-continuity of the process follows. In addition to properties established above we use the linearity of the operator on λ, which implies equality $a(\lambda/\mu) = \lambda/\mu$. This is sufficient for the given semi-Markov process to be Markovian. $\qquad\square$

20. Equation for transformed functions

THEOREM 5.9. *Let (P_x) be a continuous semi-Markov process on the line with transition generating functions satisfying differential equation [5.6] with differentiable coefficient A and continuous on x functions A' and $B(\lambda, \cdot)$ $(\lambda > 0)$. Let $\beta(x)$ be a* *continuous non-negative function on $x \in \mathbb{X}$. Then $(\forall G \in \mathfrak{A}_0)$ $(G = (a, b))$ functions $\widetilde{h}_G(x)$ and $\widetilde{g}_G(x)$ $(x \in G)$ of the form*

$$\widetilde{h}_G(x) = E_x \left(\exp \left(-\int_0^{\sigma_G} (\beta \circ X_t)\, dt \right); X_{\sigma_G} = b,\ \sigma_G < \infty \right),$$

$$\widetilde{g}_G(x) = E_x \left(\exp \left(-\int_0^{\sigma_G} (\beta \circ X_t)\, dt \right); X_{\sigma_G} = a,\ \sigma_G < \infty \right)$$

satisfy the differential equation

$$f'' + 2A(x)f' + 2B\big(\beta(x), x\big)f = 0$$

with boundary conditions

$$\widetilde{h}_G(a) = \widetilde{g}_G(b) = 0, \qquad \widetilde{h}_G(b) = \widetilde{g}_G(a) = 1.$$

Proof. It can be shown that $\widetilde{h}_G$ and $\widetilde{g}_G$ satisfy functional equations [5.1] and [5.2]. Let $G_1 = (c, d)$ and $x \in [c, d] \subset [a, b]$. We have $\sigma_G = \sigma_{G_1} \dotplus \sigma_G$ and

$$\widetilde{h}_G(x) = E_x \left(\exp \left(-\int_0^{\sigma_{G_1}} (\beta \circ X_t)\, dt \right) \exp \left(-\int_{\sigma_{G_1}}^{\sigma_{G_1} \dotplus \sigma_G} (\beta \circ X_t)\, dt \right); \right.$$

$$\left. \pi_G = b,\ \sigma_{G_1} \dotplus \sigma_G < \infty \right)$$

$$= E_x \left(\exp \left(-\int_0^{\sigma_{G_1}} (\beta \circ X_t)\, dt \right) \left(\exp \left(-\int_0^{\sigma_G} (\beta \circ X_t)\, dt \right) \circ \theta_{\sigma_{G_1}} \right); \right.$$

$$\left. \pi_G \circ \theta_{\sigma_{G_1}} = b,\ \sigma_{G_1} < \infty,\ \sigma_G \circ \theta_{\sigma_{G_1}} < \infty \right)$$

$$= E_x \left(\exp \left(-\int_0^{\sigma_G} (\beta \circ X_t)\, dt \right) E_{X_{G_1}} \left(\exp \left(-\int_0^{\sigma_G} (\beta \circ X_t)\, dt \right); \right. \right.$$

$$\left. \left. X_G = b,\ \sigma_G < \infty \right); \sigma_{G_1} < \infty \right)$$

$$= \widetilde{h}_{G_1}(x)\widetilde{h}_G(d) + \widetilde{g}_{G_1}(x)\widetilde{h}_G(c).$$

Similarly $\widetilde{g}_G(x) = \widetilde{h}_{G_1}(x)\widetilde{g}_G(d) + \widetilde{g}_{G_1}(x)\widetilde{g}_G(c)$. Evidently, $\widetilde{h}_G(a) = \widetilde{g}_G(b) = 0$ and $\widetilde{h}_G(b) = \widetilde{g}_G(a) = 1$. Let $\widetilde{h}_r(x) = \widetilde{h}_{(x-r,x+r)}(x)$, $\widetilde{g}_r(x) = \widetilde{g}_{(x-r,x+r)}(x)$. We have

$$h_r\big(\beta_2, x\big) \le \widetilde{h}_r(x) \le h_r\big(\beta_1, x\big), \qquad g_r\big(\beta_2, x\big) \le \widetilde{g}_r(x) \le g_r\big(\beta_1, x\big),$$

where

$$\beta_1 = \inf\left\{\beta(x_1) : |x_1 - x| < r\right\}, \qquad \beta_2 = \sup\left\{\beta(x_1) : |x_1 - x| < r\right\}.$$

It is easy to show (see theorem 5.1) that expansions for h_r and g_r are uniform on (λ, x) in each limited rectangle of region $(0, \infty) \times \mathbb{R}$. From here it follows that

$$\widetilde{g}_r(x) = 1/2 - A(x)r/2 + B(\beta(x), x)r^2/2 + o(r^2),$$
$$\widetilde{h}_r(x) = 1/2 + A(x)r/2 + B(\beta(x), x)r^2/2 + o(r^2)$$

uniformly on x in each finite interval. Now we can use theorem 5.3 where from semi-Markov conditions this functional equation and boundary conditions have been used.

$\square$

21. Connection with Markov processes

THEOREM 5.10. *Every continuous SM process on the line with transition generating functions g_G and h_G satisfying differential equation [5.6] with functions A, A' and $B(\lambda, \cdot)$ ($\lambda > 0$), which are continuous with respect to x, can be obtained from a Markov process $(\overline{P}_x)$, corresponding to differential equation*

$$f'' + 2A(x)f' - 2\lambda f = 0,$$

in such a way that

$$h_G(\lambda, x) = \overline{P}_x\left(\exp\int_0^{\sigma_G} B(\lambda, X_t)\,dt;\ X_{\sigma_G} = b,\ \sigma_G < \infty\right),$$
$$g_G(\lambda, x) = \overline{P}_x\left(\exp\int_0^{\sigma_G} B(\lambda, X_t)\,dt;\ X_{\sigma_G} = a,\ \sigma_G < \infty\right).$$

Proof. According to theorem 5.9, the right parts of the previous equalities satisfy differential equation [5.6] with coefficients $A(x)$, $B(\lambda, x)$ with the corresponding boundary conditions. Hence they are transition generating functions of the process with the family of distributions (P_x).

$\square$

In Chapter 6 it will be shown that this transformation corresponds to a random time change transformation of the process with the help of system of additive functionals:

$$a_t(\lambda) = -\int_0^t B(\lambda, X_{t_1})\,dt_1.$$

22. Homogenous SM process

Let us consider a SM process on the line corresponding to differential equation [5.6] with coefficients independent of x. Let $h_G(\lambda, x)$ and $g_G(\lambda, x)$ be transition generating functions of the process. Then for $x \in G = (a, b)$

$$g_G(\lambda, x) = e^{-A(x-a)} \frac{\text{sh}(b - x)\sqrt{A^2 - 2B(\lambda)}}{\text{sh}(b - a)\sqrt{A^2 - 2B(\lambda)}},$$

$$h_G(\lambda, x) = e^{A(b-x)} \frac{\text{sh}(x - a)\sqrt{A^2 - 2B(\lambda)}}{\text{sh}(b - a)\sqrt{A^2 - 2B(\lambda)}}.$$

For $A = 0$ and $B(\lambda) = -\lambda$ we obtain semi-Markov transition generating functions of the standard Wiener process [DAR 53]. If $B(\lambda, x) = B(\lambda)$ is a non-linear function of λ, this process is not Markovian. It is connected to the Wiener process (with shift) by the integral relation, considered in theorem 5.10. For example, $B(\lambda) = -\ln(1 + \lambda)$. This is a negative function. Its first derivative on λ is a completely monotone function (with minus sign). Such $B(\lambda)$ corresponds to a non-Markov semi-Markov process of diffusion type. As will be shown in the next chapter this process can be obtained from the Wiener process with a shift by random time change transformation with the help of independent Gamma-process [HAR 71b]. Besides, as it will be shown in Chapter 7, a continuous SM process can be obtained as a limit of an *a priori* given sequence of semi-Markov walks.

23. Comparison of the processes' parameters

It is interesting to compare well-known Kolmogorov's coefficients of a Markov diffusion process with coefficients of differential equation [5.6]. For this aim we will derive a formula for the resolvent kernel $R_\lambda(S \mid x)$ $(x \in \mathbb{X}, S \in \mathcal{B}(\mathbb{X}))$ (see item 3.22), which expresses this kernel in terms of transition generating functions of a SM process. In the case of $B(\lambda) = -\lambda$ this formula gives the tabulated meaning of the Laplace transformed transition function of the Markov process. It is convenient to use this for comparison of parameters.

Let (P_x) be a SM process on the line with transition generating function f_G. Let us consider a resolvent kernel R_λ, where

$$R_\lambda(S \mid x) \equiv R(\lambda; I_S \mid x) = \int_0^\infty e^{-\lambda t} P_x(X_t \in S) dt.$$

In item 3.21 a way has been obtained for the functional $R(\lambda; \varphi \mid x)$ to be evaluated as a limit of some sums suitable for piece-wise continuous functions φ. Theorem 3.4

gives another way of evaluating the resolvent kernel. For $n = 1$ this theorem implies that $\mathcal{A}_\lambda(R(\lambda, \varphi) \mid x) = -\varphi(x)$. From here

$$A_\lambda\big(R(\lambda, \varphi) \mid x\big) = \lambda^{-1}\varphi(x)A_\lambda(\mathbb{I} \mid x)$$

at points of continuity of function φ. According to item 17 and theorem 5.7, we have $A_\lambda(\mathbb{I} \mid x) = B(\lambda, x)/\gamma(x)$. Let $f(x) = R_\lambda((-\infty, b) \mid x) \equiv R(\lambda; I_{(-\infty,b)} \mid x)$. It is not difficult to show that f is bounded, continuous, differentiable everywhere and twice differentiable on intervals $(-\infty, b)$, (b, ∞). Hence it belongs to the region of determination of the operator A_λ. Therefore from item 18 we obtain a differential equation for the function f:

$$f''/2 + A(x)f' + B(\lambda, x)f = \lambda^{-1}I_{(-\infty,b)}(x)B(\lambda, x).$$

A continuous solution of this equation tending to zero as $|x| \to \infty$, the case of the coefficients A and B independent of x is represented by the following formulae: $f(x) = f_1(x)$ $(x \geq b)$, $f(x) = f_2(x)$ $(x < b)$, where

$$f_1(x) = \frac{\sqrt{A^2 - B(\lambda)/2} - A}{2\lambda\sqrt{A^2 - B(\lambda)/2}} \exp\Big(-2(x - b)\big(\sqrt{A^2 - B(\lambda)/2} + A\big)\Big),$$

$$f_2(x) = \frac{1}{\lambda} - \frac{\sqrt{A^2 - B(\lambda)/2} + A}{2\lambda\sqrt{A^2 - B(\lambda)/2}} \exp\Big(-2(b - x)\big(\sqrt{A^2 - B(\lambda)/2} - A\big)\Big).$$

Let us consider a density of the measure $R_\lambda(S \mid x)$, which is equal to a derivative of f by b:

$$p_\lambda(b \mid x) \equiv \frac{\partial}{\partial b} R_\lambda\big((-\infty, b) \mid x\big)$$

$$= \frac{-B(\lambda)/\lambda}{\sqrt{A^2 - 2B(\lambda)}} \exp\Big(A(b - x) - |b - x|\sqrt{A^2 - 2B(\lambda)}\Big).$$

For a homogenous Markov diffusion process $\xi(t)$ with an initial point x, local variance D, and parameter of shift V this density can be obtained in outcome of the Laplace transformation on t of its one-dimensional density:

$$\varphi_t(b \mid x) = \frac{1}{\sqrt{2\pi Dt}} \exp\Big(-\frac{(b - x - Vt)^2}{2Dt}\Big),$$

which has the form

$$\bar{p}_\lambda(b \mid x) = \frac{1}{\sqrt{V^2 + 2D\lambda}} \exp\Big(V(b - x)/D - |b - x|\frac{\sqrt{V^2 + 2D\lambda}}{D}\Big)$$

(see [DIT 65], etc.). A comparison of these densities reveals the sense of parameters A and $B(\lambda)$: the function $p_\lambda(b \mid x)$ is equal to $\bar{p}_\lambda(b \mid x)$ for

$$A = V/D, \qquad B(\lambda) = -\lambda/D.$$

5.2. Multi-dimensional semi-Markov processes of diffusion type

In this section, semi-Markov processes of diffusion type in d-dimensional space $(d \geq 2)$ are considered. In addition to the common properties of SM processes considered in Chapter 3, where the metric structure of space of states is used only, these processes have a lot of specific properties.

After analyzing one-dimensional SM processes of a diffusion type there is a natural desire to generalize obtained outcomes on a multi-dimensional case. In this way the qualitative saltus is obvious already with passage to two-dimensional space. Instead of a two-dimensional parameter of an elementary neighborhood of an initial point we obtain at once an infinite-dimensional one: (1) even for the simplest form of the neighborhood the arbitrary distribution on its boundary is possible, (2) the form of this neighborhood can be arbitrary. Nevertheless there is much in common in a local behavior of one-dimensional and multi-dimensional processes. Principally there is a differentiability of their transition generating functions and a differential second-order equation of elliptic type for their derivatives. Moreover, such a function has asymptotics for a sequence of small neighborhoods of an initial point which can be deduced by methods of the theory of differential equations, in particular, using methods of the Dirichlet problem. In this way the weak asymptotic expansion of the distribution density on the boundary of such a small neighborhood is obtained. This makes it possible to prove the existence of a characteristic Dynkin operator [DYN 63] as a limit of ratios of appropriate expectations, when the form of a neighborhood is arbitrary. In case of spherical neighborhoods the simple formulae for the first three coefficients of the expansion reflecting a probability sense of coefficients of the partial differential equation of elliptic type are derived.

5.2.1. *Differential equations of elliptic type*

In this section we consider continuous semi-Markov processes of diffusion type in region $S \subset \mathbb{R}^d$ $(d \geq 2)$. The transition generating function of such a process is a solutions in this region of a partial differential second-order equation of elliptic type. We look for conditions of local characterization of the processes similar to Kolmogorov's conditions, formulated for Markov processes of diffusion type. From these conditions in work [KOL 38] differential equations of parabolic type, the direct and inverse Kolmogorov equations, are derived; transition densities of the process satisfy such an equation (see [DYN 63, p. 221]). For a semi-Markov process such transition densities are useless, but it has another way of local characterization, namely, with the help of its distribution of the first exit pair from a small neighborhood of an initial point.

The basic content of the present subsection makes a solution of the inverse problem for multi-dimensional semi-Markov process of a diffusion type: for the given differential equation of elliptic type to construct the corresponding semi-Markov process and to find asymptotic local properties of this process.

24. Denotation for coordinates and derivatives

In this section we prefer to designate a point of multi-dimensional space $\mathbb{R}^d$ as $p = (p^1, \ldots, p^d)$. Designate

$$D_i u = \frac{\partial u}{\partial p^i}, \qquad D_{ij} u = \frac{\partial^2 u}{\partial p^i \partial p^j}.$$

In some integrals we designate a variable of integration $p_1, p_2, \ldots$, etc. In these cases $D_i^{(k)}$, $D_{ij}^{(k)}$, $\ldots$ means that the derivation is made on coordinates of a point p_k. To avoid a great many indexes, we will apply other designations for the first coordinates of a point $p = (x, y, z, \ldots)$, and under sign of integral $x_k, y_k, z_k, \ldots$ mean coordinates of a point p_k, a variable of an integration. As a universal designation for coordinates we will use $\pi_i(p_k) = p_k^i$, the i-th coordinate of a point p_k.

25. Differential equation

Consider in region $S \in \mathfrak{A}$ a partial differential equation of elliptic type

$$\mathcal{L}u \equiv \frac{1}{2} \sum_{i,j=1}^{d} a^{ij} D_{ij} u + \sum_{i=1}^{d} b^i D_i u - cu = 0. \qquad [5.17]$$

Here a^{ij}, b^i, c are continuous real functions, determined on S; $(\forall p \in S)$ $(a^{ij}(p))$ is a positive definite matrix; $c \geq 0$; u is an unknown differentiable function. We admit that some coefficients (and solutions as well) of this equation can depend on a parameter $\lambda \geq 0$. Without being too specific we assume that the following norming condition is fulfilled: $(\forall p \in S)$ $(\forall \lambda \geq 0)$

$$\sum_{i=1}^{d} a^{ii}(p) = d. \qquad [5.18]$$

According to standard terminology [SOB 54], equation [5.17] is a partial differential equation of elliptic type. Let $G \subset S$ and $u_G(\varphi \mid \cdot)$ be a solution of the Dirichlet problem for this equation in the bounded region G with a continuous function φ given on the boundary of the region. With the suppositions made, this solution is continuous on $G \cup \partial G$. If the function φ is determined and continuous on the whole set $S \cup \partial S$, then determining $u_G(\varphi \mid \cdot)$ we take the contraction of φ on the boundary of a set

G. We determine the solution for any $S\backslash G$ as $u_G(\varphi \mid p) = \varphi(p)$. The operator u_G defined in such a manner maps a set of continuous and bounded functions $C(S)$ in itself. Furthermore, we will understand a solution of the Dirichlet problem in this extended sense.

26. Functional equation

Show that for a family of operators $(u_G)_{G \in \mathfrak{A} \cap S}$ the following equation is fair

$$u_{G_2}(\varphi \mid p) = u_{G_1}\big(u_{G_2}(\varphi \mid \cdot) \mid p\big), \qquad\qquad [5.19]$$

where $G_1, G_2 \in \mathfrak{A}$, $G_1 \subset G_2$ and $p \in G_1$. It is the same equation that the transition generating operators of a semi-Markov process $f_G(\lambda, \cdot \mid \cdot)$ are subordinated to.

Let a function φ be determined on the boundary of G_2. In this case $u_{G_2}(\varphi \mid \cdot)$ and the function $f \equiv u_{G_1}(u_{G_2}(\varphi \mid \cdot) \mid \cdot)$ satisfy equation [5.17] inside G_1. Moreover, they coincide on the boundary of this set. According to the uniqueness theorem for a solution of the Dirichlet problem in region G_1, which is fair due to our suppositions about coefficients of the equation, we obtain $u_{G_2}(\varphi \mid \cdot) = f$ everywhere inside G_1. The equality of these functions outside of G_1 follows from a rule of extensions of a solution of the Dirichlet problem.

27. Process of diffusion type

DEFINITION 5.2. *The continuous semi-Markov process is referred to as a semi-Markov process of diffusion type in region S, if for any $\lambda \geq 0$ there exists a differential operator of elliptic type $\mathcal{L} = \mathcal{L}(\lambda)$ with appropriate coefficients (see equation [5.17]) such that the family of transition generating operators $(f_G(\lambda, \varphi \mid \cdot))$ of this process coincides with a set (u_G) of solutions of the Dirichlet problem of equation $\mathcal{L}u = 0$.*

Furthermore, it will be clear that the differential operator $\mathcal{L}(\lambda)$, defining transition generating functions of a diffusion semi-Markov process, depends on λ only through the factor $c = c(\lambda, \cdot)$.

It is known (see, e.g., [DYN 63]) that the operator of the Dirichlet problem can be represented in an integral form with some kernel $H_G(p, B)$:

$$u_G(\varphi \mid p) = \int_{\partial G} \varphi(p_1) H_G(p, dp_1).$$

This kernel for region G with piece-wise smooth boundary is determined by a density $h_G(p, p_1)$ with respect to a Lebesgue measure on the boundary:

$$H_G(p, B) = \int_{B \cap \partial G} h_G(p, p_1)\, dS(p_1) \quad (p \in G).$$

Equation [5.19] in this case passes in an integral equation

$$h_{G_2}(p, p_2) = \int_{\partial G_1} h_{G_1}(p, p_1) h_{G_2}(p_1, p_2) \, dS(p_1). \qquad [5.20]$$

For a semi-Markov process with the initial point $p \in G$ the kernel $h_G(p, \cdot) \equiv f'_G(\lambda, \cdot \mid p)$ is interpreted as a density of a transition generating function of a semi-Markov process, i.e. distribution density of the first exit pair from region G transformed by the Laplace transformation on argument $t \in \mathbb{R}_+$.

Note that the Laplace transformation image of the distribution density on argument t with a parameter λ can be interpreted as a corresponding distribution density of the process stopped at a random instant which is independent of the process and is exponentially distributed with the parameter λ (see [BOR 96]).

5.2.2. Asymptotics of distribution of a point of the first exit from arbitrary neighborhood

Weak asymptotic expansion for a distribution density function of a point of the first exit from a small neighborhood of an initial point of the process is derived. The diameter of the neighborhood is assumed to tend to zero. The form of such a neighborhood can by arbitrary. It makes it possible to use the obtained expansion as proof of existence of the characteristic Dynkin operator in its common definition.

28. Density of a point of the first exit

Consider equation [5.17] in region S. Let this region contain the origin of coordinates, point 0. Without being too specific we can assume that $(\forall i, j) \, a^{ij}(0) = \delta^{ij}$. Otherwise we could apply non-degenerate linear transformation of spaces, which reduces the equation to this form, and further we could make a transformation of obtained factors of expansion appropriate to the inverse transformation of space (see item 34).

Let G be a neighborhood of zero, a bounded region $G \subset S$, with piece-wise smooth boundary, on which a distribution density $h_G(p_0, \cdot)$ is determined. It corresponds to a semi-Markov diffusion process with the initial point $p_0 \in G$.

Let $kG = \{p : (\exists p_1 \in G) \, p = kp_1\} \, (k > 0)$ be a scale transformation of set G. Under the above supposition the density $h_{kG}(p_0, \cdot) \, (p_0 \in kG)$ will be determined. Due to rescaling the Lebesgue measure of the boundary will vary correspondingly: $|\partial(kG)| = k^{d-1}|\partial G|$. It is clear that for any reasonable asymptotics of probability density on $\partial(kG)$, this normalizing factor should be taken into account.

29. Class of regions

In [GIL 89, p. 95] the following class of regions is defined. Region G with its boundary is said to belong to class $C^{k,\alpha}$ ($k = 0, 1, 2, \ldots$, $0 \le \alpha \le 1$), if for each point $p_0 \in \partial G$ there is a ball $B = B(p_0, r)$ and one-to-one map f of this ball on $D \subset \mathbb{R}^d$ such that:

(1) $f(B \cap G) \subset \mathbb{R}_+^d \equiv \mathbb{R}^{d-1} \times \mathbb{R}_+$;

(2) $f(B \cap \partial G) \subset \partial \mathbb{R}_+^d$;

(3) $f \in C_0^{k,\alpha}(B)$, $f^{-1} \in C_0^{k,\alpha}(D)$.

In the latter case $C_0^{k,\alpha}(\cdot)$ means space of k-times differentiable functions, k-th derivatives of which are continuous on the corresponding set with Hölder index α.

Let $\mathcal{R}(G) \equiv \inf\{r > 0 : G \subset B(0, r)\}$, where $B(p, r)$ is a ball of a radius r with a center p in d-dimensional space. Furthermore, we will consider normalized neighborhoods of zero of the form $G' = (1/\mathcal{R}(G))G$, touching the surface of ball $B(0, 1)$.

For positive number $k \ge 1$ we designate $\mathfrak{A}^\circ = \mathfrak{A}^\circ(k)$ a family of normalized neighborhoods of zero such that $(\forall G \in \mathfrak{A}^\circ(k))$ there exists diffeomorphism f of a class $C_0^{2,\alpha}$ of maps $G \to B(0, 1)$ such that $(\forall p_1, p_2 \in G)$

$$k^{-1}|p_1 - p_2| \le |f(p_1) - f(p_2)| \le k|p_1 - p_2|.$$

It is clear that $\mathfrak{A}^\circ(k) \subset C^{2,\alpha}$.

30. Asymptotic expansion of a general view

Let $Q_G(p_1, p_2)$ $(p_1, p_2 \in G)$ be the Green function for the Laplace operator $\Delta \equiv \sum_{i=1}^d D_{ii}$ in region G and $K_G(p_1, p_2) = \partial Q_G(p_1, p_2)/\partial n_2$ be a derivative of the Green function according to the second argument on direction of an interior normal to the boundary of the region.

In the following theorem the family of sets $G' = RG$ is considered, where $R > 0$ $(R \to 0)$ and $G \in \mathfrak{A}^\circ$ (neighborhood of zero).

THEOREM 5.11. *Let factors b^i be continuously differentiable and a^{ij} be twice continuously differentiable in some neighborhood of zero. Also let $(\forall ij)$ $a^{ij}(0) = \delta^{ij}$ and there be positive numbers Λ_1 and Λ_2 such that $(\forall p \in \mathbb{R}^d)$*

$$\sum_{ij} a^{ij}(p_0)p^i p^j \ge \Lambda_1 \|p\|^2,$$

$$|a^{ij}(p_0)|, |b^i(p_0)|, |c(p_0)|, |a^{ij}(p_0) - \delta^{ij}|/\|p_0\| \le \Lambda_2$$

with all i, j and p_0 in some neighborhood of zero.

Then the following weak asymptotics as $R \to 0$ are true

$$R^{d-1} h_{RG}(Rp_0, Rp)$$

$$= K_G(p_0, p)$$

$$+ \int_G Q_G(p_0, p_1) \left(R \sum_i b^i(0) D_i^{(1)} K_G(p_1, p) + R X_G(p_1, p) \right.$$

$$\left. - R^2 c(0) K_G(p_1, p) + R^2 Y_G(p_1, p) \right) dV(p_1) \qquad \text{[5.21]}$$

$$+ o(R^2) \int_G Q_G(p_0, p_1) \, dV(p_1)$$

$(p_0, \ p \in G)$. *For any $k \geq 1$ the term $o(R^2)$ is uniform on all $G \in \mathfrak{A}°(k)$ and $p_0 \in G$. It is understood in a weak sense: an integral on ∂G from this term, multiplied by a function, which is continuous and bounded on this set, is a value of the order $o(R^2)$. Function $X_G(p_1, \cdot)$ is orthogonal to any constant and any linear function on the boundary, function $Y_G(p_1, \cdot)$ is orthogonal to any constant function on the boundary. Moreover*

$$X_G(p_1, p) = \sum_{ij} \sum_m a_m^{ij}(0) \pi_m(p_1) D_{ij}^{(1)} K_G(p_1, p),$$

$$Y_G(p_1, p) = \sum_{i=1}^6 Y_i(p_1, p),$$

where

$$Y_1(p_1, p) = \frac{1}{2} \sum_{ij} \sum_{mn} a_{mn}^{ij}(0) \pi_m(p_1) \pi_n(p_1) D_{ij}^{(1)} K_G(p_1, p),$$

$$Y_2(p_1, p) = \sum_i \sum_m b_m^i(0) \pi_m(p_1) D_i^{(1)} K_G(p_1, p),$$

$$Y_3(p_1, p) = \sum_{ij} \sum_m a_m^{ij}(0) \sum_{kl} \sum_n a_n^{kl}(0) \pi_m(p_1)$$

$$\times D_{ij}^{(1)} \int_G Q_G(p_1, p_2) \pi_n(p_2) D_{kl}^{(2)} K_G(p_2, p) \, dV(p_2),$$

$$Y_4(p_1, p) = \sum_i b^i(0) \sum_{kl} \sum_n a_n^{kl}(0)$$

$$\times D_i^{(1)} \int_G Q_{G'}(p_1, p_2) \pi_n(p_2) D_{kl}^{(2)} K_G(p_2, p) \, dV(p_2),$$

$$Y_5(p_1, p) = \sum_{ij} \sum_m a_m^{ij}(0) \sum_k b^k(0) \pi_m(p_1)$$

$$\times D_{ij}^{(1)} \int_G Q_G(p_1, p_2) D_k^{(2)} K_G(p_2, p) \, dV(p_2),$$

$$Y_6(p_0, p) = \sum_i b^i(0) \sum_k b^k(0)$$

$$\times D_i^{(1)} \int_G Q_G(p_1, p_2) D_k^{(2)} K_G(p_2, p) \, dV(p_2).$$

Here $a_m^{ij}, a_{mn}^{ij}, b_m^i, \ldots$ mean partial derivatives of functions $a^{ij}, a^{ij}, b^i, \ldots$ on m-th and n-th coordinates accordingly.

Proof. In what follows we will omit subscripts in designations of kernels Q_G, K_G, X_G, Y_G when it does not lead to misunderstanding.

It is sufficient to prove that for any twice continuously differentiable functions $\varphi : (0 \le \varphi \le 1)$, determined on a unit ball, the function

$$\int_{\partial G} \left(R^{d-1} h_{RG}(Rp_0, Rp) - K(p_0, p) \right.$$

$$- R \int_G Q(p_0, p_1) \left(\sum_i b^i(0) D_i^{(1)} K(p_1, p) + X(p_1, p) \right) dV(p_1)$$

$$+ R^2 \int_G Q(p_0, p_1) \left(c(0) K(p_1, p) + Y(p_1, p) \right) dV(p_1) \Big) \varphi(p) \, dS(p)$$

$$\times \left(\int_G Q(p_0, p_1) \, dV(p_1) \right)^{-1}$$

$$\tag{5.22}$$

has an order $o(R^2)$ as $R \to 0$ uniformly on all $G \in \mathfrak{A}^\circ(K)$ and $p_0 \in G$.

On the boundary of region RG we consider function φ_R, where $\varphi_R(Rp) = \varphi(p)$ ($p \in \partial G$). For the solution u_{RG} of the Dirichlet problem for equation [5.17] in region RG with meaning φ_R on the boundary of the region the following representation is true

$$u_{RG}(p_0) \equiv \int_{\partial RG} \varphi_R(p) h_{RG}(p_0, p) \, dS_R(p)$$

$$= \int_{\partial G} R^{d-1} h_{RG}(p_0, Rp) \varphi(p) \, dS(p).$$

$$\tag{5.23}$$

The equation $\mathcal{L}u = 0$ we rewrite as $\Delta u = -(\mathcal{L} - \Delta)u$. For a short we will omit argument, function φ, in designations of solutions of the Dirichlet problem: $u_G(p)$ instead of a full designation $u_G(\varphi \mid p)$. Then according to the known formula of a solution of the Dirichlet problem for the Poisson equation (see [GIL 89, SOB 54], etc.) we have

$$u_{RG}(p_0) = \psi_R(p_0) + \int_{RG} Q_R(p_0, p_1) f(p_1) \, dV_R(p_1), \qquad [5.24]$$

where

$$\psi_R(p_0) = \int_{\partial RG} K_R(p_0, p_1) \varphi_R(p_1) \, dS_R(p_1),$$

$f = (\mathcal{L} - \Delta) u_{RG}$, $dS_R(p)$, $dV_R(p)$ are elements of surface and volume of region RG. Thus, we obtain integral equation with respect to u_{RG}.

Change variables: $p_i = R\bar{p}_i$ ($\bar{p}_i \in G$, $i = 0, 1$). Using representation

$$Q_R(p_1, p_2) = \begin{cases} \dfrac{1}{(d-2)\omega_d r_{12}^{d-2}} + g(p_1, p_2), & d \geq 3 \\[2mm] \dfrac{1}{2\pi} \ln \dfrac{1}{r_{12}} + g(p_1, p_2), & d = 2 \end{cases}$$

and its uniqueness (see [SOB 54, p. 302]), where $r_{ij} = |p_i - p_j|$, $\omega_d = 2\pi^{(d/2)}/\Gamma(d/2)$ is a measure of surface of a unit ball in d-dimensional space, $g(p_1, p_2)$ is a regular harmonic function, we obtain relations

$$R^{d-2} Q_R(R\bar{p}_1, R\bar{p}_2) = Q_1(\bar{p}_1, \bar{p}_2) \equiv Q(\bar{p}_1, \bar{p}_2),$$

$$R^{d-1} K_R(R\bar{p}_1, R\bar{p}_2) = K_1(\bar{p}_1, \bar{p}_2) \equiv K(\bar{p}_1, \bar{p}_2).$$

Moreover, if $F(R\bar{p}) = \overline{F}(\bar{p})$ and $\overline{D}_i$, $\overline{D}_{ij}$ are derivations on the appropriate coordinates of a point $\bar{p}$, then $D_i F(R\bar{p}) = \overline{D}_i \overline{F}(\bar{p})/R$ and $D_{ij} F(R\bar{p}) = \overline{D}_{ij} \overline{F}(\bar{p})/R^2$. From here

$$\psi_R(Rp) \equiv \int_{\partial RG} K_R(Rp, p_1) \varphi_R(p_1) \, dS_R(p_1)$$

$$= \int_{\partial G} K_R(Rp, Rp_1) \varphi_R(Rp_1) R^{d-1} dS(p_1)$$

$$= \int_{\partial G} K(p, p_1) \varphi(p_1) \, dS(p_1) \equiv \psi(p), \qquad [5.25]$$

$$\int_{RG} Q_R(Rp, p_1)(\mathcal{L} - \Delta) u_{RG}(p_1) \, dV_R(p_1)$$

$$= \int_{G} Q_R(Rp, Rp_1)(\overline{\mathcal{L}}_R - \overline{\Delta}_R) \bar{u}_{RG}(p_1) R^d \, dV(p_1),$$

where

$$\overline{\mathcal{L}}_R \overline{u} = \sum_{ij} \overline{a}^{ij} \overline{D}_{ij} \overline{u}/R^2 + \sum_i \overline{b}^i \overline{D}_i \overline{u}/R - \overline{c} \overline{u},$$

$$\overline{\Delta}_R \overline{u} = \sum_i \overline{D}_{ii} \overline{u}/R^2.$$

According to conditions of the theorem the function $(\overline{a}^{ij}(p) - \delta^{ij})/R = (a^{ij}(Rp) - \delta^{ij})/R$ is bounded in a unit ball uniformly on all R that are small enough. Using this property, we consider an operator $\mathcal{M}_R = R \cdot (\overline{\mathcal{L}}_R - \overline{\Delta}_R)$. With the help of this operator the integral equation [5.24] can be rewritten as

$$\overline{u}_{RG}(p_0) = \psi(p_0) + R \int_G Q(p_0, p_1) \mathcal{M}_R^{(1)} \overline{u}_{RG}(p_1) \, dV(p_1), \qquad [5.26]$$

$(p_0 \in G)$. Let us estimate a norm of an operator $\mathcal{M}_R$. We have

$$\left| \mathcal{M}_R \overline{u} \right| \leq \sum_{ij} \left| \overline{a}^{ij} - \delta^{ij} \right| / R \cdot \left| \overline{D}_{ij} \overline{u} \right|$$

$$+ \sum_i \left| \overline{b}^i \right| \cdot \left| \overline{D}_i \overline{u} \right| + R \cdot \left| \overline{c} \right| \cdot \left| \overline{u} \right| \leq C_1 \cdot \left| \overline{u} \right|_{2;G},$$

where

$$|f|_{2;S} = \sup_{p \in S} \sup_{ij} \left| D_{ij} f(p) \right| + \sup_{p \in S} \sup_i \left| D_i f(p) \right| + \sup_{p \in S} \left| f(p) \right|$$

(the denotation is borrowed from [GIL 89, p. 59]), and $C_1 = C_1(d, \Lambda_2)$.

The norm $|u|_{2;S}$ for $u = u_S(\varphi \mid \cdot)$ (solution on S of the Dirichlet problem for equation $Lu = f$, $f|_{\partial S} = \varphi$) is estimated in [GIL 89, p. 100]:

$$|u|_{2;S} \leq C_2 \left(|u|_{0;S} + |\varphi|_{2;S} + |f|_{0;S} \right) \qquad [5.27]$$

with a constant $C_2 = C_2(d, \Lambda_1, \Lambda_2, K)$, where $S \in \mathfrak{A}^\circ(K)$ $(S \subset B)$. Here, as well as above, it is assumed that the function φ is determined and uniformly twice continuously differentiable on B.

Let $\mathcal{L}u = 0$ on RG and $u = \varphi_R$ on $\partial(RG)$. Then $\overline{u}$ satisfies equation $\overline{\mathcal{L}}_R \overline{u} = 0$ and $\overline{u} = \varphi$ on ∂G. From here

$$\left| \mathcal{M}_R \overline{u} \right| \leq C_1 C_2 \left(|\overline{u}|_{0;G} + |\varphi|_{2;G} \right) \leq 2 C_1 C_2 |\varphi|_{2;G}.$$

On the other hand, according to [5.25] the function ψ is a solution of the Dirichlet problem for the Laplace equation on region G with a boundary value $\psi = \varphi$ on ∂G. From here

$$\left| \mathcal{M}_R \psi \right| \leq C_1 C_3 \left(|\psi|_{0;G} + |\varphi|_{2;G} \right)$$

$(C_3 = C_3(d, K))$. Furthermore, we apply the operator $\mathcal{M}_R$ to the function $u = \int_G Q(\cdot, p) f(p)\, dV(p)$. This u is a solution of the Dirichlet problem for the Poisson equation with a right part $-f$ and with a zero boundary condition. In this case

$$|\mathcal{M}_R u| \leq C_1 C_3 (|u|_{0;G} + |f|_{0;G}) \leq C_1 C_3 (|S_G|_{0;G} + 1)|f|_{0;G},$$

where

$$S_G(p_0) = \int_G Q(p_0, p)\, dV(p).$$

The value $|S_G|_{0;G} + 1$, where S_G is the solution of the Poisson equation, is no more than a constant $C_4 = C_4(d, \Lambda_1, \Lambda_2)$ (see [GIL 89, p. 43]).

Consider an operator

$$T_R u(p) \equiv \mathcal{M}_R \int_G Q(p, p_1) u(p_1)\, dV(p_1).$$

We have received $|T_R u|_{0;G} \leq C|u|_{0;G}$ where $C = C_1 C_3 C_4$. From here under condition $RC < 1$ a solution of the integral equations [5.26] can be represented as a series

$$\overline{u}_{RG} = \psi + R \int_G Q(\cdot, p) \mathcal{M}_R \psi(p)\, dV(p)$$

$$+ \sum_{k=1}^{\infty} R^{k+1} \int_G Q(\cdot, p) (T_R^k \mathcal{M}_R) \psi(p)\, dV(p). \tag{5.28}$$

In order to evaluate terms of the first and second order we use the Taylor formula for coefficients of equation [5.17]. Let $\mathcal{M}_R = \mathcal{M}_0 + R \cdot \mathcal{N} + \mathcal{W}_R$, where

$$\mathcal{M}_0 v(p) = \sum_{ij} \sum_m a_m^{ij}(0) \pi_m(p) \overline{D}_{ij} v(p) + \sum_i b^i(0) \overline{D}_i v(p),$$

$$\mathcal{N} v(p) = \frac{1}{2} \sum_{ij} \sum_{mn} a_{mn}^{ij}(0) \pi_m(p) \pi_n(p) \overline{D}_{ij} v(p)$$

$$+ \sum_i \sum_m b_m^i(0) \pi_m(p) \overline{D}_i v(p) - c(0) v(p),$$

$$\mathcal{W}_R v(p) = \sum_{ij} \alpha_R^{ij}(p) \overline{D}_{ij} v(p)$$

$$+ \sum_i \beta_R^i(p) \overline{D}_i v(p) - R \cdot (c(Rp) - c(0)) v(p),$$

$$\alpha_R^{ij}(p) = \left(a^{ij}(Rp) - \delta^{ij}\right)/R - \sum_m a_m^{ij}(0)\pi_m(p)$$

$$- \frac{1}{2}R\sum_{mn} a_{mn}^{ij}(0)\pi_m(p)\pi_n(p),$$

$$\beta_R^i(p) = b^i(Rp) - b^i(0) - R\sum_m b_m^i(0)\pi_m(p).$$

Factors $\alpha_R^{ij}(p)$, $\beta_R^i(p)$, $R(c(Rp) - c(0))$ obviously have an order $o(R)$ with $R \to 0$ uniformly on all $p \in B$. Taking into account an estimate [5.27], we obtain $\mathcal{W}_R\psi(p) = o(R)$ for any function φ from the given class uniformly on all $G \in \mathfrak{A}^\circ(K)$ and uniformly on all $p \in G$. From here in expansion of the function $u_{RG}(Rp_0)$ on degrees of R the coefficient of the first degree is equal to

$$\int_G Q(p_0, p)\mathcal{M}_0\psi(p)\,dV(p),$$

and that of the second degree

$$\int_G Q(p_0, p)\mathcal{N}\psi(p)\,dV(p)\int_G Q(p_0, p_1)\mathcal{M}_0^{(1)}\int_G Q(p_1, p)\mathcal{M}_0\psi(p)\,dV(p)\,dV(p_1).$$

The remaining terms of the sum give the function, having an order $o(R)$ for any function φ from a class of twice continuously differentiable functions on B uniformly on all $G \in \mathfrak{A}^\circ(K)$ and uniformly on all $p \in G$. Further, we use integral representations of $\overline{u}_{RG}$ of the form [5.23] and $\psi(p)$ of the form [5.25]. Substituting them in coefficients of expansion of the function $\overline{u}_{RG}$ and changing an order of integrations, we obtain representation [5.22], equivalent to a statement of the theorem. An assertion about additional terms of first and second order to be orthogonal to any constant follows from a property $(\forall p_1 \in G)$

$$\int_{\partial G} K(p_1, p_2)\,dS(p_2) = 1$$

[SOB 54]. The orthogonality of X_1 to any linear function on the boundary follows from a property $(\forall p_1 \in G)$

$$\int_{\partial G} K(p_1, p_2)\left(\pi_k(p_2) - \pi_k(p_1)\right)dS(p_2) = 0,$$

which is a direct corollary of the Green formula (see [GIL 89, SOB 54], etc.).   □

5.2.3. Neighborhood of spherical form

31. Exit from neighborhood of spherical form

The coefficients of expansion of a density derived in theorem 5.11 receive a concrete meaning with known analytical expression of the kernels $Q(p_1, p_2)$ and $K(p_1, p_2)$.

Let $G = B \equiv \{p : \|p\| < 1\}$ be a unit ball.

THEOREM 5.12. *Under the same conditions on coefficients of equation [5.17] as in theorem 5.11 the following weak asymptotic expansion of a density of the first exit from a small spherical neighborhood holds:*

$$R^{d-1} h_{RB}(0, Rp) = \frac{1}{\omega_d} \left(1 + R \left(\frac{1}{2} \sum_{i=1}^{d} b^i(0) \pi_i(p) + X(0, p) \right) \right.$$

$$\left. + R^2 \left(-c(0) \frac{1}{2d} + Y(0, p) \right) + o(R^2) \right) \quad (R \longrightarrow 0),$$

[5.29]

where the function X is orthogonal to any constant and any linear function on the boundary, the function Y is orthogonal to any constant function on the boundary, $\omega_d = 2\pi^{(d/2)}/\Gamma(d/2)$ is a measure of a surface of a unit ball in d-dimensional space. Moreover for a process that is homogenous in space $X = 0$ and

$$Y(0, p) = \frac{1}{8\omega_d} \left(\sum_{ij}^{d} b^i b^j \pi_i(p) \pi_j(p) - \frac{1}{d} \sum_{i}^{d} (b^i)^2 \right), \quad d \geq 2.$$

Proof. It is known (see [GIL 89, MIK 68, SOB 54], etc.) that for a ball $B = \{p \in \mathbb{R}^d : \|p\| < 1\}$ the Green function Q_B is equal to

$$Q_B(p_1, p_2) \equiv Q(p_1, p_2) = \begin{cases} \dfrac{1}{(d-2)\omega_d} \left(\dfrac{1}{r_{12}^{d-2}} - \dfrac{1}{\bar{r}_{12}^{d-2}} \right), & d \geq 3 \\[3mm] \dfrac{1}{2\pi} \left(\ln \dfrac{1}{r_{12}} - \ln \dfrac{1}{\bar{r}_{12}} \right), & d = 2, \end{cases}$$

where $\bar{r}_{12}^2 = (1 - r_1^2)(1 - r_2^2) + r_{12}^2$. For $G = B$ the kernel K, called the Poisson kernel, is equal to

$$K_B(p_1, p_2) \equiv K(p_1, p_2) = \frac{1 - r_1^2}{\omega_d r_{12}^d},$$

where $r_i = \|p_i\|$, $r_{ij} = \|p_i - p_j\|$. From here the *i*-th factor in the first term of the asymptotics is equal to

$$\int_B Q(0, p_1) D_i^{(1)} K(p_1, p_2) \, dV(p_1)$$

$$= \begin{cases} \displaystyle\int_B \frac{1}{(d-2)\omega_d} \left(\frac{1}{r_1^{d-2}} - 1 \right) D_i^{(1)} \frac{1 - r_1^2}{\omega_d r_{12}^d} \, dV(p_1), & d \geq 3, \\[4mm] \displaystyle\int_B \frac{1}{2\pi} \ln \frac{1}{r_1} D_i^{(1)} \frac{1 - r_1^2}{2\pi r_{12}^2} \, dV(p_1), & d = 2. \end{cases}$$

Integrating by parts on the i-th component and taking into account that $Q(\cdot, p_1) = 0$ with $p_1 \in \partial B$, we obtain the last expression equal to

$$-\int_B \frac{1}{(d-2)\omega_d} D_i^{(1)} \left(\frac{1}{r_1^{d-2}} - 1 \right) \frac{1 - r_1^2}{\omega_d r_{12}^d} \, dV(p_1), \quad d \geq 3,$$

$$-\int_B \frac{1}{2\pi} \left(D_i^{(1)} \ln \frac{1}{r_1} \right) \frac{1 - r_1^2}{2\pi \, r_{12}^2} \, dV(p_1), \quad d = 2.$$

With appropriate $d \geq 2$ both these expressions are equal to the following expression:

$$\frac{1}{\omega_d^2} \int_B \frac{\pi_i(p_1)}{r_1^d} \frac{1 - r_1^2}{r_{12}^d} \, dV(p_1). \tag{5.30}$$

The evaluation of integral [5.30] for $d = 2$ is not difficult to perform. For arbitrary dimension $d \geq 3$ we make a change of variables: $p \mapsto \bar{p} = \mathcal{A}p$, where $\mathcal{A}$ is an orthogonal transformation, determined by a matrix (of the same label). Here the point p_2 under this transformation passes in a point $(1, 0, \ldots, 0)$, where $\|p_2\| = 1$. Let $(\Phi_1, \ldots, \Phi_{d-1})$ be the spatial polar (spherical) coordinates of a point p_2. Then a matrix of such a transformation can be as follows

$$\mathcal{A} = \begin{pmatrix} C_1 & S_1 C_2 & S_1 S_2 C_3 & \cdots & S_1 S_2 \cdots S_{d-2} C_{d-1} & S_1 S_2 \cdots S_{d-1} \\ -S_1 & C_1 C_2 & C_1 S_2 C_3 & \cdots & C_1 S_2 \cdots S_{d-2} C_{d-1} & C_1 S_2 \cdots S_{d-1} \\ 0 & -S_2 & C_2 C_3 & \cdots & C_2 S_3 \cdots S_{d-2} C_{d-1} & C_2 S_3 \cdots S_{d-1} \\ \cdot & \cdot & \cdot & \cdots & \cdot & \cdot \\ 0 & 0 & 0 & \cdots & C_{d-2} C_{d-1} & C_{d-2} S_{d-1} \\ 0 & 0 & 0 & \cdots & -S_{d-1} & C_{d-1} \end{pmatrix}.$$

Here $C_i = \cos \Phi_i$, $S_i = \sin \Phi_i$. This transformation can be received as an outcome of sequential $d-1$ elementary turns: $\mathcal{A} = \mathcal{A}_1 \cdots \mathcal{A}_{d-1}$, where

$$\mathcal{A}_k = \begin{pmatrix} 1 & 0 & \cdots & 0 & & & & & \\ 0 & 1 & \cdots & 0 & & 0 & & 0 & \\ \cdot & \cdot & \cdots & \cdot & & & & & \\ 0 & 0 & \cdots & 1 & & & & & \\ & & & & C_k & S_k & & & \\ & 0 & & & -S_k & C_k & & 0 & \\ & & & & & & 1 & 0 & \cdots & 0 \\ & 0 & & & & 0 & 0 & 1 & \cdots & 0 \\ & & & & & & \cdot & \cdot & \cdots & \cdot \\ & & & & & & 0 & 0 & \cdots & 1 \end{pmatrix}.$$

Furthermore, we have

$$\frac{1}{\omega_d^2} \int_B \frac{\pi_i(p_1)(1 - r_1^2)}{r_1^d r_{12}^d} \, dV(p_1) = \frac{1}{\omega_d^2} \int_B \frac{\pi_i(\mathcal{A}^{-1}\bar{p}_1)(1 - r_1^2)}{r_1^d r_{12}^d} \, dV(\bar{p}_1),$$

where in the second expression $r_{12}^2 = 1 + r_1^2 - 2r_1c_1$ and c_1 is cosine of angle between vectors $\bar{p}_1$ and $(1, 0, \ldots, 0)$. It is easiest to calculate this integral for first coordinates. So

$$\pi_1\left(\mathcal{A}^{-1}p\right) = C_1x - S_1y, \qquad \pi_2\left(\mathcal{A}^{-1}p\right) = S_1C_2x + C_1C_2y - S_2z$$

etc., the k-th coordinate being represented by the $k + 1$-th summand. Fortunately, in our formulae a common case can be reduced to the case of first numbers with the help of variable change. Let $T : \mathbb{R}^d \to \mathbb{R}^d$ be an orthogonal transformation, which consists of the permutation of coordinates $Tp = (p^{i_1}, \ldots, p^{i_d})$, where $(i_1, \ldots, i_d)$ is a permutation of numbers $(1, 2, \ldots, d)$. Then

$$J\left(i_1, p_2\right) \equiv \int_B \frac{\pi_{i_1}\left(p_1\right)}{F\left(\cos\langle p_1, p_2\rangle\right)} \, dV\left(p_1\right)$$

$$= \int_B \frac{\pi_{i_1}\left(T^{-1}\bar{p}_1\right)}{F\left(\cos\langle T^{-1}\bar{p}_1, p_2\rangle\right)} \, dV\left(\bar{p}_1\right)$$

$$= \int_B \frac{\pi_1\left(\bar{p}_1\right)}{F\left(\cos\langle \bar{p}_1, Tp_2\rangle\right)} \, dV\left(\bar{p}_1\right) \equiv J\left(1, Tp_2\right),$$

where $\langle p_1, p_2\rangle$ is an angle between vectors p_1 and p_2. Let us calculate an integral with $i = 1$.

$$\frac{1}{\omega_d^2} \int_B \frac{x_1\left(1 - r_1^2\right)}{r_1^d r_{12}^d} \, dV\left(p_1\right)$$

$$= \frac{1}{\omega_d^2} \int_B \frac{\left(C_1x - S_1y\right)\left(1 - r^2\right)}{r^d l^d} \, dV\left(p\right)$$

$$= \frac{1}{\omega_d^2} \int_0^1 \int_0^{2\pi} \int_0^\pi \cdots \int_0^\pi \frac{C_1rc_1\left(1 - r^2\right)}{r^d l^d} r^{d-1} s_1^{d-2} \cdots s_{d-2}^1 \, dr \, d\varphi_1 \cdots d\varphi_{d-1}$$

$$= \frac{\omega_{d-1}}{\omega_d^2} \int_0^1 \int_0^\pi \frac{C_1c_1\left(1 - r^2\right)}{l^d} s_1^{d-2} \, dr \, d\varphi_1,$$

where $l^2 = 1 + r^2 - 2rc_1$, $(r, \varphi_1, \ldots, \varphi_{d-1})$ are polar coordinates of the point p and $c_i = \cos\varphi_i$, $s_i = \sin\varphi_i$.

This integral and the similar are possible to calculate on the basis of equality

$$1 = \int_B K\left(p_1, p_2\right) dS\left(p_2\right)$$

$$= \int_0^{2\pi} \int_0^\pi \cdots \int_0^\pi \frac{1 - r^2}{\omega_d l^d} s_1^{d-2} \cdots s_{d-2}^1 \, d\varphi_1 \cdots d\varphi_{d-1}$$

$$= \omega_{d-1} \int_0^\pi \frac{1 - r^2}{\omega_d l^d} s_1^{d-2} \, d\varphi_1,$$

from here

$$\int_0^\pi \frac{c_1 s_1^{d-2}}{l^d} \, d\varphi_1 = \frac{\omega_d}{\omega_{d-1}(1-r^2)}.$$ [5.31]

Integrating by parts the intrinsic integral and using this formula, we obtain the first coefficient

$$\frac{\omega_{d-1}}{\omega_d^2} \int_0^1 \int_0^\pi \frac{C_1 c_1 (1-r^2)}{l^d} s_1^{d-2} \, dr \, d\varphi_1$$

$$= \frac{C_1 \omega_{d-1}}{\omega_d^2} \int_0^1 (1-r^2) \int_0^\pi \frac{c_1 s_1^{d-2}}{l^d} \, dr \, d\varphi_1$$

$$= \frac{C_1 \omega_{d-1}}{\omega_d^2} \int_0^1 (1-r^2) \cdot \frac{r\omega_d}{\omega_{d-1}(1-r^2)} \, dr = \frac{C_1}{2\omega_d}.$$

The last expression can also be rewritten as $x_2/(2\omega_d)$. So, in this case $J(1, p_2) = \pi_1(p_2)/(2\omega_d)$. For arbitrary $k \in \{1, \ldots, d\}$ we choose a permutation operator T in such a manner that $k = i_1$. In this case $J(k, p_2) \equiv J(i_1, p_2) = J(1, Tp_2) = \pi_1(Tp_2)/(2\omega_d) = \pi_k(p)/(2\omega_d)$. From here coefficient of a principal term of the first order is equal to

$$\frac{1}{2\omega_d} \sum_i b^i(0) \pi_i(p).$$

Now calculate a coefficient of a principal term of the second order.

$$\int_B Q(0, p_1) K(p_1, p_2) \, dV(p_1)$$

$$= \int_B \frac{1}{(d-2)\omega_d} \left(\frac{1}{r_1^{d-2}} - 1 \right) \frac{1-r_1^2}{\omega_d r_{12}^d} \, dV(p_1)$$

$$= \frac{\omega_{d-1}}{(d-2)\omega_d^2} \int_0^1 \left(\frac{1}{r^{d-2}} - 1 \right)(1-r^2) r^{d-1} \int_0^\pi \frac{s_1^{d-2}}{l^d} \, d\varphi_1 dr$$

$$= \frac{\omega_{d-1}}{(d-2)\omega_d^2} \int_0^1 (1-r^2)(1-r^{d-2}) r \cdot \frac{\omega_d}{\omega_{d-1}(1-r^2)} \, dr$$

$$= \frac{1}{(d-2)\omega_d} \int_0^1 (1-r^{d-2}) r \cdot dr = \frac{1}{2d\omega_d},$$

where $d \geq 3$. The same formula is fair for $d = 2$ despite the other method of evaluation which we omit in view of a lack of any complexities.

32. *Evaluation of orthogonal term* X

Show on an example $d \geq 3$ how the orthogonal term X can be calculated. We have

$$\int_{\partial B} X(0, p_2) \varphi(p_2) \, dS(p_2)$$

$$= \sum_{ij} \sum_m a_m^{ij}(0) \int_B Q(0, p_1) \pi_m(p_1) D_{ij}^{(1)} \psi(p_1) \, dV(p_1).$$

Here the integral is equal to

$$\int_B \frac{1}{(d-2)\omega_d} \left(\frac{1}{r_1^{d-2}} - 1 \right) \pi_m(p_1) D_{ij}^{(1)} \psi(p_1) \, dV(p_1)$$

$$= - \int_B \frac{1}{(d-2)\omega_d} D_i^{(1)} \left(\left(\frac{1}{r_1^{d-2}} - 1 \right) \pi_m(p_1) \right) D_j^{(1)} \psi(p_1) \, dV(p_1).$$

Again we accept $m = 1$ and $x_1 = \pi_1(p_1)$. The further formulae differ for cases of possible equality of indexes $i, j, 1$. There are only 5 variants:

(1) $i \neq j, 1 \neq j, 1 \neq i$;

(2) $i \neq j, 1 = i$;

(3) $i \neq j, 1 = j$;

(4) $i = j, 1 \neq i$;

(5) $i = j = 1$.

In the first case we have

$$- \int_B \frac{1}{(d-2)\omega_d} D_i^{(1)} \left(\left(\frac{1}{r_1^{d-2}} - 1 \right) x_1 \right) D_j^{(1)} \psi(p_1) \, dV(p_1)$$

$$= \int_B \frac{x_1}{\omega_d} \frac{\pi_i(p_1)}{r_1^d} D_j^{(1)} \psi(p_1) \, dV(p_1).$$

Here again we apply an integration by parts, but boundary values of the first function here are not equal to zero. We obtain the previous expression equal to

$$= \int_{B^j} \left[\frac{x_1}{\omega_d} \frac{\pi_i(p_1)}{r_1^d} \psi(p_1) \right]_{j_a}^{j_b} dV^j(p_1^{(j)}) - \int_B D_j^{(1)} \left(\frac{x_1}{\omega_d} \frac{\pi_i(p_1)}{r_1^d} \right) \cdot \psi(p_1) \, dV(p_1),$$

where integration on a d-dimensional ball B we have replaced by iterated integration:

(1) outside on the $(d-1)$-dimensional ball B^j (with the exception of the j-th coordinate) with a variable of integration $p_1^{(j)}$ and with the element of volume $dV^j(p_1^{(j)})$,

(2) inside on the j-th coordinate on an interval (j_a, j_b), where $-j_a = j_b = \sqrt{1 - \|p_1^{(j)}\|^2}$.

We see that on the boundary of B function $\psi = \varphi$, and also $r_1 = 1$ and $j_b = j_1^+ \equiv \pi_j(p_1^+)$ (the value of the j-th coordinate of p_1, when it is situated on the "upper" boundary), and $j_a = j_1^- \equiv \pi_j(p_1^-)$ (the value of the j-th coordinate of p_1, when it is situated on the "lower" boundary). Thus,

$$\int_{B^j} \left[\frac{x_1 \, \pi_i(p_1)}{\omega_d} \frac{1}{r_1^d} \psi(p_1) \right]_{j_a}^{j_b} dV^j\left(p_1^{(j)}\right)$$

$$= \int_{B^j} \frac{x_1 \pi_i(p_1)}{\omega_d r_1^d} \left(\varphi(p_1^+) - \varphi(p_1^-) \right) dV^j\left(p_1^{(j)}\right).$$

To simplify this integral, we consider an integral on a surface of a ball

$$\int_{\partial B} F(p_1) \, dS(p_1) = \int_{B^j} \left(F(p_1^+) + F(p_1^-) \right) \frac{1}{\sqrt{1 - \|p_1^{(j)}\|^2}} \, dV^j\left(p_1^{(j)}\right)$$

$$= \int_{B^j} \left(F(p_1^+)/j_1^+ - F(p_1^-)/j_1^- \right) dV^j\left(p_1^{(j)}\right).$$

We note that this integral is equal to preceding one when

$$F(p_1) = \frac{x_1 \pi_i(p_1) \pi_j(p_1)}{\omega_d} \varphi(p_1).$$

The second integral is equal to

$$\int_B (-1) \frac{x_1 \pi_i(p_1) \pi_j(p_1) d}{\omega_d r_1^{d+2}} \psi(p_1) \, dV(p_1),$$

Remembering possibilities of a permutation operator, we consider this integral for the first three coordinates of a point p_1 (to simplify matters we assume that $d \geq 5$):

$$\int_B (-1) \frac{x_1 y_1 z_1 d}{\omega_d r_1^{d+2}} \psi(p_1) \, dV(p_1)$$

$$= \int_{\partial B} \varphi(p_2) \left(\int_B (-1) \frac{x_1 y_1 z_1 d}{\omega_d r_1^{d+2}} \frac{1 - r_1^2}{\omega_d r_{12}^d} dV(p_1) \right) dS(p_2).$$

The interior integral is equal to

$$-\frac{d}{\omega_d^2} \int_B (C_1 x - S_1 y)(S_1 C_2 x + C_1 C_2 y - S_2 z)$$

$$\times (S_1 S_2 C_3 x + C_1 S_2 C_3 y + C_2 C_3 z - S_3 w) \frac{1 - r^2}{r^{d+2} l^d} \, dV(p),$$

where $p = (x, y, z, w, \ldots)$, $l^2 = 1 + r^2 - 2rc_1$, c_1 is the cosine of an angle between p and $(1, 0, \ldots, 0)$. The last expression is equal to

$$-\frac{w_{d-3}d}{w_d^2} \int_0^1 \int_0^\pi \int_0^\pi \int_0^\pi (1 - r^2)(C_1 S_1^2 C_2 S_2 C_3 x^3 + C_1(C_1^2 - 2S_1^2)C_2 S_2 C_3 x y^2$$

$$- C_1 C_2 S_2 C_3 x z^2)\frac{1}{l^d} s_1^{d-2} s_2^{d-3} s_3^{d-4} \, d\varphi_1 \, d\varphi_2 \, d\varphi_3 \, dr.$$

Using the method applied above of evaluation of integrals (formula [5.31]), having expression l^d as a denominator of the ratio, we discover that this integral is equal to $dw_d x_2 y_2 z_2/4$ (we omit details). Together with the first term this gives us $\pi_i(p_2)\pi_j(p_2)x_2(d+4)/4$. So, in the first case we obtain a common view of summand

$$a_1^{ij}\pi_i(p_2)\pi_j(p_2)x_2\frac{d+4}{4} \quad (p_2 \in \partial B),$$

which are summarized to the term Xw_d. The values of summands of the remaining four types, which compose the term X, are calculated similarly.

33. Evaluation of orthogonal term Y

The same method of evaluation is applicable to a term Y. We calculate a component of this term, which exists in the case of homogenous in space process, when the coefficients of the differential equation are constant. In this case

$$Y(0, p) = \sum_i b^i \sum_k b^k \int_B Q(0, p_1)D_i^{(1)} \int_B Q(p_1, p_2)D_k^{(2)}K(p_2, p) \, dV(p_2) \, dV(p_1).$$

We have

$$\int_B Q(0, p_1)D_i^{(1)} \int_B Q(p_1, p_2)D_k^{(2)}K(p_2, p) \, dV(p_2) \, dV(p_1)$$

$$= -\int_B D_i^{(1)}Q(0, p_1) \int_B Q(p_1, p_2)D_k^{(2)}K(p_2, p) \, dV(p_2) \, dV(p_1)$$

$$= \int_B D_k^{(2)}K(p_2, p) \int_B \frac{\pi_i(p_1)}{w_d r_1^d}Q(p_1, p_2) \, dV(p_1) \, dV(p_2).$$

An interior integral for $d \geq 3$ is equal to

$$\int_B \frac{\pi_i(p_1)}{w_d r_1^d}\frac{1}{(d-2)w_d}\left(\frac{1}{r_{12}^{d-2}} - \frac{1}{\bar{r}_{12}^{d-2}}\right) dV(p_1)$$

and for $d = 2$ it is equal to

$$\int_B \frac{\pi_i(p_1)}{2\pi r_1^2}\frac{1}{2\pi}\left(\ln\frac{1}{r_{12}} - \ln\frac{1}{\bar{r}_{12}}\right) dV(p_1).$$

Using permutation of coordinates we substitute π_i by x; further, we make replacement variables with the help of orthogonal transformation and partial integration. In the case of $d \geq 3$ we have the interior integral

$$
\frac{C_1 \omega_{d-1}}{\omega_d^2 (d-2)} \left(\int_0^{r_2} \int_0^\pi \frac{c_1 s_1^{d-2}}{r_2^{d-2} l^{d-2}(r/r_2)} d\varphi_1 dr + \int_{r_2}^1 \int_0^\pi \frac{c_1 s_1^{d-2}}{r^{d-2} l^{d-2}(r_2/r)} d\varphi_1 dr \right.
$$
$$
\left. - \int_0^1 \int_0^\pi \frac{c_1 s_1^{d-2}}{l^{d-2}(rr_2)} d\varphi_1 dr \right) \quad (d \geq 3),
$$

where $l(r) = 1 + r^2 - 2rc_1$. For $d = 2$ the interior integral is equal to

$$
\frac{C_1}{(2\pi)^2} \int_0^1 \int_0^{2\pi} c_1 \left(\ln \frac{1}{r_{12}} - \ln \frac{1}{r_{12}} \right) d\varphi_1 dr_1
$$
$$
= \frac{C_1}{(2\pi)^2} \int_0^1 \int_0^{2\pi} s_1^2 r_1 r_2 \left(\frac{1}{r_{12}^2} - \frac{1}{r_{12}^2} \right) d\varphi_1 dr_1
$$
$$
= \frac{C_1}{(2\pi)^2} \left(\int_0^{r_2} \frac{r_1}{r_2} \int_0^{2\pi} \frac{s_1^2}{l(r_1/r_2)} d\varphi_1 dr_1 + \int_{r_2}^1 \frac{r_2}{r_1} \int_0^{2\pi} \frac{s_1^2}{l(r_2/r_1)} d\varphi_1 dr_1 \right.
$$
$$
\left. - \int_0^1 r_1 r_2 \int_0^{2\pi} \frac{s_1^2}{l(r_1 r_2)} d\varphi_1 dr_1 \right).
$$

After integration with the use of formula [5.31] we discover that the interior integral is equal to

$$
\frac{C_1 r_2}{2(d-2)^2 \omega_d} \left(\frac{1}{r_2^{d-2}} - 1 \right), \quad d \geq 3,
$$
$$
\frac{C_1 r_2}{2 \cdot 2\pi} \ln \frac{1}{r_2}, \quad d = 2,
$$

where according to the denotation $C_1 r_2 = x_2$. We return to an evaluation of the outside integral. Using equality to zero of an interior integral for $p_2 \in \partial B$, we obtain

$$
\begin{cases} -\int_B \frac{1}{2(d-2)^2 \omega_d^2} D_k^{(2)} \left[x_2 \left(\frac{1}{r_2^{d-2}} - 1 \right) \right] \frac{1 - r_2^2}{r_{23}^d} dV(p_2), & d \geq 3, \\ \\ -\int_B \frac{1}{2(2\pi)^2} D_k^{(2)} \left(x_2 \ln \frac{1}{r_2} \right) \frac{1 - r_2^2}{r_{23}^2} dV(p_2), & d = 2. \end{cases}
$$

When $\pi_k \neq x$ we obtain for $d \geq 3$:

$$-\int_B \frac{x_2}{2(d-2)\omega_d}\left(-\frac{\pi_k(p_2)}{r_2^d}\right)\frac{1-r_2^2}{\omega_d r_{23}^d}\,dV(p_2)$$

$$= \frac{1}{2(d-2)\omega_d^2}\int_B \frac{x_2\pi_k(p_2)\left(1-r_2^2\right)}{r_2^d r_{23}^d}\,dV(p_2).$$

Once again, to simplify matters we substitute an arbitrary coordinate $\pi_k(p_2)$ with second y_2; furthermore, we apply an orthogonal transformation and partial integration; again we use [5.31]. As an outcome we discover that for $d \geq 3$ the previous integral is equal to

$$\frac{x_3 y_3 \omega_{d-2}}{2(d-2)\omega_d^2}\int_0^1 r(1-r^2)\int_0^\pi\int_0^\pi \frac{\left(c_1^2 - s_1^2 c_2^2\right)s_1^{d-2}s_2^{d-3}}{l^d}\,d\varphi_1 d\varphi_2 dr = \frac{x_3 y_3}{8(d-2)\omega_d}.$$

For $d = 2$ the similar expression is equal to $x_3 y_3/(16\pi)$.

When $\pi_k(p_2) = x_2$ we obtain for $d \geq 3$:

$$-\int_B \frac{1}{2(d-2)^2\omega_d^2}\left(\frac{1}{r_2^{d-2}} - 1 - \frac{x_2^2(d-2)}{r_2^d}\right)\frac{1-r_2^2}{r_{23}^d}\,dV(p_2)$$

$$= -\frac{1}{2(d-2)^2\omega_d^2}\left(\frac{\omega_d}{2d} - C_1^2\omega_{d-1}\int_0^1 (1-r^2)r\int_0^\pi \frac{c_1^2 s_1^{d-2}}{l^d}\,d\varphi_1 dr\right.$$

$$\left. - S_1^2\omega_{d-2}\int_0^1 (1-r^2)r\int_0^\pi\int_0^\pi \frac{c_2^2 s_1^d s_2^{d-3}}{l^d}\,d\varphi_1\varphi_2 dr\right)$$

$$= \frac{1}{8(d-2)\omega_d}\left(x_3^2 - \frac{1}{d}\right).$$

and similarly for $d = 2$ we obtain $(x_3^2 - 1/2)/(16\pi)$.

Making inverse permutation of indexes finally we discover that in the case of the homogenous differential equation

$$Y_6(0, p) = \frac{1}{8\,\omega_d}\left(\sum_{iqj}^d b^i b^j \pi_i(p)\pi_j(p) - \frac{1}{d}\sum_i^d \left(b^i\right)^2\right), \quad d \geq 2.$$

The evaluation Y in an inhomogenous case is rather awkward in connection with the necessity of taking into account the coincidence of indexes. $\quad\square$

5.2.4. *Characteristic operator*

Consider the characteristic Dynkin operator [DYN 63, p. 201]

$$\mathfrak{A}\varphi(p) \equiv \lim_{G \downarrow p} \frac{E_p\big(\varphi \circ X_{\sigma_G}, \ \sigma_G < \infty\big) - \varphi(p)}{m_G(p)}.$$

In this definition a sequence of open sets G is supposed to tend to a point $p \in G$ in an arbitrary manner, namely, the diameters of these sets tend to zero. We will not consider a problem of determination of characteristic operators in general. We will investigate a family of neighborhoods $RG + p$ ($R > 0$, $R \to 0$), converging to the point p, where G is some neighborhood of zero-point in $\mathbb{R}^d$ with a smooth enough boundary. The existence of the characteristic operator of a Markov process on the set of twice differentiable functions is proved in a case when on this set there exists an infinitesimal operator. However, the Markov property is not necessary for the characteristic operator $A_\lambda \varphi$ or its partial case $A_0 \varphi = \mathfrak{A}\varphi$ to exist. A proof of this fact based on theorem 5.11, which proved for a matrix of coefficients of second derivatives of special view.

34. *Arbitrary matrix of coefficients*

Let us consider equation [5.17] with arbitrary non-degenerate positive definite continuous matrix (a^{ij}) of coefficients of the second derivatives, satisfying the norming condition [5.18]. For a fixed initial point $p = 0$ let us deduce this equation to the form, when the matrix of coefficients of the second derivatives is equal to (δ^{ij}) at point 0. In order to do this let us represent the matrix $\mathbf{a} \equiv (a^{ij}(0))$ of the form $\overline{L}^T \overline{L}$, where $\overline{L} \equiv (\overline{L}^{ij})$ is a matrix of non-degenerate linear transformation, and $\overline{L}^T$ is the transposed matrix. It is well-known that the possibility of such a presentation is a necessary and sufficient condition of positive definition of the original matrix. Let us transform all sets of points of the original space $\mathbb{X} \equiv \mathbb{R}^d$ according to the map $y = Lx$, where L is the inverse matrix for $\overline{L}$: $L\overline{L} = \overline{L}L = E$, where $E = (\delta^{ij})$ (i.e. $\overline{L} = L^{-1}$) (we designate a linear map and its matrix with the same letter). The map L induces a map of the functional space $\mathcal{D}$ in itself, which we designate by the same letter. Thus, according to this definition, $(L\xi)(t) = L(\xi(t))$ ($\xi \in \mathcal{D}$, $\sqcup \geq \iota$). In this case $\sigma_{LA}(L\xi) = \sigma_A(\xi)$, $X_{\sigma_{LA}}(L\xi) = LX_{\sigma_A}(\xi)$. For the given semi-Markov family of measures (P_x) on $\mathcal{F}$, map L induces the family of measures (P_x^L), where $P_{Lx}^L(LS) = P_x(S)$ ($S \in \mathcal{F}$), and the family of semi-Markov transition and transition generating functions: F_Δ^L, f_Δ^L, where $F_{LA}^L(dt \times LS \mid Lx) = F_\Delta(dt \times S \mid x)$, $f_{LA}^L(\lambda, LS \mid Lx) = f_\Delta(\lambda, S \mid x)$. Evidently, the linear map of the space in itself preserves the semi-Markov property of a process and its diffusion type. The integral function $f_\Delta^L(\lambda, \varphi \mid x)$ of this process satisfies the transformed differential equation

$$\frac{1}{2} \sum_{ij} \widetilde{a}^{ij} u''_{ij} + \sum_{i=1}^{d} \widetilde{b}^i u'_i - \overline{c}u = 0. \qquad [5.32]$$

This is a solution of the Dirichlet problem for this equation on the set Δ with the meaning φ on the boundary of this set. Forms of coefficients of this equation follow from a rule of differentiation. For $y = Lx$ let us designate $u(x) = u(L^{-1}y) = \overline{u}(y) = \overline{u}(Lx)$. In particular, $\overline{c}(y) = c(x)$. Furthermore,

$$u_i'(x) = \sum_{k=1}^d \overline{u}_k'(y)L^{ki}, \quad \cdot \quad u_{ij}''(x) = \sum_{kl} \overline{u}_{kl}''(y)L^{ki}L^{lj}.$$

From here

$$\sum_{i=1}^d b^i(x)u_i'(x) = \sum_{ik} L^{ki}b^i(x)\overline{u}_k'(y) = \sum_{ik} L^{ki}\overline{b}^i(y)\overline{u}_k'(y).$$

Hence, $\widetilde{b}^k = \sum_{i=1}^d L^{ki}\overline{b}^i$, where $\overline{b}^i(y) = b^i(x)$. Similarly, $\widetilde{a}^{kl} = \sum_{ij} L^{ki}L^{lj}\overline{a}^{ij}$, where $\overline{a}^{ij}(y) = a^{ij}(x)$. In this case

$$\widetilde{a}^{kl}(0) = \sum_{ij} L^{ki}L^{lj}a^{ij}(0) = \sum_{ij} L^{ki}L^{lj}\sum_{m=1}^d \overline{L}^{mi}\overline{L}^{mj}$$

$$= \sum_{m=1}^d \delta^{km}\delta^{lm} = \delta^{kl}.$$

Hence, our investigation of asymptotics for small parameter meanings relates to the density of the first exit point $h_\Delta^L(0, p)$ of the transformed process. We have to find an asymptotics for the original process. Replacing variables in the integral we obtain

$$\int_{\partial\Delta} \varphi(x)h_\Delta(0, x)\,dS(x) = \int_{L\partial\Delta} \varphi(L^{-1}y)h_\Delta(0, L^{-1}y)\left(J_\Delta^L(L^{-1}y)\right)^{-1}dS(y),$$

where $J_\Delta^L(x)$ is a coefficient of variation of a surface element $\partial\Delta$ size under map $x \mapsto y = Lx$. This coefficient depends on the orientation of this element in space. It is a Jacobian analogy (see item 38). In particular, for $Lx = Rx$ (multiplying of vector x by scalar R) we have $J_\Delta^L(x) = R^{d-1}$ (see formula [5.40]). The exact meaning of this coefficient for L of a common view is not required here because in our deposition there is the product $h_\Delta(0, L^{-1}y) \cdot (J_\Delta^L(L^{-1}y))^{-1}$ which represents the investigated density $h_{L\Delta}^L(0, y)$. In particular, for $L\Delta = RG$ ($R > 0$) (i.e. for $\Delta = L^{-1}RG = RL^{-1}G$) we have

$$\int_{\partial RL^{-1}G} \varphi(x)h_{RL^{-1}G}(0, x)\,dS(x)$$

$$= \int_{\partial RG} \varphi(L^{-1}y)h_{RG}^L(0, y)\,dS(y) \qquad [5.33]$$

$$= \int_{\partial G} \varphi(L^{-1}Ry)h_{RG}^L(0, Ry)R^{d-1}\,dS(y).$$

Let φ be twice differentiable in a neighborhood of point 0. A Taylor expansion in a neighborhood of point 0 (up to the second order) for function $\varphi(L^{-1}Ry)$ as $R \to 0$ has a view $\varphi(0) + R\sum_{i=1}^{d}\sum_{k=1}^{d}\varphi'_k(0)\overline{L}^{ki}y^i + (R^2/2)\sum_{ij}\sum_{kl}\varphi''_{kl}(0)\overline{L}^{ki}\overline{L}^{lj}y^iy^j + o(R^2)$. According to formula [5.21] the term $h_{RG}^L(0, Ry)R^{d-1}$ in integral [5.33] has expansion

$$K_G(0,y) + \int_G Q_G(0,p_1)\left(R\sum_i \widetilde{b}^i(0)D_i^{(1)}K_G(p_1,y) + RX_G(p_1,y)\right.$$

$$\left. - R^2c(0)K_G(p_1,y) + R^2Y_G(p_1,y)\right)dV(p_1)$$

$$+ o(R^2)\int_G Q_G(0,p_1)\,dV(p_1).$$

Furthermore, we use properties of orthogonality of kernels X_G, Y_G, and also that of the kernel $K_G(p_0,p)$, which implies besides integrals of 1 and $\pi_i(p) - \pi_i(p_0)$ the following property:

$$\int_{\partial G} K_G(p_0,p)(\pi_i(p) - \pi_i(p_0))(\pi_j(p) - \pi_j(p_0))\,dS(p)$$

$$= 2\delta^{ij}\int_G Q_G(p_0,p)\,dV(p),$$

which also follows from the Green formula (see [GIL 89, MIK 68, SOB 54], etc.). Hence, integral [5.33] is equal to

$$\varphi(0) + R^2\left(\frac{1}{2}\sum_{kl}\varphi''_{kl}(0)\sum_i \overline{L}^{ki}\overline{L}^{li} + \sum_k \varphi'_k(0)\sum_i \overline{L}^{ki}\widetilde{b}^i(0)\right.$$

$$\left. - c(0)\varphi(0) + o(1)\right)\int_G Q_G(0,p_1)\,dV(p_1).$$

Let us note that $\sum_i \overline{L}^{ki}\overline{L}^{li} = a^{kl}(0)$, and also

$$\sum_i \overline{L}^{ki}\widetilde{b}^i(0) = \sum_i\sum_j \overline{L}^{ki}L^{ij}\overline{b}^j(0) = \sum_j \delta^{kj}\overline{b}^j = \overline{b}^k(0) = b^k(0).$$

We also denote that $G = L\Delta$. Hence the following expansion is true

$$\left(\int_{\partial R\Delta}\varphi(x)h_{R\Delta}(0,x)\,dS(x) - \varphi(0)\right)/R^2$$

$$= \left(\frac{1}{2}\sum_{ij}a^{ij}(0)\varphi''_{ij}(0) + \sum_i b^i(0)\varphi'_i(0) - c(0)\varphi(0) + o(1)\right)2C(L\Delta),$$

[5.34]

where $C(L\Delta) = \int_{L\Delta}Q_{L\Delta}(0,p_1)\,dV(p_1)$.

35. Coefficients of the equation

Using formula [5.34] we will prove the following property of coefficients: if a solution of the Dirichlet problem $u_{RG}(\varphi \mid 0)$ is an integral with respect to density h_{RG} of a transition generating function of some semi-Markov process (in this case parameter λ is the parameter of the Laplace transformation), then coefficients a^{ij} and b^i do not depend on λ.

Let φ be a continuous bounded positive function equal to $1 + m\pi_i$ in some neighborhood of point 0, where m is some value differing from zero (we determine it below). Substituting this function in formula [5.34], we obtain that the right part is equal to $1 + (mb^i(0) - c(0) + o(1))C(L\Delta)$. Assuming that the product $b^i(0)C(L\Delta)$ depends on λ, we get a contradiction: on the one hand, the left part of the equality does not increase on the whole half-line as a non-decreasing function of the Laplace transformation, while on the other hand, by choosing m appropriately it is possible to make the right part of the equality increase on some interval from the half-line. Let $\varphi = 1 + m\pi_i\pi_j$. In this case the right part is equal to $1 + (ma^{ij}(0) - c(0) + o(1))C(L\Delta)$. In the same manner we are convinced that the product $a^{ij}(0)C(L\Delta)$ does not depend on λ, but in this case a stronger assertion is true, because coefficients $a^{ij}(0)$ satisfy the norming condition [5.18]. Hence, as well as the integral $C(L\Delta)$, all the coefficients $b^i(0)$ and $a^{ij}(0)$ do not depend on parameter λ.

36. Dimensionless characteristic operator

Formula [5.34] prompts us to use R^2 as a norming factor while treating behavior of the process in a small neighborhood of its initial point. In this case it is convenient to choose some standard decreasing sequence of sets, namely, the sequence of concentric balls (RB) $(R \to 0)$. We define "dimensionless" characteristic operator as

$$\lim_{R \to 0} A_\lambda^{o,R}(\varphi \mid x) = A_\lambda^o(\varphi \mid x),$$

where

$$A_\lambda^{o,R}(\varphi \mid x) = \frac{1}{R^2}\left(f_R(\lambda, \varphi \mid x) - \varphi(x)\right).$$

Evidently this operator is a differential second-order operator of elliptic type with coefficients different from those of equation [5.17] with additional factor $C(LB)$. Let us show that under norming condition [5.18] the integral $C(LB)$ does not depend on

$L.$ We have

$$
\begin{aligned}
1 &= \int_{\partial LB} K_{LB}(0,x)\,dS(x) = \int_{\partial B} K_{LB}(0,Lx)J_{LB}^{L^{-1}}(Lx)\,dS(x) \\
&= \sum_{i=1}^{d} \int_{\partial B} (x^i)^2 K_{LB}(0,Lx)J_{LB}^{L^{-1}}(Lx)\,dS(x) \\
&= \sum_{i=1}^{d} \int_{\partial LB} (\pi_i(\overline{L}x))^2 K_{LB}(0,x)\,dS(x) \\
&= \sum_{i=1}^{d} \int_{\partial LB} \sum_{kl} \overline{L}^{ik}\overline{L}^{il} x^k x^l K_{LB}(0,x)\,dS(x) \\
&= \sum_{i=1}^{d} \int_{\partial LB} \sum_{k=1}^{d} (\overline{L}^{ik})^2 (x^k)^2 K_{LB}(0,x)\,dS(x) \\
&= \sum_{i=1}^{d}\sum_{k=1}^{d} (\overline{L}^{ik})^2 2\,C(LB) = \sum_{i=1}^{d} a^{ii}(0)2\,C(LB) = 2d\,C(LB).
\end{aligned}
$$

Thus, $C(LB) = 1/(2d)$. In the case of $a^{ij} = \delta^{ij}$ this formula can be proved by immediate integrating. We have proved the following statement.

THEOREM 5.13. *Let the SM transition generating function satisfy equation [5.17] with an arbitrary non-degenerate positive determined continuous matrix (a^{ij}) of coefficients of the second derivatives satisfying norming condition [5.18]. Then the dimensionless lambda-characteristic operator, A_λ^o, is determined on the corresponding set of functions and*

$$
A_\lambda^o(\varphi \mid x) = \frac{1}{d}\left(\frac{1}{2}\sum_{ij} a^{ij}(0)\varphi_{ij}''(0) + \sum_{i} b^i(0)\varphi_i'(0) - c(0)\varphi(0) + o(1) \right). \quad [5.35]
$$

Proof. It immediately follows from the above formula and [5.34].   □

37. Strict lambda-characteristic operator

Let us find an asymptotics of an expectation of the first exit time from a small neighborhood of zero-point:

$$
m_G(p) \equiv E_p(\sigma_G;\ \sigma_G < \infty) \quad (p \in G),
$$

which can be represented as

$$-\frac{\partial}{\partial \lambda} E_p\left(e^{-\lambda \sigma_G};\ \sigma_G < \infty\right)\big|_{\lambda=0} = -\frac{\partial}{\partial \lambda} y_G(\lambda, p)|_{\lambda=0},$$

where

$$y_G(\lambda, p) \equiv f_G(\lambda, 1 \mid p) \equiv \int_{\partial G} f_G\left(\lambda, p_1 \mid p\right) dS(p_1)$$

(see item 27). The function $f_G(\lambda, 1 \mid p)$ is a solution of the Dirichlet problem for equation $\mathcal{L}u = 0$ on G with a boundary condition $u = 1$ on ∂G. In this case we consider an arbitrary matrix of coefficients of second derivatives with limitation introduced in item 34. Moreover, between all coefficients only $c(p) = c(\lambda, p)$ (dependent on λ); we assume that $-\partial c(\lambda, p)/\partial \lambda$ is a completely monotone function of λ. It is a necessary condition for function $y_G(\lambda, p)$ to be completely monotone. As noted above (see item 34), $y_G(\lambda, p) = y_{LG}^L(\lambda, Lp)$, where the right part relates to the transformed process with transition generating functions investigated in theorem 5.17. Let us rewrite equation [5.26] for the given boundary condition

$$y_{RLG}^L(\lambda, p_0) = 1 - R^2 \int_{LG} Q_{LG}(p_0, p_1) \bar{c}(\lambda, Rp_1) y_{RLG}^L(\lambda, p_1)\, dV(p_1). \quad [5.36]$$

In order to find an equation that m_{RLG}^L satisfies, it is sufficient to differentiate both parts of this equation by λ. Since $\mathbf{a} = \overline{L}^T \overline{L}$ does not depend on λ, it is evident that the matrix L can be chosen independently of λ. Differentiation on λ of integrands for $\lambda = 0$, like those in theorem 5.7, is justified by the ArcellTheorem theorem. In order to apply this theorem it is sufficient to have monotone convergents of ratios $(\bar{c}(\lambda, p) - \bar{c}(0, p))/\lambda$ and $(y_{RG}^L(\lambda, p) - y_{RG}^L(0, p))/\lambda$ to their limits. This property follows from non-negativeness of second derivatives on λ of functions $-\bar{c}(\lambda, p)$ and $y_{RG}^L(\lambda, p)$. From here

$$m_{RLG}^L(Rp_0) = M_{RLG}^L(Rp_0)$$
$$- R^2 \int_{LG} Q_{LG}(p_0, p_1) \bar{c}(0, Rp_1) m_{RLG}^L(p_1)\, dV(p_1), \quad [5.37]$$

where

$$M_{RLG}^L(Rp_0) = R^2 \int_{LG} Q_{LG}(p_0, p_1) \gamma(Rp_1)\, dV(p_1),$$

$$\gamma(p) = \bar{c}'(0, p) \equiv \frac{\partial}{\partial \lambda} \bar{c}(\lambda, p)\big|_{\lambda=0}.$$

From equations [5.36] and [5.37] the existence and uniqueness of $m_{RLG}^L(Rp)$ follows, for which the following asymptotics is fair

$$m_{RLG}^L(Rp_0) = \int_{LG} Q_{LG}(p_0, p_1)\, dV(p_1)\left(R^2 \gamma(0) + o(R^2)\right) \quad [5.38]$$

uniformly on all $G \in \mathfrak{A}^{\circ}(K)$ and $p_0 \in LG$. In particular,

$$m_{RLG}^{L}(0) = C(LG)\left(R^2\gamma(0) + o(R^2)\right).$$

Evidently, $m_{RG}(0) = m_{RLG}^{L}(0)$. Hence under conditions of theorem 5.17 a strict lambda-characteristic operator is determined. According to formula [5.34] on the set of twice differentiable functions φ at point 0, it is expressed as

$$A_\lambda(\varphi \mid 0) = \frac{1}{\gamma(0)}\left(\sum_{ij} a^{ij}(0)\varphi_{ij}''(0) + \sum_{i} b^i(0)\varphi_i'(0) - c(0)\varphi(0)\right). \qquad [5.39]$$

38. Coefficient of transformation of surface element size

Let us consider a non-degenerate linear map $L : \mathbb{R}^d \to \mathbb{R}^d$ and region G, for which at point $p \in \partial G$ the outside normal to its surface is determined: $n = n(p)$. Let $W(p, n)$ be a tangent hyperplane to region G at point p, and $S(p, n)$ be some $(d - 1)$-dimensional region in this hyperplane. Under map L this "planar" region passes in the other planar region $LS(p, n)$. A ratio of $(d - 1)$-dimensional Lebesgue measures $|LS(p, n)|/|S(p, n)|$ evidently does not depend on the form and size of the planar region $S(p, n)$, and depends only on the orientation of this region, i.e. on n. From here it follows that this ratio is the main term of asymptotics of ratios of Lebesgue measures of small "surface" neighborhood $\delta S(p) \subset \partial G$ of the point p and its image $L\delta S(p)$. This main term depends on the normal to the surface at point p. This limit ratio is said to be a coefficient of transformation of surface element size and is denoted by $J_G^L(p)$. Let us express this coefficient in terms of matrix L and normal n.

Evidently, the coefficient of transformation of a planar figure size does not vary under parallel translation of this figure. It is convenient to assume that this planar figure is a parallelogram, having zero-point (origin of coordinates) as some of its angles. In addition, this planar figure is perpendicular to the normal $n = (n^1, \ldots, n^d)$. Let $\{x_1, \ldots, x_{d-1}\}$ be a system of linear independent vectors, perpendicular to vector n, and $|S|$ be Lebesgue measures of parallelogram S formed by these vectors in the corresponding hyperplane. It is well-known that a volume of d-dimensional parallelepiped "spanned" on d linear independent vectors is numerically equal to an absolute meaning of determinant of the matrix, composed with coordinates of these vectors. On the other hand, a volume of parallelepiped corresponding to the system of vectors $\{x_1, \ldots, x_{d-1}, n\}$ is equal to $S|n| = S$ (because $|n| = 1$ and n is perpendicular to S). Hence $S = |\sum_{k=1}^{d} n^k X_{dk}|$, where X_{dk} is an adjunct of the k-th elements of the d-th row of the matrix, composed with vectors $\{x_1, \ldots, x_{d-1}, n\}$. Let us consider vector n_1 of unit length with coordinates

$$n_1^k = X_{dk} \left/ \sqrt{\sum_{k=1}^{d} \left(X_{dk}\right)^2}\right..$$

From the theory of determinants it is well-known that $\sum_{k=1}^{d} x_i^k X_{dk} = 0$ $(1 \leq i \leq d-1)$. Hence, the vector n_1 is perpendicular to hyperplane $W(p,n)$ and therefore $n_1 = n$. From here $X_{dk} = |S|n^k$ and

$$|S| = \left| \sum_{k=1}^{d} n_1^k X_{dk} \right| = \sqrt{\sum_{k=1}^{d} (X_{dk})^2}.$$

Let us consider the system of vectors $\{y_1, \ldots, y_{d-1}\}$, where $y_k = Lx_k$. According to the previous consideration, the Lebesgue measure of parallelogram LS in corresponding hyperplane with these composing vectors is equal to

$$|LS| = \sqrt{\sum_{k=1}^{d} (Y_{dk})^2},$$

where Y_{dk} is an adjunct of the k-th element of the d-th row of the matrix composed with vectors $\{y_1, \ldots, y_{d-1}, y\}$ (y is an arbitrary vector, determining the d-th row of the matrix). Assuming $y_l^k = \sum_{i=1}^{d} L^{ki} x_l^i$, we have

$$Y_{dk} = \begin{vmatrix} y_1^1 & \cdots & y_1^{k-1} & y_1^{k+1} & \cdots & y_1^d \\ \vdots & & \vdots & \vdots & & \vdots \\ y_{d-1}^1 & \cdots & y_{d-1}^{k-1} & y_{d-1}^{k+1} & \cdots & y_{d-1}^d \end{vmatrix}.$$

From here the determinant is equal to

$$\sum L^{1a_1} \cdots L^{k-1,a_{k-1}} L^{k+1,a_{k+1}} \cdots L^{da_d}$$

$$\times \begin{vmatrix} x_1^{a_1} & \cdots & x_1^{a_{k-1}} & x_1^{a_{k+1}} & \cdots & x_1^{a_d} \\ \vdots & & \vdots & \vdots & & \vdots \\ x_{d-1}^{a_1} & \cdots & x_{d-1}^{a_{k-1}} & x_{d-1}^{a_{k+1}} & \cdots & x_{d-1}^{a_d} \end{vmatrix},$$

where the summands are determined for all collections $\{a_1, \ldots, a_{k-1}, a_{k+1}, \ldots, a_d\}$ of $d-1$ elements, where each element accepts any meaning from 1 to d. Because the determinant with repeating columns is equal to zero, the previous sum can be considered only for collections with pair-wise different elements. In this case it is possible to group summands with respect to absent elements (one from d possible). From here

$$Y_{dk} = \sum_{j=1}^{d} \sum L^{1b_1} \cdots L^{k-1,b_{k-1}} L^{k+1,b_{k+1}} \cdots L^{db_d}$$

$$\times \begin{vmatrix} x_1^{b_1} & \cdots & x_1^{b_{k-1}} & x_1^{b_{k+1}} & \cdots & x_1^{b_d} \\ \vdots & & \vdots & \vdots & & \vdots \\ x_{d-1}^{b_1} & \cdots & x_{d-1}^{b_{k-1}} & x_{d-1}^{b_{k+1}} & \cdots & x_{d-1}^{b_d} \end{vmatrix},$$

where summands are determined on all permutations

$$\{b_1, \ldots, b_{k-1}, b_{k+1}, \ldots, b_d\}$$

of numbers $\{1, \ldots, j-1, j+1, \ldots, d\}$. Let z be an index of such a permutation. Then according to our denotation and rule of the column permutation, the determinant in a common term of the previous sum is equal to $(-1)^z X'_{dj}$, where $X'_{dj} = (-1)^{d+j} X_{dj}$ is a minor of a term with index dj. Further, according to the rule of calculation of the determinant we have

$$\sum_{b_1,\ldots,b_{k-1},b_{k+1},\ldots,b_d} (-1)^z L^{1b_1} \cdots L^{k-1,b_{k-1}} L^{k+1,b_{k+1}} \cdots L^{db_d} = L'_{kj},$$

where L'_{kj} is a minor determinant of the element L^{kj} in matrix L. As a result we obtain

$$Y_{dk} = \sum_{j=1}^{d} X'_{dj} L'_{kj} = (-1)^{d+k} \sum_{j=1}^{d} X_{dj} L_{kj} = (-1)^{d+k} |S| \sum_{j=1}^{d} n^j L_{kj}.$$

At last we have

$$J_G^L(p) = |LS|/|S| = \sqrt{\sum_{k=1}^{d} \left(\sum_{j=1}^{d} n^j L_{kj}\right)^2}. \qquad [5.40]$$

Chapter 6

Time Change and Semi-Markov Processes

A time change is a transformation of the process which preserves distributions of all points of the first exit (see [VOL 58, VOL 61, DEL 72, ITO 65, BLU 62, LAM 67, ROD 71, SER 71], etc.). Many other properties of the process are also preserved. Investigation of all such properties is equivalent to the study of random sequences of states or traces (see [KOL 72]). It can be explained as follows. Let us consider a class of trajectories where two elements differ from each other only by a time change. There thus exists a correspondence between a trajectory and such a class of trajectories.[1]

This title, as well as the title random process, will refer to an admissible family of measures with parameters $x \in \mathbb{X}$ corresponding to initial points of random trajectories. The join of measures in a family may be justified if for this family there are nontrivial times of regeneration. For a random trace it can be only intrinsic Markov times [HAR 76a, CHA 79, CHA 81, JAN 81]. Among them are all the first exit times from open sets and their iterations.

The possibility of constructing a chain of the first exit points for a fixed sequence of open sets when the rank of this sequence is as small as desired makes it possible to describe almost all properties of the random trace by properties of such random chains. With the help of chains of the first exit points the following theorem is proved:

---

1. For this object there is a name "map of ways" [DYN 63, ITO 65], although it seems to be less suitable than "trace".

For distributions of two random traces to coincide, it is necessary and sufficient that there exist two continuous inverse random time changes transforming these processes in some third process. If it is admitted for the inverse time change to be discontinuous then for any two random processes with identical distributions of random trace there exists a random time change transforming one process into the other.

In semi-Markov process theory, the time change plays the special role. The time change keeping the semi-Markov property of the process is tightly connected with additive functionals [HAR 80a]. An example of the additive functional is an integral function determining the transformation of measures of a random process, like that of item 5.21, theorem 5.10.

Analysis of a general semi-Markov process would be significally simpler if this process is known to be obtained by a time change from a Markov process. Under rather weak suppositions a stepped SM process possesses this property [YAC 68] although there exist stepped SM processes with properties that are absent in an ordinary Markov process, and these properties are preserved under time change. Possibility of such Markov representation of a semi-Markov process is discussed in Chapter 8.

Studying a random trace of an SM process leads to an object "time run along the trace", when time taken in order to make a trace is considered as an additive functional on the trace. Such decomposition of a process on trace- and time-components is natural for a step-process. It is also useful for general SM processes. In this way the Lévy expansion and representation for a conditional distribution of time run along the trace were obtained. They are suitable to be expressed in terms of so called curvilinear integrals. Parameters of the Lévy expansion are tightly connected with a lambda-characteristic operator of the process. In both cases (continuous and discontinuous) the form of dependence on the parameter λ serves as an indicator of Markovness of the SM process. In particular, when a measure of the Poisson component in the Lévy expansion is not equal to zero (hence dependence on λ is not linear) the process contains intervals of constancy with non-predicted position in space, i.e. is a typically non-Markov process.

Curvilinear integrals appear to be very useful for constructing a theory of stochastic integrals, determined on the trace of a semi-Markov process. This theory is constructed in order to solve some internal problems of semi-Markov processes, but it can be useful in common theory of random processes because it gives a new perspective on the theory of stochastic integration.

6.1. Time change and trajectories

The style of representation in this section is analogous to that of Chapter 2. When dealing with time change we do not use an abstract probability space but construct a measure on a product of two spaces: the space of sample functions (trajectories of the process) and the space of time change (functions of special view). Such construction makes it possible to see in detail the construction of a measure of a transformed process and its connection with the original one. Such a method of enlargement of the original space we will use later. In this section we consider direct and inverse time changes. The name "direct time change" is connected with the sense of map ψ which transforms a point of "old" time into a point or an interval of points of "new" time. When $\psi(t)$ is equal to an interval it means in the new time scale a transformed function preserves a constant value equal to the meaning the function had before transformation at the instant t. With the help of direct time change it is convenient to express values of such Markov times that do not lose their meaning after the change of time scale, for example, the first exit time from an open set. In this case if ξ and $\widetilde{\xi}$ are original and transformed functions under the direct time change ψ, then $\sigma_\Delta(\widetilde{\xi}) = \psi(\sigma_\Delta(\xi))$. Values of the transformed function are able to be expressed in terms of inverse time change: $\widetilde{\xi} = \xi \circ \varphi$, where φ is inverse (in natural sense) for ψ.

The connection of Markov and semi-Markov processes and the presence in trajectories of semi-Markov process of intervals of constancy (see theorem 3.9) means it is necessary to study the not necessarily continuous and strictly monotone time change. By supposing such functions as a time change (direct and inverse here are equivalent), it is necessary to choose one of two modifications: continuous from the left (Ψ_1) and continuous from the right (Ψ_2). Continuity from the right (see [BLU 68], etc.) attracts us not only due to its similarity to $\mathcal{D}$, but also due to important property: ($\forall \xi \in \mathcal{D}$) ($\forall \psi \in \Psi_2$) $\xi \circ \psi \in \mathcal{D}$. However, in this case it is necessary to admit functions $\psi \in \Psi_2$, for which $\psi(0) > 0$. It especially is inconvenient that ($\forall t \geq 0$) $\theta_t \psi(0) = \psi(t) - \psi(t) = 0$ (see item 3), i.e. $\widetilde{\theta}_t \psi$ belongs to a narrower class than all Ψ_2. With a continuity from the left ($\exists \xi \in \mathcal{D}$) ($\exists \psi \in \Psi_1$) $\xi \circ \psi \notin \mathcal{D}$, but always $\psi(0) = 0$. Since we will use only those time changes which do not change a sequence of states (see item 14), and for these time changes the property $\xi \circ \psi \in \mathcal{D}$ is fulfilled, in view of a useful property $\psi(0) = 0$ we choose continuity from the left. The basic content of the section consists of a proof of the theorems on measurability of different sets and maps connected with time change.

1. Direct and inverse time change

Let Ψ_1 be a set of all non-decreasing functions continuous from the left: $\varphi : \mathbb{R}_+ \to \mathbb{R}_+$, $\varphi(0) = 0$, $\varphi(t) \to \infty$ $(t \to \infty)$; Φ a set of all continuous functions $\varphi \in \Psi_1$; and Ψ a set of all strictly increasing functions $\varphi \in \Psi_1$. Let $(\cdot)^* : \Psi_1 \to \Psi_1$ be a map of the form: ($\forall t \in \mathbb{R}_+$) ($\forall \varphi \in \Psi_1$) $\varphi^*(t) = \inf \varphi^{-1}[t, \infty)$. Obviously, ($\forall \varphi \in \Phi$) $\varphi^* \in \Psi$ and ($\forall \varphi \in \Psi$) $\varphi^* \in \Phi$, and ($\forall \varphi \in \Psi_1$) $(\varphi^*)^* = \varphi$, and also ($\forall \varphi_1, \varphi_2 \in \Psi_1$) $\varphi_1 \circ \varphi_2 \in \Psi_1$. We will suppose ($\forall \psi \in \Psi_1$) $\psi(\infty) = \infty$.

The set of functions Ψ_1 together with the sigma-algebra of its subsets defined below determines the set of time changes. The names "direct" and "inverse" time change for $\varphi \in \Psi_1$ depends on the role φ plays with respect to a function $\xi \in \mathcal{D}$: either it transforms a special Markov time by scheme $\tau(\xi) \mapsto \varphi(\tau(\xi))$ (direct change), or it transforms values of a function by scheme $\xi(t) \mapsto \xi(\varphi(t))$ (inverse change). As a rule a role of a direct change is played by strictly increasing function $\varphi \in \Psi$, and that of an inverse change by continuous function $\varphi \in \Phi$. Between these classes there is one-to-one correspondence determined by map $(\cdot)^*$.

2. Properties of time changes

Let

$$\Gamma_1(\varphi) = \left\{ (t, s) \in \mathbb{R}_+^2 : s = \varphi(t) \right\},$$
$$\Gamma_2(\varphi) = \left\{ (t, s) \in \mathbb{R}_+^2 : t = \varphi^*(s) \right\}$$

be graphs of function $\varphi \in \Psi_1$ and the appropriate inverse function. Let $\Gamma_0(\varphi) = \Gamma_1(\varphi) \cup \Gamma_2(\varphi)$ be the union of the graphs.

PROPOSITION 6.1. *The following properties are fair*

(1) $(\forall \varphi_1, \varphi_2 \in \Phi)\ (\varphi_1 \circ \varphi_2^*)^* = \varphi_2 \circ \varphi_1^*$,

(2) $(\forall \varphi \in \Psi_1)\ (\exists \varphi_1, \varphi_2 \in \Phi)\ \varphi = \varphi_1 \circ \varphi_2^*$.

Proof. (1) The following equalities are obvious:

$$\left(\varphi_1 \circ \varphi_2^* \right)^* (t) = \inf \left(\varphi_1 \circ \varphi_2^* \right)^{-1} [t, \infty) = \inf \left(\varphi_2^* \right)^{-1} \varphi_1^{-1} [t, \infty)$$
$$= \inf \left(\varphi_2^* \right)^{-1} \left[\inf \varphi_1^{-1}[t, \infty), \infty \right) = \varphi_2 \left(\inf \varphi_1^{-1}[t, \infty) \right)$$
$$= \varphi_2 \left(\varphi_1^*(t) \right).$$

(2) Obviously, in a frame (t, s) the line $\{(t, s) : s + t = c\}$ $(c \geq 0)$ intersects a set $\Gamma_0(\varphi)$ only at one point. Let $(t_c, s_c) \in \Gamma_0(\varphi)$, where $s_c + t_c = c$, and also $\varphi_1(c) = s_c$, $\varphi_2(c) = t_c$. Obviously, $\varphi_1, \varphi_2 \in \Phi$. Let us prove that $\varphi = \varphi_1 \circ \varphi_2^*$. We obtain $(\forall a \in \mathbb{R}_+)\ a = \varphi_2(a + \varphi(a))$, since $(a, \varphi(a)) \in \Gamma_0(\varphi)$. In this case according to the left-continuity of φ $\varphi_2^*(a) = a + \varphi(a)$. Furthermore, $\varphi_1(a + \varphi(a)) = \varphi(a)$, hence $\varphi(a) = \varphi_1(\varphi_2^*(a))$. $\qquad\square$

Similar facts are established by Meyer [MEY 73].

3. Basic maps

Let $\widehat{T}$ be a set of all maps $\tau : \Psi_1 \times \mathcal{D} \to \overline{\mathbb{R}}_+$. We will consider maps:

$\widetilde{X}_t : \Psi_1 \to \mathbb{R}_+ \ (t \in \mathbb{R}_+), \ \widetilde{X}_t\psi = \psi(t)$, meaning of the process at fixed instant;

$\widetilde{\theta}_t : \Psi_1 \to \Psi_1 \ (t \in \mathbb{R}_+), \ \widetilde{X}_s\widetilde{\theta}_t\psi = \widetilde{X}_{s+t}\psi - \widetilde{X}_t\psi \ (s \in \mathbb{R}_+)$, shift of the process;

$\widetilde{\alpha}_t : \Psi_1 \to \Psi_1 \ (t \in \overline{\mathbb{R}}_+), \ \widetilde{X}_s\widetilde{\alpha}_t\psi = \widetilde{X}_{s\wedge t}\psi + (0 \vee (s - t))$, stopping at the fixed instant with linear continuation;

$\widehat{X}_t : \Psi_1 \times \mathcal{D} \to \mathbb{R}_+ \times \mathbb{X} \ (t \in \mathbb{R}_+), \ \widehat{X}_t(\psi, \xi) = (\widetilde{X}_t\psi, X_t\xi)$, meaning of the two-dimensional process at the fixed instant;

$\widehat{\theta}_t : \Psi_1 \times \mathcal{D} \to \Psi_1 \times \mathcal{D} \ (t \in \mathbb{R}_+), \ \widehat{\theta}_t(\psi, \xi) = (\widetilde{\theta}_t\psi, \theta_t\xi)$, shift of the two-dimensional process on the fixed time;

$\widehat{\alpha}_t : \Psi_1 \times \mathcal{D} \to \Psi_1 \times \mathcal{D} \ (t \in \overline{\mathbb{R}}_+), \ \widehat{\alpha}_t(\psi, \xi) = (\widetilde{\alpha}_t\psi, \alpha_t\xi)$, stopping of the two-dimensional process at the fixed instant;

$\widehat{X}_\tau : \{\psi, \xi : \tau(\psi, \xi) < \infty\} \to \mathbb{R}_+ \times \mathbb{X} \ (\tau \in \widehat{T}), \ \widehat{X}_\tau(\psi, \xi) = \widehat{X}_{\tau(\psi, \xi)}(\psi, \xi)$, meaning of the two-dimensional process at the non-fixed instant;

$\widehat{\theta}_\tau : \{\psi, \xi : \tau(\psi, \xi) < \infty\} \to \Psi_1 \times \mathcal{D} \ (\tau \in \widehat{T}), \ \widehat{\theta}_\tau(\psi, \xi) = \widehat{\theta}_{\tau(\psi, \xi)}(\psi, \xi)$, shift of the two-dimensional process on the non-fixed time;

$\widehat{\alpha}_\tau : \Psi_1 \times \mathcal{D} \to \Psi_1 \times \mathcal{D} \ (\tau \in \widehat{T}), \ \widehat{\alpha}_\tau(\psi, \xi) = \widehat{\alpha}_{\tau(\psi, \xi)}(\psi, \xi)$, stopping of the two-dimensional process on the non-fixed time;

$p : \Psi_1 \times \mathcal{D} \to \Psi_1, \ p(\psi, \xi) = \psi$; the pair consisted of a time change and trajectory;

$q : \Psi_1 \times \mathcal{D} \to \mathcal{D}, \ q(\psi, \xi) = \xi$; the pair consisted of a time change and trajectory.

4. Relations of maps

PROPOSITION 6.2. *The basic maps are connected by the following relations:*

(1) $\widetilde{X}_s \circ \widetilde{\theta}_t = \widetilde{X}_{s+t} - \widetilde{X}_t$;

(2) $\widetilde{X}_s \circ \widetilde{\alpha}_t = \widetilde{X}_{s\wedge t} + (0 \vee (s - t))$;

(3) $\widetilde{\alpha}_s \circ \widetilde{\alpha}_t = \widetilde{\alpha}_{s\wedge t}$;

(4) $\widetilde{\theta}_s \circ \widetilde{\theta}_t = \widetilde{\theta}_{s+t}$;

(5) $\widetilde{\alpha}_s \circ \widetilde{\theta}_t = \widetilde{\theta}_t \circ \widetilde{\alpha}_{s+t}$;

(6) $\widehat{X}_s \circ \widehat{\theta}_t = (\widetilde{X}_s \circ \widetilde{\theta}_t, X_s \circ \theta_t)$;

(7) $\widehat{X}_s \circ \widehat{\alpha}_t = (\widetilde{X}_s \circ \widetilde{\alpha}_t, X_s \circ \alpha_t)$;

(8) $\widehat{\alpha}_s \circ \widehat{\alpha}_t = \widehat{\alpha}_{s\wedge t}$;

(9) $\widehat{\theta}_s \circ \widehat{\theta}_t = \widehat{\theta}_{s+t}$;

(10) $\widehat{\alpha}_s \circ \widehat{\theta}_t = \widehat{\theta}_t \circ \widehat{\alpha}_{s+t}$;

(11) $p \circ \widehat{\alpha}_t = \widetilde{\alpha}_t \circ p$;

(12) $p \circ \widehat{\theta}_t = \widetilde{\theta}_t \circ p$;

(13) $q \circ \widehat{\alpha}_t = \alpha_t \circ q$;

(14) $q \circ \widehat{\theta}_t = \theta_t \circ q$.

Proof. It immediately follows from the definition (see theorem 2.1).   □

5. Time change in stopped and shifted functions

PROPOSITION 6.3. *Let* $\varphi \in \Psi_1$, $\xi \in D$. *Then* $(\forall (t, s) \in \Gamma_1(\varphi) \cap \Gamma_2(\varphi))$ $\widetilde{\alpha}_t \varphi = (\widetilde{\alpha}_s \varphi^*)^*$, $\widetilde{\theta}_t \varphi = (\widetilde{\theta}_s \varphi^*)^*$ *and, besides*

(1) $(\forall t \in \overline{\mathbb{R}}_+) \, \alpha_t(\xi \circ \varphi) = (\alpha_{\varphi(t)} \xi) \circ \varphi$,

(2) $(\forall (t, s) \in \Gamma_1(\varphi) \cap \Gamma_2(\varphi)) \, \theta_t(\xi \circ \varphi) = \theta_s \xi \circ \widetilde{\theta}_t \varphi$.

Proof. The statement about $\widetilde{\alpha}_t \varphi$ and $\widetilde{\theta}_t \varphi$ is obvious. Further $(\forall t_1 \in \mathbb{R}_+)$ the following equalities are evident

$$X_{t_1} \alpha_t(\xi \circ \varphi) = X_{t_1 \wedge t}(\xi \circ \varphi) = \xi\big(\varphi(t_1 \wedge t)\big) = \xi\big(\varphi(t_1) \wedge \varphi(t)\big)$$
$$= X_{\varphi(t_1)} \alpha_{\varphi(t)} \xi = X_{t_1}\big(\alpha_{\varphi(t)} \xi \circ \varphi\big),$$
$$X_{t_1} \theta_t(\xi \circ \varphi) = X_{t_1 + t}(\xi \circ \varphi) = \xi\big(\varphi(t_1 + t)\big),$$
$$X_{t_1}\big(\theta_s \xi \circ \widetilde{\theta}_t \varphi\big) = X_{(\widetilde{\theta}_t \varphi)(t_1)} \theta_s \xi = \xi\big(s + \big(\widetilde{\theta}_t \varphi\big)(t_1)\big)$$
$$= \xi\big(s + \varphi(t + t_1) - \varphi(t)\big).$$

Since $s = \varphi(t)$ we have $X_{t_1} \theta_{t_1}(\xi \circ \varphi) = X_{t_1}(\theta_s \xi \circ \widetilde{\theta}_t \varphi)$.   □

6. Streams of sigma-algebras

Let $\mathfrak{G} = \sigma(\widehat{X}_t, \, t \in \mathbb{R}_+)$, $\mathfrak{B} = \mathfrak{G} \otimes \mathcal{F}$, $\mathfrak{G}_t = \sigma(\widetilde{X}_s, \, s \leq t) \, (t \in \mathbb{R}_+)$, $\mathfrak{B}_t = \mathfrak{G}_t \otimes \mathcal{F}_t$.

It is obvious that $\Phi \in \mathfrak{G}$ and $\Psi \in \mathfrak{G}$, and also the following properties of a stream of sigma-algebras (filtration) $(\mathfrak{B}_t)_0^\infty$ take place:

(1) $t_1 < t_2 \Rightarrow \mathfrak{B}_{t_1} \subset \mathfrak{B}_{t_2}$;

(2) $\mathfrak{B} = \mathfrak{B}_\infty = \sigma(\mathfrak{B}_t, \, t \in \mathbb{R}_+)$;

(3) $\widehat{X}_s \in \mathfrak{B}_t / \mathcal{B}(\mathbb{R}_+ \times \mathbb{X}) \, (s \leq t)$.

7. About representation of sigma-algebras

PROPOSITION 6.4. *The following properties are fair*

(1) $(\forall t \in \overline{\mathbb{R}}_+)\ \widetilde{\alpha}_t \in \mathfrak{G}/\mathfrak{G}$, $\mathfrak{G}_t = \widetilde{\alpha}_t^{-1}\mathfrak{G}$,

(2) $(\forall t \in \overline{\mathbb{R}}_+)\ \widehat{\alpha}_t \in \mathfrak{B}/\mathfrak{B}$, $\mathfrak{B}_t = \widehat{\alpha}_t^{-1}\mathfrak{B}$.

Proof. Let $S \in \mathcal{B}(\mathbb{R}_+)$, $B_1 \in \mathfrak{G}$, $B_2 \in \mathcal{F}$. Then

(1) We have

$$\widetilde{\alpha}_t^{-1}\widetilde{X}_s^{-1}S = \{\psi : \widetilde{X}_s\widetilde{\alpha}_t\psi \in S\}$$
$$= \{\psi : \widetilde{X}_{s \wedge t}\psi + (0 \vee (s - t)) \in S\} \in \mathfrak{G}_t \subset \mathfrak{G}.$$

Hence, $\widetilde{\alpha}_t$ is measurable. Besides

$$\widetilde{\alpha}_t^{-1}\mathfrak{G} = \widetilde{\alpha}_t^{-1}\sigma(\widetilde{X}_s, s \in \mathbb{R}_+) = \sigma(\widetilde{X}_s \circ \widetilde{\alpha}_t, s \in \mathbb{R}_+)$$
$$= \sigma(\widetilde{X}_{s \wedge t} + (0 \vee (s - t)), s \in \mathbb{R}_+) = \sigma(\widetilde{X}_{s \wedge t}, s \in \mathbb{R}_+)$$
$$= \sigma(\widetilde{X}_s, s \leq t) = \mathfrak{G}_t.$$

(2) We have

$$\widehat{\alpha}_t^{-1}(B_1 \times B_2) = \widetilde{\alpha}_t^{-1}B_1 \times \alpha_t^{-1}B_2 \in \mathfrak{G}_t \otimes \mathcal{F}_t \subset \mathfrak{B}.$$

Hence, $\widehat{\alpha}_t$ is measurable. Furthermore

$$\widehat{\alpha}_t^{-1}\mathfrak{B} = \widehat{\alpha}_t^{-1}\sigma(\widetilde{X}_{s_1} \circ p, X_{s_2} \circ q, s_1, s_2 \in \mathbb{R}_+)$$
$$= \sigma(\widetilde{X}_{s_1} \circ p \circ \widehat{\alpha}_t, X_{s_2} \circ q \circ \widehat{\alpha}_t, s_1, s_2 \in \mathbb{R}_+)$$
$$= \sigma(\widetilde{X}_{s_1} \circ \widetilde{\alpha}_t \circ p, X_{s_2} \circ \alpha_t \circ q, s_1, s_2 \in \mathbb{R}_+)$$
$$= \mathfrak{G}_t \otimes \mathcal{F}_t. \qquad \square$$

8. Set of Markov times

Let $\widehat{MT}$ be a set of Markov times concerning the stream of sigma-algebras $(\mathfrak{B}_t)_0^\infty$:

$$\tau \in \widehat{MT} \Longleftrightarrow \tau \in \widehat{T}, \quad (\forall t \in \mathbb{R}_+)\ \{\tau \leq t\} \in \mathfrak{B}_t.$$

Let $\mathfrak{B}_\tau$ $(\tau \in \widehat{MT})$ be the sigma-algebra of preceding events for a moment τ:

$$\mathfrak{B}_\tau = \{B \in \mathfrak{B} : (\forall t \in \mathbb{R}_+)\ B \cap \{\tau \leq t\} \in \mathfrak{B}_t\}.$$

9. Properties of the sigma-algebra

THEOREM 6.1. *The following properties are fair*

(1) $\tau \in \mathrm{MT} \Rightarrow \tau \circ q \in \widehat{\mathrm{MT}}$,

(2) $\mathfrak{B}_{\tau \circ q} = \widehat{\alpha}^{-1}_{\tau \circ q} \mathfrak{B} \ (\tau \in \mathrm{MT})$.

Proof. Let $\tau \in \mathrm{MT}$. Then $(\forall t \in \mathbb{R}_+)$

$$\{\tau \circ q \leq t\} = q^{-1}\{\tau \leq t\} = \Psi_1 \times \{\tau \leq t\} \in \mathfrak{B}_t,$$

and the first statement is proved. Let

$$\mathfrak{B}^1_{\tau \circ q} = \{B \in \mathfrak{B} : B = \widehat{\alpha}^{-1}_{\tau \circ q} B\},$$

$$\mathfrak{B}^2_{\tau \circ q} = \{B \in \mathfrak{B} : (\forall t \in \mathbb{R}_+)\ B \cap \{\tau \circ q = t\} \in \mathfrak{B}_t\}.$$

Then $B \in \mathfrak{B}_{\tau \circ q} \Leftrightarrow (\forall t \in \mathbb{R}_+)\ B \cap \{\tau \circ q \leq t\} \in \mathfrak{B}_t$ and

$$\{\tau \circ q = t\} = q^{-1}\{\tau = t\} = \Psi_1 \times \{\tau = t\} \in \mathfrak{B}_t.$$

From here $(\forall B \in \mathfrak{B}_{\tau \circ q})\ (\forall t \in \mathbb{R}_+)\ B \cap \{\tau \circ q = t\} \in \mathfrak{B}_t$, $B \in \mathfrak{B}^2_{\tau \circ q}$ and $\mathfrak{B}_{\tau \circ q} \subset \mathfrak{B}^2_{\tau \circ q}$. Further, by proposition 6.4(2)

$$B \in \mathfrak{B}^2_{\tau \circ q} \Longrightarrow B = \bigcup_{t \in \overline{\mathbb{R}}_+} (B \cap \{\tau \circ q = t\})$$

$$= \bigcup_{t \in \overline{\mathbb{R}}_+} \widehat{\alpha}^{-1}_{\tau \circ q} B \cap \{\tau \circ q = t\}$$

$$= \widehat{\alpha}^{-1}_{\tau \circ q} B.$$

From here $\mathfrak{B}^2_{\tau \circ q} \subset \mathfrak{B}^1_{\tau \circ q}$. If $B \in \mathfrak{B}^1_{\tau \circ q}$, then $B \in \mathfrak{B}$, $B = \widehat{\alpha}^{-1}_{\tau \circ q} B$ and $B \in \widehat{\alpha}^{-1}_{\tau \circ q} \mathfrak{B}$. Hence, $\mathfrak{B}^1_{\tau \circ q} \subset \widehat{\alpha}^{-1}_{\tau \circ q} \mathfrak{B}$. We have

$$\left(X_s \circ q \circ \widehat{\alpha}_{\tau \circ q}\right)^{-1} \in \sigma\left(\widetilde{X}_t \circ p \circ \widehat{\alpha}_{\tau \circ q},\ X_s \circ q \circ \widehat{\alpha}_{\tau \circ q},\ s, t \in \mathbb{R}_+\right),$$

$$X_s \circ q \circ \widehat{\alpha}_{\tau \circ q} = X_s \circ \alpha_\tau \circ q.$$

From here

$$\left(X_s \circ \alpha_\tau \circ q\right)^{-1} B = \Psi_1 \times \alpha_\tau^{-1} X_s^{-1} B \in \mathfrak{G}_0 \otimes \left(\mathcal{F}_\tau \cap \{\tau \geq 0\}\right).$$

Let $\tau = \lim_{n\to\infty} \tau_n$, where $\tau_n(\xi) = (k-1)/n$, if $(k-1)/n \le \tau(\xi) < k/n$ $(k, n \in \mathbb{N})$. In this case $\tau_n \le \tau$ and for anyone $\psi \in \Psi$ and $\xi \in \mathcal{D}$ (by left-continuity of ψ) $\tilde{X}_{\tau_n(\xi)}\psi \to \tilde{X}_{\tau(\xi)}\psi$, i.e.

$$\left(\tilde{X}_t \circ p \circ \hat{\alpha}_{\tau \circ q}\right)(\psi, \xi) = \tilde{X}_t \tilde{\alpha}_{\tau(\xi)}\psi = \tilde{X}_{t \wedge \tau(\xi)}\psi + \left(0 \vee (t - \tau(\xi))\right)$$

$$= \lim_{n\to\infty} \tilde{X}_{t \wedge \tau_n(\xi)}\psi + \left(0 \vee (t - \tau(\xi))\right).$$

We have

$$\left(0 \vee (t - \tau(\xi))\right) \in \mathcal{F}_\tau / \mathcal{B}(\mathbb{R}_+), \quad \mathcal{F}_\tau = \mathcal{F}_\tau \cap \{\tau \ge 0\},$$

$$\{\psi, \xi : \tilde{X}_{t \wedge \tau_n(\xi)}\psi \in S\} = \bigcup_{k=0}^{\infty} \left\{\psi, \xi : \tilde{X}_{t \wedge k/n}\psi \in S,\ \frac{k}{n} \le \tau(\xi) < \frac{k+1}{n}\right\}$$

$$= \bigcup_{k=0}^{\infty} \left(\tilde{X}_{t \wedge k/n}^{-1} S\right) \times \left\{\frac{k}{n} \le \tau(\xi) < \frac{k+1}{n}\right\}$$

$$\in \sigma\left(\mathfrak{G}_{k/n} \otimes \left(\mathcal{F}_\tau \cap \left\{\tau \ge \frac{k}{n}\right\}\right),\ k \in \mathbb{N}\right) \subset \mathfrak{B}^0_{\tau \circ q},$$

where

$$\mathfrak{B}^0_{\tau \circ q} = \sigma\left(\mathfrak{G}_t \otimes \left(\mathcal{F}_\tau \cap \{\tau \ge t\}\right),\ t \in \mathbb{R}_+\right).$$

Hence $\left(\tilde{X}_t \circ p \circ \hat{\alpha}_{\tau \circ q}\right)^{-1} S \in \mathfrak{B}^0_{\tau \circ q}$. Consequently it follows that $\hat{\alpha}^{-1}_{\tau \circ q} \mathfrak{B} \subset \mathfrak{B}^0_{\tau \circ q}$. Let $B_1 \in \mathfrak{G}_t$, $B_2 \in \mathcal{F}_\tau \cap \{\tau \ge t\}$. Then

$$\left(B_1 \times B_2\right) \cap \{\tau \circ q \le s\} = B_1 \times \left(B_2 \cap \{\tau \le s\}\right).$$

If $s < t$, then $B_2 \cap \{\tau \le s\} = \emptyset$. If $s \ge t$, then $B_2 \cap \{\tau \le s\} \in \mathcal{F}_\tau$. Since $B_2 \in \mathcal{F}_\tau$ and $B_1 \in \mathfrak{G}_s$, then $B_1 \times (B_2 \cap \{\tau \le s\}) \in \mathfrak{B}_s$. From here $\mathfrak{B}^0_{\tau \circ q} \subset \mathfrak{B}_{\tau \circ q}$.   □

10. Measurability of shift operators

PROPOSITION 6.5. *The following properties are fair*
  (1) $\tilde{\theta}_t \in \mathfrak{G}/\mathfrak{G}$ $(t \in \mathbb{R}_+)$;
  (2) $\hat{\theta}_t \in \mathfrak{B}/\mathfrak{B}$;
  (3) $\hat{\theta}_{\tau \circ q} \in \mathfrak{B} \cap \{\tau \circ q < \infty\}/\mathfrak{B}$ $(\tau \in MT)$.

Proof. Assertions (1) and (2) are obvious.

(3) Let $B_1 \in \mathfrak{G}$, $B_2 \in \mathcal{F}$, $S \in \mathcal{B}(\mathbb{R}_+)$. We have

$$\widehat{\theta}_{\tau o q}^{-1}(B_1 \times B_2) = \{\psi, \xi : \widetilde{\theta}_{\tau(\xi)}\psi \in B_1,\ \theta_\tau \xi \in B_2,\ \tau < \infty\}$$

and $\{\theta_\tau \xi \in B_2\} \in \mathcal{F} \cap \{\tau < \infty\}$. Let $\tau_n(\xi) = (k-1)/n$ on $\{(k-1)/n \leq \tau < k/n\}$ $(k, n \in \mathbb{N})$. If $B_1 = \widetilde{X}_s^{-1}S$ then

$$\{\psi, \xi : \widetilde{\theta}_{\tau_n(\xi)}\psi \in \widetilde{X}_s^{-1}S\}$$

$$= \{\psi, \xi : \widetilde{X}_s \widetilde{\theta}_{\tau_n(\xi)}\psi \in S\}$$

$$= \{\psi, \xi : \widetilde{X}_{s+\tau_n(\xi)}\psi - \widetilde{X}_{\tau_n(\xi)}\psi \in S\}$$

$$= \bigcup_{k=0}^{\infty} \left\{\psi, \xi : \widetilde{X}_{s+k/n}\psi - \widetilde{X}_{k/n}\psi \in S,\ \frac{k}{n} \leq \tau(\xi) < \frac{k+1}{n}\right\} \in \mathfrak{B}$$

and since $(\forall \psi, \xi \in \Psi_1 \times \mathcal{D})\ (\forall s \in \mathbb{R}_+)\ \widetilde{X}_{s+\tau_n(\xi)}\psi \to \widetilde{X}_{s+\tau(\xi)}\psi$, then

$$(\forall B_1 \in \mathfrak{G})\quad \{\widetilde{\theta}_{\tau(\xi)}\psi \in B_1\} \in \mathfrak{B}. \qquad\qquad \square$$

6.2. Intrinsic time and traces

The name "intrinsic time" is connected with properties of a trajectory, not time-dependent. The basic property of these Markov times is borrowed from a time of the first exit σ_Δ ($\Delta \in \mathfrak{A}$). The class of intrinsic time is closed with respect to operations $\dotplus$, combination and passage to a limit.

The equivalence relation is considered when two trajectories belong to one class, or have the same sequence of states (trace), if they can be reduced to the same trajectory by a time change [KOL 72, CHA 79]. A sigma-algebra $\mathcal{F}^*$ of subsets invariant with respect to time change is defined. It can be represented in terms of intrinsic times and traces.

11. Intrinsic time

Let

$$\mathcal{T}_c = \{\tau \in \mathcal{T} : (\forall \varphi \in \Phi)\ (\forall \xi \in \mathcal{D})\ \tau(\xi \circ \varphi) = \varphi^*(\tau(\xi))\}.$$

If $\tau(\xi \circ \varphi) \in \varphi^{-1}\{\tau(\xi)\}$, then, obviously, $X_\tau(\xi \circ \varphi) = X_\tau \xi$ on $\{\tau < \infty\}$. From here $(\forall \tau \in \mathcal{T}_c)\ X_\tau(\xi \circ \varphi) = X_\tau \xi$ $(\tau < \infty)$. Let us designate $\mathrm{IMT} = \mathrm{MT} \cap \mathcal{T}_c$ and call it as class "of intrinsic Markov times". This class is rather narrower than a similar class from [CHA 81].

Let $\mathcal{F}^*$ be a set of all $B \in \mathcal{F}$ such that $(\forall \varphi \in \Phi)$ $(\forall \xi \in \mathcal{D})$

$$\xi \in B \Longleftrightarrow \xi \circ \varphi \in B.$$

Obviously, $\mathcal{F}^*$ is a sigma-algebra.

12. Properties of IMT class

THEOREM 6.2. *The following properties are fair:*

(1) $\tau_1, \tau_2 \in \mathrm{IMT} \Rightarrow \tau_1 \dotplus \tau_2 \in \mathrm{IMT}$,

(2) If $\tau_n \in \mathrm{IMT}$ *($\forall n \in \mathbb{N}$) and* $\tau_n = \tau$ *on* $B_n \in \mathcal{F}_{\tau_n} \cap \mathcal{F}^*$*, where* $\bigcup_{n=1}^{\infty} B_n = \mathcal{D}$, *then* $\tau \in \mathrm{IMT}$.

(3) If $(\forall n \in \mathbb{N})$ $\tau_n \in \mathrm{IMT}$ *and* $\tau_n \uparrow \tau$*, then* $\tau \in \mathrm{IMT}$.

(4) $T(\sigma_\Delta, \; \Delta \in \mathfrak{A}) \subset \mathrm{IMT}$ *(see item 2.14).*

Proof. (1) Let $\xi \in \mathcal{D}$, $\varphi \in \Phi$, $\tau_1, \tau_2 \in \mathcal{T}_c$ and $\tau_1(\xi) < \infty$. Then

$$\left(\tau_1 \dotplus \tau_2\right)(\xi \circ \varphi) = \tau_1(\xi \circ \varphi) + \tau_2 \theta_{\tau_1}(\xi \circ \varphi).$$

According to theorem 6.3(2)

$$\theta_{\tau_1(\xi \circ \varphi)}(\xi \circ \varphi) = \theta_{\varphi^*(\tau_1(\xi))}(\xi \circ \varphi) = \theta_{\tau_1(\xi)}\xi \circ \left(\widetilde{\theta}_{\tau_1(\xi)}\varphi^*\right)^*.$$

From here

$$\tau_2\theta_{\tau_1}(\xi \circ \varphi) = \left(\widetilde{\theta}_{\tau_1(\xi)}\varphi^*\right)\left(\tau_2\theta_{\tau_1(\xi)}\xi\right)$$

$$= \varphi^*\left(\tau_1(\xi) + \tau_2\theta_{\tau_1(\xi)}\xi\right) - \varphi^*\left(\tau_1(\xi)\right)$$

$$= \varphi^*\left(\left(\tau_1 \dotplus \tau_2\right)(\xi)\right) - \varphi^*\left(\tau_1(\xi)\right),$$

$$\left(\tau_1 \dotplus \tau_2\right)(\xi \circ \varphi) = \varphi^*\left(\left(\tau_1 \dotplus \tau_2\right)(\xi)\right).$$

If $\tau_1(\xi) = \infty$, then $\tau_1(\xi \circ \varphi) = \infty$. From here also

$$\left(\tau_1 \dotplus \tau_2\right)(\xi \circ \varphi) = \varphi^*\left(\left(\tau_1 \dotplus \tau_2\right)(\xi)\right),$$

if we set $\varphi^*(\infty) = \infty$. Since $\tau_1, \tau_2 \in \mathrm{MT} \Rightarrow \tau_1 \dotplus \tau_2 \in \mathrm{MT}$ (see theorems 2.5(1); 2.6(4); 2.8(1); 2.9(1)) the first statement is proved.

(2) We have

$$\{\tau \leq t\} = \bigcup_{n=1}^{\infty}\{\tau \leq t\} \cap B_n = \bigcup_{n=1}^{\infty}\{\tau_n \leq t\} \cap B_n \in \mathcal{F}_t.$$

From here $\tau \in \text{MT}$. Furthermore

$$\tau(\xi \circ \varphi) = \sum_{n=1}^{\infty} \tau(\xi \circ \varphi) J_n(\xi) = \sum_{n=1}^{\infty} \tau(\xi \circ \varphi) J_n(\xi \circ \varphi)$$

$$= \sum_{n=1}^{\infty} \tau_n(\xi \circ \varphi) J_n(\xi \circ \varphi) = \sum_{n=1}^{\infty} \varphi^* \big(\tau_n(\xi)\big) J_n(\xi)$$

$$= \sum_{n=1}^{\infty} \varphi^* \big(\tau(\xi)\big) J_n(\xi) = \varphi^* \big(\tau(\xi)\big),$$

where $\overline{B}_i = \mathcal{D} \setminus B_i$, $J_n(\xi) = I(B_n \overline{B}_{n-1} \cdots \overline{B}_1 \mid x)$, $I(B \mid \xi) \equiv I_B(\xi)$ is indicator of the set B. From here $\tau \in \mathcal{T}_c$.

(3) If $\tau_n \in \text{IMT}$ and $\tau_n \uparrow \tau$, then $\tau \in \text{MT}$ (see [BLU 68, p. 32]) and, in addition, $(\forall \xi \in \mathcal{D}) \, (\forall \varphi \in \Phi)$

$$\tau(\xi \circ \varphi) = \lim_{n \to \infty} \tau_n(\xi \circ \varphi) = \lim_{n \to \infty} \varphi^* \big(\tau_n(\xi)\big) = \varphi^* \big(\tau(\xi)\big)$$

by a continuity of φ^* from the left. Hence, $\tau \in \mathcal{T}_c$.

(4) Let $\Delta \in \mathfrak{A}$, $\xi \in \mathcal{D}$, $\varphi \in \Phi$. We have

$$\sigma_\Delta(\xi \circ \varphi) = \inf \varphi^{-1} \xi^{-1} (\mathbb{X} \setminus \Delta)$$

$$= \inf \varphi^{-1} \big(\inf \xi^{-1}(\mathbb{X} \setminus \Delta) \big)$$

$$= \varphi^* \big(\sigma_\Delta(\xi)\big).$$

Since $\sigma_\Delta \in \text{MT}$ and $T(\sigma_\Delta, \ \Delta \in \mathfrak{A})$ is a semigroup with respect to $\dot{+}$, generated by a class $\{\sigma_\Delta, \ \Delta \in \mathfrak{A}\}$, then according to item 1 we obtain proof of the fourth assertion. $\square$

13. About non-constancy of the function on the left of intrinsic time

PROPOSITION 6.6. *If $\tau \in \text{IMT}$, then $(\forall \varepsilon > 0) \, (\forall \xi \in \mathcal{D}) \, \xi$ is not constant on $[\tau(\xi) - \varepsilon, \tau(\xi)]$.*

Proof. Let $\tau \in \text{IMT}$ and $(\exists \varepsilon < 1) \, \tau(\xi) = t_0 > 0$ and ξ be constant on $[t_0 - \varepsilon, t_0]$. Since IMT $\subset$ ST $\subset \mathcal{T}_a$ (see theorem 2.8) and $\tau = \tau \circ \alpha_\tau$ (see theorem 2.2), then $\tau(\xi) = \tau(\alpha_{t_0} \xi)$. Let

$$\varphi(t) = \begin{cases} t, & t \in [0, t_0 - \varepsilon], \\ t_0 - \varepsilon, & t \in [t_0 - \varepsilon, t_0 - \varepsilon + 1], \\ t - 1, & t \in [t_0 - \varepsilon + 1, \infty); \end{cases}$$

Then $\alpha_{t_0}\xi = \alpha_{t_0}(\xi \circ \varphi)$ (up to $t_0 - \varepsilon$ $\varphi(t) = t$, with $t \in [t_0 - \varepsilon, t_0]$ $(\xi \circ \varphi)(t) = \xi(t_0 - \varepsilon) = \xi(t_0) = \xi(t)$). From here $\tau(\xi \circ \varphi) = \tau(\alpha_{t_0}(\xi \circ \varphi)) = \tau(\alpha_{t_0}\xi) = t_0$. On the other hand, $\varphi^*(\tau(\xi)) = \varphi^*(t_0) = t_0 + 1 > \tau(\xi \circ \varphi)$. It is a contradiction.   □

14. Sequence of states (trace)

Let $\mathcal{M}$ be a set of equivalence classes of set $\mathcal{D}$ with respect to an equivalence relation:

$$\xi_1 \sim \xi_2 \Longleftrightarrow (\exists \varphi_1, \varphi_2 \in \Phi) \quad \xi_1 \circ \varphi_1 = \xi_2 \circ \varphi_2$$

The map $\ell : \mathcal{D} \to \mathcal{M}$ corresponding to each function its class of equivalency is said to be a sequence of states (or trace) of this function. Thus, $\ell(\xi)$ is the trace of the trajectory $\xi \in \mathcal{D}$. In [KOL 72, p. 109] this object received the name "curve in metric space". This name seems to be unsuitable for such a trace as sequence of isolated points (in case of step-functions). In the case of continuous trajectories we will use this name in aspect of "curvilinear integrals".

15. Representation of the sigma-algebra

PROPOSITION 6.7. *The following property is fair (see item 11):*

$$B \in \mathcal{F}^* \Longleftrightarrow (\exists B' \subset \mathcal{M}) \quad B = \ell^{-1}B', \; B \in \mathcal{F}.$$

Proof. Let $B \in \mathcal{F}^*$. We have $B \subset \ell^{-1}\ell B$ for $\ell B \subset \mathcal{M}$. Furthermore

$$\xi \in \ell^{-1}\ell B \Longrightarrow \ell\xi \in \ell B \Longrightarrow (\exists \xi' \in B) \quad \ell\xi = \ell\xi'$$
$$\Longrightarrow (\exists \varphi, \varphi' \in \Phi)(\exists \xi' \in B) \quad \xi \circ \varphi = \xi' \circ \varphi' \in B$$
$$\Longrightarrow \xi \in B.$$

Hence, $B = \ell^{-1}\ell B$. Let $B' \subset \mathcal{M}$. We have

$$\xi \in \ell^{-1}B' \Longleftrightarrow \ell\xi \in B' \Longleftrightarrow (\forall \varphi \in \Phi) \quad \ell(\xi \circ \varphi) = \ell\xi \in B'$$
$$\Longleftrightarrow (\forall \varphi \in \Phi) \quad \xi \circ \varphi \in \ell^{-1}B'.$$

If $\ell^{-1}B' \in \mathcal{F}$, then $\ell^{-1}B' \in \mathcal{F}^*$.   □

16. Class IMT′ and its relation with IMT and MT

Let us consider a class IMT′ of measurable functions $\tau \in \mathcal{T}$ such that $(\forall t_1, t_2 \in \mathbb{R}_+)$ $(\forall \xi_1, \xi_2 \in \mathcal{D})$

$$\left(\ell\alpha_{t_1}\xi_1 = \ell\alpha_{t_2}\xi_2, \; \tau(\xi_1) \leq t_1\right) \Longrightarrow \tau(\xi_2) \leq t_2.$$

PROPOSITION 6.8. *The following property holds*

$$\mathrm{IMT}' \subset \mathrm{MT}.$$

Proof. According to theorem 2.9 it is enough to prove that $\mathrm{IMT}' \subset \mathcal{T}_a$. Let $\tau \in \mathrm{IMT}'$ and $\tau(\xi) \le t$. Then $\ell\alpha_t\xi = \ell\alpha_t\alpha_t\xi$ and, hence, $\tau\alpha_t\xi \le t$. Let $\tau\alpha_t\xi \le t$. Then $\tau(\xi) \le t$, i.e.

$$\{\tau \le t\} = \{\tau \circ \alpha_t \le t\}. \qquad \square$$

17. Necessary condition of class IMT

Let us consider a class of time (see item 2.2)

$$\mathcal{T}_b = \left\{\tau \in \mathcal{T} : (\forall t \in \mathbb{R}_+)\, \{\tau \le t\} = \alpha_t^{-1}\{\tau < \infty\}\right\}.$$

PROPOSITION 6.9. *The following property is fair*

$$\mathrm{IMT} \subset \mathcal{T}_b.$$

Proof. According to item 11, $\mathrm{IMT} \subset MT$. From here $(\forall \tau \in \mathrm{IMT})\, (\forall t \in \mathbb{R}_+)$

$$\tau(\xi) \le t \Longrightarrow \tau\alpha_t\xi < \infty.$$

Function $\alpha_t\xi$ is constant on $[t, \infty)$. By theorem 6.6, function $\tau(\alpha_t\xi)$ cannot be more than t. Therefore, from condition $\tau\alpha_t\xi < \infty$ it follows that $\tau\alpha_t\xi \le t$. Hence, $\{\tau\circ\alpha_t \le t\} = \{\tau \le t\} = \{\tau \circ \alpha_t < \infty\}$. $\qquad \square$

18. The main property of class IMT

THEOREM 6.3. *The following property is fair*

$$\mathrm{IMT} = \mathrm{IMT}'.$$

Proof. We divide the proof into two parts.

(1) Let $\tau \in \mathrm{IMT}$. Then, by theorem 6.9, we have $\tau(\alpha_t\xi) < \infty \Rightarrow \tau(\xi) \le t$. Hence, if $\tau(\alpha_{t_1}\xi_1) = \tau(\alpha_{t_2}\xi_2) < \infty$, then $\tau(\xi_1) = \tau(\xi_2) \le t_1 \wedge t_2$. Let $\ell\alpha_{t_1}\xi_1 = \ell\alpha_{t_2}\xi_2$ and $\tau(\xi_1) \le t_1 < \infty$. Then $(\exists \varphi_1, \varphi_2 \in \Phi)$

$$\alpha_{t_1}\xi_1 \circ \varphi_1 = \alpha_{t_2}\xi_2 \circ \varphi_2$$

and, by theorem 6.3(1), $\alpha_{\varphi_1^*(t_1)}(\xi_1 \circ \varphi_1) = \alpha_{\varphi_2^*(t_2)}(\xi_2 \circ \varphi_2)$. Hence,

$$\tau(\alpha_{t_1}\xi_1 \circ \varphi_1) = \varphi_1^*\big(\tau(\alpha_{t_1}\xi_1)\big) = \tau\big(\alpha_{\varphi_1^*(t_1)}(\xi_1 \circ \varphi_1)\big)$$

and since $\tau(\alpha_{t_1}\xi) = \tau(\xi_1) < \infty$,

$$\tau\big(\alpha_{\varphi_1^*(t_1)}(\xi_1 \circ \varphi_1)\big) = \tau\big(\alpha_{\varphi_2^*(t_2)}(\xi_2 \circ \varphi_2)\big)$$
$$= \varphi_1^*\big(\tau(\alpha_{t_1}\xi_1)\big)$$
$$= \varphi_1^*\big(\tau(\xi_1)\big) < \infty.$$

From here $\tau(\xi_1 \circ \varphi_1) = \tau(\xi_2 \circ \varphi_2) \le \varphi_1^*(t_1) \wedge \varphi_2^*(t_2)$, i.e.

$$\tau(\xi_2 \circ \varphi_2) \le \varphi_2^*(t_2) \implies \varphi_2^*(\tau(\xi_2)) \le \varphi_2^*(t_2)$$
$$\implies \tau(\xi_2) \le t_2.$$

Hence, $\tau \in \text{IMT}'$.

(2) Let $\tau \in \text{IMT}'$. Then by theorem 6.3(1) $(\forall \xi \in \mathcal{D})\ (\forall \varphi \in \Phi)\ (\forall t \in \mathbb{R}_+)$

$$\ell\alpha_t(\xi \circ \varphi) = \ell\big(\alpha_{\varphi(t)}\xi \circ \varphi\big) = \ell\alpha_{\varphi(t)}\xi'$$
$$\big(\tau(\xi \circ \varphi) \le t\big) \iff \big(\tau(\xi) \le \varphi(t)\big).$$

We have

$$\big(\tau(\xi) \le \varphi(t)\big) \implies \varphi^*\big(\tau(\xi)\big) \le \varphi^*\big(\varphi(t)\big) = \inf \varphi^{-1}\big(\varphi(t)\big) \le t.$$

On the other hand, on a continuity of φ

$$\big(\tau(\xi) > \varphi(t)\big) \implies \big(\varphi^*\big(\tau(\xi)\big) > \sup \varphi^{-1}\big(\varphi(t)\big) \ge t\big).$$

From here $\big(\tau(\xi) \le \varphi(t)\big) \iff \big(\varphi^*(\tau(\xi)) \le t\big)$, i.e.

$$\big(\tau(\xi \circ \varphi) \le t\big) \iff \big(\varphi^*\big(\tau(\xi)\big) \le t\big).$$

Hence, $\tau(\xi \circ \varphi) = \varphi^*(\tau(\xi))$ and $\tau \in \text{IMT}$. $\qquad\qquad$ □

6.3. Canonical time change for trajectories with identical traces

According to the definition two functions have one trace if with the help of two time changes they can be transformed into the same function. The property of coincidence of traces of two functions can be checked by a comparison of values of these functions in an countable set of time of the first exit and their iterations. In a common case coincidence of all these values is not yet enough for equality of sequences of states. It is necessary to check up simultaneous presence or lack of intervals of constancy before time of discontinuity and on infinity. These intervals cannot be discovered on a finite set of first exit times. Therefore, it is desirable to have effectively checked conditions of their lack (see theorems 3.11 and 3.12). The construction of canonical time changes for two functions with identical sequences of states serves as preparation for the further constructions connected with random processes and time changes.

19. Trace values in a point

Let $\tau \in T$, $\delta = (\Delta_1, \Delta_2, \ldots)$, $(\Delta_i \subset \mathbb{X})$, $\overline{\mathbb{X}} = \mathbb{X} \cup \{\overline{\infty}\}$ where $\overline{\infty} \notin \mathbb{X}$. We will use denotations

$$\gamma_\tau : \mathcal{D} \to \overline{\mathbb{X}} = \mathbb{X} \cup \{\overline{\infty}\} \text{ is the map}$$

$$\gamma_\tau \xi = \begin{cases} \overline{\infty}, & \tau(\xi) = \infty, \\ X_\tau \xi, & \tau(\xi) < \infty. \end{cases}$$

$\ell_\delta : \mathcal{D} \to \overline{\mathbb{X}}^\infty$ where $\ell_\delta \xi = (\gamma_0 \xi, \gamma_{\sigma_\delta^1} \xi, \gamma_{\sigma_\delta^2} \xi, \ldots)$, $\gamma_0 \xi = X_0 \xi$.

20. Preservation of an order

We use $\mathrm{DS}(\mathfrak{A}_{r_n})$, a class of deducing sequences composed of elements of an open covering $\mathfrak{A}_{r_n}$ of a rank r_n.

PROPOSITION 6.10. *Let* $\xi_1, \xi_2 \in \mathcal{D}$ *and* $(\exists (\delta_n)_1^\infty,\ \delta_n \in \mathrm{DS}(\mathfrak{A}_{r_n}),\ r_n \to 0)\ (\forall n \in \mathbb{N})$

$$\ell_{\delta_n} \xi_1 = \ell_{\delta_n} \xi_2.$$

Then for any sequence $\delta = (\Delta_1, \Delta_2, \ldots)$ $(\Delta_i \in \widetilde{\mathfrak{A}}$, *i.e. it is a closed set*) $\ell_\delta \xi_1 = \ell_\delta \xi_2$ *and for any* $\tau_1, \tau_2 \in T(\sigma_\Delta, \Delta \in \widetilde{\mathfrak{A}})$

(1) $\tau_1 \xi_1 < \tau_2 \xi_1 \Rightarrow \tau_1 \xi_2 \leq \tau_2 \xi_2$;

(2) $\tau_1 \xi_2 < \tau_2 \xi_2 \Rightarrow \tau_1 \xi_1 \leq \tau_2 \xi_1$.

Proof. (1) We have $(\forall n \in \mathbb{N})\ (\forall \tau \in T(\sigma_\Delta, \Delta \in \widetilde{\mathfrak{A}}))\ \tau L_{\delta_n} \xi_1 = \tau L_{\delta_n} \xi_2 = \infty$ or $(\exists k \in \mathbb{N})\ \tau L_{\delta_n} \xi_1 = \sigma_{\delta_n}^{k-1} \xi_1 < \infty$, $\tau L_{\delta_n} \xi_2 = \sigma_{\delta_n}^{k-1} \xi_2 < \infty$ and $X_\tau L_{\delta_n} \xi_1 = X_\tau L_{\delta_n} \xi_2$. From the definition of $\tau \in T(\sigma_\Delta, \Delta \in \widetilde{\mathfrak{A}})$ it follows that $(\forall \xi \in \mathcal{D})\ \tau L_{\delta_n} \xi \downarrow \tau(\xi)$ $(n \to \infty)$ and $X_\tau L_{\delta_n} \xi \to X_\tau \xi$ for $\tau \xi < \infty$. From here it follows that either times $\tau \xi_1$, $\tau \xi_2$ are infinite, or both are finite. In the last case $X_\tau \xi_1 = X_\tau \xi_2$. This means that $(\forall \tau \in T(\sigma_\Delta, \Delta \in \widetilde{\mathfrak{A}}))\ \gamma_\tau \xi_1 = \gamma_\tau \xi_2$ and for any $\delta = (\Delta_1, \Delta_2, \ldots)$ $(\Delta_i \in \widetilde{\mathfrak{A}})$ $\ell_\delta \xi_1 = \ell_\delta \xi_2$.

(2) If $\tau_1, \tau_2 \in T(\sigma_\Delta, \Delta \in \widetilde{\mathfrak{A}})$, then $\tau_1 L_{\delta_n} \xi_1 \downarrow \tau_1 \xi_1$, $\tau_2 L_{\delta_n} \xi_1 \downarrow \tau_2 \xi_1$; $\tau_1 L_{\delta_n} \xi_2 \downarrow \tau_1 \xi_2$, $\tau_2 L_{\delta_n} \xi_2 \downarrow \tau_2 \xi_2$ $(n \to \infty)$, but for $\tau_1 L_{\delta_n} \xi_1$ and $\tau_1 L_{\delta_n} \xi_2$ there is the same corresponding number k_n in a sequence $\ell_{\delta_n} \xi_1 = \ell_{\delta_n} \xi_2$. Also, for $\tau_2 L_{\delta_n} \xi_1$ and $\tau_2 L_{\delta_n} \xi_2$, there is a corresponding common number s_n. Let $\tau_1 \xi_1 < \tau_2 \xi_1$. Then $k_n < s_n$ for rather large n and, hence, $\tau_1 \xi_2 \leq \tau_2 \xi_2$. The same is true for $\tau_1 \xi_2 < \tau_2 \xi_2$. $\qquad\square$

21. *Interval of constancy on the left of* t

Let us say that ξ has no intervals of constancy on the left of $t - 0$ $(0 < t \le \infty)$ if $(\forall t_1 < t)$ ξ is not constant on $[t_1, t)$. Let $\mathcal{D}^\tau$ be a set of all functions $\xi \in \mathcal{D}$ not having an interval of constancy on the left of $\tau(\xi) - 0$ $(\tau \in MT_+)$. Let $\mathcal{D}^\infty$ be a set of all functions not having an infinite interval of constancy (see item 3.46).

Learning intervals of constancy before random instants makes sense for processes having a non-trivial trace, i.e. a trace not reducing to a sequence of isolated points. Evidently, a step-process does not belong to $\mathcal{D}^\tau$ for any jump time τ (any jump time has an interval of constancy just before any jump time). According to proposition 6.6, for any $\tau \in MT$, which is not a point of discontinuity of the function ξ, $\xi \in \mathcal{D}^\tau$.

22. *Measurability*

Let us denote

$$\mathcal{T}_n = T(\sigma_\Delta, \ \Delta \in \mathfrak{A}_{r_n}), \qquad \mathcal{T}_0 = \bigcup_{n=1}^{\infty} \mathcal{T}_n.$$

PROPOSITION 6.11. *The following property is fair*

$$\tau \in \mathrm{IMT} \Longrightarrow \mathcal{D}^\tau \in \mathcal{F}^* \cap \mathcal{F}_\tau.$$

Proof. We have

$$\{\tau < \infty\} \cap \mathcal{D}^\tau = \bigcap_{n=1}^{\infty} \ \bigcup_{\tau_1 \in \mathcal{T}_0} \{\xi : \tau(\xi) - 1/n < \tau_1(\xi) < \tau(\xi) < \infty\},$$

where $\mathcal{T}_0 = \bigcup_{n=1}^{\infty} \mathcal{T}_n$, $\mathcal{T}_n = T(\sigma_\Delta, \ \Delta \in \mathfrak{A}_{r_n})$, $r_n \downarrow 0$. From here $\{\tau < \infty\} \cap \mathcal{D}^\tau \in \mathcal{F}_\tau$ (see, e.g., [BLU 68, p. 33]). Furthermore,

$$\{\tau = \infty\} \cap \mathcal{D}^\tau = \{\tau = \infty\} \cap \mathcal{D}^\infty \in \mathcal{F}_\tau \cap \{\tau = \infty\} \subset \mathcal{F}_\tau.$$

From here $\mathcal{D}^\tau \in \mathcal{F}_\tau$. Let $\varphi \in \Phi$. Because $\tau, \tau_1 \in \mathcal{T}_c$, we have

$$\tau(\xi) - \frac{1}{n} < \tau_1(\xi) < \tau(\xi) \Longleftrightarrow \varphi^*(\tau(\xi) - 1/n) < \varphi^*(\tau_1(\xi))$$

$$< \varphi^*(\tau(\xi)) \Longleftrightarrow \tau(\xi \circ \varphi) - \varepsilon_n < \tau_1(\xi \circ \varphi) < \tau(\xi \circ \varphi),$$

where $\varepsilon_n = \varphi^*(\tau(\xi)) - \varphi^*(\tau(\xi) - 1/n)$, $\varepsilon_n \to 0$ $(n \to \infty)$. From here

$$\xi \in \mathcal{D}^\tau \cap \{\tau < \infty\} \Longleftrightarrow \xi \circ \varphi \in \mathcal{D}^\tau \cap \{\tau < \infty\}.$$

Similarly, it can be proved that

$$\xi \in \mathcal{D}^{\tau} \cap \{\tau = \infty\} \Longleftrightarrow \xi \circ \varphi \in \mathcal{D}^{\tau} \cap \{\tau = \infty\}.$$

Hence $\mathcal{D}^{\tau} \in \mathcal{F}^{*}$. □

23. Criterion of equivalence of two trajectories

Let $(\forall n \in \mathbb{N})$ $\delta_n = (\Delta_{n_1}, \Delta_{n_2}, \dots) \in \mathrm{DS}(\mathfrak{A}_{r_n})$ and $\ell_{\delta_n} \xi_1 = \ell_{\delta_n} \xi_2$ $(r_n \downarrow 0)$. We denote $[\delta_n] = ([\Delta_{n_1}], [\Delta_{n_2}], \dots)$ and $\sigma^k_{[\delta_n]} = \tau^k_n$. According to proposition 6.6, $(\forall k, n \in \mathbb{N})$ $\gamma_{\tau^k_n} \xi_1 = \gamma_{\tau^k_n} \xi_2$, and the set of pairs $(\tau^k_n \xi_1, \tau^k_n \xi_2)$ is linear ordered by the relation $(a, b) < (c, d) \Leftrightarrow (a < c, b \leq d) \vee (a \leq c, b < d)$. We identify equal pairs. For any $m \in \mathbb{N}$ we consider pairs $(\tau^k_n \xi_1, \tau^k_n \xi_2)$ $(n \leq m, k \in \mathbb{N})$. We unite points $(\tau^k_n \xi_1, \tau^k_n \xi_2)$ of quadrant $\mathbb{R}_+ \times \mathbb{R}_+$ with broken line Γ_m with vertice in these points and beginning at the point $(0, 0)$. Let it pass through these points according to the given order from small to large. If there is only a finite number of such points in quadrant $\mathbb{R}_+ \times \mathbb{R}_+$, then the last infinite link is a ray drawing from the last point with inclination coefficient 1. Consider the point (t^m_c, s^m_c) of intersection of the broken line Γ_m with line $t + s = c$. Let $(\overline{t}^m_c, \overline{s}^m_c)$ and let $(\widehat{t}^m_c, \widehat{s}^m_c)$ be the greatest vertex not exceeding (t^m_c, s^m_c), and smallest (if it exists) not less than (t^m_c, s^m_c) correspondingly. Since sums $\overline{t}^m_c + \overline{s}^m_c$ do not decrease and are bounded from above by value c, and sums $\widehat{t}^m_c + \widehat{s}^m_c$ do not increase and are bounded from below by value c, there exist limits $\overline{c} = \lim_{m \to \infty}(\overline{t}^m_c + \overline{s}^m_c)$ and $\widehat{c} = \lim_{m \to \infty}(\widehat{t}^m_c + \widehat{s}^m_c)$. Then according to ordering any sequence $\overline{t}^{m_n}_c$ tends to the same limit $\overline{t}_c$, hence $\overline{t}^m_c \uparrow \overline{t}_c$ and correspondingly $\overline{s}^m_c \uparrow \overline{s}_c$, $\widehat{t}^m_c \downarrow \widehat{t}_c$, $\widehat{s}^m_c \downarrow \widehat{s}_c$. Therefore, the limits $t_c = \lim_{m \to \infty} t^m_c$, $s_c = \lim_{m \to \infty} s^m_c$ exist. Let $\Gamma = \Gamma(\xi_1, \xi_2)$ be the curve determined by family of points $\{(t_c, s_c)\}$ $(c \geq 0)$. We will call it the *time run comparison curve* for functions ξ_1 and ξ_2.

According to construction $c_1 < c_2 \Rightarrow (t_{c_1}, s_{c_1}) < (t_{c_2}, s_{c_2})$, and since $t_c + s_c = c$, then t_c and s_c are continuous $((t_{c_2} - t_{c_1}) + (s_{c_2} - s_{c_1}) = c_2 - c_1)$. If $I(\mathcal{D}^{\infty} \mid \xi_1) = I(\mathcal{D}^{\infty} \mid \xi_2)$, then either simultaneously $\sup\{\tau^k_n(\xi_1) : \tau^k_n(\xi_1) < \infty\} = \sup\{\tau^k_n(\xi_2) : \tau^k_n(\xi_2) < \infty\} = \infty$, or both these limits are less than infinity. In this case if a point of intersection (t^m_c, s^m_c) for all m belongs to a last infinite link, then existence of limit (t_c, s_c) follows from the existence of the limit $(\overline{t}_c, \overline{s}_c)$. Let us designate $v_1(c \mid \xi_1, \xi_2) = t_c$, $v_2(c \mid \xi_1, \xi_2) = s_c$. Under our suppositions $t_c, s_c \to \infty$ as $c \to \infty$ and hence $v_1(\cdot \mid \xi_1, \xi_2), v_2(\cdot \mid \xi_1, \xi_2) \in \Phi$. We refer to these functions as *canonical inverse time changes* corresponding to two functions with identical traces.

THEOREM 6.4. *Let $\xi_1, \xi_2 \in \mathcal{D}$. The following criterion for equivalence of functions $\ell\xi_1 = \ell\xi_2$ is fair if and only if the following conditions are fulfilled:*

(1) $(\forall \tau \in \mathcal{T}_0)$ $\gamma_\tau \xi_1 = \gamma_\tau \xi_2$;

(2) $(\forall \tau \in \mathcal{T}_0)$ $I(\mathcal{D}^{\tau} \mid \xi_1) = I(\mathcal{D}^{\tau} \mid \xi_2)$;

(3) $I(\mathcal{D}^{\infty} \mid \xi_1) = I(\mathcal{D}^{\infty} \mid \xi_2)$.

Proof. Sufficiency. Let us consider the time run comparison curve $\Gamma(\xi_1, \xi_2)$. We will prove that $\xi_1 \circ v_1 = \xi_2 \circ v_2$.

Let t_c, s_c be points of continuity of ξ_1 and ξ_2 correspondingly. Then $(\exists k \in \mathbb{N})$

$$t_c^n \in \left[\tau_n^k \xi_1, \tau_n^{k+1} \xi_1\right), \qquad s_c^n \in \left[\tau_n^k \xi_2, \tau_n^{k+1} \xi_2\right),$$

$$\rho\big(\xi_1(t_c^n), \xi_1(\tau_n^k \xi_1)\big) < r_n, \qquad \rho\big(\xi_2(s_c^n), \xi_2(\tau_n^k \xi_2)\big) \leq r_n$$

and since $\xi_1(\tau_n^k(\xi_1)) = \xi_2(\tau_n^k(\xi_2))$, then $\rho(\xi_1(t_c^n), \xi_2(s_c^n)) \leq 2r_n$ and $\rho(\xi_1(t_c), \xi_2(s_c)) = 0$, i.e. in this case $(\xi_1 \circ v_1)(c) = (\xi_2 \circ v_2)(c)$.

Let t_c be a point of discontinuity of ξ_1. Then $(\exists n, k \in \mathbb{N}) \, t_c = \tau_n^{k-1} \xi_1$. However, because of equality $(\forall n \in \mathbb{N}) \, \ell_{\delta_n} \xi_1 = \ell_{\delta_n} \xi_2$ we find that $\tau_n^{k-1} \xi_2$ is also a point of discontinuity of function ξ_2. Consider a set of points of the curve Γ with the first coordinate t_c. If this set consists of one point, then

$$\xi_2(s_c) = \xi_2\big(\tau_n^{k-1} \xi_2\big) = \xi_1\big(\tau_n^{k-1} \xi_1\big) = \xi_1(t_c).$$

If this set consists of an interval with edges (t_c, s_1), (t_c, s_2), $(s_1 < s_2)$, then from equality $I(\mathcal{D}^{\tau_n^{k-1}} \mid \xi_1) = I(\mathcal{D}^{\tau_n^{k-1}} \mid \xi_2)$ it follows that $s_1 = \tau_n^{k-1} \xi_2$. Besides, if $\tau_m^i \xi_1 > \tau_m^{k-1} \xi_1$, then $\tau_m^i \xi_2 > \tau_m^{k-1} \xi_2$. Hence on the interval (s_1, s_2) the function ξ_2 does not contain points $\tau_m^i \xi_2$ $(\forall i, m \in \mathbb{N})$, i.e. it is constant and, consequently, it is equal to $\xi_2(\tau_n^{k-1} \xi_2)$. Assuming $\xi_2(s_2) \neq \xi_2(\tau_n^{k-1} \xi_2)$, we obtain $(\exists m, l \in \mathbb{N}) \, s_2 = \tau_m^{l-1} \xi_2$, and hence $t_c = \tau_m^{l-1} \xi_1$. The latter contradicts the supposition, from which it follows that

$$\xi_1(t_c) = \xi_1\big(\tau_n^{k-1} \xi_1\big) = \xi_2\big(\tau_n^{k-1} \xi_2\big) \neq \xi_2\big(\tau_m^{l-1} \xi_2\big) = \xi_1\big(\tau_m^{l-1} \xi_1\big).$$

Hence $\xi_2(s_c) = \xi_1(t_c)$ and in this case $(\xi_1 \circ v_1)(c)(\xi_2 \circ v_2)(c)$.

Necessity. Let $\ell \xi_1 = \ell \xi_2$ and $\xi_1 \circ \varphi_1 = \xi_2 \circ \varphi_2$ $(\varphi_1, \varphi_2 \in \Phi)$. We have $(\forall \tau \in \mathrm{MT})$

$$\tau(\xi \circ \varphi) = \infty \Longleftrightarrow \tau(\xi) = \infty, \qquad \tau(\xi) < \infty \Longrightarrow X_\tau(\xi \circ \varphi) = X_\tau(\xi)$$

(see item 11). From here $\gamma_\tau \xi_1 = \gamma_\tau \xi_2$ and since $\mathcal{D}^\tau \in \mathcal{F}^*$, then $I(\mathcal{D}^\tau \mid \xi_1) = I(\mathcal{D}^\tau \mid \xi_2)$. Because $\mathcal{T}_0 \subset \mathrm{IMT}$, necessity is proved. $\qquad \square$

The similar result is placed in [CHA 81].

24. Notes to criterion of equivalence

(1) According to theorem 6.2, if $\tau \in \mathrm{IMT}$ and $\xi \notin \mathcal{D}^\tau$, then either $\tau(\xi)$ is a point of discontinuity of ξ or $\tau(\xi) = \infty$ and ξ has an infinite interval of constancy. Therefore,

condition $(\forall \tau \in \mathcal{T}_0)\ I(\mathcal{D}^\tau \mid \xi_1) = I(\mathcal{D}^\tau \mid \xi_2)$ is meaningful only for points of discontinuity of functions ξ_1, ξ_2, and also for point ∞. The statement of theorem 6.4 is also fair in a case when $\mathcal{T}_0$ is replaced by $\widetilde{\mathcal{T}}_0$, the similar set of the first exit time from closed sets; however, condition $(\forall \tau \in \widetilde{\mathcal{T}}_0)\ I(\mathcal{D}^\tau \mid \xi_1) = I(\mathcal{D}^\tau \mid \xi_2)$ should be supplemented by the requirement: either $\tau(\xi_1)$, $\tau(\xi_2)$ are points of discontinuity of the appropriate functions or $\tau(\xi_1) = \tau(\xi_2) = \infty$.

(2) Similar statements are fair, when Markov times $\sigma_{r_n}^k$ $(r_n \downarrow 0)$ or appropriate time of the first exit from the closed balls are considered.

(3) For two functions with identical sequences of states $(\ell\xi_1 = \ell\xi_2)$ we have constructed a time run comparison curve $\Gamma \subset \mathbb{R}_+ \times \mathbb{R}_+$. This curve contains segments parallel to axes of ordinates with edges (t, s_1), (t, s_2) if and only if ξ_2 is constant on an interval $[s_1, s_2)$ and point t does not belong to an interval of constancy of function ξ_1. If ξ_2 has no intervals of constancy on the whole set $\mathbb{R}_+$, then, obviously, $\xi_2 \circ \Gamma(\cdot) = \xi_1$, where $\Gamma(\cdot) \in \Phi$ is a single-valued function, defined by curve Γ.

25. First exit time from a closed set

PROPOSITION 6.12. *Let $\ell\xi_1 = \ell\xi_2$ and $\Gamma = \Gamma(\xi_1, \xi_2)$. Then $(\forall \tau \in T(\sigma_\Delta,\ \Delta \in \widetilde{\mathfrak{A}}))$ either $(\tau(\xi_1), \tau(\xi_2)) \in \Gamma$, or $\tau(\xi_1) = \tau(\xi_2) = \infty$.*

Proof. The curve Γ is constructed for a sequence $(\delta_n)_1^\infty$ $(\delta_n \in \mathrm{DS}(\mathfrak{A}_{r_n}))$, and for any $m, k \in \mathbb{N}$ $(\tau_m^k \xi_1, \tau_m^k \xi_2) \in \Gamma$, where $\tau_n^k = \sigma_{[\delta_n]}^k$. If $\tau \in T(\sigma_\Delta,\ \Delta \in \widetilde{\mathfrak{A}})$, then either $\tau L_{[\delta_n]}\xi_1 = \tau L_{[\delta_n]}\xi_2 = \infty$ or $(\tau L_{[\delta_n]}\xi_1, \tau L_{[\delta_n]}\xi_2) \in \Gamma$. Since $(\forall \xi \in \mathcal{D})\ \tau L_{[\delta_n]}\xi \downarrow \tau(\xi)$ and Γ is a continuous curve, then $(\tau(\xi_1), \tau(\xi_2)) \in \Gamma$. $\square$

26. Measurability of canonical direct time changes

Let $d = \{(\xi_1, \xi_2) \in \mathcal{D} \times \mathcal{D} : \ell\xi_1 = \ell\xi_2\}$. From sigma-compactness of space $\mathbb{X}$, it follows that $d \in \mathcal{F} \otimes \mathcal{F}$. We call d a diagonal of traces.

PROPOSITION 6.13. *The following property is fair:*

$$v_1^*, v_2^* \in (\mathcal{F} \otimes \mathcal{F}) \cap d/\mathfrak{G},$$

where v_1, v_2 are canonical inverse time changes determined in item 24 (map $(\cdot)^*$ (see item 1)).

Proof. We have $v_1^*, v_2^* \in \Psi$ according to construction. Map $(\cdot)^* : \Psi_1 \to \Psi_1$ is measurable with respect to $\mathcal{G}/\mathcal{G}$. Therefore it is enough to prove that $v_i \in (\mathcal{F} \otimes \mathcal{F}) \cap d/\mathcal{G}$, but $v_1(c \mid \xi_1, \xi_2) = t_c = \lim_{m \to \infty} t_c^m$ and $v_2(c \mid \xi_1, \xi_2) = s_c = \lim_{m \to \infty} s_c^m$ (in terms of theorem 6.4). The measurability of t_c^m, s_c^m follows from that of vertices $(\bar{t}_c^m, \bar{s}_c^m)$ and $(\bar{\bar{t}}_c^m, \bar{\bar{s}}_c^m)$ of the broken line Γ_m as functions of (ξ_1, ξ_2). $\qquad\square$

27. Equivalence under shift and stopping

PROPOSITION 6.14. *Let $\tau \in \mathrm{IMT}$ and $\ell\xi_1 = \ell\xi_2$. Then*

(1) $\tau(\xi_1) < \infty \Rightarrow \ell\theta_\tau\xi_1 = \ell\theta_\tau\xi_2$;

(2) $\ell\alpha_\tau\xi_1 = \ell\alpha_\tau\xi_2$.

Proof. (1) Let $\tau \in \mathrm{IMT}$ and $\ell\xi_1 = \ell\xi_2$. Then $\tau(\xi_1) < \infty \Leftrightarrow \tau(\xi_2) < \infty$. Let $\tau(\xi_1) < \infty$, $\xi_1 \circ \varphi_1 = \xi_2 \circ \varphi_2$ ($\varphi_1, \varphi_2 \in \Phi$), $\tau_1 \in \mathrm{IMT}$. Then, under theorem 6.2(1),

$$\tau_1\theta_\tau\xi_1 < \infty \iff (\tau \dot{+} \tau_1)(\xi_1) < \infty \iff \varphi_1^*((\tau \dot{+} \tau_1)(\xi_1)) < \infty.$$

Since $\varphi_1^*((\tau \dot{+} \tau_1)(\xi_1)) = \varphi_2^*((\tau \dot{+} \tau_1)(\xi_2))$, then

$$\tau_1\theta_\tau\xi_1 < \infty \iff \tau_1\theta_\tau\xi_2 < \infty.$$

Besides, from condition $\tau_1\theta_\tau\xi_1, \tau_1\theta_\tau\xi_2 < \infty$ it follows that

$$X_{\tau_1}\theta_\tau\xi_1 = X_{\tau \dot{+} \tau_1}\xi_1 = X_{\tau \dot{+} \tau_1}(\xi_1 \circ \varphi_1)$$

and because of $X_{\tau \dot{+} \tau_1}(\xi_1 \circ \varphi_1) = X_{\tau \dot{+} \tau_1}(\xi_2 \circ \varphi_2)$,

$$X_{\tau_1}\theta_\tau\xi_1 = X_{\tau_1}\theta_\tau\xi_2.$$

From here for any $\tau_1 \in \mathcal{T}_0$ (see item 22) $\gamma_{\tau_1}\theta_\tau\xi_1 = \gamma_{\tau_1}\theta_\tau\xi_2$. Furthermore, for $\tau_1 < \infty$ we have

$$\theta_\tau\xi_1 \in \mathcal{D}^{\tau_1} \iff (\forall \varepsilon > 0)\,(\exists \tau_2 \in \mathcal{T}_0)\quad 0 < \tau_1\theta_\tau\xi_1 - \tau_2\theta_\tau\xi_1 < \varepsilon$$
$$\iff (\forall \varepsilon > 0)\,(\exists \tau_2 \in \mathcal{T}_0)\quad 0 < (\tau \dot{+} \tau_1)(\xi_1) - (\tau \dot{+} \tau_2)(\xi_1) < \varepsilon$$
$$\iff (\forall \varepsilon > 0)\,(\exists \tau_2 \in \mathcal{T}_0)\quad 0 < \varphi_1^*((\tau \dot{+} \tau_1)\xi_1) - \varphi_1^*((\tau \dot{+} \tau_2)\xi_1) < \varepsilon,$$

and because of $\varphi_1^*((\tau \dot{+} \tau_1)\xi_1) = \varphi_2^*((\tau \dot{+} \tau_1)\xi_2)$,

$$\theta_\tau\xi_1 \in \mathcal{D}^{\tau_1} \iff \theta_\tau\xi_2 \in \mathcal{D}^{\tau_1}.$$

Hence, $(\forall \tau_1 \in \mathcal{T}_0)\,(\tau_1 < \infty)$

$$I(\mathcal{D}^{\tau_1} \mid \theta_\tau\xi_1) = I(\mathcal{D}^{\tau_1} \mid \theta_\tau\xi_2).$$

Similarly, it can be proved that $I(\mathcal{D}^\infty \mid \theta_\tau \xi_1) = I(\mathcal{D}^\infty \mid \theta_\tau \xi_2)$. From here under theorem 6.4, $\ell\theta_\tau\xi_1 = \ell\theta_\tau\xi_2$.

(2) We have $\alpha_\tau(\xi_1 \circ \varphi_1) = \alpha_\tau(\xi_2 \circ \varphi_2)$ and according to proposition 6.3(1),

$$\alpha_{\tau(\xi_1 \circ \varphi_1)}(\xi_1 \circ \varphi_1) = \alpha_{\varphi_1^*(\tau(\xi_1))}(\xi_1 \circ \varphi_1) = (\alpha_\tau \xi_1) \circ \varphi_1.$$

From here $(\alpha_\tau \xi_1) \circ \varphi_1 = (\alpha_\tau \xi_2) \circ \varphi_2$ and consequently $\ell\alpha_\tau\xi_1 = \ell\alpha_\tau\xi_2$.   □

28. Preservation of order under time change

PROPOSITION 6.15. *Let $\xi \in \mathcal{D}$, $\varphi \in \Phi$ and $\tau_1 \in T(\sigma_\Delta,\ \Delta \in \widetilde{\mathfrak{A}})$. Then*

(1) $\tau_1(\xi \circ \varphi) \in \{\inf \varphi^{-1}\{\tau_1(\xi)\}, \sup \varphi^{-1}\{\tau_1(\xi)\}\}$;

(2) if $\tau \in \mathrm{IMT}$, then the following two equivalences are fair:

$$\tau_1(\xi) < \tau(\xi) \Longleftrightarrow \tau_1(\xi \circ \varphi) < \tau(\xi \circ \varphi),$$
$$\tau_1(\xi) > \tau(\xi) \Longleftrightarrow \tau_1(\xi \circ \varphi) > \tau(\xi \circ \varphi).$$

Proof. (1) We have $\sigma_\emptyset(\xi \circ \varphi) \equiv 0 \in \varphi^{-1}\{\sigma_\emptyset \xi\} = \varphi^{-1}\{0\}$. Let for some $\tau \in T(\sigma_\Delta, \Delta \in \widetilde{\mathfrak{A}})$ $\tau(\xi \circ \varphi) \in \varphi^{-1}\{\tau(\xi)\}$. Then for $\Delta \in \mathfrak{A}$ we have

$$(\tau \dotplus \sigma_\Delta)(\xi) = \inf\left[\tau(\xi), \infty\right) \cap \xi^{-1}(\mathbb{X} \setminus \Delta),$$
$$(\tau \dotplus \sigma_\Delta)(\xi \circ \varphi) = \inf\left[\tau(\xi \circ \varphi), \infty\right) \cap \varphi^{-1}\xi^{-1}(\mathbb{X} \setminus \Delta)$$
$$\geq \inf \varphi^{-1}\left(\left[\tau(\xi), \infty\right) \cap \xi^{-1}(\mathbb{X} \setminus \Delta)\right)$$
$$\geq \inf \varphi^{-1}\left(\inf\left[\tau(\xi), \infty\right) \cap \xi^{-1}(\mathbb{X} \setminus \Delta)\right)$$
$$= \inf \varphi^{-1}\{(\tau \dotplus \sigma_\Delta)(\xi)\}.$$

On the other hand, $(\forall t_n \in (\tau(\xi), \infty) \cap \xi^{-1}(\mathbb{X} \setminus \Delta))$ it is fair that $(\tau \dotplus \sigma_\Delta)(\xi \circ \varphi) \leq \inf \varphi^{-1}\{t_n\}$, and if $t_n \downarrow \inf[\tau(\xi), \infty) \cap \xi^{-1}(\mathbb{X} \setminus \Delta)$, then

$$\inf \varphi^{-1}\{t_n\} \downarrow \sup \varphi^{-1}\{\inf\left[\tau(\xi), \infty\right) \cap \xi^{-1}(\mathbb{X} \setminus \Delta)\} = \sup \varphi^{-1}\{(\tau \dotplus \sigma_\Delta)(\xi)\},$$
$$(\tau \dotplus \sigma_\Delta)(\xi \circ \varphi) \leq \sup \varphi^{-1}\{(\tau \dotplus \sigma_\Delta)\xi\}.$$

From here $(\tau \dotplus \sigma_\Delta)(\xi \circ \varphi) \in \varphi^{-1}\{(\tau \dotplus \sigma_\Delta)\xi\}$. Hence, $(\forall \tau \in T(\sigma_\Delta, \Delta \in \widetilde{\mathfrak{A}}))$

$$\tau(\xi \circ \varphi) \in \varphi^{-1}\{\tau(\xi)\}.$$

On interval $(\inf \varphi^{-1}\{\tau(\xi)\}, \sup \varphi^{-1}\{\tau(\xi)\})$ the function $\xi \circ \varphi$ is evidently constant and equal to $\xi(\tau(\xi))$, but $(\forall \tau \in T(\sigma_\Delta, \Delta \in \widetilde{\mathfrak{A}}))$ point $\tau(\xi)$ cannot belong an interval of constancy of the function ξ. It implies the first assertion.

(2) Let $\tau_1 \in T(\sigma_\Delta,\ \Delta \in \widetilde{\mathfrak{A}}),\ \tau \in$ IMT. Then $\tau(\xi \circ \varphi) = \varphi^*(\tau(\xi))$. However, for $\tau_1(\xi) < \tau(\xi)$

$$\sup \varphi^{-1}\{\tau_1(\xi)\} < \inf \varphi^{-1}\{\tau(\xi)\} = \varphi^*(\tau(\xi)),$$

i.e. $\tau_1(\xi \circ \varphi) < \tau(\xi \circ \varphi)$. If $\tau_1(\xi) > \tau(\xi)$, then

$$\tau_1(\xi \circ \varphi) = \inf \varphi^{-1}\{\tau_1(\xi)\} > \inf \varphi^{-1}\{\tau(\xi)\} = \tau(\xi \circ \varphi),$$

and if $\tau_1(\xi) = \tau(\xi)$, then

$$\tau(\xi \circ \varphi) \leq \tau_1(\xi \circ \varphi). \qquad \qquad \square$$

29. Intrinsic Markov time and curve Γ

PROPOSITION 6.16. *If* $\ell\xi_1 = \ell\xi_2,\ \tau \in$ IT *and* $\tau(\xi_1) < \infty$, *then*

$$\big(\tau(\xi_1), \tau(\xi_2)\big) \in \Gamma$$

(definition of Γ; see item 23).

Proof. Let $\xi_1 \circ \varphi_1 = \xi_2 \circ \varphi_2$ $(\varphi_1, \varphi_2 \in \Phi)$ and $\tau \in$ IMT. In this case there can be two possibilities.

If $\tau(\xi_1)$ is a point of discontinuity of ξ_1, then $\tau(\xi_1 \circ \varphi_1) = \tau(\xi_2 \circ \varphi_2)$ is a point of discontinuity of $\xi_1 \circ \varphi_1 = \xi_2 \circ \varphi_2$, i.e. $\tau(\xi_2)$ is a point of discontinuity of ξ_2. From here $(\exists \tau_1, \tau_2 \in T(\sigma_\Delta,\ \Delta \in \widetilde{\mathfrak{A}}))$

$$\tau(\xi_1) = \tau_1(\xi_1), \qquad \tau(\xi_2) = \tau_2(\xi_2).$$

Let $(\tau_1(\xi_1), \tau_1(\xi_2)) \neq (\tau_2(\xi_1), \tau_2(\xi_2))$. Both these points by virtue of theorem 6.12 belong to Γ. Consider the case $(\tau_1(\xi_1), \tau_1(\xi_2)) < (\tau_2(\xi_1), \tau_2(\xi_2))$. Because $\tau_2(\xi_1)$ is a point of discontinuity of function ξ_1, by virtue of a condition $I(\mathcal{D}^{\tau_2} \mid \xi_1) = I(\mathcal{D}^{\tau_2} \mid \xi_2)$ would be simultaneously $\tau_1(\xi_1) < \tau_2(\xi_1)$ and $\tau_1(\xi_2) < \tau_2(\xi_2)$. Then, by proposition 6.15(2),

$$\tau_1(\xi_2) < \tau_2(\xi_2) = \tau(\xi_2) \iff \tau_1(\xi_2 \circ \varphi_2) < \tau(\xi_2 \circ \varphi_2)$$
$$\iff \tau_1(\xi_1 \circ \varphi_1) < \tau(\xi_1 \circ \varphi_1)$$
$$\iff \tau_1(\xi_1) < \tau(\xi_1).$$

We obtain a contradiction.

Let $\tau(\xi_1),\ \tau(\xi_2)$ be points of continuity of ξ_1 and ξ_2. Then on the left of points $\tau(\xi_1), \tau(\xi_2)$ for functions ξ_1, ξ_2 accordingly there are no intervals of a constancy (theorem 6.4) and $\tau(\xi_1), \tau(\xi_2)$ are limits from the left for the set of points $\tau_n(\xi_1), \tau_n(\xi_2)$

accordingly, where $\tau_n \in T(\sigma_\Delta, \ \Delta \in \widetilde{\mathfrak{A}})$. Let $\tau_n(\xi_1) \uparrow \tau(\xi_1)$. Then $\tau_n(\xi_2) \uparrow a \leq \tau(\xi_2)$. If $a < \tau(\xi_2)$, we obtain an inconsistency: according to proposition 6.15(2) $(\exists \tau_0(\xi_2) \in (a, \tau(\xi_2)), \ \tau_0 \in T(\sigma_\Delta, \ \Delta \in \widetilde{\mathfrak{A}}))$

$$\tau_0(\xi_2) < \tau(\xi_2) \Longrightarrow \tau_0(\xi_1) < \tau(\xi_1),$$

$$\tau_0(\xi_2) > a \Longrightarrow \tau_0(\xi_1) \geq \tau_n(\xi_1) \Longrightarrow \tau_0(\xi_1) \geq \tau(\xi_1).$$

From here $(\tau(\xi_1), \tau(\xi_2)) = \lim_{n \to \infty} (\tau_n(\xi_1), \tau_n(\xi_2)) \in \Gamma$.   □

30. Direct values and inverse time changes

PROPOSITION 6.17. *Let* $\tau \in \mathrm{IMT}$, $\ell\xi_1 = \ell\xi_2$, $i = 1, 2$. *Then*

(1) $v_i(\tau(\xi_1) + \tau(\xi_2) \mid \xi_1, \xi_2) = \tau(\xi_i)$;

(2) $v_i^*(\tau(\xi_i) \mid \xi_1, \xi_2) = \tau(\xi_1) + \tau(\xi_2)$ *(v_1, v_2; see item 23).*

Proof. Assertion (1) follows from the definition of v_1, v_2 and proposition 6.16. Let us prove assertion (2). According to the definition, $(\forall t \in \mathbb{R}_+) \ v_i^*(t \mid \xi_1, \xi_2) = \inf v_i^{-1}(\{t\} \mid \xi_1, \xi_2)$. There can be two possibilities.

Let $\tau(\xi_1)$, $\tau(\xi_2)$ be points of discontinuity. According to condition $I(\mathcal{D}^\tau \mid \xi_1) = I(\mathcal{D}^\tau \mid \xi_2)$, for any point $(t, s) \in \Gamma$ it is fair that $(t, s) < (\tau(\xi_1), \tau(\xi_2)) \Rightarrow t < \tau(\xi_1)$, $s < \tau(\xi_2)$. From here $\tau(\xi_1) + \tau(\xi_2) = \inf v_1^{-1}(\{\tau(\xi_1)\} \mid \xi_1, \xi_2) = \inf v_2^{-1}(\{\tau(\xi_2)\} \mid \xi_1, \xi_2)$.

Let $\varphi \in \Phi$ and $t > 0$. Let for any $\varepsilon > 0$ φ be not constant on $[t - \varepsilon, t]$, i.e. $(\forall t' < t) \ \varphi(t') < \varphi(t)$. In general case $\varphi^*(\varphi(t)) \leq t$, but if $\varphi^*(\varphi(t)) < t$, we obtain a contradiction: $\varphi(\varphi^*(\varphi(t))) = \varphi(t) < \varphi(t)$. Let $\tau(\xi_1)$, $\tau(\xi_2)$ be points of continuity of corresponding functions and, hence, both ξ_1, and ξ_2 have no intervals of constancy on the left of corresponding points. Hence,

$$v_i^*(\tau(\xi_i) \mid \xi_1, \xi_2) = v_i^*(v_i(\tau(\xi_1) + \tau(\xi_2) \mid \xi_1, \xi_2) \mid \xi_1, \xi_2)$$
$$= \tau(\xi_1) + \tau(\xi_2).$$ □

31. Permutability of time change with shift and stopping operators

THEOREM 6.5. *Let* $\tau \in \mathrm{IMT}$, $\ell\xi_1 = \ell\xi_2$. *Then*

(1) if $\tau(\xi_1) < \infty$, *then*

$$v_1^*(\cdot \mid \theta_\tau \xi_1, \theta_\tau \xi_2) = \widetilde{\theta}_{\tau(\xi_1)} v_1^*(\cdot \mid \xi_1, \xi_2);$$

(2) if $\tau(\xi_1) < \infty$, *then*

$$v_1(\cdot \mid \theta_\tau \xi_1, \theta_\tau \xi_2) = \widetilde{\theta}_{\tau(\xi_1) + \tau(\xi_2)} v_1(\cdot \mid \xi_1, \xi_2);$$

(3) $(\forall s \in \mathbb{R}_+)$

$$v_1\big(s \mid \alpha_\tau \xi_1, \alpha_\tau \xi_2\big) = v_1\big(s \wedge \big(\tau(\xi_1) + \tau(\xi_2)\big) \mid \xi_1, \xi_2\big)$$
$$+ \big(0 \vee \big(s - \tau(\xi_1) - \tau(\xi_2)\big)\big)/2;$$

(4) $(\forall s \in \mathbb{R}_+)$

$$v_1^*\big(s \mid \alpha_\tau \xi_1, \alpha_\tau \xi_2\big) = v_1^*\big(s \wedge \tau(\xi_1) \mid \xi_1, \xi_2\big) + 2\big(0 \vee \big(s - \tau(\xi_1)\big)\big);$$

(5)

$$v_2\big(v_1^*\big(\cdot \mid \theta_\tau \xi_1, \theta_\tau \xi_2\big) \mid \theta_\tau \xi_1, \theta_\tau \xi_2\big) = \widetilde{\theta}_{\tau(\xi_1)} v_2\big(v_1^*\big(\cdot \mid \xi_1, \xi_2\big) \mid \xi_1, \xi_2\big);$$

(6)

$$v_2\big(v_1^*\big(\cdot \mid \alpha_\tau \xi_1, \alpha_\tau \xi_2\big) \mid \alpha_\tau \xi_1, \alpha_\tau \xi_2\big) = \widetilde{\alpha}_{\tau(\xi_1)} v_2\big(v_1^*\big(\cdot \mid \xi_1, \xi_2\big) \mid \xi_1, \xi_2\big).$$

The same is fair with replacing indexes $(1, 2) \mapsto (2, 1)$ *(see item 23).*

Proof. (1) We have $\ell\theta_\tau \xi_1 = \ell\theta_\tau \xi_2$ (proposition 6.14). Since any point of non-constancy of v_1 on the left is a limiting point for the set of points $\tau_n(\xi_1) + \tau_n(\xi_2)$ $(\tau_n \in \mathrm{IMT})$, it is sufficient to check equality at points $\tau_n(\theta_\tau \xi_1)$. We have

$$v_1^*\big(\tau_n \theta_\tau \xi_1 \mid \theta_\tau \xi_1, \theta_\tau \xi_2\big) = \tau_n\big(\theta_\tau \xi_1\big) + \tau_n\big(\theta_\tau \xi_2\big)$$
$$= \big(\tau \dotplus \tau_n\big)(\xi_1) - \tau(\xi_1) + \big(\tau \dotplus \tau_n\big)(\xi_2) - \tau(\xi_2)$$
$$= \widetilde{X}_{\tau_n(\theta_\tau \xi_1)} \widetilde{\theta}_{\tau(\xi_1)} v_1^*\big(\cdot \mid \xi_1, \xi_2\big)$$
$$= v_1^*\big(\big(\tau \dotplus \tau_n\big)(\xi_1) \mid \xi_1, \xi_2\big) - v_1^*\big(\tau(\xi_1) \mid \xi_1, \xi_2\big)$$

(see item 3, proposition 6.2(1), theorem 6.2, proposition 6.17).

(2) We have

$$v_1\big(\tau_n\big(\theta_\tau \xi_1\big) + \tau_n\big(\theta_\tau \xi_2\big) \mid \theta_\tau \xi_1, \theta_\tau \xi_2\big)$$
$$= \tau_n\big(\theta_\tau \xi_1\big) = \big(\tau \dotplus \tau_n\big)(\xi_1) - \tau(\xi_1).$$

On the other hand,

$$\widetilde{X}_{\tau_n(\theta_\tau \xi_1) + \tau_n(\theta_\tau \xi_2)} \widetilde{\theta}_{\tau(\xi_1) + \tau(\xi_2)} v_1\big(\cdot \mid \xi_1, \xi_2\big)$$
$$= \widetilde{X}_{(\tau \dotplus \tau_n)(\xi_1) + (\tau \dotplus \tau_n)(\xi_2)} v_1\big(\cdot \mid \xi_1, \xi_2\big)$$
$$- X_{\tau(\xi_1) + \tau(\xi_2)} v_1\big(\cdot \mid \xi_1, \xi_2\big)$$
$$= \big(\tau \dotplus \tau_n\big)(\xi_1) - \tau(\xi_1).$$

From here it follows that

$$\widetilde{\theta}_{\tau(\xi_1)+\tau(\xi_2)} v_1\big(\cdot \mid \xi_1, \xi_2\big) = v_1\big(\cdot \mid \theta_\tau \xi_1, \theta_\tau \xi_2\big).$$

(3) $(\forall s \le \tau(\xi_1) + \tau(\xi_2))$ $v_1(s \mid \alpha_\tau \xi_1, \alpha_\tau \xi_2) = v_1(s \mid \xi_1, \xi_2)$ according to definition of v_1 (see item 23). Since a coefficient of inclination of the curve $\Gamma(\alpha_\tau \xi_1, \alpha_\tau \xi_2)$ after the point $(\tau(\xi_1), \tau(\xi_2))$ is equal to 1, $(\forall a \ge 0)$

$$v_1\big(\tau(\xi_1) + \tau(\xi_2) + 2a \mid \alpha_\tau \xi_1, \alpha_\tau \xi_2\big) = v_1\big(\tau(\xi_1) + \tau(\xi_2) \mid \xi_1, \xi_2\big) + a.$$

From here

$$v_1\big(s \mid \alpha_\tau \xi_1, \alpha_\tau \xi_2\big) = v_1\big(s \wedge \big(\tau(\xi_1) + \tau(\xi_2)\big) \mid \xi_1, \xi_2\big)$$
$$+ \big(0 \vee \big(s - \tau(\xi_1) - \tau(\xi_2)\big)\big)/2.$$

(4) Since $v_1(\tau(\xi_1) + \tau(\xi_2) \mid \alpha_\tau \xi_1, \alpha_\tau \xi_2) = \tau(\xi_1)$ and up to that point $(\tau(\xi_1),$ $\tau(\xi_2))$ curves $\Gamma(\xi_1, \xi_2)$ and $\Gamma(\alpha_\tau \xi_1, \alpha_\tau \xi_2)$ coincide, $(\forall s \le \tau(\xi_1))$

$$v_1^*\big(s \mid \alpha_\tau \xi_1, \alpha_\tau \xi_2\big) = v_1^*\big(s \mid \xi_1, \xi_2\big).$$

Since coefficients of inclination of the rectilinear part after the point $\tau(\xi_1) + \tau(\xi_2)$ of the function $v_1(\cdot \mid \alpha_\tau \xi_1, \alpha_\tau \xi_2)$ and that of the function $v_1^*(\cdot \mid \alpha_\tau \xi_1, \alpha_\tau \xi_2)$ after point $\tau(\xi_1)$ are mutually reciprocal,

$$v_1^*\big(s \mid \alpha_\tau \xi_1, \alpha_\tau \xi_2\big) = v_1^*\big(s \wedge \tau(\xi_1) \mid \xi_1, \xi_2\big) + 2\big(0 \vee \big(s - \tau(\xi_1)\big)\big).$$

(5) According to item (1) of the proof, we have

$$v_2\big(v_1^*\big(s \mid \theta_\tau \xi_1, \theta_\tau \xi_2\big) \mid \theta_\tau \xi_1, \theta_\tau \xi_2\big) = v_2\big(\widetilde{\theta}_{\tau(\xi_1)} v_1^*\big(s \mid \xi_1, \xi_2\big) \mid \theta_\tau \xi_1, \theta_\tau \xi_2\big)$$
$$= \widetilde{\theta}_{\tau(\xi_1)+\tau(\xi_2)} v_2\big(v_1^*\big(\tau(\xi_1) + s \mid \xi_1, \xi_2\big) - \tau(\xi_1) - \tau(\xi_2) \mid \xi_1, \xi_2\big)$$
$$= v_2\big(v_1^*\big(\tau(\xi_1) + s \mid \xi_1, \xi_2\big) \mid \xi_1, \xi_2\big) - v_2\big(\tau(\xi_1) + \tau(\xi_2) \mid \xi_1, \xi_2\big)$$
$$= v_2\big(v_1^*\big(\tau(\xi_1) + s \mid \xi_1, \xi_2\big) \mid \xi_1, \xi_2\big) - \tau(\xi_2).$$

On the other hand,

$$\widetilde{X}_s \widetilde{\theta}_{\tau(\xi_1)} v_2\big(v_1^*\big(\cdot \mid \xi_1, \xi_2\big) \mid \xi_1, \xi_2\big)$$
$$= v_2\big(v_1^*\big(\tau(\xi_1) + s \mid \xi_1, \xi_2\big) \mid \xi_1, \xi_2\big)$$
$$- v_2\big(v_1^*\big(\tau(\xi_1) \mid \xi_1, \xi_2\big) \mid \xi_1, \xi_2\big)$$
$$= v_2\big(v_1^*\big(\tau(\xi_1) + s \mid \xi_1, \xi_2\big) \mid \xi_1, \xi_2\big) - \tau(\xi_2).$$

(6) According to item (3) of the proof, we have

$$v_2\big(v_1^*\big(s \mid \alpha_\tau\xi_1, \alpha_\tau\xi_2\big) \mid \alpha_\tau\xi_1, \alpha_\tau\xi_2\big)$$

$$= v_2\big(v_1^*\big(s \mid \alpha_\tau\xi_1, \alpha_\tau\xi_2\big) \wedge \big(\tau(\xi_1) + \tau(\xi_2)\big) \mid \xi_1, \xi_2\big)$$
$$+ \big(0 \vee \big(v_1^*\big(s \mid \alpha_\tau\xi_1, \alpha_\tau\xi_2\big) - \tau(\xi_1) - \tau(\xi_2)\big)\big)/2$$

$$= v_2\big(v_1^*\big(s \mid \xi_1, \xi_2\big) \wedge \big(\tau(\xi_1) + \tau(\xi_2)\big) \mid \xi_1, \xi_2\big) + \big(0 \vee 2\big(s - \tau(\xi_1)\big)\big)/2$$

$$= v_2\big(v_1^*\big(s \wedge \tau(\xi_1) \mid \xi_1, \xi_2\big) \mid \xi_1, \xi_2\big) + 0 \vee \big(s - \tau(\xi_1)\big)$$

$$= \widetilde{X}_s\widetilde{\alpha}_{\tau(\xi_1)}v_2\big(v_1^*\big(\cdot \mid \xi_1, \xi_2\big) \mid \xi_1, \xi_2\big). \qquad \square$$

6.4. Coordination of function and time change

The direct time change which is not changing a sequence of states need not be a strictly growing function. In this case intervals of constancy of a direct time change should be chosen in a special manner. Such intervals cut out some segments from a time scale of a transformed function. On the other hand, they must not change their sequence of states. This can only be the case when deleted segments are closed intervals of constancy of the transformed function or its periods, if any. To avoid anomalies connected with periodicity, the definition of coordination of a direct time change and a function concerns to the stopped functions. The measurability of a set of coordinated pairs: "direct time change – function" follows from outcomes of section 6.3 (see proposition 6.13).

32. Coordinated functions

We shall state that a time change $\psi \in \Psi_1$ is coordinated with function $\xi \in \mathcal{D}$, when $(\forall t \in \mathbb{R}_+)\, \alpha_t\xi \circ \psi^* \in \mathcal{D}$ and $\ell(\alpha_t\xi \circ \psi^*) = \ell\alpha_t\xi$. Let $\mathcal{H}$ be the set of the coordinated pairs $(\psi, \xi) \in \Psi_1 \times \mathcal{D}$. Obviously, $\Psi \times \mathcal{D} \subset \mathcal{H} \neq \Psi_1 \times \mathcal{D}$ and $(\forall(\psi, \xi) \in \mathcal{H})$ $\ell(\xi \circ \psi) = \ell\xi$.

An interval open or closed at some of its ends at least is said to be an interval of constancy of the function $f : \mathbb{R}_+ \to A$ if f is constant on this interval. Let us designate by $\mathrm{IC}(f)$ the set of intervals of constancy of the function f.

33. Criterion of coordination

As mentioned earlier we denote $[\Delta]$ a closure of an interval Δ.

PROPOSITION 6.18. *The following equivalence is fair*

$$(\psi, \xi) \in \mathcal{H} \Longleftrightarrow (\forall\Delta \in \mathrm{IC}(\psi))\quad [\Delta] \in \mathrm{IC}(\xi).$$

Proof. Necessity. Let $(t_1, t_2) \in IC(\psi)$, and this interval be not contained in any interval of constancy of the form $(s, t]$. We have $\psi^* \psi(t_2) = t_1$ and $\psi(\psi(t_2) + c) \to t_2$ $(c \downarrow 0)$. Let $(\exists t \in (t_1, t_2]) \, \xi(t_1) \neq \xi(t)$. Then $\xi(t_1) = (\alpha_t \xi)(t_1) \neq (\alpha_t \xi)(t_2)$. Hence,

$$\big((\alpha_t \xi) \circ \psi^*\big)\big(\psi(t_2)\big) \neq \lim_{c \downarrow 0} \big((\alpha_t \xi) \circ \psi^*\big)\big(\psi(t_2) + c\big), \quad \alpha_t \xi \circ \psi^* \notin \mathcal{D}.$$

Sufficiency. Let $\xi = \alpha_t \xi$, $\tau \in IMT$, $\tau(\xi) < \infty$. According to proposition 6.6, $(\forall \varepsilon > 0) \, [\tau(\xi) - \varepsilon, \tau(\xi)] \notin IC(\xi)$. Then, according to the condition, $(\forall \varepsilon > 0)$ $(\tau(\xi) - \varepsilon, \tau(\xi)] \notin IC(\psi)$. Hence, $\psi^* \psi(\tau(\xi)) = \tau(\xi)$. From here $(\forall \Delta \in \mathfrak{A}, \sigma_\Delta \xi < \infty)$

$$\big(\xi \circ \psi^*\big)\big(\psi(\sigma_\Delta(\xi))\big) = \xi\big(\sigma_\Delta(\xi)\big) \notin \Delta$$

and $(\forall t < \psi(\sigma_\Delta(\xi))) \, (\exists t_1 < \sigma_\Delta(\xi))$

$$\big(\xi \circ \psi^*\big)(t) = \xi(t_1) \in \Delta.$$

Hence, $\psi(\sigma_\Delta(\xi)) = \sigma_\Delta(\xi \circ \psi^*)$ and $X_{\sigma_\Delta}(\xi \circ \psi^*) = X_{\sigma_\Delta} \xi$.

Let $(\exists \tau \in IMT, \tau(\xi) < \infty)$

$$\psi\big(\tau(\xi)\big) = \tau\big(\xi \circ \psi^*\big), \qquad X_\tau\big(\xi \circ \psi^*\big) = X_\tau \xi.$$

Then $(\forall \Delta \in \mathfrak{A})$

$$\begin{aligned}
\big(\tau \dotplus \sigma_\Delta\big)\big(\xi \circ \psi^*\big) &= \tau\big(\xi \circ \psi^*\big) + \sigma_\Delta \theta_\tau\big(\xi \circ \psi^*\big) \\
&= \psi\big(\tau(\xi)\big) + \sigma_\Delta \theta_{\psi(\tau(\xi))}\big(\xi \circ \psi^*\big) \\
&= \psi\big(\tau(\xi)\big) + \sigma_\Delta\big(\theta_{\tau(\xi)} \xi \circ \widetilde{\theta}_{\psi(\tau(\xi))} \psi^*\big) \\
&= \psi\big(\tau(\xi)\big) + \big(\widetilde{\theta}_{\tau(\xi)} \psi\big)\big(\sigma_\Delta \theta_{\tau(\xi)} \xi\big) \\
&= \psi\big(\tau(\xi)\big) + \psi\big(\tau(\xi) + \sigma_\Delta \theta_{\tau(\xi)} \xi\big) - \psi\big(\tau(\xi)\big) \\
&= \psi\big((\tau \dotplus \sigma_\Delta)\xi\big),
\end{aligned}$$

since $\widetilde{\theta}_{\tau(\xi)} \psi = \big(\widetilde{\theta}_{\psi(\tau(\xi))} \psi^*\big)^*$ with respect to $\theta_{\tau(\xi)} \xi$ satisfies the same condition as ψ with respect to ξ. Consequently,

$$X_{\tau \dotplus \sigma_\Delta}\big(\xi \circ \psi^*\big) = X_{\tau \dotplus \sigma_\Delta}(\xi).$$

From here it follows that $\ell_\delta \xi = \ell_\delta(\xi \circ \psi^*)$ for any sequence $\delta = (\Delta_1, \Delta_2, \ldots)$ $(\Delta_i \in \mathfrak{A})$. Furthermore, since $(\forall \tau \in IMT) \, (\forall \varepsilon > 0)$

$$\big(\tau(\xi) - \varepsilon, \tau(\xi)\big] \in IC(\psi),$$

$I(\mathcal{D}^\tau \mid \xi) = I(\mathcal{D}^\tau \mid \xi \circ \psi^*)$. For example, if $I(\mathcal{D}^\tau \mid \xi) = 0$, then $(\exists \varepsilon > 0)$ $(\forall \delta \in \mathrm{DS})$ $(\forall n \in \mathbb{N})$

$$\sigma^n_\delta(\xi) \leq \tau(\xi) - \varepsilon \quad \text{or} \quad \sigma^n_\delta(\xi) \geq \tau(\xi).$$

Correspondingly $(\exists \varepsilon_1 > 0)$

$$\psi\big(\sigma^n_\delta(\xi)\big) \leq \psi\big(\tau(\xi)\big) - \varepsilon_1 \quad \text{or} \quad \psi\big(\sigma^n_\delta(\xi)\big) \geq \psi\big(\tau(\xi)\big),$$

therefore $I(\mathcal{D}^\tau \mid \xi \circ \psi^*) = 0$. $\qquad\qquad\qquad\qquad\qquad\qquad\qquad$ □

34. Measurability of a set of coordinated pairs

PROPOSITION 6.19. *The following property is fair*

$$\mathcal{H} \in \mathfrak{B}.$$

Proof. In proposition 6.18 it is proved that $(\forall(\psi, \xi) \in \mathcal{H})$ $(\forall \tau \in \mathrm{IMT})$

$$\psi^*\big(\psi(\tau(\xi))\big) = \tau(\xi),$$

i.e. $(\forall \varepsilon > 0)$ $(\tau(\xi) - \varepsilon, \tau(\xi)) \notin \mathrm{IC}(\psi)$.

Let $r_n \downarrow 0$ and $(\forall \delta_n \in \mathrm{DS}(r_n))$ $(\forall k, n, m \in \mathbb{N}, \; 1/m \leq \sigma^k_{\delta_n} \xi < \infty)$

$$\big(\sigma^k_{\delta_n} \xi - 1/m, \sigma^k_{\delta_n}\big] \in \mathrm{IC}(\psi).$$

Then $[a, b] \notin \mathrm{IC}(\xi) \Leftrightarrow (\exists k, n \in \mathbb{N})$ $\sigma^k_{\delta_n} \xi \in (a, b]$, therefore $(\exists m \in \mathbb{N}, \; \sigma^k_{\delta_n} \xi - 1/m \in (a, b])$

$$\big(\sigma^k_{\delta_n} \xi - 1/m, \sigma^k_{\delta_n} \xi\big] \notin \mathrm{IC}(\psi) \implies (a, b] \notin \mathrm{IC}(\psi) \implies (\psi, \xi) \in \mathcal{H}.$$

Hence,

$$\mathcal{H} = \bigcap_{k,m,n \in \mathbb{N}} \big(\big\{(\psi, \xi) : \psi\big(\sigma^k_{\delta_n} \xi - 1/m\big) < \psi\big(\sigma^k_{\delta_n} \xi\big)\big\}$$

$$\cup \big\{(\psi, \xi) : \sigma^k_{\delta_n}(\xi) < 1/m\big\} \cup \big\{(\psi, \xi) : \sigma^k_{\delta_n}(\xi) = \infty\big\}\big).$$

Evidently, two last sets belong to $\mathfrak{B}$. Besides,

$$\big\{(\psi, \xi) : \psi\big(\tau(\xi) - 1/m\big) < \psi\big(\tau(\xi)\big)\big\}$$

$$= \bigcup_{t_1, t_2 \in \mathbb{R}'_+, \; t_1 < t_2} \big\{(\psi, \xi) : \psi\big(\tau(\xi) - 1/m\big) \leq t_1, t_2 < \psi\big(\tau(\xi)\big)\big\},$$

where $\mathbb{R}'_+$ is the set of all rational $t \in \mathbb{R}_+$, and

$$\{(\psi, \xi) : \psi(\tau(\xi) + a) > t\}$$

$$= \bigcup_{k,n \in \mathbb{N}} \left\{ (\psi, \xi) : \psi\left(\frac{k}{n} + a\right) > t, \ \tau(\xi) \in \left[\frac{k}{n}, \frac{k+1}{n}\right) \right\}$$

$$= \bigcup_{k,n \in \mathbb{N}} (p^{-1} \widetilde{X}^{-1}_{k/n+a}(t, \infty)) \left(q^{-1} \left\{ \tau \in \left[\frac{k}{n}, \frac{k+1}{n}\right) \right\} \right) \in \mathfrak{B}.$$
□

35. Transformation of trajectory under time change

Let $u : \mathcal{H} \to \mathcal{D}$ be the map $u(\psi, \xi) = \xi \circ \psi^*$ transformation of a trajectory generated by a time change.

PROPOSITION 6.20. *The following property is fair*

$$u \in \mathfrak{B} \cap \mathcal{H}/\mathcal{F}.$$

Proof. Let $\tau \in T(\sigma_\Delta, \ \Delta \in \mathfrak{A}), t \in \mathbb{R}_+, S \in \mathcal{B}(\mathbb{X})$. Then

$$u^{-1} \beta_\tau^{-1} \{\infty\} = \{(\psi, \xi) \in \mathcal{H} : \beta_\tau(\xi \circ \psi^*) = \infty\}$$
$$= \{(\psi, \xi) \in \mathcal{H} : \tau(\xi \circ \psi^*) = \infty\}$$
$$= \{(\psi, \xi) \in \mathcal{H} : \tau(\xi) = \infty\} \in \mathfrak{B}.$$

Furthermore,

$$u^{-1} \beta_\tau^{-1} ([0, t] \times S) = \{(\psi, \xi) \in \mathcal{H} : \tau(\xi \circ \psi^*) \le t, \ X_\tau(\xi \circ \psi^*) \in S\}$$
$$= \{(\psi, \xi) \in \mathcal{H} : \psi(\tau(\xi)) \le t, \ X_\tau \xi \in S\}.$$

From the proof of proposition 6.19 it follows that

$$\{(\psi, \xi) \in \mathcal{H} : \psi(\tau(\xi)) \le t\} \in \mathfrak{B}.$$

Measurability of the other component is obvious.
□

36. Correspondence of time change and trajectory

PROPOSITION 6.21. *Let* $\varphi_1, \varphi_2 \in \Phi, \ \xi_1, \xi_2 \in \mathcal{D}$ *and* $\xi_1 \circ \varphi_1 = \xi_2 \circ \varphi_2$. *Then*

$$(\varphi_2 \circ \varphi_1^*, \xi_1) \in \mathcal{H}.$$

Proof. We have $(\forall s \in \mathbb{R}_+)$

$$X_{s \wedge \varphi_1^*(t)}\big(\xi_1 \circ \varphi_1\big) = X_{s \wedge \varphi_1^*(t)}\big(\xi_2 \circ \varphi_2\big)$$

$$\Longrightarrow X_{\varphi_1(s \wedge \varphi_1^*(t))}\xi_1 = X_{\varphi_2(s \wedge \varphi_1^*(t))}\xi_2$$

$$\Longrightarrow X_{\varphi_1(s) \wedge t}\xi_1 = X_{\varphi_2(s) \wedge \varphi_2\varphi_1^*(t)}\xi_2$$

$$\Longrightarrow X_{\varphi_1(s)}\alpha_t\xi_1 = X_{\varphi_2(s)}\alpha_{\varphi_2\varphi_1^*(t)}\xi_2$$

$$\Longrightarrow X_s\big(\alpha_t\xi_1 \circ \varphi_1\big) = X_s\big(\alpha_{\varphi_2\varphi_1^*(t)}\xi_2 \circ \varphi_2\big).$$

Hence, $\alpha_t\xi_1 \circ \varphi_1 = \big(\alpha_{\varphi_2\varphi_1^*(t)}\xi_2\big) \circ \varphi_2$. From here $(\forall t \in \mathbb{R}_+)\ \alpha_t\xi_1 \circ \varphi_1 \circ \varphi_2^* \in \mathcal{D}$. According to propositions 6.18 and 6.19, it is sufficient to show that $(\forall \tau \in \mathrm{IMT})$ $\psi^*\psi(\tau(\xi_1)) = \tau(\xi_1)$, where $\psi = \varphi_2 \circ \varphi_1^*$, $\xi_1 = \alpha_t\xi_1$, $\xi_2 = \alpha_{\psi(t)}\xi_2$. According to proposition 6.1 we have

$$\big(\psi^* \circ \psi\big)\big(\tau(\xi_1)\big) = \big(\varphi_1 \circ \varphi_2^* \circ \varphi_2 \circ \varphi_1^*\big)\big(\tau(\xi_1)\big)$$

$$= \big(\varphi_1 \circ \varphi_2^* \circ \varphi_2\big)\big(\tau(\xi_1 \circ \varphi_1)\big)$$

$$= \big(\varphi_1 \circ \varphi_2^* \circ \varphi_2\big)\big(\tau(\xi_2 \circ \varphi_2)\big)$$

$$= \big(\varphi_1 \circ \varphi_2^* \circ \varphi_2 \circ \varphi_2^*\big)\big(\tau(\xi_2)\big)$$

$$= \big(\varphi_1 \circ \varphi_2^*\big)\big(\tau(\xi_2)\big) = \varphi_1\big(\tau(\xi_2 \circ \varphi_2)\big)$$

$$= \varphi_1\big(\tau(\xi_1 \circ \varphi_1)\big) = \big(\varphi_1 \circ \varphi_1^*\big)\big(\tau(\xi_1)\big) = \tau(\xi_1). \qquad \square$$

37. Property IMT with respect to coordinated pairs

PROPOSITION 6.22. $(\forall \tau \in \mathrm{IMT})\ (\forall(\psi,\xi) \in \mathcal{H})$

$$\tau\big(\xi \circ \psi^*\big) = \psi\big(\tau(\xi)\big).$$

Proof. Since $(\forall k \in \mathbb{N})\ (\forall \delta \in \mathrm{DS})$

$$\psi\big(\sigma_\delta^k(\xi)\big) = \sigma_\delta^k\big(\xi \circ \psi^*\big),$$

from theorem 6.4 and proposition 6.18 it follows that

$$\Gamma\big(\xi, \xi \circ \psi^*\big) = \Gamma_1(\psi) \cup \Gamma_2(\psi)$$

(see items 2 and 23). From proposition 6.16 it follows that $(\forall \tau \in \mathrm{IMT})$

$$\big(\tau(\xi), \tau(\xi \circ \psi^*)\big) \in \Gamma\big(\xi, \xi \circ \psi^*\big).$$

From proposition 6.6 it follows that

$$\tau(\xi \circ \psi^*) = \inf\{s : (\tau(\xi), s) \in \Gamma(\xi, \xi \circ \psi^*)\} = \psi(\tau(\xi)). \qquad \square$$

6.5. Random time changes

Constructed in item 24, canonical time changes are used to prove the following theorem: *for two random functions with identical distributions of points of the first exit and indicators of sets D^τ there exist two random time changes, transforming these functions into one.* The reversion of this theorem is trivial. We consider general random time changes, when the joint distribution Q of direct time changes and trajectories is given. This joint distribution with the help of a map $(\psi, \xi) \to u(\psi, \xi) \in D$ induces a measure of a transformed random function. Transformation of distribution of an original random function $Q \circ q^{-1}$ (marginal distribution corresponding to map $q(\psi, \xi) = \xi$) in distribution of a random function $Q \circ u^{-1}$ we call as a time change transformation for distributions of random functions or random processes.

38. Basic maps

Let us introduce the following denotations:

$p_1 : D \times D \mapsto D$, $p_1(\xi_1, \xi_2) = \xi_1$ the first coordinate of a two-dimensional vector;

$p_2 : D \times D \mapsto D$, $p_2(\xi_1, \xi_2) = \xi_2$ the second coordinate of a two-dimensional vector;

$w_1 : d \mapsto \Psi \times D$, $w_1(\xi_1, \xi_2) = (v_1^*(\cdot \mid \xi_1, \xi_2), \xi_1)$ the first pair with a direct time change (d see item 26);

$w_2 : d \mapsto \Psi \times D$, $w_2(\xi_1, \xi_2) = (v_2^*(\cdot \mid \xi_1, \xi_2), \xi_2)$ the second pair with a direct time change;

$z_1 : d \mapsto \mathcal{H}$, $z_1(\xi_1, \xi_2) = (v_2(\cdot \mid \xi_1, \xi_2) \circ v_1^*(\cdot \mid \xi_1, \xi_2), \xi_1)$ the first coordinated pair;

$z_2 : d \mapsto \mathcal{H}$, $z_2(\xi_1, \xi_2) = (v_1(\cdot \mid \xi_1, \xi_2) \circ v_2^*(\cdot \mid \xi_1, \xi_2), \xi_2)$ the second coordinated pair.

39. Existence of canonical time change

PROPOSITION 6.23. *Let $\mathbb{P}$ be a probability measure on $\mathcal{F} \otimes \mathcal{F}$, $\mathbb{P}(d) = 1$, $P_1 = \mathbb{P} \circ p_1^{-1}$, $P_2 = \mathbb{P} \circ p_2^{-1}$. Then*

(1) there exist probability measures Q_1 and Q_2 on $\mathfrak{B}$ such that $Q_i(\Psi \times D) = 1$, $P_i = Q_i \circ q^{-1}$ ($i = 1, 2$) and $Q_1 \circ u^{-1} = Q_2 \circ u^{-1}$;

(2) there exist probability measures $\overline{Q}_1$ and $\overline{Q}_2$ on $\mathfrak{B}$ such that $\overline{Q}_i(\mathcal{H}) = 1$, $P_i = \overline{Q}_i \circ q^{-1}$ ($i = 1, 2$), $P_2 = \overline{Q}_1 \circ u^{-1}$, $P_1 = \overline{Q}_2 \circ u^{-1}$.

Proof. (1) From proposition 6.13 it follows that $w_i \in (\mathcal{F} \otimes \mathcal{F}) \cap d/\mathfrak{B}$. Let $Q_i = \mathbb{P} \circ w_i^{-1}$ $(i = 1, 2)$. We have $p_i = q \circ w_i$ (see item 3). From here $P_i = \mathbb{P} \circ w_i^{-1} \circ q^{-1} = Q_i \circ q^{-1}$. According to construction of v_1 and v_2, for given ξ_1, ξ_2 (see item 23) it is fair

$$(u \circ w_1)(\xi_1, \xi_2) = \xi_1 \circ v_1(\cdot \mid \xi_1, \xi_2)$$
$$= \xi_2 \circ v_2(\cdot \mid \xi_1, \xi_2)$$
$$= (u \circ w_2)(\xi_1, \xi_2).$$

From here $Q_1 \circ u^{-1} = \mathbb{P} \circ w_1^{-1} \circ u^{-1} = \mathbb{P} \circ w_2^{-1} \circ u^{-1} = Q_2 \circ u^{-1}$.

(2) We have $p_i = q \circ z_i$ and $z_i \in (\mathcal{F} \otimes \mathcal{F}) \cap d/\mathfrak{B}$. Let $\overline{Q}_i = \mathbb{P} \circ z_i^{-1}$. From here $P_i = \mathbb{P} \circ z_i^{-1} \circ q^{-1} = \overline{Q}_i \circ q^{-1}$. In this case

$$(u \circ z_1)(\xi_1, \xi_2) = u\big(v_2(v_1^*(\cdot \mid \xi_1, \xi_2) \mid \xi_1, \xi_2), \xi_1\big)$$
$$= \xi_1 \circ v_1(\cdot \mid \xi_1, \xi_2) \circ v_2^*(\cdot \mid \xi_1, \xi_2)$$
$$= \xi_2 = p_2(\xi_1, \xi_2).$$

From here

$$\overline{Q}_1 \circ u^{-1} = \mathbb{P} \circ z_1^{-1} \circ u^{-1} = \mathbb{P} \circ p_2^{-1} = P_2.$$

The same with replacement $(1, 2) \mapsto (2, 1)$. $\qquad\square$

40. Existence of measure on diagonal of traces

Let us designate

$$\mathcal{E} = \{A \subset \mathcal{M} : \ell^{-1} A \in \mathcal{F}\}.$$

Evidently, $\mathcal{E}$ is a sigma-algebra, $\ell \in \mathcal{F}/\mathcal{E}$. From proposition 6.7 it immediately follows that $\mathcal{F}^* = \ell^{-1}\mathcal{E}$. This means that any set $B \in \mathcal{F}^*$ can be represented as $B = \ell^{-1} B'$, where $B' \subset \mathcal{E}$, but it follows that B is representable as $B = \ell^{-1}\ell B$. Actually, $\ell^{-1}\ell B = \ell^{-1}\ell\ell^{-1} B'$, and since $\ell\ell^{-1}$ means identity map on $\mathcal{M}$, we obtain required relation. In other words, $\mathcal{F}^*$ is a set of all $B \in \mathcal{F}$ representable as $B = \ell^{-1}\ell B$.

In the following theorem and below we will use a sigma-algebra of subsets of the set $\mathcal{D}$, which all functions and subsets determining the trace are measurable with respect to. Let us designate

$$\mathcal{F}^\circ = \sigma(\gamma_\tau, \mathcal{D}^\tau, \ \tau \in \mathcal{T}_0).$$

Evidently $\mathcal{F}^\circ \subset \mathcal{F}^*$. From the Blackwell theorem [CHA 81] it follows inverse inclusion as well, hence $\mathcal{F}^\circ = \mathcal{F}^*$. We do not show a proof of this theorem. In all our constructions, touching traces and time changes, it is sufficient to use the sigma-algebra $\mathcal{F}^\circ$.

PROPOSITION 6.24. *Let P_1, P_2 be probability measures on $\mathcal{F}$ such that $(\forall \ell, k \in \mathbb{N})$ $(\forall \tau_i, \tau_i' \in \mathcal{T}_0) (\forall S_i \in \mathcal{B}(\overline{\mathbb{X}}))$*

$$P_1 \left(\bigcap_{i=1}^{k-1} \gamma_{\tau_i}^{-1} S_i \bigcap_{i=1}^{\ell-1} \mathcal{D}^{\tau_i'} \right) = P_2 \left(\bigcap_{i=1}^{k-1} \gamma_{\tau_i}^{-1} S_i \bigcap_{i=1}^{\ell-1} \mathcal{D}^{\tau_i'} \right)$$

($\mathcal{T}_0$ see item 22). Then there exists a probability measure $\mathbb{P}$ on $\mathcal{F} \otimes \mathcal{F}$ such that $\mathbb{P}(d) = 1$, $P_1 = \mathbb{P} \circ p_1^{-1}$, $P_2 = \mathbb{P} \circ p_2^{-1}$.

Proof. According to a given condition and the theorem on extension of measure [SHI 80, p. 167], $P_1 = P_2$ on $\mathcal{F}^\circ$. A probability measure $\mathbb{P}$ on $\mathcal{F} \otimes \mathcal{F}$ we define as an extension of the function of sets determining on sets $A_1 \times A_2$ ($A_1, A_2 \in \mathcal{F}$) by the formula

$$\mathbb{P}(A_1 \times A_2) = \int_D P_1(A_1 \mid \mathcal{F}^\circ) P_2(A_2 \mid \mathcal{F}^\circ) P_1(d\xi).$$

Let $A_1, A_2 \in \mathcal{F}^\circ$ and $A_1 \cap A_2 = \emptyset$. Then P_1-a.s. $P_i(A_i \mid \mathcal{F}^\circ) = I(A_i \mid \cdot)$ and

$$\mathbb{P}(A_1 \times A_2) = \int_D I(A_1 \mid \xi) I(A_2 \mid \xi) P_1(d\xi) = 0.$$

According to theorem 6.4 and due to the sigma-compactness of $\mathbb{X}$, set $(\mathcal{D} \times \mathcal{D}) \setminus d$ is exhausted by countable set of rectangles $A_1 \times A_2$ ($A_1, A_2 \in \mathcal{F}$, $A_1 \cap A_2 = \emptyset$). From here $\mathbb{P}(d) = 1$. In this case $(\forall A \in \mathcal{F})$

$$\mathbb{P} \circ p_1^{-1}(A) = \int P_1(A \mid \mathcal{F}^\circ) P_1(d\xi) = \mathbb{P}(A \times \mathcal{D}) = P_1(A),$$

$$\mathbb{P} \circ p_2^{-1}(A) = \int P_2(A \mid \mathcal{F}^\circ) P_1(d\xi) = \int P_2(A \mid \mathcal{F}^\circ) P_2(d\xi) = P_2(A). \qquad \square$$

41. Preservation of random trace under time change

PROPOSITION 6.25. *For any probability measure Q on $\mathfrak{B} \cap \mathcal{H}$ $P \circ \ell^{-1} = \widetilde{P} \circ \ell^{-1}$, where $P = Q \circ q^{-1}$, $\widetilde{P} = Q \circ u^{-1}$.*

Proof. Measurability of u is proved in proposition 6.20; q is measurable as a projection. We have $\ell q(\psi, \xi) = \ell \xi = \ell(\xi \circ \psi^*) = \ell u(\psi, \xi)$ and, hence, $\ell \circ q = \ell \circ u$. Therefore

$$P \circ \ell^{-1} = Q \circ q^{-1} \circ \ell^{-1} = Q \circ u^{-1} \circ \ell^{-1} = \widetilde{P} \circ \ell^{-1}. \qquad \square$$

42. Note on construction of measure on the diagonal

According to proposition 6.11 and the evident $\mathcal{F}^*$-measurability of the function γ_τ ($\tau \in$ IMT), we have $\mathcal{F}^\circ \subset \mathcal{F}^*$. Therefore, from condition $P_1 \circ \ell^{-1} = P_2 \circ \ell^{-1}$, fulfillment of proposition 6.24 follows (see item 40). On the other hand, if the condition of proposition 6.24 is fulfilled, then from propositions 6.23 – 6.25 it follows that $P_1 \circ \ell^{-1} = P_2 \circ \ell^{-1}$ and $P_1 = P_2$ on $\mathcal{F}^*$. Let us consider a probability measure $\mathbb{P}^*$ on $\mathcal{F} \otimes \mathcal{F}$ such that $(\forall A_1, A_2 \in \mathcal{F})$

$$\mathbb{P}^*(A_1 \times A_2) = P_1(P_1(A_1 \mid \mathcal{F}^*) P_2(A_2 \mid \mathcal{F}^*)).$$

Then since $\mathcal{F}^\circ \subset \mathcal{F}^*$ and $P_1 = P_2$ on $\mathcal{F}^*$,

$$\begin{aligned}
\mathbb{P}(A_1 \times A_2) &= P_1(P_1(A_1 \mid \mathcal{F}^\circ) P_2(A_2 \mid \mathcal{F}^\circ)) \\
&= P_1(P_1(A_1 \mid \mathcal{F}^*) P_2(A_2 \mid \mathcal{F}^*)) \\
&= \mathbb{P}^*(A_1 \times A_2),
\end{aligned}$$

i.e. $\mathbb{P} = \mathbb{P}^*$.

43. Criterion of identity of traces

Let (Q_x) be a family of probability measures on $\mathfrak{B}$. This family we call admissible if $(\forall x \in \mathbb{X})$ $Q_x(\mathcal{H}) = 1$, $(\forall B \in \mathfrak{B})$ $Q_x(B)$ is a $\mathcal{B}(\mathbb{X})$-measurable function of x and the marginal distributions (P_x) $(P_x = Q_x \circ q^{-1})$ composes an admissible family of probability measures on $\mathcal{F}$ (see item 2.42).

THEOREM 6.6. *Let* (P_x^1) *and* (P_x^2) *be two admissible family of probability measures on* $\mathcal{F}$. *Then condition* $(\forall x \in \mathbb{X})$

$$P_x^1 \circ \ell^{-1} = P_x^2 \circ \ell^{-1}$$

is equivalent to either of the two conditions:

(1) there exist two admissible families of probability measures (Q_x^1) *and* (Q_x^2) *on* $\mathfrak{B}$ *such that* $(\forall x \in \mathbb{X})$ $Q_x^i(\Psi \times \mathcal{D}) = 1$, $P_x^i = Q_x^i \circ q^{-1}$ $(i = 1, 2)$

$$Q_x^1 \circ u^{-1} = Q_x^2 \circ u^{-1} \equiv \tilde{P}_x;$$

(2) there exist two admissible families of probability measures $(\overline{Q}_x^1)$ *and* $(\overline{Q}_x^2)$ *on* $\mathfrak{B}$ *such that* $(\forall x \in \mathbb{X})$ $P_x^i = \overline{Q}_x^i \circ q^{-1}$ $(i = 1, 2)$, $P_x^2 = \overline{Q}_x^1 \circ u^{-1}$, $P_x^1 = \overline{Q}_x^2 \circ u^{-1}$.

Proof. It follows from propositions 6.23, 6.24 and 6.25. □

44. Note on criterion

If either of two processes with identical traces, for example (P_x^2), does not contain any intervals of constancy, then, according to item 24 we have $P_x^1 = \mathbb{P}_x \circ p_1^{-1} = \mathbb{P}_x \circ \widetilde{w}_2^{-1} \circ u^{-1} = \widetilde{Q}_x^2 \circ u^{-1}$, where $\widetilde{w}_2(\xi_1, \xi_2) = ((\Gamma(\cdot \mid \xi_1, \xi_2))^*, \xi_2), \Gamma(\cdot \mid \xi_1, \xi_2)$ is a function corresponding to time run comparison curve Γ from item 23, and $\Gamma(\cdot \mid \xi_1, \xi_2) \in \Phi$. Here also $\widetilde{Q}_x^2 \circ q^{-1} = P_x^2$, i.e. $\widetilde{Q}_x^2$ is an analogy of $\overline{Q}_x^2$.

45. Maps on set of traces

Let $\tau \in \mathrm{IMT}$. We define maps on the set $\mathcal{M}$:

$$\overline{X}_\tau : \ell\{\tau < \infty\} \mapsto \mathbb{X}, X_\tau = \overline{X}_\tau \circ \ell;$$
$$\overline{\theta}_\tau : \ell\{\tau < \infty\} \mapsto \mathcal{M}, \ell \circ \theta_\tau = \overline{\theta}_\tau \circ \ell;$$
$$\overline{\alpha}_\tau : \mathcal{M} \mapsto \mathcal{M}, \ell \circ \alpha_\tau = \overline{\alpha}_\tau \circ \ell.$$

Maps $\overline{X}_\tau$, $\overline{\theta}_\tau$ and $\overline{\alpha}_\tau$ are determined to be single-valued, because if $\ell\xi_1 = \ell\xi_2$, then $\ell\alpha_\tau\xi_1 = \ell\alpha_\tau\xi_2$ and $\ell\theta_\tau\xi_1 = \ell\theta_\tau\xi_2$ ($\tau < \infty$) (see proposition 6.14), therefore $X_0\theta_\tau\xi_1 = X_\tau\xi_1 = X_\tau\xi_2 = X_0\theta_\tau\xi_2$. From this definition it easily follows that

$$\overline{X}_\tau \in \mathcal{E} \cap \ell\{\tau < \infty\}/\mathcal{B}(\mathbb{X}), \qquad \overline{\theta}_\tau \in \mathcal{E} \cap \ell\{\tau < \infty\}/\mathcal{E}, \qquad \overline{\alpha}_\tau \in \mathcal{E}/\mathcal{E}.$$

46. Regeneration times of random traces

Let $(\forall B \in \mathcal{E}) R_x(B)$ be a $\mathcal{B}(\mathbb{X})$-measurable function of x and $R_x(\overline{X}_0 = x) = 1$ (admissible family). We will state that admissible family $(R_x)_{x \in \mathbb{X}}$ of probability measures on $\mathcal{E}$ has a Markov time $\tau \in \mathrm{IMT}$ as its regenerative time ($\tau \in \mathrm{RT}(R_x)$), if $(\forall x \in \mathbb{X}) R_x$-a.s. on the set $\ell\{\tau < \infty\}$

$$R_x(\overline{\theta}_\tau^{-1}B \mid \overline{\alpha}_\tau^{-1}\mathcal{E})(\mu) = R_{\overline{X}_\tau\mu}(B) \quad (B \in \mathcal{E}, \mu \in \mathcal{M}).$$

THEOREM 6.7. *For any admissible family of measures (P_x) on $\mathcal{F}$ it is fair*

$$\mathrm{IMT} \cap \mathrm{RT}(P_x) \subset \mathrm{RT}(R_x),$$

where $R_x = P_x \circ \ell^{-1}$.

Proof. For $\tau \in \mathrm{IMT} \cap \mathrm{RT}(P_x)$ and $A, B \in \mathcal{E}$ we have

$$R_x(\overline{\theta}_\tau^{-1}B \cap \overline{\alpha}_\tau^{-1}A) = P_x(\ell^{-1}\overline{\theta}_\tau^{-1}B \cap \ell^{-1}\overline{\alpha}_\tau^{-1}A)$$
$$= P_x(\theta_\tau^{-1}\ell^{-1}B \cap \alpha_\tau^{-1}\ell^{-1}A)$$
$$= P_x(P_{X_\tau}(\ell^{-1}B); \{\tau < \infty\} \cap \alpha_\tau^{-1}\ell^{-1}A),$$

according to item 45 and because of $\ell^{-1}B$, $\ell^{-1}A \in \mathcal{F}$. The last expression is equal to

$$P_x\left(P_{\overline{X}_\tau \circ \ell}\left(\ell^{-1}B\right); \; \ell^{-1}\ell\{\tau < \infty\} \cap \ell^{-1}\bar{\alpha}_\tau^{-1}A\right)$$
$$= R_x\left(R_{\overline{X}_\tau}(B); \; \ell\{\tau < \infty\} \cap \bar{\alpha}_\tau^{-1}A\right),$$

according to definition of $\overline{X}_\tau$ and because of

$$\{\tau < \infty\} = \ell^{-1}\ell\{\tau < \infty\}. \qquad \Box$$

While investigating time change in families of measures, a question can arise: does time change preserve regeneration time? It is interesting because a family of distributions of traces for both a transformed and original process can possess a set of regeneration time reach enough. Simple examples show that even in the case of stepped SM processes, some time change can cause a loss of the semi-Markov property. For this aim it is enough to determine a conditional distribution of a sequence of sojourn times as that of a sequence of dependent random values. Preservation of regeneration time under time change requires a special class of time changes.

6.6. Additive functionals and time change preserving the semi-Markov property

A time change preserving the semi-Markov property of the process can be constructed with the help of the Laplace family of additive functionals which is enough to be determined on the set of non-constancy of the process, i.e. the random set which does not contain intervals of constancy of trajectories. On the other hand, any time change transforming one semi-Markov process into another determines a Laplace family of additive functionals [HAR 80a]. As an example the time change with the help of independent strictly increasing homogenous process with independent increments is considered [GIH 73, ITO 60, HAR 76a, SER 71].

47. Random set

In the given context the word "random" has the sense "depending on trajectory".

Let $M(\cdot): \mathcal{D} \mapsto \mathcal{B}(\mathbb{R}_+)$ be a function of the following view:

(1) $(\forall \xi \in \mathcal{D})$ $M(\xi)$ is a measurable set of point on the half-line $\mathbb{R}_+$ $(M(\xi) \in \mathcal{B}(\mathbb{R}_+))$; in addition

$$\mathfrak{T} \equiv \left\{(t, \xi) \in \mathbb{R}_+ \times \mathcal{D}, \; t \in M(\xi)\right\} \in \mathcal{B}(\mathbb{R}_+) \otimes \mathcal{F};$$

(2) $(\forall \xi \in \mathcal{D})$ $0 \in M(\xi)$;

(3) $(\forall \xi \in \mathcal{D})$ $t_1, t_1 + t_2 \in M(\xi), t_2 \geq 0 \Rightarrow t_2 \in M(\theta_{t_1}\xi)$;

(4) $(\forall \xi \in \mathcal{D})$ $(\forall t \in \mathbb{R}_+)$ $M(\xi) \cap [0, t] = M(\alpha_t\xi) \cap [0, t]$.

We call such a function a random set. Let $\mathfrak{L}$ be a class of such functions. The trivial example of random set is $M(\xi) \equiv \mathbb{R}_+$. A simple non-trivial example is

$$M_0(\xi) = \{\tau(\xi) : \tau \in \mathcal{T}_0\}, \qquad [6.1]$$

where $\mathcal{T}_0$ is defined in item 22. Measurability of M_0 follows from countability of set of parameters and $\mathcal{B}(\mathbb{R}_+) \otimes \mathcal{F}$-measurability of the set $\{(\tau(\xi), \xi), \xi \in \mathcal{D}\}$. Conditions (2) and (4) are evident. Condition (3) follows from closure of $\mathcal{T}_0$ with respect to operation $\dotplus$ (see item 2.14).

48. Additive functional on random set

Let $M \in \mathfrak{L}$. Consider functional $a_t(\xi) \equiv a(t, \xi)$ on $\mathfrak{X}$ of the following view:

(1) $a \in (\mathcal{B}(\mathbb{R}_+) \otimes \mathcal{F}) \cap \mathfrak{X}/\mathcal{B}(\mathbb{R})$, measurability;

(2) $(\forall \xi \in \mathcal{D})\ a_0(\xi) = 0$, zero initial value;

(3) $(\forall \xi \in \mathcal{D})\ t_1, t_1 + t_2 \in M(\xi), t_2 \geq 0 \Rightarrow a_{t_1+t_2}(\xi) = a_{t_1}(\xi) + a_{t_2}(\theta_{t_1}\xi)$, additivity;

(4) $(\forall \xi \in \mathcal{D})\ (\forall t \in \mathbb{R}_+)\ a_t(\xi) = a_t(\alpha_t \xi)$, independence of future.

We call $(a(\cdot, \cdot), M(\cdot))$ as *additive functional on a random set*. If $M(\xi) \equiv \mathbb{R}_+$, then the words "random set" can be omitted.

49. Family of additive functionals

Consider a family of additive functionals on a random set $(a(\lambda \mid \cdot, \cdot), M(\cdot))_{\lambda \geq 0}$. We will also designate functional $a(\lambda \mid t, \xi)$ as $a_t(\lambda \mid \xi)$, and even $a(\lambda)$ if two arguments are absent. The family $a(\lambda)$ is said to be Laplacian if $(\forall \xi \in \mathcal{D})\ (\forall t \in M(\xi))$ $a_t(\lambda \mid \xi) \geq 0$ and $\exp(-a_t(\lambda \mid \xi))$ is a completely monotone function of $\lambda \geq 0$. Hence in this case there exists a probability measure $G_t(\cdot \mid \xi)$ on $[0, \infty)$ such that

$$\exp\left(-a_t(\lambda \mid \xi)\right) = \int_0^\infty e^{-\lambda s} G_t(ds \mid \xi).$$

From the evident $\mathcal{F}_t$-measurability of the function $a_t(\lambda \mid \cdot)$, determined on a measurable set $\{\xi \in \mathcal{D} : t \in M(\xi)\} \in \mathcal{F}_t$, and from the formula of inversion of the Laplace transformation [FEL 67] it follows that $(\forall s \in \mathbb{R}_+)$ the function $G_t([0, s) \mid \cdot)$ is $\mathcal{F}_t$-measurable.

50. Coordinated extension of function

We denote by $[A]_\uparrow$ a closure of the set $A \in \mathbb{R}_+$ with respect to convergence from the left. For the random set $S \in \mathfrak{L}$ we define class $\Psi_1(S)$ of functions $\psi : S \mapsto \mathbb{R}_+$ by following properties:

(1) $\psi(0) = 0$;

(2) ψ does not decrease on S;

(3) $\psi(t) \to \infty$ $(t \to \infty, t \in S)$;

(4) ψ is continuous from the left on S;

(5) if $t_1 \in [S]_\uparrow$, $t_2 \in M_0(\xi)$ and $t_1 < t_2$, then $\psi(t_1) < \psi(t_2)$.

PROPOSITION 6.26. *Let $\xi \in \mathcal{D}$ and $S \in \mathfrak{L}$ be a random set such that $M_0 \subset S$ (see formula [6.1]). A function $\psi : S \mapsto \mathbb{R}_+$ can be extended up to the function $\psi_1 \in \Psi_1$, coordinated with ξ (i.e. $(\psi_1, \xi) \in \mathcal{H}$, see item 32) if and only if $\psi \in \Psi_1(S)$.*

Proof. Necessity of conditions (1)–(4) is evident. If $t_1 < t_2$ and $t_2 \in M_0(\xi)$, then $[t_1, t_2] \notin \mathrm{IC}(\xi)$ and therefore $(\forall \psi \in \Psi_1, (\psi, \xi) \in \mathcal{H})$ $(t_1, t_2] \notin \mathrm{IC}(\xi)$, i.e. $\psi(t_1) < \psi(t_2)$, condition (5).

Sufficiency. Define $\psi_1 = \psi$ on S, $\psi(t) = \psi(t - 0)$ for $t \in [S]_\uparrow$,

$$\psi_1(t) = \psi(t' - 0) + \frac{t - t'}{t'' - t'} \left(\psi(t'' + 0) - \psi(t' - 0) \right)$$

for $t \in (t', t'']$, where $t' = \sup(t_1 \in S, t_1 < t)$, $t'' = \inf(t_1 \in S, t_1 \geq t)$ and $t'' < \infty$; if $t'' = \infty$, $\psi_1(t) = \psi(t' - 0) + t - t'$. Then $\psi_1(0) = 0$, ψ_1 does not decrease on $\mathbb{R}_+$, $\psi_1(t) \to \infty$ $(t \to \infty)$, ψ_1 is continuous from the left. Let $t_1 < t_2$ and $\psi_1(t_1) = \psi_1(t_2))$. From construction it follows that $M_0(\xi) \cap (t_1, t_2] = \emptyset$, i.e. $[t_1, t_2] \in \mathrm{IC}(\xi)$. From here by 6.18, $(\psi_1, \xi) \in \mathcal{H}$. □

51. Time change determined by functional

Consider a Laplace family of additive functionals on a random set $(a(\lambda), M_1(\cdot))_{\lambda \geq 0}$ where $M_1 \in \mathfrak{L}$, $(\forall \xi \in \mathcal{D})$ $M_0(\xi) \subset M_1(\xi)$. Let P be a probability measure on $\mathcal{F}$. Let us construct on $\mathcal{B}(\mathbb{R}_+^n) \otimes \mathcal{F}$ a projective family of measures $(Q_{\tau_1, \dots, \tau_n})$ $(n \in \mathbb{N}$, $\tau_i \in \mathcal{T}_0)$ of the form

$$Q_{\tau_1, \dots, \tau_n} \left(\bigcap_{i=1}^{n} \left\{ k_1 \left(\widehat{X}_{\tau'_{ni} \circ q} \circ \widehat{\theta}_{\tau_{ni-1} \circ q} \right) \in S_i \right\} \cap q^{-1} B \cap \left\{ \tau_{nn} \circ q < \infty \right\} \right)$$

$$= P \left(\prod_{i=1}^{n} G_{\tau'_{ni}} (S_i) \circ \theta_{\tau_{ni-1}}, B \cap \left\{ \tau_{nn} < \infty \right\} \right),$$

where $k_1(t, x) = t$ (either the first coordinate of the pair or ∞; see item 4.2), $S_i \in \mathcal{B}(\mathbb{R}_+)$, $B \in \mathcal{F}$, τ_{ni} is the i-th order statistics of a collection $(\tau_1, \dots, \tau_n)$, $\tau_{ni} =$

$\tau_{ni-1} \dotplus \tau'_{ni}$ (e.g., $\tau'_{ni} = \sigma_{\Delta_{ni}}$); $(\forall \xi \in \mathcal{D}, \tau_{ni}(\xi) < \infty)$ $\tau'_{ni}(\theta_{\tau_{ni}}\xi) \in M_1(\theta_{\tau_{ni}}\xi)$ and $(\forall \lambda > 0)$

$$\exp\left(-a_{\tau'_{ni}}(\lambda \mid \xi)\right) = \int_0^\infty e^{-\lambda t} G_{\tau'_{ni}}(dt \mid \xi),$$

$G_\tau(dt \mid \xi) \equiv G_{\tau(\xi)}(dt \mid \xi)$ is a probability measure on $(0, \infty)$; see item 49.

THEOREM 6.8. *Let $(a(\lambda), M_1(\cdot))_{\lambda \geq 0}$ be a Laplace family of additive functionals on a random set where $M_1 \in \mathfrak{L}$, $(\forall \xi \in \mathcal{D})$ $M_0(\xi) \subset M_1(\xi)$ and $(\forall \lambda > 0)$ $(\forall \xi \in \mathcal{D})$ the functional $a_t(\lambda \mid \xi)$ as a function of $t \in M_1$ belongs to class $\Psi_1(M_1)$. Then the constructed above projective family of measures $(Q_{\tau_1,\dots,\tau_n})$ can be extended up to a probability measure Q on $\mathfrak{B}$, concentrated on $\mathcal{H}$, and, in addition, $P = Q \circ q^{-1}$.*

Proof. Since

$$k_1\left(\widehat{X}_{\tau' \circ q} \circ \widehat{\theta}_{\tau \circ q}\right)(\psi, \xi) = \psi\left((\tau \dotplus \tau')\xi\right) - \psi\left(\tau(\xi)\right),$$

and map $\xi \mapsto G_{\tau(\xi)}(\cdot \mid \xi)$ is measurable, it is sufficient to show that $(\forall \xi \in \mathcal{D})$ the projective family $(G_{\tau_1,\dots,\tau_n}(\cdot \mid \xi))$ $(n \in \mathbb{N}, \tau_i \in M_1(\xi))$ of the form

$$G_{\tau_1,\dots,\tau_n}\left(\bigcap_{i=1}^n \{\psi(\tau_{ni}(\xi)) - \psi(\tau_{ni-1}(\xi)) \in S_i\} \mid \xi\right)$$

$$= \prod_{i=1}^n G_{\tau'_{ni}}\left(S_i \mid \theta_{\tau_{ni-1}}\xi\right)$$

can be extended up to a probability measure G on $\mathcal{B}(\Psi_1(\xi))$, where $\Psi_1(\xi)$ is a cross-section of set $\mathcal{H}$ on coordinate ξ. Note that since functionals $a_\tau(\lambda \mid \xi) \equiv a(\lambda \mid \tau(\xi), \xi)$ are additive on $M_1(\xi)$, the distribution $G_{\tau_2}(\cdot \mid \theta_{\tau_1}\xi)$ is that of increments on interval $[\tau_1(\xi), (\tau_1 \dotplus \tau_2)(\xi)]$ of a process with independent increments determined on set $M_1(\xi)$. Properties of trajectories of a process with independent increments are well-known to coincide, in general, with logarithm properties of the generating function of the process (see [SKO 61]). Show that from conditions on $a_\tau(\lambda \mid \xi)$ follows the trajectories of the constructed process with independent increments extended on $\mathbb{R}_+$ to be a.s. coordinated with ξ. Let $\mathcal{U}(\cdot \mid \xi)$ be a distribution of this process. Obviously, $\mathcal{U}(\psi(0) = 0 \mid \xi) = 1$. Let $N_n \in \sigma(\widetilde{X}_t, t \in M_1(\xi))$, $N_n \supset N_{n+1}$ and $\mathcal{U}(N_n \mid \xi) \geq \varepsilon_1 > 0$. Without loss of generality, suppose $N_n = \bigcap_{i=1}^n \{\psi(t_i) \in S_i^n\}$, where $M_0(\xi) \subset (t_1, t_2, \dots) \subset M_1(\xi)$, all S_i^n are compact, and in addition, $(\forall i \in \mathbb{N})$ $S_i^n \supset S_i^{n+1}$.

Since $(\forall (t_n) \subset M_1(\xi))$ $t_n \uparrow \infty \Rightarrow (\forall l \in \mathbb{N})$ $G_{t_n}([l, \infty) \mid \xi) \to 1$ (it follows from property (3) of item 50), for any sequence $(\varepsilon(l))_1^\infty$ $(\varepsilon(l) > 0)$ there exists a sequence $(a(l))_1^\infty$ $(a(l) \uparrow \infty)$ such that $(\forall n \in \mathbb{N})$

$$t_n \geq a(l) \Longrightarrow G_{t_n}[l, \infty) \geq 1 - \varepsilon(l).$$

Let $\varepsilon(l) = \varepsilon \cdot 2^{-l-1}$, $(t_{n1}, \ldots, t_{nn})$ be an order statistics of collection $\{t_1, \ldots, t_n\} \subset M_1(\xi)$, $\ell(t_{ni}) = l \Leftrightarrow a(l) \leq t_{ni} < a(l+1)$ $(l \in \mathbb{N}_0, a(0) = 0)$;

$$K_n = \bigcap_{i=1}^{n} \{\psi(t_{ni}) \geq \ell(t_{ni})\} = \bigcap_{l=0}^{\infty} \bigcap_{\ell(t_{ni})=l} \{\psi(t_{ni}) \geq l\}$$

$$= \bigcap_{l=0}^{\infty} \{\psi(t_{ni(l)}) \geq l\},$$

where $t_{ni(l)} = \min(t_{ni} : t_{ni} \geq a(l))$. In this case

$$\mathcal{U}(K_n \mid \xi) \geq 1 - \sum_{l=0}^{\infty} \mathcal{U}(\psi(t_{ni(l)}) < l \mid \xi)$$

$$= 1 - \sum_{l=0}^{\infty} G_{t_{ni(l)}}((0, l) \mid \xi) \geq 1 - \varepsilon$$

and $K_n \supset K_{n+1}$. If $\psi \in \bigcap_{n=1}^{\infty} K_n$, $t_n' \uparrow \infty$ and $(t_n') \subset M_1(\xi)$, evidently, $\psi(t_n') \uparrow \infty$.

Since $(\forall (t_n) \subset M_1(\xi))$

$$t_n \uparrow t, \ t \in M_1(\xi) \Longrightarrow (\forall l \in N) \quad G'_{t_n, t}(0, l^{-1}] \longrightarrow 1$$

(this property follows from (4) of item 50), for any $(\varepsilon(t, l))_{l=1}^{\infty}$ $(\varepsilon(t, l) > 0)$ there exists a sequence $(b(t, l))_{l=1}^{\infty}$, $(b(t, l) \downarrow 0)$ such that

$$1 - b(t, l) \leq t_n < t \Longrightarrow G'_{t_n, t}(0, l^{-1}] \geq 1 - \varepsilon(t, l).$$

Here and in what follows $G'_{t_1, t_2}(S \mid \xi) = G_{t_2'}(S \mid \theta_{t_1} \xi)$, where $t_2 = t_1 \dotplus t_2'$ $(t_1, t_2 \in M_1(\xi), t_2' \in M_1(\theta_{t_1} \xi))$. Let $\varepsilon(t_n, l) = \varepsilon 2^{-n-l-1}$,

$$L_n = \bigcap_{i=1}^{n} \bigcap_{j=0}^{i-1} \{\psi(t_{ni}) - \psi(t_{nj}) \leq (\ell(t_{ni}, t_{nj}))^{-1}\}$$

$$= \bigcap_{i=1}^{n} \bigcap_{l=0}^{\infty} \bigcap_{j:\ell(t_{ni}, t_{nj})=l} \{\psi(t_{ni}) - \psi(t_{nj}) \leq l^{-1}\}$$

$$= \bigcap_{i=1}^{n} \bigcap_{l=0}^{\infty} \{\psi(t_{ni}) - \psi(t_{nj(l)}) \leq l^{-1}\},$$

where $\ell(t_{ni}, t_{nj}) = l \Leftrightarrow t_{ni} - b(t_{ni}, l) \leq t_{nj} < t_{ni} - b(t_{ni}, l+1)$, $t_{nj(l)} = \min(t_{nj} : t_{nj} \geq t_{ni} - b(t_{ni}, l))$ $(l \in \mathbb{N}_0, b(t_{ni}, 0) = \infty)$. Then

$$\mathcal{U}(L_n \mid \xi) \geq 1 - \sum_{i=1}^{n} \sum_{l=0}^{\infty} G'_{t_{nj(l)}, t_{ni}}(l^{-1}, \infty) \geq 1 - \varepsilon,$$

$(\forall n \in N)$ $L_n \supset L_{n+1}$ and any function $\psi \in \bigcap_{n=1}^{\infty} L_n$ is continuous from the left on $(t_1, t_2, \ldots) \subset M_1(\xi)$.

Since $(\forall (t_n) \subset M_1(\xi))$

$$t_n \uparrow t' < t, \quad t \in M_0(\xi) \Longrightarrow G'_{t'-0,t}(0, \infty) = 1$$

(follows from property (5) of item 50), for any sequence $(\varepsilon(t, l))_{l=1}^{\infty}$ $(\varepsilon(t, l) > 0)$ there exists a sequence of numbers $(c(t, l))_{l=1}^{\infty}$ $(c(t, l) > 0)$ such that

$$t - l^{-1} \le t_n < t - 1/(l+1) \Longrightarrow G'_{t_n, t}[c(t, l), \infty) \ge 1 - \varepsilon(t, l).$$

Let $\epsilon(t'_n, \ell) = \epsilon 2^{-n-\ell-1}$, where t'_n is the n-th element of the set $M_0(\xi)$ in the sequence (t_n). Let the condition $\ell(t_{ni}, t_{nj}) = \ell$ be equivalent to the condition $t_{ni} - 1/\ell \le t_{nj} < t_{ni} - 1/(\ell+1)$ $(\ell \in N_0)$, and

$$M_n = \bigcap_{\substack{i \le n, \\ t_{ni} \in M_0(\xi)}} \bigcap_{j=0}^{i-1} \{\psi(t_{ni}) - \psi(t_{nj}) \ge c(t_{ni}, \ell(t_{ni}, t_{nj}))\}$$

$$= \bigcap_{i} \bigcap_{l=0}^{\infty} \bigcap_{j:\ell(t_{ni}, t_{nj})=l} \{\psi(t_{ni}) - \psi(t_{nj}) \ge c(t_{ni}, l)\}$$

$$= \bigcap_{i} \bigcap_{l=0}^{\infty} \{\psi(t_{ni}) - \psi(t_{nj(l)}) \ge c(t_{ni}, l)\},$$

where $t_{nj(l)} = \max\{t_{nj} : t_{ni} - l^{-1} \le t_{nj} < t_{ni} - 1/(l+1)\}$. Then

$$\mathcal{U}(M_n \mid \xi) \ge 1 - \sum_{i} \sum_{l=0}^{\infty} G'_{t_{nj(l)}, t_{ni}}(0, c(t_{ni}, l)) \ge 1 - \varepsilon,$$

$(\forall n \in N)$ $M_n \supset M_{n+1}$ and for any function $\psi \in \bigcap_{n=1}^{\infty} M_n$ on (t_n) property (5) of item 50 is fulfilled.

Let $R_n = K_n L_n M_n N_n$. Then $(\forall n \in N)$ $R_n \supset R_{n+1} \in \sigma(\widetilde{X}_t, t \in M_1(\xi))$ and $\mathcal{U}(R_n) \ge \varepsilon_1 - 3\varepsilon \ge \varepsilon_1/2$, if $\varepsilon = \varepsilon_1/6$. Since $\mathcal{U}$ is a probability measure on $\mathcal{B}(\mathbb{R}_+^{M_1(\xi)})$, $\bigcap_{n=1}^{\infty} R_n \ne \emptyset$, and, in addition, for any ψ from this conjunction the conditions of proposition 6.26 are fulfilled. Hence the probability measure $\mathcal{U}$ and the canonical extension of the function ψ up to the function ψ' on $\mathbb{R}_+$, constructed in proposition 6.26, induces a probability measure $\mathcal{U}'(\cdot \mid \xi)$ on $\mathcal{B}(\Psi_1(\xi))$. Here $(\forall n \in N)$ $(\forall t_i \in \mathbb{R}_+)$ $(\forall S_t \in \mathcal{B}(\mathbb{R}_+)$ the function $\mathcal{U}'(\bigcap_{i=1}^{n}\{\varphi(t_i) \in S_i\} \mid \xi)$ is measurable with respect to $\mathcal{F}$. The distribution Q on $\mathfrak{B}$ is determined by the relation

$$Q\left(\bigcap_{i=1}^{n}\{\psi(t_i) \in S_i\} \cap q^{-1}B\right) = P\left(\mathcal{U}'\left(\bigcap_{i=1}^{n}\{\psi(t_i) \in S_i\}\right); B\right)$$

(a canonical extension of distributions $(Q_{\tau_1,\ldots,\tau_n})$) and is concentrated on $\mathcal{H} = \bigcup_{\xi \in \mathcal{D}} \{(\psi, \xi) : \psi \in \Psi_1(\xi)\}$. Evidently,

$$P = Q \circ q^{-1}.$$ $\square$

52. Note on uniqueness

Note that the uniqueness of extension of Q is guaranteed only for the case when $(\forall \xi \in \mathcal{D})\ M_0(\xi)$ is dense in $\mathbb{R}_+$. However, the transformed measure $\widetilde{P} = Q \circ u^{-1}$ is always unique in spite of possible non-uniqueness of Q. This measure is determined by the expression $(\forall n \in \mathbb{N})\ (\forall \lambda_i > 0)\ (\forall \varphi_i \in C)\ (\forall \tau_i \in \mathcal{T}_0)$

$$\widetilde{P}\left(\prod_{i=1}^{n} e^{-\lambda_i \tau_i} (\varphi_i \circ X_{\tau_i}) \right) = P\left(\prod_{i=1}^{n} e^{-a_{\tau_i}(\lambda_i)} (\varphi_i \circ X_{\tau_i}) \right).$$

Actually, $(\forall B \in \mathcal{F}^*)\ u^{-1}B \cap \mathcal{H} = q^{-1}B \cap \mathcal{H}$ (by 6.22 and item 11). Let $\tau_i \in \mathcal{T}_0$, $\tau_1 < \cdots < \tau_n$, $\tau_i = \tau_{i-1} \dotplus \tau_i'$ ($\tau_0 = 0$). We have

$$\left(\tau_i' \circ \theta_{\tau_{i-1}} \circ u \right)(\psi, \xi) = \tau_i' \theta_{\psi(\tau_{i-1}(\xi))} \left(\xi \circ \psi^* \right)$$

$$= \tau_i' \left(\theta_{\tau_{i-1}} \xi \circ \widetilde{\theta}_{\psi(\tau_{i-1}(\xi))} \psi^* \right)$$

$$= \left(\widetilde{\theta}_{\psi(\tau_{i-1}(\xi))} \psi^* \right)^* \left(\tau_i' \left(\theta_{\tau_{i-1}} \xi \right) \right)$$

$$= \left(\widetilde{\theta}_{\tau_{i-1}(\xi)} \psi \right) \left(\tau_i' \left(\theta_{\tau_{i-1}} \xi \right) \right)$$

$$= \psi\left(\tau_{i-1}(\xi) + \tau_i' \left(\theta_{\tau_{i-1}} \xi \right) \right) - \psi\left(\tau_{i-1}(\xi) \right)$$

$$= k_1 \left(\widehat{X}_{\tau_i' \circ q} \circ \widehat{\theta}_{\tau_{i-1} \circ q} \right)(\psi, \xi).$$

Then

$$\widetilde{P}\left(\prod_{i=1}^{n} e^{-\lambda_i \tau_i' \circ \theta_{\tau_{i-1}}}; B \right) = Q\left(\prod_{i=1}^{n} e^{-\lambda_i \tau_i' \circ \theta_{\tau_{i-1}} \circ u};\ u^{-1} B \cap \mathcal{H} \right)$$

$$= Q\left(\prod_{i=1}^{n} e^{-\lambda_i k_1 (\widehat{X}_{\tau_i' \circ q} \circ \widehat{\theta}_{\tau_{i-1} \circ q})};\ q^{-1} B \right)$$

$$= P\left(\prod_{i=1}^{n} e^{-a_{\tau_i'}(\lambda_i) \circ \theta_{\tau_{i-1}}};\ B \right).$$

The measure $\widetilde{P}$ is determined uniquely with collection of these integrals.

53. Preservation of regeneration times

THEOREM 6.9. *Let $(a(\lambda), M(\cdot))_{\lambda \geq 0}$ be a Laplace family of additive functionals satisfying the conditions of proposition 6.26; (P_x) be an admissible family of probability measures on $\mathcal{F}$ and $(\widetilde{P}_x) = Q_x \circ u^{-1}$ be a family of transformed probability measures on $\mathcal{F}$, determined by corresponding family of probability measures (Q_x), constructed in theorem 6.8. Then*

$$\mathrm{IMT} \cap \mathrm{RT}\,(P_x) \subset \mathrm{RT}\,(\widetilde{P}_x).$$

Proof. Let $\tau \in \mathrm{IMT}$. Then $X_\tau \in \mathcal{F}^*/\mathcal{B}(\mathbb{X})$. Furthermore, let $\tau_n = \sigma_{\delta_n}^k$ for $\sigma_{\delta_n}^k \leq \tau < \sigma_{\delta_n}^{k+1}$, where $\delta_n \in \mathrm{DS}(\mathfrak{A}_n)$, $\mathfrak{A}_n$ is a covering of $\mathbb{X}$ of rank r_n ($r_n \downarrow 0$). Since $M_0(\xi)$ is dense in $M(\xi)$, $\tau_n \uparrow \tau$ and $a_{\tau_n}(\lambda) \uparrow a_\tau(\lambda)$, where $a_\tau(\lambda \mid \xi)$ is an extension by continuity of the function $a_t(\lambda \mid \xi)$ with $M_0(\xi)$ on $M(\xi) = \{t = \tau(\xi), \tau \in \mathrm{IMT}\}$. In addition $\{\sigma_{\delta_n}^k \leq \tau < \sigma_{\delta_n}^{k+1}\} \in \mathcal{F}^*$. We have

$$\widetilde{P}_x\big(e^{-\lambda\tau}(\varphi \circ X_\tau)\big) = \lim_{n\to\infty} \widetilde{P}_x\big(e^{-\lambda\tau_n}(\varphi \circ X_\tau)\big)$$

$$= \lim_{n\to\infty} \sum_{k=0}^{\infty} P_x\Big(e^{-a_{\sigma_{\delta_n}^k}(\lambda)}(\varphi \circ X_\tau); \{\sigma_{\delta_n}^k \leq \tau < \sigma_{\delta_n}^{k+1}\}\Big)$$

$$= \lim_{n\to\infty} P_x\big(e^{-a_{\tau_n}(\lambda)}(\varphi \circ X_\tau)\big) = P_x\big(e^{-a_\tau(\lambda)}(\varphi \circ X_\tau)\big).$$

Hence,

$$\widetilde{P}_x\big(e^{-\lambda\tau}(\varphi \circ X_\tau)\big) = P_x\big(e^{-a_\tau(\lambda)}(\varphi \circ X_\tau)\big).$$

Similarly we prove that $(\forall n \in \mathbb{N})\,(\forall \tau_i \in \mathrm{IMT})\,(\forall \lambda_i > 0)\,(\forall \varphi_i \in C)$

$$\widetilde{P}_x\left(\prod_{i=1}^{n} e^{-\lambda_i \tau_i}(\varphi_i \circ X_{\tau_i})\right) = P_x\left(\prod_{i=1}^{n} e^{-a_{\tau_i}(\lambda_i)}(\varphi_i \circ X_{\tau_i})\right).$$

Furthermore, $(\forall n, m \in \mathbb{N})\,(\forall \tau \in \mathrm{IMT})\,(\forall \lambda_i, \lambda_i' > 0)\,(\forall \varphi_i, \varphi_i' \in C)\,(\forall \tau_i, \tau_i' \in \mathcal{T}_0)$

$$\widetilde{P}_x\left(\left(\left(\prod_{i=1}^{n} e^{-\lambda_i \tau_i}(\varphi_i \circ X_{\tau_i})\right) \circ \theta_\tau\right)\left(\left(\prod_{i=1}^{m} e^{-\lambda_i' \tau_i'}(\varphi_i' \circ X_{\tau_i'})\right) \circ \alpha_\tau\right); \tau < \infty\right)$$

$$= P_x\left(\left(\prod_{i=1}^{n} e^{-a_{\tau_i}(\lambda_i)\circ\theta_\tau}(\varphi_i \circ X_{\tau_i}) \circ \theta_\tau\right)\right.$$

$$\left. \times \left(\prod_{i=1}^{m} e^{-a_{\tau_i'\circ\alpha_\tau}(\lambda_i')}(\varphi_i' \circ X_{\tau_i'} \circ \alpha_\tau)\right); \tau < \infty\right).$$

We note that the second factor is $\mathcal{F}_\tau$-measurable and $\{\tau < \infty\} \in \mathcal{F}^*$. Then if $\tau \in$ MP(P_x), the last expression is equal to

$$P_x\left(P_{X_\tau}\left(\prod_{i=1}^n e^{-a_{\tau_i}(\lambda_i)}(\varphi_i \circ X_{\tau_i})\right)\left(\prod_{i=1}^m e^{-a_{\tau_i' \circ a_{\tau_i}}(\lambda_i')}(\varphi_i' \circ X_{\tau_i'} \circ \alpha_\tau)\right); \tau < \infty\right)$$

$$= \widetilde{P}_x\left(\widetilde{P}_{X_\tau}\left(\prod_{i=1}^n e^{-\lambda_i \tau_i}(\varphi_i \circ X_{\tau_i})\right)\left(\prod_{i=1}^m e^{-\lambda_i' \tau_i'}(\varphi_i' \circ X_{\tau_i'})\right) \circ \alpha_\tau; \tau < \infty\right).$$

From here it follows that $\tau \in$ MP$(\widetilde{P}_x)$.   □

54. Process with independent increments

A probability measure $\mathcal{U}$ on $\mathfrak{G}$ is said to be a measure of a homogenous process with independent increments, if $\mathcal{U}$-a.s. $(\forall t \in \mathbb{R}_+)(\forall B \in \mathfrak{G})$

$$\mathcal{U}(\widetilde{\theta}_t^{-1}B \mid \mathfrak{G}_t) = \mathcal{U}(B).$$

Let us determine a Laplace family of additive functionals $(a(\lambda))$: $(\forall t \in \mathbb{R}_+)(\forall \lambda \geq 0)$ which is independent of ξ

$$e^{-a_t(\lambda)} = \mathcal{U}(e^{-\lambda \widetilde{X}_t}).$$

From homogenity of the process it follows that $a_t(\lambda) = t\alpha(\lambda)$, where $\alpha(\lambda)$ is a positive function with a derivative completely monotone on λ. Thus, the constructed Laplace family of additive functionals satisfies conditions of theorem 6.8 and determines a time change transforming the SM process (P_x) in SM process $(\widetilde{P}_x)$ by formula

$$\widetilde{P}_x(e^{-\lambda \tau}\varphi \circ X_\tau; \tau < \infty) = P_x(e^{-a_\tau(\lambda)}\varphi \circ X_\tau; \tau < \infty),$$

where $\tau \in$ IMT.

The time change with the help of an independent process with independent increments is also investigated in [SER 71].

55. Strong semi-Markov process

A process (P_x) is said to be *strong semi-Markov process* if IMT $\subset$ RT(P_x).

In theorem 6.9 we considered a Laplace family of additive functionals which determines a time change preserving strong semi-Markov property. Thus, a method of construction of strong semi-Markov processes is given. As an original process can be considered a strong Markov process which contains all intrinsic Markov times in its class of regeneration times. An example of Laplace family of additive functionals preserving the strong semi-Markov property, but not preserving Markovness, is the family considered above, connected to homogenous process with independent positive increments, if it is not linear on λ.

56. On linearity of the Laplace family

By analogy with the criterion of Markovness for semi-Markov processes (see theorems 3.6, 3.8) it can be predicted that linearity of the Laplace family of additive functionals on λ is necessary and sufficient for the corresponding time change to preserve the Markov property of the process. As it follows from results of [BLU 62] (see also [BLU 68, p. 234]), it is fair under additional supposition that the map $t \mapsto a_t(\lambda \mid \cdot)$ is continuous. Let $(P_x)_{x \in \mathbb{R}_+}$ be a family of degenerate distributions of a Markov process for which $(\forall t \in \mathbb{R}_+)\, P_x(X_t = x + t) = 1$ (movement to the right with unit velocity). Let $\xi_x(t) = x + t\ (t, x \in \mathbb{R}_+)$ and

$$a_t(\lambda \mid \xi_x) = \begin{cases} \lambda t, & 1 \notin [x, t + x), \\ \lambda t + \ln f(\lambda), & 1 \in [x, t + x), \end{cases}$$

where $f(\lambda) = \int_0^\infty e^{-\lambda t} F(dt)$, F is some probability distribution on $\mathcal{B}(\mathbb{R}_+)$ and $F(\{0\}) = 0$. Then $(\forall x \in \mathbb{R}_+)\, P_x$-a.s. $a(\lambda)$ is a Laplace family of additive functionals:

$$\begin{aligned} a_{t_1 + t_2}(\lambda \mid \xi_x) &= a_{t_1}(\lambda \mid \xi_x) + a_{t_2}(\lambda \mid \theta_{t_1} \xi_x) \\ &= a_{t_1}(\lambda \mid \xi_x) + a_{t_2}(\lambda \mid \xi_{x + t_1}) \\ &= \begin{cases} \lambda(t_1 + t_2), & 1 \notin [x, x + t_1 + t_2), \\ \lambda(t_1 + t_2) + \ln f(\lambda) & \text{otherwise.} \end{cases} \end{aligned}$$

This family of additive functionals, in general, is not linear on λ. However, for $f(\lambda) = a/(a + \lambda)\ (a > 0)$, the corresponding time change preserves the Markovness of the family. In the transformed family trajectories since the first hitting time into the unit level has an interval of constancy distributed exponentially. For any other $f(\lambda)$, including degenerated distributions when $f(\lambda) = e^{-a\lambda}$, Markovness of the family is not preserved. Note that in the latter case $(\forall x \in \mathbb{R}_+)\, P_x$-a.s. the family $(a(\lambda))$ is linear on λ. We will return to the question about linearity of a family of additive functionals while analyzing a time run along the trace.

6.7. Distribution of a time run along the trace

For any random process there exists a special class of additive functionals which is determined by the process itself. It is called an additive functional of a time run along the trace. Consider a function $\xi \in \mathcal{D}$ and its trace $\ell\xi$. From other functions $\xi_1 \in \ell^{-1}\ell\xi$ the original function differs with its individual distribution of time along the trace. If we consider functions $\tau \in$ IMT to be marks on a trace, this distribution represents an additive functional $(T_\tau(\xi))_{\tau \in \text{IMT}}$, where $T_\tau(\xi) \equiv \tau(\xi)$. Its main properties follow from properties of functions ξ themselves and those of intrinsic times. Thus, the time run is a "twice random" non-negative additive functional on a trace. The first "randomness" is dependent on a trace. The second "randomness" distinguishes an individual trajectory between trajectories with the same trace. It follows that the distribution of a random process can be represented with the help of two distributions: a distribution on set of traces, and depending on the trace a conditional distribution on the set of time runs. The second distribution, factually, is a distribution of some random process. For a semi-Markov process this secondary process possesses the simplest properties. It is a process with independent positive increments on some linearly ordered set. A distribution of this process can be determined with the help of a so called Laplace family of additive functionals of a time run. As well as in a usual monotone process with independent increments determined on a segment of the line, for these functionals the Lévy formula is fair. A measure of a Poisson component in this formula determines jumps of the time run process. It is natural to interpret these jumps as intervals of constancy of a trajectory.

57. Representation of sigma-algebra

In item 40 we have defined the sigma-algebra $\mathcal{F}^\circ = \sigma(\gamma_\tau, \mathcal{D}^\tau, \tau \in \mathcal{T}_0)$. We assume that $\mathcal{T}_0$ contains $\tau \equiv \infty$ (e.g., $\tau = \sigma_{\mathbb{X}}$). Thus, using the definition, $\mathcal{D}^\infty \in \mathcal{F}^\circ$. This sigma algebra also contains other useful sets such as $\{\tau_1 < \tau_2\}$, $\{\tau < \infty\}$, $\{\sup(\tau_n) < \tau\}$, $\{\sup(\tau_n) < \infty\}$, where $\tau, \tau_i \in \mathcal{T}_0$ (see below).

PROPOSITION 6.27. *For any $\tau \in \mathcal{T}_0$ the following representation is fair:*

$$\mathcal{F}^\circ = \sigma\left(\alpha_\tau^{-1}\mathcal{F}^\circ, \theta_\tau^{-1}\mathcal{F}^\circ\right).$$

Proof. Let $\tau_1 \in \mathcal{T}_0 : \tau_1 = \sigma_{(\Delta_1',\ldots,\Delta_k')}$, $\tau = \sigma_{(\Delta_1,\ldots,\Delta_n)}$ $(\Delta_i, \Delta_i' \in \mathcal{P}(\mathfrak{A}_0))$. With times τ and τ_1 we consider all preceding times, i.e. chains of times:

(1) $0 \leq \sigma_{\Delta_1'} \leq \sigma_{(\Delta_1',\Delta_2')} \leq \cdots \leq \sigma_{(\Delta_1',\ldots,\Delta_k')}$;

(2) $0 \leq \sigma_{\Delta_1} \leq \cdots \leq \sigma_{(\Delta_1,\ldots,\Delta_n)}$.

Let

$$z = \left\{(\varnothing), ((\Delta_1'),\ldots,(\Delta_1',\ldots,\Delta_k')), ((\Delta_1),\ldots,(\Delta_1,\ldots,\Delta_n))\right\}$$

be the corresponding full collection (see item 4.1), $I(z)$ be the set of all indexes of intersection for the given z (item 4.11) and $J(z)$ be the set of all sequences of indexes of intersection (item 4.12). One and only one variational series, composed with times

$$\left(\sigma_{\Delta_1'},\ldots,\sigma_{(\Delta_1',\ldots,\Delta_k')},\sigma_{\Delta_1},\ldots,\sigma_{(\Delta_1,\ldots,\Delta_n)}\right),$$

corresponds to any sequence $(\alpha) \in J(z)$. This sequence $(\alpha) = (\alpha_1,\ldots,\alpha_{N+1})$ $(N \geq 0)$ depends only on corresponding values of $X_0, X_{\tau_{\alpha_1}},\ldots,X_{\tau_{\alpha_N}}$ and, correspondingly, is $\mathcal{F}^\circ$-measurable. In this case the set $\{\tau_1 < \tau\}$ is a union of a finite number of sets $B(\alpha)$, where $B(\alpha)$ is a set of all $\xi \in \mathcal{D}$ for which a variational series for the collection

$$\left(\sigma_{\Delta_1'}\xi,\ldots,\sigma_{(\Delta_1',\ldots,\Delta_k')}\xi,\sigma_{\Delta_1}\xi,\ldots,\sigma_{(\Delta_1,\ldots,\Delta_n)}\xi\right)$$

has the sequence of indexes of intersection (α) and hence $\{\tau_1 < \tau\} \in \mathcal{F}^\circ$. It is sufficient to check $\mathcal{F}^\circ$-measurability of maps α_τ and θ_τ on sets of the form $\{\tau_1 = \infty\}$, $\{\tau_1 < \infty,\, X_{\tau_1} \in S\}$ $(S \in \mathcal{B}(\mathbb{X}))$, $\mathcal{D}^{\tau_1}$, $\mathcal{D}^\infty \in \mathcal{F}^\circ$. We have

$$\alpha_\tau^{-1}\{\tau_1 = \infty\} = \{\tau_1 \circ \alpha_\tau = \infty\} = \{\tau_1 > \tau\} \cup \{\tau_1 = \tau = \infty\} \in \mathcal{F}^\circ,$$

$$\alpha_\tau^{-1}\{\tau_1 < \infty,\, X_{\tau_1} \in S\} = \{\tau_1 \circ \alpha_\tau < \infty,\, X_{\tau_1 \wedge \tau} \in S\}$$

$$= \{\tau_1 \leq \tau,\, \tau_1 < \infty,\, X_{\tau_1} \in S\} \in \mathcal{F}^\circ,$$

$$\alpha_\tau^{-1}\mathcal{D}^{\tau_1} = \{\tau \geq \tau_1\} \cap \mathcal{D}^{\tau_1} \in \mathcal{F}^\circ,$$

$$\alpha_\tau^{-1}\mathcal{D}^\infty = \{\xi : \alpha_\tau\xi \in \mathcal{D}^\infty\} = \{\tau = \infty\} \cap \mathcal{D}^\infty \in \mathcal{F}^\circ,$$

and also

$$\theta_\tau^{-1}\{\tau_1 = \infty\} = \{\tau < \infty\} \cap \{\tau \dotplus \tau_1 = \infty\} \in \mathcal{F}^\circ,$$

$$\theta_\tau^{-1}\{\tau_1 < \infty,\, X_{\tau_1} \in S\} = \{\tau < \infty\} \cap \{\tau \dotplus \tau_1 < \infty,\, X_{\tau \dotplus \tau_1} \in S\} \in \mathcal{F}^\circ,$$

$$\theta_\tau^{-1}\mathcal{D}^{\tau_1} = \{\tau < \infty\} \cap \mathcal{D}^{\tau \dotplus \tau_1} \in \mathcal{F}^\circ,$$

$$\theta_\tau^{-1}\mathcal{D}^\infty = \{\tau < \infty\} \cap \mathcal{D}^\infty \in \mathcal{F}^\circ.$$

Hence, $\sigma(\alpha_\tau^{-1}\mathcal{F}^\circ,\, \theta_\tau^{-1}\mathcal{F}^\circ) \subset \mathcal{F}^\circ$. On the other hand,

$$\{\tau_1 = \infty\} = \alpha_\tau^{-1}\{\tau_1 = \tau = \infty\}$$

$$\cup \bigcup_{i=1}^k \left(\alpha_\tau^{-1}\{\sigma_{(\Delta_1',\ldots,\Delta_{i-1}')} \leq \tau < \sigma_{(\Delta_1',\ldots,\Delta_i')}\}\right.$$

$$\left.\cap\, \theta_\tau^{-1}\{\sigma_{(\Delta_i',\ldots,\Delta_k')} = \infty\}\right) \in \sigma\left(\alpha_\tau^{-1}\mathcal{F}^\circ,\, \theta_\tau^{-1}\mathcal{F}^\circ\right),$$

$$\{\tau_1 < \infty,\, X_{\tau_1} \in S\} = \alpha_\tau^{-1}\{\tau_1 \leq \tau < \infty,\, X_{\tau_1} \in S\}$$

$$\cup \bigcup_{i=1}^k \alpha_\tau^{-1}\{\sigma_{(\Delta_1',\ldots,\Delta_{i-1}')} \leq \tau < \sigma_{(\Delta_1',\ldots,\Delta_i')}\}$$

$$\cap\, \theta_\tau^{-1}\{\sigma_{(\Delta_i',\ldots,\Delta_k')} < \infty,\, X_{\sigma_{(\Delta_i',\ldots,\Delta_k')}} \in S\} \in \sigma\left(\alpha_\tau^{-1}\mathcal{F}^\circ,\, \theta_\tau^{-1}\mathcal{F}^\circ\right),$$

$$\mathcal{D}^{\tau_1} = \alpha_\tau^{-1}\big(\{\tau_1 < \tau\} \cap \mathcal{D}^{\tau_1}\big) \cup \bigcup_{i=1}^{k} \alpha_\tau^{-1}\big\{\sigma_{(\Delta_1', \ldots, \Delta_{i-1}')} \le \tau < \sigma_{(\Delta_1', \ldots, \Delta_i')}\big\}$$

$$\cap \, \theta_\tau^{-1} \mathcal{D}^{\sigma(\Delta_i', \ldots, \Delta_k')} \in \sigma\big(\alpha_\tau^{-1}\mathcal{F}^\circ, \theta_\tau^{-1}\mathcal{F}^\circ\big),$$

$$\mathcal{D}^\infty = \alpha_\tau^{-1}\big(\{\tau = \infty\} \cap \mathcal{D}^\infty\big) \cup \big(\alpha_\tau^{-1}\{\tau < \infty\} \cap \theta_\tau^{-1} \mathcal{D}^\infty\big) \in \sigma\big(\alpha_\tau^{-1}\mathcal{F}^\circ, \theta_\tau^{-1}\mathcal{F}^\circ\big),$$

Hence, $\mathcal{F}^\circ \subset \sigma\big(\alpha_\tau^{-1}\mathcal{F}^\circ, \theta_\tau^{-1}\mathcal{F}^\circ\big).$ $\qquad\square$

58. Joint measurability of conditional probabilities

Furthermore, we will consider conditional probabilities, connected with different original measures, as functions of ξ and of a parameter of a family of measures (in the given case $x \in \mathbb{X}$). A question arises about joint measurability.

Let $\mathcal{F}'$ be a sigma algebra being a subset of $\mathcal{F}$. Consider a conditional distribution $P_x(\cdot \mid \mathcal{F}')$ $(x \in \mathbb{X})$. It is well-known (see [BLU 68, p. 22] and also [HAL 53]) that for a countably-generated sigma-algebra $\mathcal{F}'$ and $\mathcal{B}(\mathbb{X})$-measurable probabilities $P_x(B)$ $(B \in \mathcal{F})$ (as functions of x) there exists a $\mathcal{B}(\mathbb{X}) \otimes \mathcal{F}'$-measurable version of a conditional probability, depending on a parameter (i.e. $(\forall B \in \mathcal{F})$ $P_x(B \mid \mathcal{F}')(\xi)$ is a $\mathcal{B}(\mathbb{X}) \otimes \mathcal{F}'$-measurable function). A proof of this fact follows from the property that $\mathcal{F}'$ can be represented as $\sigma(\bigcup_{n=1}^\infty \mathcal{F}_n')$, where $\mathcal{F}_n'$ is a finite algebra of subsets of the set $\mathcal{D}$, and the family of conditional probabilities $(P_x(B \mid \mathcal{F}_n'))_{n=1}^\infty$ composes a martingale [SHI 80]. If function X_0 is $\mathcal{F}'$-measurable, a superposition of this function with the conditional probabilities $P_{X_0}(B \mid \mathcal{F}')$ (or with expectation $E_{X_0}(f \mid \mathcal{F}')$) is a $\mathcal{F}'$-measurable function for any sets $B \in \mathcal{F}$ (or $\mathcal{F}$-measurable functions f).

59. Representation of conditional probability

PROPOSITION 6.28. *Let (P_x) be an admissible family of measures, and also $\tau \in \mathcal{T}_0 \cap \mathrm{RT}(P_x)$. Then $(\forall x \in \mathbb{X})$ P_x-a.s. for any $\mathcal{F}$-measurable functions f and g on the set $\{\tau < \infty\}$ it is fair*

$$E_x\big((f \circ \alpha_\tau)(g \circ \theta_\tau) \mid \mathcal{F}^\circ\big) = E_x\big(f \circ \alpha_\tau \mid \alpha_\tau^{-1}\mathcal{F}^\circ\big) E_x\big(g \circ \theta_\tau \mid \theta_\tau^{-1}\mathcal{F}^\circ\big).$$

Proof. Let $A, B \in \mathcal{F}^\circ$. We have

$$E_x\big((f \circ \alpha_\tau)(g \circ \theta_\tau)\big); \ \alpha_\tau^{-1} A \cap \theta_\tau^{-1} B\big)$$

$$= E_x\big(E_x\big((f \circ \alpha_\tau)(g \circ \theta_\tau) \mid \mathcal{F}^\circ\big); \ \alpha_\tau^{-1} A \cap \theta_\tau^{-1} B\big).$$

On the other hand, according to the property of regeneration, the first expression is equal to

$$E_x\big(E_{X_\tau}(g; B)(f \circ \alpha_\tau); \ \alpha_\tau^{-1} A \cap \{\tau < \infty\}\big).$$

Using the definition of conditional probability and taking into account the evident $\alpha_\tau^{-1}\mathcal{F}^\circ$-measurability of function X_τ and set $\{\tau < \infty\}$, we obtain

$$E_x\big(E_x\big(f \circ \alpha_\tau \mid \alpha_\tau^{-1}\mathcal{F}^\circ\big)E_{X_\tau}(g; B);\ \alpha_\tau^{-1}A \cap \{\tau < \infty\}\big). \qquad [6.2]$$

Note that P_x-a.s.

$$E_{X_\tau}(g; B) = E_x\big(g \circ \theta_\tau;\ \theta_\tau^{-1}B \mid X_\tau\big),$$

where conditional probability, as usual, is with respect to the sigma-algebra $X_\tau^{-1}\mathcal{B}(\mathbb{X})$, generated by the map X_τ on set $\{\tau < \infty\}$. Since X_τ can be represented as $X_0 \circ \theta_\tau$, $X_\tau^{-1}\mathcal{B}(\mathbb{X}) \subset \theta_\tau^{-1}\mathcal{F}^\circ$. Using well-known property of conditional distributions (see, e.g., [SHI 80, p. 230]), we obtain

$$E_x\big(g \circ \theta_\tau;\ \theta_\tau^{-1}B \mid X_\tau\big) = E_x\big(E_x\big(g \circ \theta_\tau \mid \theta_\tau^{-1}\mathcal{F}^\circ\big);\ \theta_\tau^{-1}B \mid X_\tau\big).$$

Substituting the last expression in formula [6.2] and again using the property of regeneration we obtain

$$E_x\big(E_x\big(f \circ \alpha_\tau \mid \alpha_\tau^{-1}\mathcal{F}^\circ\big)E_x\big(g \circ \theta_\tau \mid \theta_\tau^{-1}\mathcal{F}^\circ\big);\ \theta_\tau^{-1}B \cap \alpha_\tau^{-1}A\big).$$

From here, using the theorem on extension of measure [GIH 71, p. 53], P_x-a.s. on $\{\tau < \infty\}$

$$E_x\big((f \circ \alpha_\tau)(g \circ \theta_\tau) \mid \mathcal{F}^\circ\big)$$
$$= E_x\big(f \circ \alpha_\tau \mid \alpha_\tau^{-1}\mathcal{F}^\circ\big)E_x\big(g \circ \theta_\tau \mid \theta_\tau^{-1}\mathcal{F}^\circ\big). \qquad \square$$

60. Additive functional of time run

Let (P_x) be an admissible family of measures. Consider $\mathcal{F}^\circ$-measurable function $E_x\big(e^{-\lambda\tau} \mid \alpha_\tau^{-1}\mathcal{F}^\circ\big)$ (see item 40). According to the definition T_0 is countable. The sigma-algebra $\mathcal{B}(\mathbb{X})$ is countably-generated too. It is generated by a system of open balls $\{S(x, r)\}$, $(x \in \mathbb{X}', r \in \mathbb{R}'_+)$, where $\mathbb{X}'$ is a countable set which is dense everywhere in $\mathbb{X}$, $\mathbb{R}'_+$ is the set of all positive rational numbers. From here $\mathcal{F}^\circ$ and $\alpha_\tau^{-1}\mathcal{F}^\circ$ are countably-generated sigma-algebras. Since X_0 is $\mathcal{F}^\circ$-measurable function, there exists $\mathcal{F}^\circ$-measurable modification of the function $E_{X_0}\big(e^{-\lambda\tau} \mid \alpha_\tau^{-1}\mathcal{F}^\circ\big)$. In the following theorem we consider $\mathcal{F}^\circ$-measurable functional

$$b_\tau(\lambda) = -\log E_{X_0}\big(e^{-\lambda\tau} \mid \alpha_\tau^{-1}\mathcal{F}^\circ\big)$$

on a random set $M_0(\xi)$ (see item 47, formula (6.1)), where $b_\tau(\lambda)(\xi) \equiv b(\lambda \mid \tau(\xi), \xi)$. Note that, by proposition 6.27, $(\forall x \in \mathbb{X})$ P_x-a.s.

$$E_{X_0}\big(e^{-\lambda\tau} \mid \alpha_\tau^{-1}\mathcal{F}^\circ\big) = E_{X_0}\big(e^{-\lambda\tau} \mid \mathcal{F}^\circ\big).$$

We will call the functional $b_\tau(\lambda)$ a functional of *time run*.

THEOREM 6.10. *Let* $(P_x) \in SM$. *Then* $(\forall x \in \mathbb{X})$ P_x-*a.s. the family of functionals* $(b(\lambda))$ $(\lambda \geq 0)$ *is a Laplace family of additive functionals on a random set* $M_0(\xi)$ *satisfying the conditions of theorem 6.8.*

Proof. Let us prove additivity of functional $b(\lambda)$. It is sufficient to establish that $(\forall A_1, A_2 \in \mathcal{F})$ $(\forall \tau_1, \tau_2 \in \mathcal{T}_0)$ $(\forall \lambda > 0)$ $(\forall x \in \mathbb{X})$

$$E_x \left(\exp \left(- b_{\tau_1 \dotplus \tau_2}(\lambda) \right); A \right)$$
$$= E_x \left(\exp \left(- \left(b_{\tau_1}(\lambda) + b_{\tau_2}(\lambda) \circ \theta_{\tau_1} \right) \right); A \right),$$

where $A = \alpha_{\tau_1}^{-1} A_1 \cap \theta_{\tau_1}^{-1} A_2$. According to definition of $b_\tau(\lambda)$, we have

$$E_x \left(\exp \left(- b_{\tau_1 \dotplus \tau_2}(\lambda) \right); A \right)$$
$$= E_x \left(E_{X_0} \left(\exp \left(- \lambda(\tau_1 \dotplus \tau_2) \right) \mid \mathcal{F}^\circ \right); A \right)$$
$$= E_x \left(E_x \left(\exp \left(- \lambda(\tau_1 \dotplus \tau_2) \right) \mid \mathcal{F}^\circ \right) P_x \left(A \mid \mathcal{F}^\circ \right) \right)$$
$$= E_x \left(\exp \left(- \lambda(\tau_1 \dotplus \tau_2) \right) P_x \left(A \mid \mathcal{F}^\circ \right) \right)$$
$$= E_x \left(e^{-\lambda \tau_1} e^{-\lambda \tau_2 \theta_{\tau_1}} P_x \left(\alpha_{\tau_1}^{-1} A_1 \mid \alpha_{\tau_1}^{-1} \mathcal{F}^\circ \right) P_x \left(\theta_{\tau_1}^{-1} A_2 \mid \theta_{\tau_1}^{-1} \mathcal{F}^\circ \right) \right).$$

According to the well-known property of conditional probabilities (see, e.g., Neveu [NEV 69]), conditional probability $P_x(\theta_{\tau_1}^{-1} A_2 \mid \theta_{\tau_1}^{-1} \mathcal{F}^\circ)$ can be represented as $f \circ \theta_{\tau_1}$, where f is a $\mathcal{F}^\circ$-measurable function. Using the regeneration property of family (P_x), the preceding expression can be written as

$$E_x \left(e^{-\lambda \tau_1} P_x \left(\alpha_{\tau_1}^{-1} A_1 \mid \alpha_{\tau_1}^{-1} \mathcal{F}^\circ \right) E_{X_{\tau_1}} \left(e^{-\lambda \tau_2} f \right); \tau < \infty \right).$$

On the other hand, for any $x \in \mathbb{X}$ it is fair

$$E_x \left(e^{-\lambda \tau_2} f \right) = E_x \left(E_x \left(e^{-\lambda \tau_2} \mid \mathcal{F}^\circ \right) f \right)$$
$$= E_x \left(\exp \left(- b_{\tau_2}(\lambda) \right) f \right).$$

From here $E_{X_{\tau_1}} \left(e^{-\lambda \tau_2} f \right) = E_{X_{\tau_1}} \left(\exp(-b_{\tau_2}(\lambda)) f \right)$ and also by the regeneration condition we obtain

$$E_x \left(\exp \left(- \lambda \tau_1 \right) \left(\exp \left(- b_{\tau_2}(\lambda) \right) \circ \theta_{\tau_1} \right) E_x \left(A \mid \mathcal{F}^\circ \right) \right)$$
$$= E_x \left(E_{X_0} \left(e^{-\lambda \tau_1} \mid \mathcal{F}^\circ \right) \left(\exp \left(- b_{\tau_2}(\lambda) \right) \circ \theta_{\tau_1} \right); A \right)$$
$$= E_x \left(\exp \left(- \left(b_{\tau_1}(\lambda) + b_{\tau_2}(\lambda) \circ \theta_{\tau_1} \right) \right); A \right).$$

Since $\mathcal{F} = \sigma(\alpha_\tau^{-1} \mathcal{F}, \theta_\tau^{-1} \mathcal{F})$ (see item 2.24, theorem 2.14), the equality

$$E_x \left(\exp \left(- b_{\tau_1 \dotplus \tau_2}(\lambda) \right); A \right)$$
$$= E_x \left(\exp \left(- \left(b_{\tau_1}(\lambda) + b_{\tau_2}(\lambda) \circ \theta_{\tau_1} \right) \right); A \right)$$

is fair for all $A \in \mathcal{F}$. Evidently, $b_0(\lambda) = 0$. By proposition 6.28, we have

$$\exp\left(-b_\tau(\lambda)\right) = E_{X_0}\left(e^{-\lambda\tau} \mid \alpha_\tau^{-1}\mathcal{F}^\circ\right),$$

i.e. $b_\tau(\lambda)$ does not depend on the future. Obviously,

$$e^{-b_\tau(\lambda)} = \int_0^\infty e^{-\lambda t} P_x\left(\tau \in ds \mid \mathcal{F}^\circ\right),$$

i.e. P_x-a.s. $(b(\lambda),\ M_0(\cdot))$ is a Laplace family of additive functionals on a random set. For this family to determine a time change it is sufficient that the conditions of theorem 6.8 are fulfilled. From them it remains to check continuity from the left and properties on infinity, but these properties immediately follow from the definition of the functional. For example, if P_x-a.s. $\tau_n \uparrow \tau \in \mathcal{T}_0$,

$$P_{X_0}\left(\exp\left(-\lambda\tau_n\right) \mid \mathcal{F}^\circ\right) \longrightarrow P_{X_0}\left(\exp(-\lambda\tau) \mid \mathcal{F}^\circ\right). \qquad \square$$

From the proven theorem it follows that for two SM processes with coinciding distributions of traces there exists an additive functional determining a time change which transforms one process into the other, namely if (P_x^1), (P_x^2) are two families of measures such that $P_x^1 \circ \ell^{-1} = P_x^2 \circ \ell^{-1}$, then $b^2(\lambda,\tau) = P_{X_0}^2(\exp(-\lambda\tau) \mid \mathcal{F}^\circ)$ is such a Laplace family of additive functionals which, by theorem 6.10, determines the time change transforming family (P_x^1) into (P_x^2).

61. Composition of sequences

In order to conduct further investigation, deducing sequences of a special view will be useful for us. Let $(\mathfrak{A}_n)_{n=1}^\infty$ be a sequence of countable open coverings of the space $\mathbb{X}$ with decreasing ranks: $(\forall \Delta \in \mathfrak{A}_n)$ diam $\Delta \leq 1/n$, and let $\delta_n \in \mathrm{DS}(\mathfrak{A}_n)$. Define a composition of k deducing sequences $\delta_1 \times \cdots \times \delta_k$, where $\delta_i = (\Delta_{i1}, \Delta_{i2}, \ldots)$. This is a sequence of the order-type ω^k with elements $(\Delta_{1i_1} \cap \cdots \cap \Delta_{ki_k})\,(i_1, \ldots, i_k) \in \mathbb{N}^k$ (index of dimension k). A rank of this sequence is equal to the smallest rank of the component ranks. It is convenient to use sequences of compositions $(\delta_1 \times \cdots \times \delta_k)_{k=1}^\infty$ because $(\forall \xi \in \mathcal{D})$, a sequence of the first exit times $(\tau_{i_1,\ldots,i_k}(\xi))$, determined by k-th composition is a sub-sequence of sequence $(\tau_{i_1,\ldots,i_{k+1}}(\xi))$, determined by the $(k+1)$-th composition. The definition of times $\tau(i_1, \ldots, i_k)$ is based on the following property: for any $\Delta \in \mathfrak{A}$ and $\delta \in \mathrm{DS}$

$$\sigma_{\Delta \cap \Delta_1} \dotplus \cdots \dotplus \sigma_{\Delta \cap \Delta_k} \longrightarrow \sigma_\Delta$$

as $k \to \infty$, where $\delta = (\Delta_1, \Delta_2, \ldots)$, and if $\sigma_\Delta \xi < \infty$, then $(\exists n \in \mathbb{N})\ \sigma_\Delta \xi = (\sigma_{\Delta \cap \Delta_1} \dotplus \cdots \dotplus \sigma_{\Delta \cap \Delta_n})(\xi)$. It is not difficult to note that we again turn to the situation of Chapter 4 where construction of a projective family of measures with the help of an admissible family of semi-Markov transition functions has been considered. Every system of initial segments of k deducing sequences determines a full collection $z \in \mathcal{Z}_0(\mathfrak{A}_0)$ (see item 4.1). This full collection determines a finite set $I(z)$ of indexes of intersection, on which a lexicographical ordering in alphabet $\mathbb{N}_0$ is introduced (see item 4.11). Here we will not show in detail the rule of index change while increasing the dimension, i.e. under addition of new deducing sequences to a given finite system. It is important to note that to each index $\alpha = (i_1, \ldots, i_k) \neq 0$ there corresponds: firstly, the set (it can be empty) $\Delta_\alpha = \bigcap_{j=1}^k \Delta_{ji_j}$ and, secondly, the immediately preceding index α'. In this case $\tau_\alpha = \tau_{\alpha'} + \sigma_{\Delta_\alpha}$. Every index of dimension k is at the same time an index of dimension m $(m > k)$, in which $m - k$ coordinates are equal to zero. Thus, if $\alpha < \beta$, $\tau_\alpha \leq \tau_\beta$ $(\alpha, \beta \in \mathcal{A}_k)$. The set of indexes $\mathcal{A}_k$ of dimension k is countable. A rank of a composition decreases as $1/k$ with growth of dimension.

62. Lévy formula

Formula In the following theorem a decomposition of $b(\lambda)$, analogous to the Lévy formula [ITO 65, p. 49], for non-decreasing process with independent increments is proved.

THEOREM 6.11. *Let* $(P_x) \in$ SM, $(b(\lambda))$ $(\lambda \geq 0)$ *be a Laplace family of additive functionals of time run, defined in item 60. Then* $(\forall x \in \mathbb{X})$ $(\forall \tau \in \mathcal{T}_0)$ P_x*-a.s. the following expansion is fair:*

$$b_\tau(\lambda) = \lambda a_\tau + \int_{0+}^\infty \left(1 - e^{-\lambda u}\right) n_\tau(du)$$
$$- \sum_{\tau_i < \tau} \log E_{X_0}\left(e^{-\lambda \sigma_0} \mid \mathcal{F}^\circ\right) \circ \theta_{\tau_i},$$

[6.3]

where a_τ *and* $(\forall B \in \mathcal{B}(0, \infty))$ $n_\tau(B)$ *(Lévy measure) are* $\mathcal{F}^\circ$*-measurable additive functionals on the random set* M_0*;* $(\tau_i + \sigma_0)$ *is a finite or countable infinite collection of* $\mathcal{F}^\circ$*-measurable Markov times of the class* MT$_+$*; and* σ_0 *is the first exit time from the initial point of a trajectory.*

Proof. Let $\tau = \tau_\beta$ for some $m \in \mathbb{N}$ and $\beta \in \mathcal{A}_m$. Consider partitioning of the set $\mathcal{T}_0$ by all Markov times of the form τ_α $(\alpha \in \mathcal{A}_n)$ for some $(n \geq m)$. Then, by theorem 6.10, and due to representation of τ in the form of a finite iteration, corresponding to index β, $(\forall \lambda > 0)$ $(\forall x \in \mathbb{X})$ P_x-a.s. the following representation of $b_\tau(\lambda) \equiv b(\lambda, \tau)$ in the form of a finite sum is fair:

$$b_\tau(\lambda) = \sum^{(n)} b\left(\lambda, \sigma_{\Delta_\alpha}\right) \circ \theta_{\alpha'},$$

where sum is made up on all $\alpha \in \mathcal{A}_n$ such that $\alpha < \beta$. We will sum up separately small and large terms of this sum. Let $\varepsilon > 0$,

$$b_\varepsilon^n(\lambda, \tau) = \sum_1^{(n)} b(\lambda, \sigma_{\Delta_\alpha}) \circ \theta_{\alpha'},$$

where all terms not more than ε are summed up;

$$B_\varepsilon^n(\lambda, \tau) = \sum_2^{(n)} b(\lambda, \sigma_{\Delta_\alpha}) \circ \theta_{\alpha'},$$

where all summands exceeding ε are summed up. As $n \to \infty$, evidently, the first sum does not decrease, the second sum does not increase. On the other hand, as $\varepsilon \to 0$ the first sum does not increase, and the second sum does not decrease. Hence, P_x-a.s. $b(\lambda, \tau) = b_0(\lambda, \tau) + B(\lambda, \tau)$, where

$$b_0(\lambda, \tau) = \lim_{\varepsilon \to 0} \lim_{n \to \infty} b_\varepsilon^n(\lambda, \tau),$$

$$B(\lambda, \tau) = \lim_{\varepsilon \to 0} \lim_{n \to \infty} B_\varepsilon^n(\lambda, \tau).$$

The first part we call *continuous*, and the second *purely discontinuous* component of the additive functional with respect to some natural $\mathcal{F}^\circ$-measurable metric on the trace. Both functions, $\exp(-b_0(\lambda, \tau))$ and $\exp(-B(\lambda, \tau))$, are completely monotone on λ as limits of completely monotone functions.

Consider the continuous component. We have

$$\frac{\partial}{\partial \lambda} E_x\left(e^{-\lambda \tau} \mid \mathcal{F}^\circ\right) = \frac{\partial}{\partial \lambda} e^{-b(\lambda, \tau)} = -b'(\lambda, \tau) e^{-b(\lambda, \tau)}$$

$$= -E_x\left(\tau e^{-\lambda \tau} \mid \mathcal{F}^\circ\right),$$

from here

$$b'(\lambda, \tau) = \frac{E_{X_0}\left(\tau e^{-\lambda \tau} \mid \mathcal{F}^\circ\right)}{E_{X_0}\left(e^{-\lambda \tau} \mid \mathcal{F}^\circ\right)}.$$

Hence,

$$\frac{\partial}{\partial \lambda} b_0(\lambda, \tau) = \lim_{\varepsilon \to 0} \lim_{n \to \infty} \sum_1^{(n)} b'\left(\lambda, \sigma_{\Delta_\alpha}\right) \circ \theta_{\alpha'}$$

$$= \lim_{\varepsilon \to 0} \lim_{n \to \infty} \left(\sum_1^{(n)} E_{X_0}\left(\sigma_{\Delta_\alpha} e^{-\lambda \sigma_{\Delta_\alpha}} \mid \mathcal{F}^\circ\right) \circ \theta_{\alpha'} \right.$$

$$\left. + \sum_1^{(n)} \left(b'\left(\lambda, \sigma_{\Delta_\alpha}\right) \circ \theta_{\alpha'}\right)\left(1 - E_{X_0}\left(e^{-\lambda \sigma_{\Delta_\alpha}} \mid \mathcal{F}^\circ\right) \circ \theta_{\alpha'}\right) \right).$$

Evidently, $b'(\lambda, \tau)$ is an additive functional too. Therefore

$$\sum_1^{(n)} \left(b'\left(\lambda, \sigma_{\Delta_\alpha}\right) \circ \theta_{\alpha'}\right)\left(1 - e^{-b(\lambda, \sigma_{\Delta_\alpha}) \circ \theta_{\alpha'}}\right) \leq b'(\lambda, \tau)\varepsilon.$$

From here it follows that the derivatives of $b_0(\lambda, \tau)$ on λ is a completely monotone function as a limit of a sequence of completely monotone functions. Hence, using Bernstein's theorem, there exists a measure $m(du, \tau)$ on $\mathbb{R}_+$ such that

$$\frac{\partial b_0(\lambda, \tau)}{\partial \lambda} = \int_0^\infty e^{-\lambda u} m(du, \tau) = m(\{0\}, \tau) + \int_{0+}^\infty e^{-\lambda u} m(du, \tau).$$

Therefore

$$b_0(\lambda, \tau) = \lambda m(\{0\}, \tau) + \int_{0+}^\infty \left(1 - e^{-\lambda u}\right) \frac{1}{u} m(du, \tau).$$

Denoting $a_\tau = m(\{0\}, \tau)$, $n_\tau(du) = (1/u) m(du, \tau)$, we obtain the required representation of the continuous component. Evidently, a_τ and $n_\tau(du)$ are additive on the random set M_0 and they are $\mathcal{F}^\circ$-measurable.

Consider the purely discontinuous component. According to the chosen way of construction of sequences of partitions of set $\mathcal{T}_0$ for any $\varepsilon > 0$, limit $\lim_{n\to\infty} B_\varepsilon^n(\lambda, \tau)$ is a sum of a finite number of terms. As $\varepsilon \to 0$ only some new terms are added to the sum. To every term there exists a corresponding sequence of included intervals having non-empty intersection. Let $[\tau_{\alpha'_n}, \tau_{\alpha_n}]$ be such an interval where $\tau_{\alpha_n} = \tau_{\alpha'_n} \dot{+} \sigma_{\Delta_{\alpha_n}}$; $\tau_1 = \lim_{n\to\infty} \tau_{\alpha'_n}$ and $\tau_2 = \lim_{n\to\infty} \tau_{\alpha_n}$. In this case $\tau_1 \leq \tau_2$. Hence, τ_{α_n} can be represented as $\tau_1 \dot{+} \sigma_{\Delta_{\alpha_n}}$, where $P_{X(\tau_1)}(\sigma_0 > 0) > 0$, and since diam $\Delta_{\alpha_n} \to 0$, $\tau_2 = \lim_{n\to\infty}(\tau_1 \dot{+} \sigma_{\Delta_{\alpha_n}}) = \tau_1 \dot{+} \sigma_0$. Note that for a semi-Markov process the law of 0 or 1 is not fulfilled, and hence the values $0 < P_x(\sigma_0 > 0) < 1$ are admissible. Thus, the situation $\tau_1 = \tau_2$ is possible. By continuity of $b(\lambda, \tau)$ from the left we obtain $b(\lambda, \tau_{\alpha'_n}) \to b(\lambda, \tau_1)$. Since trajectories of the process are continuous from the right and have limits from the left, *a priori* we can only assert that $X(\tau_{\alpha'_n}) \to X(\tau_1 - 0)$. If for any n $\tau_{\alpha_n} > \tau_1$, the point τ_1 is not a point of discontinuity. Hence $X(\tau_{\alpha'_n}) \to X(\tau_1)$. If beginning with some n all τ_{α_n} are equal to τ_2, this τ_2 is a point of discontinuity. In this case the situation $\tau_1 = \tau_2$ is interpreted as an event $\sigma_0 = 0$, which in the given case has a positive probability, equal to the probability of event $\xi \in \mathcal{D}^{\tau_2}$. We denote a limit of the sequence of values $b(\lambda, \sigma_{\alpha_n}) \circ \theta_{\tau(\alpha'_n)}$ as $b(\lambda, \sigma_0) \circ \theta_{\tau_1}$, in spite of τ_1, in general, is not a Markov time. A time $\tau_2 = \tau_1 \dot{+} \sigma_0$ is a Markov time from the set MT_+ as a lower boundary of Markov times $\tau_{\alpha_n} \in \mathrm{MT}$. This limit can be written as $-\log E_{X_0}(\exp(-\lambda\sigma_0) \mid \mathcal{F}^\circ) \circ \theta_{\tau_1}$. So, the second component can be written in the form

$$B(\lambda, \tau) = -\sum_{\tau_i < \tau} \log E_{X_0}\left(\exp\left(-\lambda\sigma_0\right) \mid \mathcal{F}^\circ\right) \circ \theta_{\tau_i}$$

for corresponding random times τ_i, where $\tau_i \dot{+} \sigma_0 \in \mathrm{MT}_+$. Evidently, indexes of these times can be chosen depending only on γ_τ ($\tau \in \mathcal{T}_0$). $\qquad\square$

We call this formula *a Lévy expansion* for semi-Markov processes and $n_\tau(du)$ a *Lévy measure*. From this expansion of the function $b(\lambda, \tau)$ it follows a representation of τ:

$$\tau = a_\tau + \int_{0+}^{\infty} u \mathcal{P}_\tau(du) + \sum_{\tau_i < \tau} \sigma_0 \circ \theta_{\tau_i}, \tag{6.4}$$

where $\mathcal{P}_\tau(du)$ is $\mathcal{F}^\circ$-measurable conditional Poisson random measure on $\mathbb{R}_+ \times [0, \tau)$, where $[0, \tau)$ is a segment of a trace, and in addition, where $E_x(\mathcal{P}_\tau(du) \mid \mathcal{F}^\circ) = n_\tau(du)$. A proof of this representation, similar to the proof of corresponding to Lévy representation in the theory of processes with independent increments, consists of construction of a process by the given conditional distribution $P_x(\cdot \mid \mathcal{F}^\circ)$, determined by four independent components: a non-random function a_τ, a Poisson random measure $\mathcal{P}_\tau$ on $[0, \tau) \times (0, \infty)$, a (non-random) countable collection of elements (τ_i) and a sequence of independent non-negative random values $(\sigma_0 \theta_{\tau_i})$. Here $\tau \in \text{IMT}$ is considered as a mark on the trace (see below). Atoms (t_i, s_i) of measure $\mathcal{P}$ are interpreted as lengths (t_i) and initial points (s_i) as intervals of constancy of a trajectory with a given trace. They are so called conditional Poisson intervals of constancy. Measure $n_\tau(du)$ is a measure of intensity of random measure $\mathcal{P}_\tau(du)$, its expectation with respect to measure $P_x(\cdot \mid \mathcal{F}^\circ)$. A time τ_i in Lévy expansion is the initial point of an interval of constancy of another kind. We call them conditionally fixed intervals of constancy with random lengths $\sigma_0 \circ \theta_{\tau_i}$, which with positive probability are more than zero: $P_x(\sigma_0 \circ \theta_{\tau_i} > 0 \mid \mathcal{F}^\circ) > 0$. The time $\tau_i \dotplus \sigma_\Delta$ relates to a set of the first exit times of function

$$P_{X(\tau_i)}\left(\exp\left(-\lambda \sigma_\Delta\right) \mid \mathcal{F}^\circ\right) = P_{X(\tau_i)}\left(\exp\left(-\lambda \sigma_\Delta\right) \mid \alpha_{\sigma_\Delta}^{-1} \mathcal{F}^\circ\right)$$

beyond a level ε ($\varepsilon > 0$) and therefore it is a Markov time with respect to the stream of sigma-algebras $(\widetilde{\mathcal{F}}_t)$ ($t \geq 0$), where $\widetilde{\mathcal{F}}_t$ is a completion of the sigma algebra $\mathcal{F}_t$ with respect to measure P_x for all $x \in \mathbb{X}$. In what follows we will deal with processes for which P_x-a.s. $P_{X(\tau)}(\exp(-\lambda \sigma_0) \mid \mathcal{F}^\circ) = P_{X(\tau)}(\exp(-\lambda \sigma_0))$. The first exit time of the value of such a function beyond a given level is already a Markov time in common sense. Moreover it is an intrinsic Markov time, and under some regularity conditions of the semi-Markov process it is its regeneration time.

6.8. Random curvilinear integrals

Up to now we have considered a trace as an equivalence class on the Skorokhod space which is invariant with respect to a time change. In this section we will interpret the trace as a linearly ordered set of points taken from the metric space of states. From such a point of view it is natural to consider an additive functional along the trace and an integral with respect to this additive functional. It is a curvilinear integral, a well-known object of mathematical analysis [KOL 72]. We investigate a random curvilinear integral which turns out to be a useful tool for studying properties of semi-Markov processes. As an example of application of the random curvilinear integral we consider a problem of a semi-Markov process having an interval of constancy at a fixed time $t > 0$.

63. Trace as a linearly ordered set

Obviously, IMT is a partly ordered set. Besides, we have proved (see proposition 6.27) that $(\forall \tau, \tau_1 \in \mathrm{IMT})$ the set $\{\tau < \tau_1\}$ belongs to $\mathcal{F}^\circ$. According to the definition, the sets $\{X_\tau \in S, \ \tau < \infty\}$ $(S \in \mathcal{B}(\mathbb{X}))$ and $\mathcal{D}^\tau$ also belong to $\mathcal{F}^\circ$. Theorem 6.4 (criterion of equality of traces of two functions) prompts a thought to characterize the trace by properties of some maps:

(1) $\zeta : \mathrm{IMT} \to \overline{\mathbb{X}} \equiv \mathbb{X} \cup \{\infty\}$ corresponds to meaning of trajectory at the time τ;

(2) $\chi : \mathrm{IMT} \to \{0,1\}$ corresponds to an indicator of set $\mathcal{D}^\tau$.

The trace must also correspond to relation of linear ordering, determined on IMT (which, in turn, must be coordinated with ζ and χ).

64. Realization of linear ordering

Let a structure of linear ordering, and maps ζ and χ be determined in the IMT set. We say that function $\xi \in \mathcal{D}$ realizes this structure and these maps if for any $(\forall \tau, \tau_1 \in \mathrm{IMT})$ the following relations are fulfilled

$$\tau_1 < \tau_2 \iff \tau_1(\xi) < \tau_2(\xi),$$

$$\tau_1 = \tau_2 \iff \tau_1(\xi) = \tau_2(\xi),$$

$$\zeta(\tau) = \gamma_\tau \xi, \qquad \chi(\tau) = J_\tau \xi,$$

where $J_\tau = I_{\mathcal{D}^\tau}$ (see item 23).

PROPOSITION 6.29. *For a structure of linear ordering given on* IMT *and maps* $\zeta : \mathrm{IMT} \to \overline{\mathbb{X}}$ *and* $\chi : \mathrm{IMT} \to \{0,1\}$ *there exists a function* $\xi \in \mathcal{D}$, *realizing this structure and maps, if and only if the following conditions are fulfilled:*

(1) $\tau \dotplus \sigma_\Delta < \infty \Rightarrow \zeta(\tau \dotplus \sigma_\Delta) \notin \Delta$;

(2) $\tau_1 \le \tau < \tau_1 \dotplus \sigma_\Delta \Rightarrow \zeta(\tau) \in \Delta$;

(3) $\tau < \infty \Rightarrow (\exists n \in \mathbb{N}) \ \sigma_\delta^n \ge \tau$;

(4) $\chi(\infty) = 0 \Rightarrow (\exists n \in \mathbb{N}) \ \sigma_\delta^n = \infty$;

(5) $\chi(\tau) = 0, \ \tau < \infty \Rightarrow (\exists \varepsilon > 0) \ (\forall \delta \in \mathrm{DS}, \ \mathrm{rank}\, \delta \le \varepsilon) \ (\exists n \in \mathbb{N}) \ \tau = \sigma_\delta^n$;

(6) $(\forall k, \ell, m, n \in \mathbb{N} : \sigma_{\delta_n}^{k-1} \le \sigma_{\delta_m}^{\ell} < \sigma_{\delta_n}^{k}) \ (\exists \varepsilon > 0) \ (\forall r : 0 < r \le \varepsilon)$

$$\{(s,t) \in \mathbb{N}^2 : \sigma_{\delta_n}^{k-1} < \sigma_{\delta_t}^s < \sigma_{\delta_m}^\ell, \ \zeta(\sigma_{\delta_t}^s) \in \Delta_{nk} \setminus \Delta_{nk}^{-r}\} = \emptyset,$$

where $\delta_n = (\Delta_{n1}, \Delta_{n2}, \ldots)$;

(7) $\chi(\sigma_{\delta_n}^k) = 0 \Rightarrow \zeta(\tau'_{nk}) \in \Delta_{nk}$, where $\tau'_{nk} = \sup\{\sigma_{\delta_m}^\ell : \sigma_{\delta_m}^\ell < \sigma_{\delta_n}^k\}$;

(8) $\chi(\infty) = 1 \Rightarrow \{\sigma_{\delta_n}^k : \sigma_{\delta_n}^k < \infty\}$, *and it has no maximal term.*

The logical quantors

$$\left(\forall \tau, \tau_1 \in \text{IMT}\right) \left(\forall \Delta \in \mathfrak{A}\right) \left(\forall \delta \in \text{DS}\right)$$

precede conditions (1)–(5).

The logical quantor

$$\left(\exists \left(\delta_n\right)_1^\infty, \ \delta_n \in \text{DS}, \ \text{rank} \, \delta_n \longrightarrow 0\right)$$

precedes conditions (6)–(8).

Proof. Necessity. Let $\xi \in \mathcal{D}$, $\tau \in \text{IMT}$, $\zeta(\tau) = \gamma_\tau \xi$, $\chi(\tau) = J_\tau \xi$ and a linear order on IMT be determined by values of $\tau(\xi)$. Then properties (1), (2), (3), (4) and (8) are obvious. Property (5) follows from an interval of constancy (IC) on the left of $\tau - 0$ to be possible only in the case where $\tau(\xi)$ is a point of discontinuity. Its right part is an evidently necessary and sufficient condition for discontinuity at point $\tau(\xi)$. Property (6) follows since for any function $\xi \in \mathcal{D}$ there is always a rather rich collection of sets $\Delta \in \mathfrak{A}$ which ξ exits "correctly" from. In particular, from any system of concentric balls no more than a countable subsystem of balls can be chosen for which property $\sigma_{\Delta-r}(\xi) \to \sigma_\Delta(\xi)$ $(r \to 0)$ is not fulfilled. Property (7) follows from $\tau'_{nk}(\xi) \in [\sigma_{\delta_n}^{k-1}(\xi), \ \sigma_{\delta_n}^k(\xi))$.

Sufficiency. We will construct a function ξ with required properties (from theorem 6.4 it follows that any other such function has the same trace). Consider the sequence $(\delta_n)_1^\infty$ from conditions (6)–(8). We will choose for all $\sigma_{\delta_n}^k < \infty$ points $\kappa(\sigma_{\delta_n}^k)$ on axis $\mathbb{R}_+$ according to the linear order on IMT and values of function χ. Without loss of generality, we can assume that $\zeta(\sigma_{\delta_n}^{k-1}) \in \Delta_{nk}$ for any n and k. From here by condition (1) it follows: $\sigma_{\delta_n}^{k-1} < \sigma_{\delta_n}^k$. Let $\chi(\infty) = 1$. Choose points for δ_1: $\kappa(\sigma_{\delta_1}^n) = n+m$, where $\sigma_{\delta_1}^n < \infty$ and m is a number of those $k \leq n$, for which $\chi(\sigma_{\delta_1}^k) = 0$. A double enlargement of a distance between points corresponds to an interval of constancy of unit length situated before the appropriate time point. A choice of remain points is being made on parts of the axis without these IC. Let points for $\delta_1, \ldots, \delta_n$ be already located. If $\kappa(\sigma_{\delta_k}^\ell) < \kappa(\sigma_{\delta_m}^c)$ are two neighboring points in this construction and $\sigma_{\delta_{n+1}}^{s=1} < \cdots < \sigma_{\delta_{n+1}}^{s=1+L}$ are times from the following sequence, located between $\sigma_{\delta_k}^\ell$ and $\sigma_{\delta_m}^c$, then we assume

$$\kappa\left(\sigma_{\delta_{n+1}}^{s+i}\right) = \kappa\left(\sigma_{\delta_k}^\ell\right) + \frac{i+j}{L+M+1} t,$$

where $t = \kappa(\sigma_{\delta_m}^c) - \kappa(\sigma_{\delta_k}^\ell)$, if $\chi(\sigma_{\delta_m}^c) = 1$, and $t = (\kappa(\sigma_{\delta_m}^c) - \kappa(\sigma_{\delta_k}^\ell))/2$, if $\chi(\sigma_{\delta_m}^c) = 0$; M is a number of those $i \leq L$, for which $\chi(\sigma_{\delta_{n+1}}^{s+i}) = 0$; j is a number of those $d \leq i$, for which $\chi(\sigma_{\delta_{n+1}}^{s+d}) = 0$. A double reduction of an interval between neighboring points corresponds to putting IC before the point $\kappa(\sigma_{\delta_m}^c)$. If there exists the last point after the n-th step of construction, $\kappa(\sigma_{\delta_k}^\ell) < \infty$, and $\sigma_{\delta_{n+1}}^{s+1} < \sigma_{\delta_{n+1}}^{s+2}$ are times next to $\sigma_{\delta_k}^\ell$, then their location on $[\kappa(\sigma_{\delta_k}^\ell),\infty)$ with gaps 1 or 2 is being made according to the first construction rule. Thus, the rule of allocation of all points $\kappa(\sigma_{\delta_n}^k)$ $(k, n \in \mathbb{N})$ is determined. In addition, $\sup \kappa(\sigma_{\delta_n}^k) = \infty$ (by condition (8)). If $\chi(\infty) = 1$, we locate points $(\kappa(\sigma_{\delta_1}^k))$ (which are of a finite number, by condition (4)) uniformly on interval $[0, 1]$ taking into account the corresponding IC, as in the case of $\chi(\infty) = 1$ on the $(n + 1)$-th step. The construction of $\kappa(\sigma_{\delta_n}^k)$ is finished. Further-more, in accordance with ζ we construct a sequence of functions. After n-th stage we determine a function

$$\xi_n : \xi_n(t) = \zeta(\sigma_{\delta_k}^\ell) \quad (\kappa(\sigma_{\delta_k}^\ell) \leq t < \kappa(\sigma_{\delta_m}^c)),$$

where extreme terms of the inequality are neighboring points (we assume $\sigma_{\delta_k}^0 = 0$ and $\kappa(0) = 0$, and also $\kappa(\infty) = \infty$ for the case when there are no points of the n-th construction on the right of t). From conditions (2), (3) and (4) it follows that the sequence (ξ_n) converges uniformly, since rank $\delta_n \to 0$. According to the con-struction $(\forall n \in \mathbb{N})$ $\xi_n \in \mathcal{D}$. From here $\xi = \lim \xi_n \in \mathcal{D}$. Again according to the construction $\xi(\kappa(\sigma_{\delta_n}^k)) = \zeta(\sigma_{\delta_n}^k)$, from here $\sigma_{\delta_n}^k(\xi) \leq \kappa(\sigma_{\delta_n}^k)$. However, from con-ditions (6) and (7) it follows that $\sigma_{\delta_n}^k(\xi) = \kappa(\sigma_{\delta_n}^k)$. Actually, it is true for $k = 0$. Let $\sigma_{\delta_n}^{k-1}(\xi) = \kappa(\sigma_{\delta_n}^{k-1})$. If $t \in [\sigma_{\delta_n}^{k-1}(\xi), \kappa(\sigma_{\delta_n}^k))$ exists, for which $\xi(t) \notin \Delta$ (from uni-form convergence it follows that in this case $\xi(t) \in \partial\Delta$), then either t is some point of discontinuity (in this case t is included in $(\kappa(\sigma_{\delta_m}^\ell)) \cap [\kappa(\sigma_{\delta_n}^{k-1}), \kappa(\sigma_{\delta_n}^k))$, and for any point from the latter sub-sequence, by condition (2), $\zeta(\sigma_{\delta_m}^\ell) \in \Delta$, i.e. contradic-tion), or t is a point of continuity. If on the interval $(t, \kappa(\sigma_{\delta_n}^k))$ there is at least one point $\kappa(\sigma_{\delta_m}^\ell)$, then, by condition (6), $(\exists r > 0)$ $\xi(t) \in \Delta^{-r}$, i.e. a contradiction. If on the interval $(t, \kappa(\sigma_{\delta_n}^k))$ there are no points of view $\kappa(\sigma_{\delta_n}^\ell)$, it means that ξ is con-stant and $\xi(t) = \zeta(t') = \lim \zeta(\tau_n)$, where $\tau_n < \tau'$, $\tau_n \to \tau'$, and, by condition (7), we obtain a contradiction. According to construction ξ has jumps at the same points, where the right part of condition (5) is fulfilled, and it has IC only when $\chi(\tau) = 0$. Hence, the constructed function realizes the structure of ordering and maps ζ and χ on the subset $\mathcal{T}_2^1 = (\sigma_{\delta_n}^k)_{k,n \in \mathbb{N}}$ of the set IMT. On the other hand, any finite $\tau \in$ IMT can be determined with the help of showing a sequence of intervals containing it: $\sigma_{\delta_n}^{k_n} \leq \tau < \sigma_{\delta_n}^{k_n+1}$. In this case $\tau = \sup \sigma_{\delta_n}^{k_n}$. Let us define $\kappa(\tau) = \lim \kappa(\sigma_{\delta_n}^{k_n})$. Note that $\zeta(\tau) = \lim \zeta(\sigma_{\delta_n}^{k_n}) = \xi(\kappa(\tau))$ and $\kappa(\sigma_{\delta_n}^{k_n}) \leq \kappa(\tau) < \kappa(\sigma_{\delta_n}^{k_n+1})$. By uniqueness of correspondence of a value of $\tau \in$ IMT and a sequence of containing its intervals, determined by sequence (δ_n), it follows that $\kappa(\tau) = \tau(\xi)$, i.e. $\zeta(\tau) = \xi(\tau)$. According to the construction $\sigma_{\delta_n}^k < \infty \Leftrightarrow \kappa(\sigma_{\delta_n}^k) < \chi(\infty) = 0$. Besides:

(a) if $\chi(\infty) = 1$, then, by condition (8),

$$\sup\left\{\kappa(\sigma_{\delta_n}^k) : \sigma_{\delta_n}^k < \infty\right\} = \infty \Longleftrightarrow \sup\{\tau \in \text{IMT} : \tau < \infty\} = \infty$$

and in the left part $\sigma_{\delta_n}^k$ can be replaced by $\tau \in$ IMT;

(b) if $\chi(\infty) = 0$, then, by construction,

$$\sup\left\{\kappa\left(\sigma_{\delta_n}^k\right) : \sigma_{\delta_n}^k < \infty\right\} \le 1 \iff \sup\{\tau \in \mathrm{IMT} : \tau < \infty\} < \infty,$$

where in the left part $\sigma_{\delta_n}^k$ can be replaced by $\tau \in \mathrm{IMT}$. From here it follows that $(\forall \tau \in \mathrm{IMT})\ \tau < \infty \Leftrightarrow \tau(\xi) < \infty$ and, moreover, $\zeta(\tau) = \gamma_\tau \xi$. Due to the absence of IC in ξ before $\tau \in \mathrm{IMT}$, which is a point of continuity of ξ, we have $(\forall \tau \in \mathrm{IMT})$

$$\chi(\tau) = J_\tau \xi. \qquad \qquad \square$$

65. Curvilinear integral

We have defined the trace (item 14), we have proved a criterion of equality of the traces of two functions ξ_1, $\xi_2 \in \mathcal{D}$ (theorem 6.4), and we have also shown that the trace can be determined in terms of linear ordering on IMT and that of map ζ. Function χ only makes more precise linear ordering in points of discontinuity and on infinity (proposition 6.29). Now we define one kind of integral along the trace, in particular, called a curvilinear integral along the trace. While constructing it we need not consider *a priori* given linearly-ordered structure and functions ζ and χ on IMT, for which we would check the conditions of proposition 6.29. Furthermore, we will propose that in constructions relating to a trace and the class IMT there already exists a realizing function $\xi \in \mathcal{D}$.

Let $\mathrm{IMT}(\xi)$ be the set of all contraction of $\tau \in \mathrm{IMT}$ on $\ell\xi$, in which an order is determined according to $\tau(\xi)$ (which is corrected in a point of discontinuity of ξ and at infinity). Thus, if $\xi_1, \xi_2 \in \ell^{-1}\ell\xi$, then

$$\tau_1(\xi_1) < \tau_2(\xi_1) \implies \tau_1(\xi_2) < \tau_2(\xi_2),$$
$$\tau_1(\xi_1) = \tau_2(\xi_1) \implies \tau_1(\xi_2) = \tau_2(\xi_2).$$

This linearly ordered set will serve as a set of parameters while defining a curvilinear integral. We will name it a set of indexes of state of the corresponding trace $\ell\xi$. Set $\mathcal{T}_0(\xi)$ of all contractions $\tau \in \mathcal{T}_0$ on set $\ell\xi$ plays a role of a countable everywhere dense set in $\mathrm{IMT}(\xi)$. The last aspect can be corrected with the help of a metric on $\mathrm{IMT}(\xi)$.

For any $\tau_1, \tau_2 \in \mathcal{T}_0$ and $\xi \in \mathcal{D}$ we define a function ρ_ℓ as follows:

$$\rho_\ell(\tau_1, \tau_2 \mid \xi) = \inf\left\{\mathrm{diam}\,\Delta : \Delta \in \mathfrak{A}_0,\ \tau_1\xi \le \tau_2\xi < (\tau_1 \dotplus \sigma_\Delta)(\xi)\right\},$$

if $\tau_1\xi \le \tau_2\xi$, and

$$\rho_\ell(\tau_1, \tau_2 \mid \xi) = \inf\left\{\mathrm{diam}\,\Delta : \Delta \in \mathfrak{A}_0,\ \tau_2\xi \le \tau_1\xi < (\tau_2 \dotplus \sigma_\Delta)(\xi)\right\},$$

if $\tau_2\xi \le \tau_1\xi$. Denote $\mathcal{F}' = \sigma(\gamma_\tau, \tau \in \mathcal{T}_0)$.

PROPOSITION 6.30. *For any* $\xi \in \mathcal{D}$ *function* $\rho_\ell(\cdot, \cdot \mid \xi)$ *is a metric on* $\mathcal{T}_0$, *and for any* $\tau_1, \tau_2 \in \mathcal{T}_0$ *it is* $\mathcal{F}'$-*measurable.*

Proof. $\mathcal{F}'$-measurability follows immediately from the definition. Fulfilment of the zero axiom and the symmetry axiom is evident. Let us check the triangle axiom. Let $\tau_1(\xi) < \tau_2(\xi)$. Investigate three possible cases of location of the third point (excluding coincidence) denoting $\rho_{ij} = \rho_\ell(\tau_i, \tau_j \mid \xi)$ and omitting the argument ξ.

If $\tau_3 < \tau_1$, then from condition $\tau_3 \dotplus \sigma_\Delta > \tau_2$ it follows that $\tau_1 \dotplus \sigma_\Delta = \tau_3 \dotplus \sigma_\Delta > \tau_2$; from here

$$\rho_{12} \leq \rho_{32} \leq \rho_{13} + \rho_{32}.$$

If $\tau_3 > \tau_2$, then from condition $\tau_1 \dotplus \sigma_\Delta > \tau_3$ it follows that $\tau_2 \dotplus \sigma_\Delta = \tau_1 \dotplus \sigma_\Delta > \tau_3$; from here

$$\rho_{12} \leq \rho_{13} \leq \rho_{13} + \rho_{32}.$$

If $\tau_1 < \tau_3 < \tau_2$, then from conditions $\tau_1 \dotplus \sigma_{\Delta_1} > \tau_3$ and $\tau_3 \dotplus \sigma_{\Delta_2} > \tau_2$ it follows that $\tau_1 \dotplus \sigma_{\Delta_1 \cup \Delta_2} = \tau_1 \dotplus \sigma_{\Delta_1} \dotplus \sigma_{\Delta_2} \dotplus \sigma_{\Delta_1 \cup \Delta_2} > \tau_2$. From here $\rho_{12} \leq \text{diam}(\Delta_1 \cup \Delta_2) \leq \text{diam}(\Delta_1) + \text{diam}(\Delta_2)$. Passing on the right part to the lower bound by all Δ_1 and Δ_2, we obtain $\rho_{12} \leq \rho_{13} + \rho_{32}$. $\square$

In this metric the IMT(ξ) set is not in general closed and connected. Each $\tau \in$ IMT(ξ), while being a point of discontinuity, is situated on a positive distance ρ_ℓ from the set $[0, \tau) \cap$ IMT(ξ) of preceding points, which does not contain a maximal element, if $J_\tau(\xi) = 1$.

An additive functional on IMT(ξ) is said to be such a function of two arguments $A(\tau_1, \tau_2)$ that if $\tau_1 \geq \tau_2$, then $A(\tau_1, \tau_2) = 0$, and if $\tau_1 \leq \tau_2 \leq \tau_3$, then $A(\tau_1, \tau_3) = A(\tau_1, \tau_2) + A(\tau_2, \tau_3)$. Furthermore, we will be interested in non-negative, left-continuous additive functionals, i.e. if $\tau_n < \tau$, $\tau_n, \tau \in$ IMT, $\rho_\ell(\tau_n, \tau) \to 0$ $(n \to \infty)$, then $A(\tau_n, \tau) \to 0$. Let f be a real function on IMT(ξ), $\delta_n \in$ DS and $\tau \in$ IMT(ξ), $\tau < \infty$. Consider a sum

$$\sum_{k=1}^{\infty} f\left(\sigma_{\delta_n}^{k-1}\right) A\left(\sigma_{\delta_n}^{k-1}, \sigma_{\delta_n}^k \wedge \tau\right).$$

If this sum tends to a limit independent of a sequence (δ_n) as rank $\delta_n \to 0$, then this limit is called a *curvilinear integral* of function f on a segment of the trace $[0, \tau) \cap$ IMT(ξ) with respect to the additive functional A. We will designate this integral as

$$\int_0^\tau f(\tau_1) A(d\tau_1).$$

66. Existence of a curvilinear integral

We call a variation of function f on the interval $[0, \tau) \cap \mathrm{IMT}(\xi)$ the functional

$$\bigvee_0^\tau (f) = \sup_{\delta \in \mathrm{DS}} \sum_{k=1}^{\infty} \sup \{ |f(\sigma_\delta^{k-1}) - f(\tau_1)| : \tau_1 \in [\sigma_\delta^{k-1}, \sigma_\delta^k \wedge \tau) \cap \mathrm{IMT}(\xi) \},$$

which, evidently, coincides with a variation of function $g_\xi(t)$, which is a naturally parameterized function $f \colon (\forall t \in \mathbb{R}_+)$

$$g_\xi(t) = f\big(\sup \{ \tau \in \mathrm{IMT}(\xi) : \tau(\xi) \leq t \} \big)$$

on the interval $[0, \tau(\xi))$. Function g can be determined in terms of values $g_\xi(\sigma_{\delta_n}^k(\xi)) = f(\sigma_{\delta_n}^k)$, because $g_\xi(t)$ is constant on all intervals of constancy of the function ξ.

PROPOSITION 6.31. *Let f be a function of bounded variation. For any $\tau \in \mathrm{IMT}(\xi)$ ($\tau < \infty$) if f is continuous from the left on $[0, \tau) \cap \mathrm{IMT}(\xi)$, then there exists a curvilinear integral on this interval (continuity is considered in the sense of the metric ρ_ℓ).*

Proof. We use the sequence $(\delta_n)_1^\infty$, where δ_n is a deducing sequence of an order-type ω^n. In this case all points of partition of n-th order are included in the set of points of partition of $(n+1)$-th order. We have

$$S_1 \leq \sum_{k \in \mathbb{N}_0^n} f(\tau_k) A(\sigma_{\delta_n}^{k'}, \sigma_{\delta_n}^k \wedge \tau) \leq S_2,$$

where k' is a preceding point for k under the lexicographical ordering in alphabet $\mathbb{N}_0^n$, and also

$$\tau_k \in [\sigma_{\delta_n}^{k'}, \sigma_{\delta_n}^k \wedge \tau) \cap \mathrm{IMT}(\xi), \qquad M_k = \sup f(\tau_k), \qquad m_k = \inf f(\tau_k),$$

$$S_1 = \sum_{k \in \mathbb{N}_0^n} m_k A(\sigma_{\delta_n}^{k'}, \sigma_{\delta_n}^k \wedge \tau), \qquad S_2 = \sum_{k \in \mathbb{N}_0^n} M_k A(\sigma_{\delta_n}^{k'}, \sigma_{\delta_n}^k \wedge \tau).$$

As $n \to \infty$ the left term of this inequality does not decrease, and the right one does not increase. Consider a sequence of embedded intervals

$$[\sigma_{\delta_1}^{k_1'}, \sigma_{\delta_1}^{k_1} \wedge \tau) \supset [\sigma_{\delta_2}^{k_2'}, \sigma_{\delta_2}^{k_2} \wedge \tau) \supset \cdots,$$

for which $M_{k_n} - m_{k_n} \geq \varepsilon$ ($\varepsilon > 0$). From boundedness of the variations of f the finiteness of number of such sequences follows. For them because of continuity from the left there exists a limit $\lim_{n \to \infty} f(\sigma_{\delta_n}^{k_n'})$ equal to $f(\tau')$, where $\tau' = \lim_{n \to \infty} \sigma_{\delta_n}^{k_n'} < \infty$.

In this case

$$f(\sigma_{\delta_n}^{k_n})A(\sigma_{\delta_n}^{k'_n}, \sigma_{\delta_n}^{k_n} \wedge \tau) \longrightarrow f(\tau')A(\tau', \tau' + 0) \quad (n \longrightarrow \infty),$$

where $A(\tau', \tau' + 0) = \lim_{n \to \infty} A(\tau', \sigma_{\delta_n}^{k_n} \wedge \tau)$, if the set $\bigcap_{n=1}^{\infty} [\sigma_{\delta_n}^{k'_n}, \sigma_{\delta_n}^{k_n} \wedge \tau)$ is not empty. Otherwise, the limit of the sequence of this terms is equal to zero. Since $\varepsilon > 0$ is arbitrary, the limit of these sums exists. This limit, evidently, does not depend on the choice of a sequence of deducing sequences (δ_n) with ranks converging to zero. It means a curvilinear integral exists. $\qquad\qquad\Box$

67. Evaluation of curvilinear integrals

It can be made with the help of a natural parametrization. Let $\tau \in \mathrm{IMT}(\xi)$. Determine an additive functional a_ξ: $a_\xi(t_1, t_2) = a_\xi(0, t_2) - a_\xi(0, t_1)$, where $t_1 \leq t_2$ and

$$a_\xi(0, t) = A\big(0, \sup\{\tau \in \mathrm{IMT}(\xi) : \tau(\xi) \leq t\}\big)$$

$$= \sup\big(A(0, \tau) : \tau \in \mathrm{IMT}(\xi),\ t(\xi), \tau(\xi) \leq t\big).$$

In this case $a_\xi(t_1, t_2) = 0$, if ξ is constant on $[t_1, t_2]$. Then on $\mathrm{IMT}(\xi)$

$$\int_0^\tau f(\tau_1) A(d\tau_1) = \int_0^{\tau(\xi)} g_\xi(t) a_\xi(dt),$$

where $g_\xi(t)$ is defined in item 66. If $f(\tau) = f_1(\zeta(\tau))$,

$$g_\xi(t) = f_1\big(\zeta\big(\sup\{\tau \in \mathrm{IMT}(\xi) : \tau(\xi) \leq t\}\big)\big) = f_1\big(\xi(t)\big).$$

This case corresponds to a usual definition of a curvilinear integral, when a curve is determined with the help of a numerical parameter. A proof consists of a check of a usual rule of replacement of variables in the integral.

68. Random curvilinear integral

Let (P_x) be an admissible family of probability measures on $\mathcal{F}$, determining a random process with trajectories from $\mathcal{D}$. For any $x \in \mathbb{X}$ and $\tau \in \mathrm{IMT}$ the distribution $P_x(\tau < t \mid \mathcal{F}^\circ)(\xi)$ $(t \in \mathbb{R}_+, \xi \in \mathcal{D})$ can be interpreted as a conditional distribution of τ under fixed $\ell\xi$, which is defined for almost all $\ell\xi$. An expectation $E_x(\tau \mid \mathcal{F}^\circ)(\xi)$, considering as a function of $\tau \in \mathrm{IMT}(\xi)$, is an example of a non-decreasing continuous from the left function, which determines an additive functional $\mu_1(\tau_1, \tau_2) = E_x(\tau_2 - \tau_1 \mid \mathcal{F}^\circ)$ on $\mathrm{IMT}(\xi)$. Continuity from the left follows from the fact that if $\tau_n \in \mathrm{IMT}(\xi)$, $\tau_n < \tau$ and $\rho_\ell(\tau_n, \tau) \to 0$, then τ is not a point of discontinuity, hence $(\forall \xi_1 \in \ell^{-1}\ell\xi)\ \tau_n(\xi_1) \to \tau(\xi_1)$. The simple example of a curvilinear integral of a continuous function f is expression

$$\int_0^\infty P_x\big(f_1(X_t);\ t < \tau \mid \mathcal{F}^\circ\big)dt = \int_0^\tau f_1(X_{\tau_1})\mu_1(d\tau_1).$$

For a semi-Markov process the additive functional μ can be written in a homogenous form: $\mu_1(\tau_1, \tau_2) = \mu(\tau_2) - \mu(\tau_1)$, where $\mu(\tau) = P_{X_0}(\tau \mid \mathcal{F}^\circ)$ and therefore

$$\mu(\tau_1 \dotplus \tau_2) = \mu(\tau_1) + \mu(\tau_2) \circ \theta_{\tau_1}$$

(homogenous additivity). In this case a linearly ordered family $\mathrm{IMT}(\xi)$ is a process with independent increments with respect to a measure $P_x(\cdot \mid \mathcal{F}^\circ)$: $(\forall \tau_1, \tau_2 \in \mathrm{IMT}(\xi),$ $\tau_1 < \tau_2)$ τ_1 and $\tau_2 - \tau_1$ are independent.

A vectorial additive (multiplicative) functional is said to be a finite system of one-dimensional additive (multiplicative) functionals ($\mathcal{F}^0$-measurable also). Furthermore, we use d-dimensional functionals ($d \geq 2$). An example of a vector additive functional is the integral $S_t^g \equiv \int_0^t g(X_s)\,dt$, where $g(x)$ is a vector-valued function, or a corresponding curvilinear integral.

69. Probability of hitting in an interval of constancy

We know that trajectories of a typical SM process contain intervals of constancy. A quantitative description of a localization and lengths of IC can be derived from Lévy formula [6.3] according to Lévy representation [6.4] (the most important outcome from the Lévy formula). There are two kinds of possible IC: Poisson and conditionally-fixed. We are interested in the probability that a point t ($t \in \mathbb{R}_+$) belongs to some IC of the function $\xi \in \mathcal{D}$. In addition, we will distinguish events $\{t \in \mathrm{ICR}\}$ and $\{t \in \mathrm{ICF}\}$ of t to hit in random (Poisson) interval of constancy (ICR), and that of fixed (ICF). Besides we will consider a function

$$R_t(\xi) = t - \sup\big\{\tau(\xi) : \tau \in \mathrm{IMT},\ \tau(\xi) \leq t\big\},$$

which is called an (inverse) sojourn time in IC (see item 1.18). Let us evaluate a functional

$$\int_0^\infty e^{-\lambda t} E_x\big(f(X_t)e^{-\lambda_1 R_t};\ A \cap \{t < \tau\} \mid \mathcal{F}^\circ\big)\,dt \quad (\lambda, \lambda_1 > 0),$$

where f is a continuous function on $\mathbb{X}$, for A there are three possibilities: $\{t \notin \mathrm{IC}\}$, $\{t \in \mathrm{ICF}\}$ or $\{t \in \mathrm{ICR}\}$ (in the first event, obviously, $R_t = 0$). Let $\tau = \sigma_\Delta$, where $\Delta \in \mathfrak{A}$ and the set $\Delta \cup \partial\Delta$ is compact.

PROPOSITION 6.32. *Let* $(P_x) \in$ SM. *Then* $(\forall x \in \mathbb{X})$ P_x-*a.s.*

$$\int_0^\infty e^{-\lambda t} E_x \big(f(X_t) e^{-\lambda_1 R_t};\ t < \sigma_\Delta,\ t \in \mathrm{ICR} \mid \mathcal{F}^\circ \big)\, dt$$

$$= \int_0^{\sigma_\Delta} f(X_\tau) e^{-b(\lambda,\tau)} B_{\lambda+\lambda_1}(d\tau),$$

$$\int_0^\infty e^{-\lambda t} E_x \big(f(X_t) e^{-\lambda_1 R_t};\ t < \sigma_0,\ t \in \mathrm{ICF} \mid \mathcal{F}^\circ \big)\, dt$$

$$= \int_0^{\sigma_\Delta} f(X_\tau) e^{-b(\lambda,\tau)} C_{\lambda+\lambda_1}(d\tau),$$

$$\int_0^\infty e^{-\lambda t} E_x \big(f(X_t);\ t < \sigma_\Delta,\ t \notin \mathrm{IC} \mid \mathcal{F}^\circ \big)\, dt$$

$$= \int_0^{\sigma_\Delta} f(X_\tau) e^{-b(\lambda,\tau)} A(d\tau),$$

where

$$A(\tau_1,\tau_2) = a_{\tau_2} - a_{\tau_1},$$

$$B_\lambda(\tau_1,\tau_2) = (1/\lambda) \int_0^\infty \big(1 - e^{-\lambda u}\big)\big(n_{\tau_2}(du) - n_{\tau_1}(du)\big),$$

$$C_\lambda(\tau_1,\tau_2) = (1/\lambda) \sum_{\tau_1 \le \tau_i < \tau_2} \big(1 - P_x\big(\exp\big(-\lambda\sigma_0 \circ \theta_{\tau_i}\big) \mid \mathcal{F}^\circ\big)\big).$$

Proof. Note that $R_t = \lim(t - \sigma_\delta^{k'})$ (rank $\delta \to 0$), where $\sigma_\delta^{k'} \le t < \sigma_\delta^k$. We have

$$\int_0^\infty e^{-\lambda t} E_x \big(f(X_t) e^{-\lambda_1 R_t};\ t < \sigma_\Delta \mid \mathcal{F}^\circ \big)\, dt$$

$$= E_x \Big(\int_0^{\sigma_\Delta} e^{-\lambda t} e^{-\lambda_1 R_t} f(X_t)\, dt \mid \mathcal{F}^\circ \Big)$$

$$= \lim_{\mathrm{rank}\,\delta \to 0} \sum_{\sigma_\delta^k < \sigma_\Delta} f(X(\sigma_\delta^{k'})) E_x \Big(\int_{\sigma_\delta^{k'}}^{\sigma_\delta^k} e^{-\lambda t} e^{-\lambda_1 (t - \sigma_\delta^k)}\, dt \mid \mathcal{F}^\circ \Big) \qquad [6.5]$$

$$= \lim_{\mathrm{rank}\,\delta \to 0} \sum_k f(X(\sigma_\delta^{k'})) \exp\big(-b(\lambda,\sigma_\delta^{k'})\big) \frac{1}{(\lambda+\lambda_1)}$$

$$\times \big(1 - \exp\big(-b(\lambda+\lambda_1,\sigma_\delta^k) + b(\lambda+\lambda_1,\sigma_\delta^{k'})\big)\big).$$

An increment of $b(\lambda, \tau)$ at the point of continuity $(\tau \neq \tau_i)$ is equal to an increment of function $\lambda a_\tau + \int_{0+}^{\infty} (1 - e^{-\lambda u}) n_\tau(du)$, and at the points of discontinuity $(\tau = \tau_i)$ it is equal to

$$- \log E_x \left(\exp \left(- \lambda \sigma_0 \circ \theta_{\tau_i} \right) \mid \mathcal{F}^\circ \right).$$

From here we discover that expression [6.5] is equal to

$$\int_0^{\sigma_\Delta} f(X_\tau) e^{-b(\lambda, \tau)} \left(A(d\tau) + B_{\lambda + \lambda_1}(d\tau) + C_{\lambda + \lambda_1}(d\tau) \right).$$

We have

$$\int_0^{\infty} e^{-\lambda t} E_x \left(f(X_t) e^{-\lambda_1 R_t}; \ t \in \mathrm{ICF}, \ t < \sigma_\Delta \mid \mathcal{F}^\circ \right) dt$$

$$= E_x \left(\int_0^{\sigma_\Delta} f(X_t) e^{-\lambda t} e^{-\lambda_1 R_t} I_{\mathrm{ICF}}(t) \, dt \mid \mathcal{F}^\circ \right)$$

$$= \sum_{\tau_i < \sigma_\Delta} E_x \left(f(X(\tau_i)) \int_{\tau_i}^{\tau_i + \sigma_0} e^{-\lambda t} e^{-\lambda_1 (t - \tau_i)} \, dt \mid \mathcal{F}^\circ \right)$$

$$= \frac{1}{\lambda + \lambda_1} \sum_{\tau_i < \sigma_\Delta} f(X(\tau_i)) E_x \left(e^{-\lambda \tau_i} \left(1 - e^{-(\lambda + \lambda_1) \sigma_0} \circ \theta_{\tau_i} \right) \mid \mathcal{F}^\circ \right).$$

Note that $\sigma_0 \circ \theta_{\tau_i} = \tau_i \dotplus \sigma_0 - \tau_i$ is independent of τ_i with respect to measure $P_x(\cdot \mid \mathcal{F}^\circ)$. This property does not immediately follow from properties of SM processes, because τ_i is not its regeneration time, but it can be obtained as a limit for $\sigma_\delta^{k'} \uparrow \tau_i, \sigma_\delta^k \downarrow \tau_i \dotplus \sigma_0$, hence $\sigma_\delta^k - \sigma_\delta^{k'} \to \sigma_0 \circ \theta_{\tau_i}$. Therefore

$$\frac{1}{\lambda + \lambda_1} \sum_{\tau_i < \sigma_\Delta} f(X(\tau_i)) e^{-b(\lambda, \tau_i)} \left(1 - E_x \left(e^{-(\lambda + \lambda_1) \sigma_0} \circ \theta_{\tau_i} \mid \mathcal{F}^\circ \right) \right)$$

$$= \int_0^{\sigma_\Delta} f(X_\tau) e^{-b(\lambda, \tau)} C_{\lambda + \lambda_1}(d\tau).$$

Evidently, $\{t \in \mathrm{ICR}\} = \bigcup_{\varepsilon > 0} \{t \in \mathrm{ICR}_\varepsilon\}$, where $\{t \in \mathrm{ICR}_\varepsilon\}$ means hitting t into a random interval of length more than ε. We have

$$\{t \notin \mathrm{ICR}_\varepsilon\} = \bigcup_{\delta \in \mathrm{DS}} \bigcup_k \{t \in [\sigma_\delta^{k'}, \sigma_\delta^k), \ \widetilde{\mathcal{P}}([\varepsilon, \infty) \times [\sigma_\delta^k, \sigma_\delta^{k'})) = 0\},$$

where $\widetilde{\mathcal{P}}(A \times [\tau_1, \tau_2]) = \mathcal{P}_{\tau_2}(A) - \mathcal{P}_{\tau_1}(A)$ and $\widetilde{\mathcal{P}}$ is a Poisson random measure on

$(0, \infty) \times \mathrm{IMT}(\xi)$. From here

$$\int_0^\infty e^{-\lambda t} E_x\big(f(X_t)e^{-\lambda_1 R_t};\ t \notin \mathrm{ICR},\ t < \sigma_\Delta \mid \mathcal{F}^\circ\big)\, dt$$

$$= \lim_{\varepsilon \to 0}\ \lim_{\mathrm{rank}\,\delta \to 0} \int_0^\infty e^{-\lambda t} \sum_k E_x\big(f(X_t)e^{-\lambda_1 R_t};\ t \in [\sigma_\delta^{k'}, \sigma_\delta^k \wedge \sigma_\Delta),$$

$$\widetilde{\mathcal{P}}\big([\varepsilon, \infty) \times [\sigma_\delta^{k'}, \sigma_\delta^k \wedge \sigma_\Delta)\big) = 0 \mid \mathcal{F}^\circ\big)\, dt$$

$$= \lim_{\varepsilon \to 0}\ \lim_{\mathrm{rank}\,\delta \to 0} \sum_k f(X_{\sigma_\delta^{k'}}) E_x\left(\int_{\sigma_\delta^{k'}}^{\sigma_\delta^k \wedge \sigma_\Delta} e^{-\lambda t} e^{-\lambda_1(t - \sigma_\delta^{k'})}\, dt;\right.$$

$$\left.\widetilde{\mathcal{P}}\big([\varepsilon, \infty) \times [\sigma_\delta^{k'}, \sigma_\delta^k \wedge \sigma_\Delta)\big) = 0 \mid \mathcal{F}^\circ\right)$$

$$= \lim_{\varepsilon \to 0}\ \lim_{\mathrm{rank}\,\delta \to 0} \sum_k f(X_{\sigma_\delta^k}) \frac{1}{(\lambda + \lambda_1)} E_x\big(\exp\big(\lambda_1 \sigma_\delta^{k'}\big)\big(\exp\big(-(\lambda + \lambda_1)\sigma_\delta^{k'}\big)$$

$$- \exp\big(-(\lambda + \lambda_1)(\sigma_\delta^k \wedge \sigma_\Delta)\big)\big);\, \widetilde{\mathcal{P}}\big([\varepsilon, \infty) \times [\sigma_\delta^{k'}, \sigma_\delta^k \wedge \sigma_\Delta)\big) = 0 \mid \mathcal{F}^\circ\big).$$

$$[6.6]$$

Using the Lévy representation for $\tau \in \mathrm{IMT}$, we discover that the conditional average in [6.6] is equal to

$$\exp\big(-b(\lambda, \sigma_\delta^{k'})\big) E_x\big(1 - \exp\big(-(\lambda + \lambda_1)(\sigma_\delta^k \wedge \sigma_\Delta - \sigma_\delta^{k'})\big);$$

$$\widetilde{\mathcal{P}}\big([\varepsilon, \infty)[\sigma_\delta^{k'}, \sigma_\delta^k \wedge \sigma_\Delta)\big) = 0 \mid \mathcal{F}^\circ\big)$$

$$= \exp\big(-b(\lambda, \sigma_\delta^{k'})\big) E_x\left(1 - \exp\left(-(\lambda + \lambda_1)\left[\big(a(\sigma_\delta^k \wedge \sigma_\Delta) - a(\sigma_\delta^{k'})\big)\right.\right.\right.$$

$$\left.\left.\left. + \int_{0+}^\varepsilon u\big(\mathcal{P}(du, \sigma_\delta^k \wedge \sigma_\Delta) - \mathcal{P}(du, \sigma_\delta^{k'})\big) + \sum_{\sigma_\delta^{k'} \le \tau_i < \sigma_\delta^k \wedge \sigma_\Delta} \sigma_0 \circ \theta_{\tau_i}\right]\right)\right);$$

$$\left. + \widetilde{\mathcal{P}}\big([\varepsilon, \infty) \times [\sigma_\delta^{k'}, \sigma_\delta^k \wedge \sigma_\Delta)\big) = 0 \mid \mathcal{F}^\circ\right)$$

$$= \exp\big(-b(\lambda, \sigma_\delta^{k'})\big) \exp\big(-\widetilde{n}\big([\varepsilon, \infty) \times [\sigma_\delta^{k'}, \sigma_\delta^k \wedge \sigma_\Delta)\big)\big)$$

$$\times \left(1 - \exp\left(-(\lambda + \lambda_1)\big(a(\sigma_\delta^k \wedge \sigma_\Delta) - a(\sigma_\delta^{k'})\big)\right.\right.$$

$$\left.\left. - \int_{0+}^\varepsilon \big(1 - e^{-(\lambda + \lambda_1)u}\big)\big(n(du, \sigma_\delta^k \wedge \sigma_\Delta) - n(du, \sigma_\delta^{k'})\big)\right)\right.$$

$$\times \prod_{\sigma_\delta^{k'} \le \tau_i < \sigma_\delta^k \wedge \sigma_\Delta} E_x\big(\exp\big(-(\lambda + \lambda_1)\sigma_0 \circ \theta_{\tau_i}\big) \mid \mathcal{F}^\circ\big)\bigg),$$

where $\tilde{n}$ is a measure of intensity of the Poisson random measure $\tilde{\mathcal{P}}$. This expression with a boundedness $t \notin \mathrm{IC}\,C_\varepsilon$ differs from the corresponding factor without this boundedness by a term $\exp(-\tilde{n}([\varepsilon, \infty) \times [\sigma_\delta^{k'}, \sigma_\delta^k \wedge \sigma_\Delta)))$ (which tends to 1 as rank $\delta \to 0$) and by an upper limit of integration on measure $n(du, \cdot)$. Let

$$B_\lambda^\varepsilon(\tau_1, \tau_2) = \frac{1}{\lambda} \int_{0+}^\varepsilon \left(1 - e^{-\lambda u}\right)\left(n(du, \tau_2) - n(du, \tau_1)\right).$$

From an evident continuity of integral $\int_{0+}^\varepsilon (1 - e^{-\lambda u})\, n(du, \tau)$ on τ (with respect to ρ_ℓ) when $\delta \to 0$, convergence of the sum to a curvilinear integral follows

$$\int_0^{\sigma_\Delta} f(X_\tau) e^{-b(\lambda, \tau)} \left(A(d\tau) + B_{\lambda+\lambda_1}^\varepsilon(d\tau) + C_{\lambda+\lambda_1}(d\tau)\right).$$

On the other hand, from convergence $\int_{0+}^\varepsilon (1 - e^{-u})n(du, \tau) \to 0$ when $\varepsilon \to 0$, convergence follows

$$\int_0^{\sigma_\Delta} f(X_\tau) e^{-b(\lambda, \tau)} B_{\lambda+\lambda_1}^\varepsilon(d\tau) \longrightarrow 0,$$

and we discover that

$$\int_0^\infty e^{-\lambda t} P_x\left(f(X_t) e^{-\lambda_1 t}; \ t \notin \mathrm{ICR}, \ t < \sigma_\Delta \mid \mathcal{F}^\circ\right) dt$$
$$= \int_0^{\sigma_\Delta} f(X_\tau) e^{-b(\lambda, \tau)} \left(A(d\tau) + C_{\lambda+\lambda_1}(d\tau)\right),$$

hence

$$\int_0^\infty e^{-\lambda t} P_x\left(f(X_t) e^{-\lambda_1 t}; \ t \notin \mathrm{IC}, \ t < \sigma_\Delta \mid \mathcal{F}^\circ\right) dt$$
$$= \int_0^{\sigma_\Delta} f(X_\tau) e^{-b(\lambda, \tau)} A(d\tau),$$

and also

$$\int_0^\infty e^{-\lambda t} P_x\left(f(X_t) e^{-\lambda_1 R_t}; \ t \in \mathrm{ICR}, \ t < \sigma_\Delta \mid \mathcal{F}^\circ\right) dt$$
$$= \int_0^{\sigma_\Delta} f(X_\tau) e^{-b(\lambda, \tau)} B_{\lambda+\lambda_1}(d\tau).$$
$\qquad\square$

6.9. Characteristic operator and curvilinear integral

In this section a connection between two directions of investigation of semi-Markov processes is considered:

(1) an analytical direction in which the principal objects of investigation are transition generating function and a lambda-characteristic operator (see Chapters 3 and 5);

(2) a probability direction, dealing with conditional distributions of a time run with respect to trace.

A supposition about existence of a rather simple connection between these directions arises due to Markov processes, which are described in terms of these directions with some extreme but similar properties.

70. Representation of operator

Consider a SM process. Assume that in point $x \in \mathbb{X}_0$ there exists a limit

$$A_\lambda \mathbb{I}(x) = \lim_{\Delta \downarrow x} \frac{E_x\left(e^{-\lambda \sigma_\Delta} - 1\right)}{E_x(\sigma_\Delta)}.$$

Evidently, function A_λ has all derivatives on λ for $\lambda > 0$ and its first derivative can be represented in the form

$$-\frac{\partial}{\partial \lambda} A_\lambda \mathbb{I}(x) = \lim_{\Delta \downarrow x} \frac{E_x\left(\sigma_\Delta e^{-\lambda \sigma_\Delta}\right)}{E_x(\sigma_\Delta)}.$$

From here it follows that this function is completely monotone. Possessing such a property this function, using Bernstein's theorem, can be represented as an outcome of the Laplace transformation of some distribution function. From here, in turn, it follows that function $-A_\lambda \mathbb{I}(x)$ itself can be represented in Lévy form (see theorem 6.11)

$$-A_\lambda \mathbb{I}(x) = \lambda \alpha(x) + \int_{0+}^{\infty} \left(1 - e^{-\lambda u}\right)\nu(du \mid x),$$

where $\alpha(x)$ is some non-negative function and $\nu(du, x)$ is some measure on $(0, \infty)$ depending on x.

On the other hand, the following limit is true

$$-A_\lambda \mathbb{I}(x) = \lim_{\Delta \downarrow x} \frac{E_x\left(e^{-b(\lambda, \sigma_\Delta)} - 1\right)}{E_x(\sigma_\Delta)},$$

where $b(\lambda, \sigma_\Delta) \to 0$ for almost all ξ. We suppose that

$$\frac{E_x\left((b(\lambda, \sigma_\Delta))^2\right)}{E_x(\sigma_\Delta)} \longrightarrow 0, \qquad [6.7]$$

P_x-a.s. (we will justify this assumption below). Under this condition

$$-A_\lambda \mathbb{I}(x) = \lim_{\Delta \downarrow x} \frac{E_x\big(b(\lambda, \sigma_\Delta)\big)}{E_x(\sigma_\Delta)} = \lim_{\Delta \downarrow x} \frac{E_x\big(b_0(\lambda, \sigma_\Delta)\big)}{E_x(\sigma_\Delta)},$$

and, hence, in the expansion of $A_\lambda \mathbb{I}(x)$ the coefficients have a sense

$$\alpha(x) = \lim_{\Delta \downarrow x} \frac{E_x\big(a(\sigma_\Delta)\big)}{E_x(\sigma_\Delta)}, \qquad \nu(B \mid x) = \lim_{\Delta \downarrow x} \frac{E_x\big(n(B, \sigma_\Delta)\big)}{E_x(\sigma_\Delta)}. \qquad [6.8]$$

These formulae show connection between parameters of Lévy expansion for the process of time run along the trace and lambda-characteristic operator. This connection can be made more precise with the help of additive functionals $a(\tau)$ and $n(B, \tau)$ represented by curvilinear integrals.

71. Continuous and discrete parts of functional

Consider a functional $\mu(\tau) = E_{X_0}(\tau \mid \mathcal{F}^\circ)$ ($\tau \in \mathrm{IMT}$), for which the following representation is fair $\mu(\tau_1 \dotplus \tau) = \mu(\tau_1) + \mu(\tau) \circ \theta_{\tau_1}$. Let $E_x(\tau) < \infty$. Then P_x-a.s. $\mu(\tau) < \infty$. Let $(\delta'_n)_{n=1}^\infty$ be a sequence of deducing sequences with decreasing ranks (rank $\delta'_n \to 0$), and $\delta_n = \delta'_1 \times \cdots \times \delta'_n$ be a composition of first n sequences, i.e. a sequence of order-type ω^n (see item 61). Consider a representation of $\mu(\tau)$ in the form of a finite sum, where $\tau = \tau_\beta < \infty$ ($\beta \in \mathcal{A}_m$)

$$\mu(\tau) = \sum_{\alpha < \beta} \mu_{\sigma_{\Delta_\alpha}} \circ \theta_{\alpha'}$$

(for denotation for multi-dimensional indexes, see item 61). We represent this sum in the form of two summands

$$\sum_{\alpha < \beta} \big(\mu_{\sigma_{\Delta_\alpha}} \circ \theta_{\alpha'}\big) I\big(\mu_{\sigma_{\Delta_\alpha}} \circ \theta_{\alpha'} \leq \varepsilon\big),$$

$$\sum_{\alpha < \beta} \big(\mu_{\sigma_{\Delta_\alpha}} \circ \theta_{\alpha'}\big) I\big(\mu_{\sigma_{\Delta_\alpha}} \circ \theta_{\alpha'} > \varepsilon\big).$$

As $n \to \infty$ and $\varepsilon \to 0$, each of these sums tends to a limit. The first limit is a continuous component of the additive functional $\mu_0(\tau)$, which is obviously equal to

$$\mu_0(\tau) = a_\tau + \int_{0+}^\infty u n_\tau(du).$$

The second limit is a discrete component which can be represented in the form

$$M(\tau) = \sum_{\tau_i < \tau} E_{X_0}\big(\sigma_0 \mid \mathcal{F}^\circ\big) \circ \theta_{\tau_i}.$$

72. Representation of parameters in the form of integrals

According to item 65, the IMT(ξ) set can be considered as a metric space with the metric ρ_ℓ. Due to continuity from the left of functionals $b(\lambda, \tau)$ and $\mu(\tau)$, they coincide with distribution functions of some measures $\widetilde{b}$ and $\widetilde{\mu}$ on Borel subsets of the set IMT(ξ). Obviously, the measure $\widetilde{b}$ and its components, including the Lévy expansion, are absolutely continuous with respect to the measure $\widetilde{\mu}$. Let

$$\beta(\lambda, \tau) = \lambda\alpha'(\tau) + \int_{0+}^{\infty} (1 - e^{-\lambda u})\nu'(du \mid \tau) + \sum_{t_i < \tau} p(\tau_i)$$

be a density of the measure $\widetilde{b}$ with respect to $\widetilde{\mu}$, i.e.

$$b(\lambda, \tau) = \int_0^\tau \beta(\lambda, \tau_1)\mu(d\tau_1),$$

$$b_0(\lambda, \tau) = \int_0^\tau \lambda\alpha'(\lambda, \tau_1)\mu(d\tau_1) + \int_0^\tau \int_{0+}^{\infty} (1 - e^{-\lambda u})\nu'(du \mid \tau_1)\mu(d\tau_1),$$

where $\alpha'(\tau)$ and $\nu'(du \mid \tau)$ correspond to a_τ and $n(du, \tau)$,

$$p(\tau_i) = -\big(\log E_{X_0}(e^{-\lambda\sigma_0} \mid \mathcal{F}^\circ)/P_{X_0}(\sigma_0 \mid \mathcal{F}^\circ)\big) \circ \theta_{\tau_i}.$$

Let $\alpha'(\tau)$ and $\int_{0+}^{\infty} (1 - e^{-\lambda u})\nu'(du \mid \tau)$ be continuous in some neighborhood of point 0. Then, by formulae [6.8],

$$\alpha(X_0) = \lim_{\Delta \downarrow x} \frac{E_{X_0}(a(\sigma_\Delta))}{E_{X_0}(\sigma_\Delta)} = \lim_{\Delta \downarrow x} \frac{E_{X_0}\big(\int_0^{\sigma_\Delta} \alpha'(\tau)\mu(d\tau)\big)}{E_{X_0}\big(\int_0^{\sigma_\Delta} \mu(d\tau)\big)} = \alpha'(0),$$

$$\nu(B \mid X_0) = \lim_{\Delta \downarrow x} \frac{E_x(n(B, \sigma_\Delta))}{E_x(\sigma_\Delta)} = \lim_{\Delta \downarrow x} \frac{E_x\big(\int_0^{\sigma_\Delta} \nu'(B \mid \tau)\mu(d\tau)\big)}{E_{X_0}\big(\int_0^{\sigma_\Delta} \mu(d\tau)\big)} = \nu'(B \mid 0).$$

In this case

$$b_0(\lambda, \tau) = \int_0^\tau (-A_\lambda \mathbb{I} \circ X_{\tau_1})\mu(d\tau_1) \qquad [6.9]$$

is a required representation of a parameter of the Lévy expansion in terms of the lambda-characteristic operator.

73. Conditional degeneration of Markov processes

Let (P_x) be a family of measures of a Markov process with pseudo-local lambda-characteristic operator in some region $\Delta \in \mathfrak{A}$. Then in this region the process does not have any fixed intervals of constancy and, besides, condition [6.7] is fulfilled. Hence, formula [6.9] is fair. Since for a Markov process $A_\lambda \mathbb{I} = -\lambda$ (see theorem 3.6(2)), we discover that for this process $b(\lambda, \tau) = \lambda \mu_\tau$ for all $\tau < \sigma_\Delta$. It means that a time run along a fixed trace represents a determinate (non-random) process. In other words, a conditional distribution of a time run along the trace in a Markov process is degenerate [HAR 89].

6.10. Stochastic integral with respect to semi-Markov process

In this section we assume $\mathbb{X} = \mathbb{R}^d$ ($d \geq 1$) and (P_x) to be a family of measures of a semi-Markov process of diffusion type. We consider a construction of stochastic integrals appropriating to a semi-Markov process as an integrating function. Stochastic integrals for processes of such a class are required in the problem of absolute continuity (see [HAR 02]). Since semi-Markov process is not Markovian (in general) the method of construction of stochastic integrals based on the classic theory of martingales [LIP 86] is not applicable. However as we know semi-Markov processes possess a partly Markov structure which can be used in order to construct some kind of stochastic integral with the help a modernized method. Of course this method can be applied to the proper Markov processes of diffusion type too. For them it brings the same results as the traditional method of stochastic integrating. In this respect the former can be considered as a generalization of the latter.

6.10.1. Semi-martingale and martingale along the trace

74. Asymptotic formulae

From items 5.34 and 5.36 it follows that neighborhoods of spherical form play a special role due to their simple asymptotical properties. Let $B = B(0, 1)$ and $B_r = B_r(x) = x + rB$ be an open ball with radius r and center x. In item 5.36 [5.35] the following asymptotical representation is proved:

$$f_{\sigma_{B_r}}(\lambda, \varphi \mid x) - \varphi(x) = r^2 A_\lambda^{(0)}(\varphi \mid x) + o(r^2) \quad (r \longrightarrow 0) \qquad [6.10]$$

where

$$A_\lambda^{(0)}(\varphi \mid x) = \frac{1}{d}\left(\frac{1}{2}\sum_{i,j=1}^{d} a^{ij}(x)D_{ij}\varphi(x) + \sum_{i=1}^{d} b^i(x)D_i\varphi(x) - c(x, \lambda)\varphi(x)\right),$$

$\lambda \geq 0$; φ is a twice continuously differentiable function.

The second asymptotical formula relates to the expectation of the first exit time from a small spherical neighborhood of the starting point of the process. In equation

(5.17) we assume $c(0, x) \equiv 0$ and $0 < \gamma(x) < \infty$ where $\gamma(x) = \partial c(\lambda, x)/\partial \lambda|_{\lambda=0}$. From results of item 5.37 the asymptotics follows:

$$E_x(\sigma_{B_r}) = (r^2 \gamma(x) + o(r^2))/d \quad (r \longrightarrow 0).$$ [6.11]

Formula [6.11] can be generalized as follows. Let us consider the integral $S_t^g = \int_0^t g(X_s)\,ds$, where g is a continuous positive function on $\mathbb{X}$. Continuity from the right of trajectories implies the following asymptotics of the expectation

$$E_x(S_{\sigma_{B_r}}^g) = (r^2 \gamma(x)g(x) + o(r^2))/d \quad (r \longrightarrow 0).$$ [6.12]

Let us assume function γ to be continuous on G. Formulae [6.10], and [6.12] imply the equivalence: as $r \to 0$

$$\frac{E_x(\exp(-\lambda\sigma_{B_r})\varphi(X_{\sigma_{B_r}})) - \varphi(x)}{E_x(S_{\sigma_{B_r}}^g)} = \frac{A_\lambda^{(0)}(\varphi \mid x)}{g(x)\gamma(x)}(1 + o(1)).$$ [6.13]

This equivalence has a key meaning while constructing a stochastic integral with respect to a semi-Markov process. As will be shown below different g functions generate different additive functionals determining curvilinear integrals.

75. Sequence of inscribed balls

Let $G \in \mathfrak{A}$ and $x \equiv \xi(0) \in G$. We denote $r(x, G) = \sup\{r > 0 : B_r(x) \subset G\}$. For given value $u > 0$ let $\sigma_{G,u}(\xi) = \sigma_{r(X_0(\xi),G)\wedge u}(\xi)$. Then $\sigma_{G,u} \in \mathcal{T}$ (Markov time) and $\sigma_G = \sigma_{G,u} \dotplus \sigma_G$. Hence for any $n \geq 1$ it is true that $\sigma_G = \sigma_{G,u}^n \dotplus \sigma_G$, where $\sigma_{G,u}^n = \sigma_{G,u} \dotplus \cdots \dotplus \sigma_{G,u}$ (in the right part of the equality identical summands are repeated n times). Let us assume $\sigma_{G,u}^0 = 0$, and also in the case $\xi(0) \notin G$ let $\sigma_{G,u}(\xi) = 0$. Hence for any ξ there exists the limit

$$\lim_{n\to\infty} \sigma_{G,u}^n \equiv \sigma_{G,u}^\infty \leq \sigma_G.$$

Evidently, $\lim x_n \notin G$, where $x_n = X_{\sigma_{G,u}^n}$, and consequently, if $\xi \in \mathcal{C}$ (continuous), then $\sigma_{G,u}^\infty = \sigma_G$.

The method of inscribed balls can be easily generalized for the case of inscribed neighborhoods of rather arbitrary form if a fixed family of neighborhoods corresponds to every point of the space of states.

76. Semi-martingale and martingale

Consider semi-Markov process of diffusion type, controlled by equation [5.17]. Let $G \in \mathfrak{A}$ be a bounded set and $x \in G$. Consider a deducing sequence from G of rank r. Let N_G^δ be the number of jumps that the stepped function L_δ makes before its first exit from G; t_k be the time of the k-th jump; x_k be the meaning of the process at the time of the k-th jump. In this case there exists $\Delta_{k+1} \in \mathfrak{A}^o$ such that $t_{k+1} = t_k \dotplus \sigma_{\Delta_{k+1}}$. Consider representation of the integral of the function $\varphi(X_{\sigma_G})$. We have

$$E_x\big(\varphi(X_{\sigma_G})\big) - \varphi(x) = E_x\left(\sum_{k=1}^{N_G^\delta} \big(\varphi(x_k) - \varphi(x_{k-1})\big)\right)$$

$$= \sum_{n=1}^{\infty} E_x\left(\sum_{k=1}^{n} \big(\varphi(x_k) - \varphi(x_{k-1})\big); \ N_G^\delta = n\right)$$

$$= \sum_{k=1}^{\infty} E_x\big(\varphi(x_k) - \varphi(x_{k-1}); \ N_G^\delta \geq k\big)$$

$$= \sum_{k=1}^{\infty} E_x\big(E_{x_{k-1}}\big(\varphi(X_{\sigma_{\Delta_k}}) - \varphi(X_0)\big); \ N_G^\delta \geq k\big).$$

Using equivalence [6.13] as $\lambda = 0$ and $g(x) = A_0^{(0)}(\varphi \mid x)/\gamma(x)$, we obtain

$$= \sum_{k=1}^{\infty} E_x\big(E_{x_{k-1}}\big(S_{\sigma_{\Delta_k}}^g\big)\big(1 + \epsilon(k,r)\big); \ N_G^\delta \geq k\big),$$

where $\epsilon(k,r) \to 0$ $(r \to 0)$ uniformly in k. From here the latter expression can be rewritten as follows

$$= \sum_{k=1}^{\infty} E_x\big(E_{x_{k-1}}\big(S_{\sigma_{\Delta_k}}^g\big); \ N_G^\delta \geq k\big)\big(1 + \epsilon(r)\big)$$

$$= \sum_{k=1}^{\infty} E_x\big(S_{t_k}^g - S_{t_{k-1}}^g; \ N_G^\delta \geq k\big)\big(1 + \epsilon(r)\big) \longrightarrow E_x\big(S_{\sigma_G}^g\big).$$

Denoting $A_0(\varphi \mid x) = A_0^{(0)}(\varphi \mid x)/\gamma(x)$, we obtain the formula

$$E_x\big(\varphi(X_{\sigma_G})\big) - \varphi(x) = E_x\left(\int_0^{\sigma_G} A_0(\varphi \mid X_t)\, dt\right). \qquad [6.14]$$

This formula is well-known in the theory of Markov processes [DYN 63], where $(A_0(\varphi \mid x))$ is the characteristic Dynkin operator. From additivity of the functional S_τ^g it follows that this formula is true after change σ_G by an arbitrary time $\tau \in T^o$. Let $\varphi(X) = X^i - Y_i$. Then $A_0^{(0)}(\varphi \mid Y) = b^i(Y)$, where $b(Y) = (b^1(Y), \ldots, b^d(Y))$ is the vector of coefficients of the first derivatives in equation [5.17]. Extending denotation S_t^g to vector functions g, we can write

$$E_x\big(X_\tau - X_0\big) = E_x\big(S_\tau^{b/\gamma}\big) = E_x\left(\int_0^\tau \frac{b(X_t)}{\gamma(X_t)}\, dt\right).$$

Taking into account this formula, we consider a vector additive functional (process)

$$M_t^{(1)} = X_t - X_0 - S_t^{b/\gamma}.$$

Additivity of this functional follows from additivity of the difference:

$$X_{\tau_1 \dotplus \tau_1} - X_0 = X_{\tau_1} - X_0 + (X_{\tau_2} - X_0) \circ \theta_{\tau_1}.$$

In order to prove the martingale property of this process we note that for any τ_1 and τ_2 from $\mathcal{T}^0$ there exists $\tau_3 \in \mathcal{T}^0$ such that on the set $\tau_1 \leq \tau_2$ $\tau_2 = \tau_1 \dotplus \tau_3$ (the proof follows from relation $\{t < \sigma_\Delta\} \Rightarrow \{\sigma_\Delta = t \dotplus \sigma_\Delta\}$; see item 2.16). Hence, it is sufficient to consider the pair of times τ_1 and $\tau_1 \dotplus \tau_2$. Using the semi-Markov property, we have

$$E_x\left(M_{\tau_1 \dotplus \tau_2}^{(1)} \mid \mathcal{F}_{\tau_1}\right) = E_x\left(M_{\tau_1}^{(1)} + M_{\tau_2}^{(1)} \circ \theta_{\tau_1} \mid \mathcal{F}_{\tau_1}\right)$$
$$= M_{\tau_1}^{(1)} + E_{X_{\tau_1}}\left(M_{\tau_2}^{(1)}\right) = M_{\tau_1}^{(1)}.$$

So the martingale property is fulfilled on the partly ordered set of Markov times $\mathcal{T}^0$, but in general is not fulfilled on the whole semi-axis, or on any other interval. Analogously with semi-Markov processes such a process would naturally be called semi-martingale (unfortunately, this term is frequently used for the union of sub- and super-martingale classes). The definition of a stochastic integral can be made in terms of semi-martingales. However, properties of semi-martingales are not sufficient to derive the Ito formula or its close analogy. For this aim we use another modification of the martingale, namely $\mathcal{F}^0$-measurable martingale along the trace. Let us note that formula [6.14] can be rewritten as follows

$$E_x\left(\varphi\left(X_{\sigma_G}\right)\right) - \varphi(x) = E_x\left(\int_0^{\sigma_G} \frac{A_0\left(\varphi \mid X_\tau\right)}{\gamma\left(X_\tau\right)} d\mu_\tau\right), \qquad [6.15]$$

On the basis of this formula we obtain the second variant of a vector additive function, which can be called a martingale on the partly ordered set $\mathcal{T}^0$, or a martingale along the trace:

$$M_\tau^{(2)} = X_\tau - X_0 - \mu_\tau^f,$$

where $f = b/\gamma$. The martingale property follows from the following relation

$$E_x\left(M_{\tau_1 \dotplus \tau_2}^{(2)} \mid \mathcal{F}_{\tau_1}\right) = M_{\tau_1}^{(2)} + E_{X_{\tau_1}}\left(M_{\tau_2}^{(2)}\right) = M_{\tau_1}^{(2)}.$$

The martingale $(M_\tau^{(2)})$ cannot be extended to a "semi-martingale". Its realizations are determined on every trace of a trajectory. The latter is considered an ordered set of points from the metric space $\mathbb{X}$. In the given example the $\mathcal{F}^0$-measurable additive functional corresponds to the first exit times from open sets. In contrast to the semi-martingale, the martingale along the trace, can also be determined in the case when the expectation of this time is infinite (see examples of additive functionals).

77. *Quadratic characteristic*

Let us show that martingale $M \equiv M^{(2)}$ is quadratically integrable, and find its quadratic characteristic. Note that martingale M (like the original process) has dimension d, hence its quadratic characteristic is a matrix with elements

$$K^{ij}(M, \tau) = E_x(M_\tau^i M_\tau^j) \quad (\tau \in \mathcal{T}^0).$$

We see that $M_0 = 0$ P_x-a.s. Consider the expression

$$K_m^{ij}(M, \sigma_G) = E_x\left(\sum_{k=1}^{N}(M_{t_k}^i - M_{t_{k-1}}^i)\sum_{k=1}^{N}(M_{t_k}^j - M_{t_{k-1}}^j)\right), \qquad [6.16]$$

where $t_k = \sigma_{\delta_m}^k$, $\delta_m \in \mathcal{S}_m$ is a deducing sequence from $G \in \mathfrak{A}$ of order type ω^m, and of rank r_m ($r_m \to 0$ as $m \to \infty$), and $N = N_G^{\delta_m}(\xi)$ is number of jumps of the quantified process $L_{\delta_m}(\xi)$ by the time σ_G (here $\sigma_G(\xi) = \sigma_{\delta_m}^N(\xi)$). Denote $\Delta M_k^s = M_{t_k}^s - M_{t_{k-1}}^s$. Then

$$K_m^{ij}(M, \sigma_G) = \sum_{n=1}^{\infty} E_x\left(M_{\sigma_G}^i M_{\sigma_G}^j; N = n\right)$$

$$= \sum_{n=1}^{\infty} E_x\left(\sum_{k=1}^{n}\Delta M_k^i \Delta M_k^j + \sum_{k=1}^{n-1}\sum_{l=k+1}^{n}\Delta M_k^i \Delta M_l^j\right.$$

$$\left. + \sum_{l=1}^{n-1}\sum_{k=l+1}^{n}\Delta M_k^i \Delta M_l^j; N = n\right).$$

By varying the order of summation:

$$\sum_{n=1}^{\infty}\sum_{k=1}^{n-1}\sum_{l=k+1}^{n} \longmapsto \sum_{k=1}^{\infty}\sum_{l=k+1}^{\infty}\sum_{n=l}^{\infty},$$

taking into account that $\{N \geq l\} \in \mathcal{F}_{\tau_{l-1}}$, and using the martingale property, we obtain

$$\sum_{n=1}^{\infty} E_x\left(\sum_{k=1}^{n-1}\sum_{l=k+1}^{n}\Delta M_k^i \Delta M_l^j + \sum_{l=1}^{n-1}\sum_{k=l+1}^{n}\Delta M_k^i \Delta M_l^j; N = n\right) = 0.$$

Consequently,

$$K_m^{ij}(M, \sigma_G) = \sum_{k=1}^{\infty} E_x(\Delta M_k^i \Delta M_k^j; \ N \geq k)$$

$$= \sum_{k=1}^{\infty} E_x\Big((x_k^i - x_{k-1}^i - \mu_{t_k}^{b^i/\gamma} + \mu_{t_{k-1}}^{b^i/\gamma})(x_k^j - x_{k-1}^j - \mu_{t_k}^{b^j/\gamma} + \mu_{t_{k-1}}^{b^j/\gamma}); \ N \geq k)$$

$$= \sum_{k=1}^{\infty} E_x\Big((X_{\sigma_{\Delta_k}}^i - X_0^i - \mu_{\sigma_{\Delta_k}}^{b^i/\gamma})(X_{\sigma_{\Delta_k}}^j - X_0^j - \mu_{\sigma_{\Delta_k}}^{b^j/\gamma}) \circ \theta_{t_{k-1}}; \ N \geq k)$$

$$= \sum_{k=1}^{\infty} E_x\Big(E_{x_{k-1}}\big((X_{\sigma_{\Delta_k}}^i - X_0^i - \mu_{\sigma_{\Delta_k}}^{b^i/\gamma})(X_{\sigma_{\Delta_k}}^j - X_0^j - \mu_{\sigma_{\Delta_k}}^{b^j/\gamma})\big); \ N \geq k)$$

$$= \sum_{k=1}^{\infty} E_x\Big(E_{x_{k-1}}\big((X_{\sigma_{\Delta_k}}^i - X_0^i)(X_{\sigma_{\Delta_k}}^j - X_0^j) - (X_{\sigma_{\Delta_k}}^i - X_0^i)\mu_{\sigma_{\Delta_k}}^{b^j/\gamma}$$

$$- (X_{\sigma_{\Delta_k}}^j - X_0^j)\mu_{\sigma_{\Delta_k}}^{b^i/\gamma} + \mu_{\sigma_{\Delta_k}}^{b^i/\gamma}\mu_{\sigma_{\Delta_k}}^{b^j/\gamma}\big); \ N \geq k\Big).$$

Estimate the order of the members. According to formula [6.13] we have

$$\Big|E_x\big((X_{\sigma_{\Delta_k}}^i - X_0^i)\mu_{\sigma_{\Delta_k}}^{b^j/\gamma}\big)\Big| \leq r_m E_x\big(\mu_{\sigma_{\Delta_k}}^{|b^j|/\gamma}\big).$$

From here it follows that the sum of these members is negligible as $m \to \infty$.

Furthermore, we will estimate members $E_x(\mu_{\sigma_{\Delta_k}}^{b^i/\gamma}\mu_{\sigma_{\Delta_k}}^{b^j/\gamma})$. We have

$$\frac{E_x\big(\mu_{\sigma_{\Delta_k}}^{b^i/\gamma}\mu_{\sigma_{\Delta_k}}^{b^j/\gamma}\big)}{E_x\big(\mu_{\sigma_{\Delta_k}}^g\big)} \sim \frac{(b^i(x)b^j(x)/\gamma^2(x))E_x\big((\mu_{\sigma_{\Delta_k}})^2\big)}{g(x)E_x\big(\mu_{\sigma_{\Delta_k}}\big)}.$$

We know that ratio $E_x((\mu_{\sigma_{\Delta_k}})^2)/E_x(\mu_{\sigma_{\Delta_k}})$ tends to zero as $m \to \infty$. So it follows that the sum of these members is also negligible.

Let us estimate members of the form $E_x((X_{\sigma_{\Delta_k}}^i - X_0^i)(X_{\sigma_{\Delta_k}}^j - X_0^j))$. Using formula [6.13] with function $\varphi(y) = (y^i - x^i)(y^j - x^j)$, when $A_0^{(0)}(\varphi \mid x) = a^{ij}(x)$, we obtain

$$E_x\big((X_{\sigma_{\Delta_k}}^i - X_0^i)(X_{\sigma_{\Delta_k}}^j - X_0^j)\big) = E_x\big(S_{\sigma_{\Delta_k}}^f\big)\big(1 + \epsilon_1(m, k)\big)$$

with $f(x) = a^{ij}(x)/\gamma(x)$, where $\epsilon_1(m, k) \to 0$ as $m \to \infty$ uniformly in k. From here

$$K_m^{ij}(M, \sigma_G) = \sum_{k=1}^{\infty} E_x \left(S_{\sigma_{\Delta_k}}^f \circ \theta_{t_{k-1}}; \; N \geq k \right) \left(1 + \epsilon(m) \right)$$

$$= E_x \left(\sum_{k=1}^{N} S_{\sigma_{\Delta_k}}^f \circ \theta_{t_{k-1}} \right) \left(1 + \epsilon(m) \right)$$

$$= E_x \left(S_{\sigma_G}^f \right) \left(1 + \epsilon(m) \right),$$

where $\epsilon(m) \to 0$. Therefore

$$K_m^{ij}(M, \sigma_G) = E_x \left(S_{\sigma_G}^f \right) = E_x \left(\int_0^{\sigma_G} \left(a^{ij}(X_t)/\gamma(X_t) \right) dt \right).$$

From the method of construction it follows that this formula is true after change σ_G by an arbitrary finite time $\tau \in \mathcal{T}^0$,

$$K^{ij}(M, \tau) \equiv E_x \left(M_\tau^i M_\tau^j \right) = E_x \int_0^\tau \left(\frac{a^{ij}(X_t)}{\gamma(X_t)} \right) dt. \qquad [6.17]$$

Positive definiteness of matrix $K^{ij}(M, \tau)$ follows from positive definiteness of matrix (a^{ij}). We now prove that (m_τ^{ij}) is a martingale along the trace, where

$$m_\tau^{ij} = M_\tau^i M_\tau^j - \mu_\tau^f.$$

We obtain

$$E_x \left(m_{\tau_1 + \tau_2}^{ij} \mid \mathcal{F}_{\tau_1} \right) = m_{\tau_1}^{ij} + E_x \left(M_{\tau_1}^i \left(M_{\tau_2}^j \circ \theta_{\tau_1} \right) + \left(M_{\tau_2}^i \circ \theta_{\tau_1} \right) M_{\tau_1}^j \right.$$

$$\left. + \left(M_{\tau_2}^i M_{\tau_2}^j \right) \circ \theta_{\tau_1} - \mu_{\tau_2}^f \circ \theta_{\tau_1} \mid \mathcal{F}_{\tau_1} \right).$$

The second member of this sum is equal to zero due to the semi-Markov property of the process, and because of $E_x(M_\tau^i) = 0$ and $E_x(M_\tau^i M_\tau^j - \mu_\tau^f) = 0$ for any $x \in G$. Hence the matrix additive functional with the components

$$\mu_\tau^{a^{ij}/\gamma} \equiv \int_0^\tau \frac{a^{ij}(X_{t_1})}{\gamma(X_{t_1})} dt_1 \qquad [6.18]$$

is an analogy of the quadratic characteristic of a one-dimensional quadratically integrable martingale.

6.10.2. *Stochastic integral with respect to a martingale*

Construction of integral

We consider a vector martingale with a standard denotation of the vector as a column. Then a^* (row) is the transposed vector a (column), and a^*b is the scalar product of two vectors of the same dimension d (in coordinate denotation: $a^*b = \sum_{i=1}^{d} a^i b^i$). Furthermore, we will consider integrals of the scalar product of a vector-valued integrand function by increments of a vector-valued integrating function. In order to define stochastic integral with respect to a general semi-martingale we consider firstly a stepped function such that there exists a sequence of Markov times $(\tau_k)_1^N$, where N is an integer-valued function with $\{N = k\} \in \mathcal{F}_{\tau_k}$, $\tau_k \in T^0$, $\tau_0 = 0 \leq \tau_1 \leq \cdots$, $\tau_N = \sigma_G$. In particular these times can be taken from the family $(\sigma_{\delta_n}^k)$. We assume that for any $t \in [\tau_{k-1} < \tau_k)$ $f(t) = f(\tau_{k-1})$, and for any k $f(\tau_{k-1})$ is a $\mathcal{F}_{\tau_{k-1}}$-measurable random element with values from $\mathbb{R}^d$. Define the stochastic integral on the interval $[0, \sigma_G)$ of function f by semi-martingale (M_t) as the sum

$$\int_0^{\sigma_G} \left(f(\tau) \right)^* dM_\tau = \sum_{k=1}^{N} \left(f(\tau_{k-1}) \right)^* \left(M_{\tau_k} - M_{\tau_{k-1}} \right).$$

Denote by $\mathcal{L}_0$ the class of such functions f. Note that for any two functions of this class there exists a sequence of Markov times with properties enumerated above, such that both functions have constant meaning on any interval between neighboring Markov times from this sequence. From here it follows that $\mathcal{L}_0$ is a linear space. In order to extend the definition of the stochastic integral on a wider class of functions we consider (according to the standard method of extension of stochastic integrals [GIH 75, p. 72]) Hilbert space $\mathcal{L}_2$, determined by the scalar product

$$(f, g) \equiv E_x \left(\int_0^{\sigma_G} \sum_{ij} f^i(t) g^j(t) \left(\frac{a^{ij}(X_t)}{\gamma(X_t)} \right) dt \right), \qquad [6.19]$$

and by the completion of space $\mathcal{L}_0$ with respect to a metric, corresponding to the norm $\|f\| = (f, f)^{1/2}$.

In this case for $f \in \mathcal{L}_0$ it is true:

$$\|f\|^2 = E_x \left(\int_0^{\sigma_G} \left(f(\tau) \right)^* dM_\tau \right)^2. \qquad [6.20]$$

Actually,

$$
E_x \left(\int_0^{\sigma_G} (f(\tau))^* \, dM_\tau \right)^2
$$

$$
= \sum_{ij} E_x \left(\int_0^{\sigma_G} f^i(\tau) \, dM_\tau^i \right) \left(\int_0^{\sigma_G} f^j(\tau) \, dM_\tau^j \right)
$$

$$
= \sum_{ij} \sum_{n=1}^\infty E_x \left(\left(\sum_{k=1}^n f^i(\tau_{k-1}) \Delta M_k^i \right) \left(\sum_{l=1}^n f^j(\tau_{l-1}) \Delta M_l^j \right); \ N = n \right)
$$

$$
= \sum_{ij} \sum_{n=1}^\infty E_x \left(\sum_{k=1}^n f^i(\tau_{k-1}) f^j(\tau_{k-1}) \Delta M_k^i \Delta M_k^j; \ N = n \right)
$$

$$
= \sum_{ij} \sum_{k=1}^\infty E_x \left(f^i(\tau_{k-1}) f^j(\tau_{k-1}) \left(M_{\tau_k'}^i M_{\tau_k'}^j \right) \circ \theta_{\tau_{k-1}}; \ N \geq k \right),
$$

where $\tau_k' \in T^0$ is an increment of the Markov time such that $\tau_k = \tau_{k-1} + \tau_k'$. According-ing to semi-Markov property the latter expression is equal to

$$
= \sum_{ij} \sum_{k=1}^\infty E_x \left(f^i(\tau_{k-1}) f^j(\tau_{k-1}) \mu_{\tau_k'}^{a^{ij}/\gamma} \circ \theta_{\tau_{k-1}}; \ N \geq k \right)
$$

$$
= \sum_{ij} E_x \sum_{n=1}^\infty \left(\sum_{k=1}^n f^i(\tau_{k-1}) f^j(\tau_{k-1}) S_{\tau_k'}^{a^{ij}/\gamma} \circ \theta_{\tau_{k-1}}; \ N = n \right)
$$

$$
= \sum_{ij} \sum_{n=1}^\infty E_x \left(\int_0^{\sigma_G} f^i(t) f^j(t) \frac{a^{ij}(X_t)}{\gamma(X_t)} \, dt; \ N = n \right)
$$

$$
= E_x \left(\int_0^{\sigma_G} \sum_{ij} f^i(t) f^j(t) a^{ij}(X_t) \frac{1}{\gamma(X_t)} \, dt \right).
$$

Consider an extension of the stochastic integral on space $\mathcal{L}_2$, consisting of limits of all sequences with elements from class $\mathcal{L}_0$, mutually converging (see [LOE 62]) with respect to scalar product [6.19]. This extension is said to be a stochastic integral (in the sense of mean square convergence) with respect to martingale (M_τ). In this case for any function $f \in \mathcal{L}_2$ formula [6.20] is true.

78. Stochastic integral with respect to a semi-Markov process

A stochastic integral with respect to a semi-Markov process of diffusion type, which can be called stochastic integral along the trace of this process, can be naturally defined as a sum of the stochastic integral with respect to the martingale plus the curvilinear integral with respect to the additive functional of bounded variation:

$$\int_0^{\sigma_G} \left(f(\tau)\right)^* dX_\tau = \int_0^{\sigma_G} \left(f(\tau)\right)^* dM_\tau$$

$$+ \int_0^{\sigma_G} \left(f(\tau)\right)^* b(X_\tau) \frac{1}{\gamma(X_t)} d\mu_\tau.$$

[6.21]

6.10.3. Ito-Dynkin's formula

79. Method of inscribed balls

In order to derive Ito's formula we will use the method of inscribed balls. Its application is connected with asymptotic formula [6.11]. For any point $x \in G$ we consider a sequence of inscribed balls $B_u(x) = x + (r(x,G) \wedge u)B$ $(u > 0)$. Let $t_k = \sigma_{G,u}^k$ and $x_k = X_{\sigma_{G,u}^k}$. We denote

$$J_u \equiv \sum_{k=1}^{\infty} \varphi(x_{k-1}) \Delta x_k^i \Delta x_k^j,$$

where $\Delta x_k^i = x_k^i - x_{k-1}^i$, and φ is a continuous function. We have

$$E_x(J_u) = E_x\left(\sum_{k=1}^{\infty} \varphi(x_{k-1}) \Delta x_k^i \Delta x_k^j\right)$$

$$= E_x\left(\sum_{k=1}^{\infty} \varphi(x_{k-1}) E_x\left(\Delta X_k^i \Delta X_k^j \mid \mathcal{F}_{t_{k-1}}\right)\right)$$

$$= E_x\left(\sum_{k=1}^{\infty} \varphi(x_{k-1}) E_{x_{k-1}}\left((X_{\sigma_{G,u}}^i - X_0^i)(X_{\sigma_{G,u}}^j - X_0^j)\right)\right)$$

$$= E_x\left(\sum_{k=1}^{\infty} E_{x_{k-1}}\left(S_{\sigma_{G,u}}^{\varphi a^{ij}/\gamma}\right)(1 + \epsilon(k,u))\right)$$

$$= E_x\left(\sum_{k=1}^{\infty} E_{x_{k-1}}\left(S_{\sigma_{G,u}}^{\varphi a^{ij}/\gamma}\right)\right)(1 + \epsilon(u))$$

$$= E_x\left(S_{\sigma_G}^{\varphi a^{ij}/\gamma}\right)(1 + \epsilon(u)) = E_x\left(\mu_{\sigma_G}^{\varphi a^{ij}/\gamma}\right)(1 + \epsilon(u)).$$

From here

$$E_x(J_u) \longrightarrow E_x\left(\mu_{\sigma_G}^{\varphi a^{ij}/\gamma}\right) \quad (u \longrightarrow 0).$$

LEMMA 6.1. *Under the assumptions made above, J_u tends in probability to the curvilinear integral*

$$J \equiv \int_0^{\sigma_G} \frac{\varphi(X_\tau) a^{ij}(X_\tau)}{\gamma(X_\tau)} \, d\mu_\tau.$$

Proof. Denote $\Delta\mu_{t_k}^{\varphi a^{ij}/\gamma} = \mu_{t_k}^{\varphi a^{ij}/\gamma} - \mu_{t_{k-1}}^{\varphi a^{ij}/\gamma}$. We have

$$E_x\left(J_u - J\right)^2 = E_x\left(\sum_{k=1}^{\infty}\left(\varphi(x_{k-1})\Delta x_k^i \Delta x_k^j - \Delta\mu_{t_k}^{\varphi a^{ij}/\gamma}\right)\right)^2$$

$$= E_x\left(\sum_{k=1}^{\infty}\left(\varphi(x_{k-1})\Delta x_k^i \Delta x_k^j - \Delta\mu_{t_k}^{\varphi a^{ij}/\gamma}\right)^2\right.$$

$$+ 2\sum_{k<l}\left(\varphi(x_{k-1})\Delta x_k^i \Delta x_k^j - \Delta\mu_{t_k}^{\varphi a^{ij}/\gamma}\right)$$

$$\left. \times \left(\varphi(x_{l-1})\Delta x_l^i \Delta x_l^j - \Delta\mu_{t_l}^{\varphi a^{ij}/\gamma}\right)\right).$$

The second member of this sum is equal to

$$E_x\left(2\sum_{k=1}^{\infty}\sum_{l=k+1}^{\infty}\left(\varphi(x_{k-1})\Delta x_k^i \Delta x_k^j - \Delta\mu_{t_k}^{\varphi a^{ij}/\gamma}\right)\right.$$

$$\left. \times \left(\varphi(x_{l-1})\Delta x_l^i \Delta x_l^j - \Delta\mu_{t_l}^{\varphi a^{ij}/\gamma}\right)\right).$$

Using the semi-Markov property we obtain

$$= 2E_x\left(\sum_{k=1}^{\infty}\sum_{l=k+1}^{\infty}\left(\varphi(x_{k-1})\Delta x_k^i \Delta x_k^j - \Delta\mu_{t_k}^{\varphi a^{ij}/\gamma}\right)\right.$$

$$\left. \times E_x\left(\varphi(x_{l-1})\Delta x_l^i \Delta x_l^j - \Delta\mu_{t_l}^{\varphi a^{ij}/\gamma} \mid \mathcal{F}_{t_{l-1}}\right)\right)$$

$$= 2E_x\left(\sum_{k=1}^{\infty}\sum_{l=k+1}^{\infty}\left(\varphi(x_{k-1})\Delta x_k^i \Delta x_k^j - \Delta\mu_{t_k}^{\varphi a^{ij}/\gamma}\right)\right.$$

$$\left. \times E_{x_{l-1}}\left(\varphi(X_0)\left(X_{\sigma_{G,u}}^i - X_0^i\right)\left(X_{\sigma_{G,u}}^j - X_0^j\right) - S_{\sigma_{G,u}}^{\varphi a^{ij}/\gamma}\right)\right)$$

$$= 2E_x \left(\sum_{k=1}^{\infty} \sum_{l=k+1}^{\infty} \left(\varphi(x_{k-1}) \Delta x_k^i \Delta x_k^j - \Delta \mu_{t_k}^{\varphi a^{ij}/\gamma} \right) \right.$$

$$\left. \times E_{x_{l-1}} \left(\varphi(X_0) (X_{\sigma_{G,u}}^i - X_0^i)(X_{\sigma_{G,u}}^j - X_0^j) - \mu_{\sigma_{G,u}}^{\varphi a^{ij}/\gamma} \right) \right)$$

$$= 2E_x \left(\sum_{k=1}^{\infty} \sum_{l=k+1}^{\infty} \left(\varphi(x_{k-1}) \Delta x_k^i \Delta x_k^j - \Delta \mu_{t_k}^{\varphi a^{ij}/\gamma} \right) E_{x_{l-1}} \left(\epsilon(u,l) \mu_{\sigma_{G,u}}^{\varphi a^{ij}/\gamma} \right) \right)$$

$$= 2\epsilon(u) E_x \left(\sum_{k=1}^{\infty} \left(\varphi(x_{k-1}) \Delta x_k^i \Delta x_k^j - \Delta \mu_{t_k}^{\varphi a^{ij}/\gamma} \right) E_{x_{l-1}} \left(\mu_{\sigma_G}^{\varphi a^{ij}/\gamma} \right) \right)$$

$$= 2\epsilon(u) E_x \left(\sum_{k=1}^{\infty} \left(\varphi(x_{k-1}) \Delta x_k^i \Delta x_k^j - \Delta \mu_{t_k}^{\varphi a^{ij}/\gamma} \right) E_{x_{l-1}} \left(S_{\sigma_G}^{\varphi a^{ij}/\gamma} \right) \right)$$

$$\leq 2\epsilon(u) \sup_{x \in G} E_x \left(S_{\sigma_G}^{\varphi a^{ij}/\gamma} \right) E_x \left(\sum_{k=1}^{\infty} \left(|\varphi(x_{k-1})| \cdot |\Delta x_k^i| \cdot |\Delta x_k^j| + \Delta \mu_{t_k}^{|\varphi| \cdot |a^{ij}|/\gamma} \right) \right)$$

We have

$$E_x \left(\sum_{k=1}^{\infty} \left(\Delta \mu_{t_k}^{|\varphi| \cdot |a^{ij}|/\gamma} \right) \right) = E_x \left(\mu_{\sigma_G}^{|\varphi| \cdot |a^{ij}|/\gamma} \right) = E_x \left(S_{\sigma_G}^{|\varphi| \cdot |a^{ij}|/\gamma} \right)$$

bounded uniformly. On the other hand

$$E_x \left(\sum_{k=1}^{\infty} \left(|\varphi(x_{k-1})| \cdot |\Delta x_k^i| \cdot |\Delta x_k^j| \right) \right)$$

$$\leq C \sup_{x \in G} |\varphi(x)| E_x \left(\sum_{k=1}^{\infty} \left(r(x_{k-1}, G) \wedge u \right)^2 \right),$$

where C is the upper bound of the ratio of the maximal axis of the ellipsoid to the minimal axis on set G. The sum in the latter expression is equal to

$$E_x \left(\sum_{k=1}^{\infty} \left(r(x_{k-1}, G) \wedge u \right)^2 \right) = d \cdot E_x \left(\sum_{k=1}^{\infty} E_{x_{k-1}} \left(S_{\sigma_{G,u}}^{1/\gamma} \right) (1 + \epsilon(u,k)) \right)$$

$$= d \cdot E_x \left(\sum_{k=1}^{\infty} E_{x_{k-1}} \left(S_{\sigma_{G,u}}^{1/\gamma} \right) \right) (1 + \epsilon(u))$$

$$= d \cdot E_x \left(S_{\sigma_{G,u}}^{1/\gamma} \right) (1 + \epsilon(u)) < \infty$$

uniformly in $x \in G$. So the second member of the sum, representing $E_x(J_u - J)^2$, tends to zero as $u \to 0$. Consider the first member. We have

$$= E_x \left(\sum_{k=1}^{\infty} \left(\varphi(x_{k-1}) \Delta x_k^i \Delta x_k^j - \Delta \mu_{t_k}^{\varphi a^{ij}/\gamma} \right)^2 \right)$$

$$= E_x \left(\sum_{k=1}^{\infty} \left(\varphi^2(x_{k-1}) \Delta^2 x_k^i \Delta^2 x_k^j - 2\Delta \mu_{t_k}^{\varphi a^{ij}/\gamma} \varphi(x_{k-1}) \Delta x_k^i \Delta x_k^j + \Delta^2 \mu_{t_k}^{\varphi a^{ij}/\gamma} \right) \right).$$

The sum of members, containing $\Delta^2 x_k^i \Delta^2 x_k^j$, tends to zero, because ratio $E_x(\Delta^2 x_k^i \times \Delta^2 x_k^j)/E_x(\sigma_{G,u})$ tends to zero due to property $A_0(g \mid x) = 0$, where $g(y) = (y^i - x^i)^2 (y^j - x^j)^2$. The sum of members containing $2\Delta \mu_{t_k}^{\varphi a^{ij}/\gamma} \varphi(x_{k-1}) \Delta x_k^i \Delta x_k^j$ tends to zero due to estimate $|\Delta x_k^i \Delta x_k^j| \leq u^2$. The sum of members containing $(\Delta \mu_{t_k}^{\varphi a^{ij}/\gamma})^2$ tends to zero due to the property of semi-Markov process $E_x(\mu_{G_r}^2)/E_x(\mu_{G_r}) \to 0$ $(G_r \downarrow x)$. The lemma is proved.   □

80. Notes

A variant of Ito's formula for the stochastic integral with respect to a semi-Markov process can be derived in a standard way from the lemma proved above. Namely, for a twice differentiable function φ, the following representation is true

$$\varphi(X_{\sigma_G}) = \varphi(X_0) + \int_0^{\sigma_G} \nabla \varphi(X_\tau)^* \, dX_\tau$$

$$+ \frac{1}{2} \sum_{ij} \int_0^{\sigma_G} D_{ij} \varphi(X_\tau) a^{ij}(X_\tau)/\gamma(X_\tau) \, d\mu_\tau.$$

[6.22]

In this representation the first integral is a stochastic integral with respect to the semi-Markov process, and the second one is a curvilinear integral with respect to the additive functional.

In this book the case of a finite $\mathcal{F}^0$-measurable additive functional μ_τ is considered. The rather similar method is applicable to the case of infiniteness of this functional.

Chapter 7

Limit Theorems for Semi-Markov Processes

A distribution of any semi-Markov process induces a probability measure of a stepped semi-Markov process of the first exits, which is appropriate to a sequence of iterated times of the first exit from balls of a radius $r > 0$ (see item 2.18). If $r \to 0$, this stepped semi-Markov process weakly converges to an initial semi-Markov process. It is natural to consider conditions under which an *a priori* given sequence of stepped semi-Markov processes weakly converges to a general semi-Markov process. These conditions are convenient for expressing in terms of a time of the first exit. The language of time of the first exit has appeared well adapted for the study of weak compactness and weak convergence of probability measures in space $\mathcal{D}$, not only for semi-Markov processes (see [ALD 78a, ALD 78b, GUT 75]). For semi-Markov processes, a simple sufficient condition of weak convergence of distributions is derived in terms of semi-Markov transition functions. The task of the analysis of weak convergence of distributions of semi-Markov walks can be divided into two parts: at first to consider convergence of distributions of inserted Markov walks, and then to make time change transformation of an obtained limit appropriate to a sequence of distributions of intervals between jumps. In this way we come to an outcome which was proved with a few other methods by Silvestrov [SIL 74]. As has been shown in [HAR 79], the limiting theorems for semi-Markov processes can be applied for a construction of a Markov process, which is not broken at a finite instant, with the given distribution of points of the first exit from open sets (see [BLU 62, KNI 64, SHI 71a, SHI 71b]).

7.1. Weak compactness and weak convergence of probability measures

In terms of time of the first exit from open sets the conditions of weak compactness and weak convergence of probability measures in space $\mathcal{D}$ are derived

[BIL 77,   GIH 71,   GIH 75,   SKO 56,   SKO 58,   HAR 76b,   HAR 77,   ALD 78a,
ALD 78b, GUT 75].

1. Definitions and denotations

Let $N_t(\xi)$ be the number of points of discontinuity of function $\xi \in \mathcal{D}$ on an interval $(0, t]$ $(t > 0)$; $N_t(\xi) \in \mathbb{N}_0 \cup \{\infty\}$. Define functionals

$$h_r^t(\xi) = \min\left\{\sigma_r^{k+1}(\xi) - \sigma_r^k(\xi) : k \in \{0, 1, \ldots, N_t(L_r\xi)\}\right\},$$

where $\xi \in \mathcal{D}$; $r, t > 0$, and for $\delta \in DS$

$$h_\delta^t(\xi) = \min\left\{\sigma_\delta^{k+1}(\xi) - \sigma_\delta^k(\xi) : k \in \{0, 1, \ldots, N_t(L_\delta\xi)\}, \sigma_\delta^{k+1} - \sigma_\delta^k > 0\right\}.$$

(the minimum is sought among positive differences). We will use functionals h_r^t and h_δ^t for a statement of two variants of the limiting theorems in terms of time of the first exit.

For any $c, t > 0$ and $\xi \in \mathcal{D}$, let

$$\Delta_c^t(\xi) = \sup\left\{\left(\rho\big(\xi(t_1), \xi(t_2)\big) \wedge \rho\big(\xi(t_2), \xi(t_3)\big)\right) \vee \rho\big(\xi(0), \xi(t_4)\big) : \right.$$
$$\left. 0 < t_2 \le t, \, 0 \vee (t_2 - c) \le t_1 \le t_2 \le t_3 \le t_2 + c, \, 0 \le t_4 \le c\right\}.$$

Let $(\mathbb{X}_t)_{t \ge 0}$ be a non-decreasing continuous function from the right set of compact sets $\mathbb{X}_t \subset \mathbb{X}$, λ_c^t be a continuous non-decreasing function on arguments c and t $(c, t > 0)$ for which $(\forall t > 0)$ $\lambda_{0+}^t = 0$. Define a set:

$$K\big(\mathbb{X}_t, \lambda_c^t\big) = \left\{\xi \in \mathcal{D} : (\forall c, t > 0) \, \xi(t) \in \mathbb{X}_t, \, \Delta_c^t(\xi) \le \lambda_c^t\right\}.$$

2. Compact sets in $\mathcal{D}$

PROPOSITION 7.1. *The following properties are fair*

(1) set $K(\mathbb{X}_t, \lambda_c^t)$ is a compact set in $(\mathcal{D}, \rho_\mathcal{D})$;

(2) for any compact set $K \subset \mathcal{D}$ there exists a set $(\mathbb{X}_t)_{t \ge 0}$ and function λ_c^t $(c, t > 0)$ with the listed above properties such that $K \subset K(\mathbb{X}_t, \lambda_c^t)$.

Proof. The proof of this outcome is similar to the proof of theorem 1 in [GIH 71, p. 502] about a form of a compact set in space $\mathcal{D}[0, 1]$ with modifications connected with a replacement of the Skorokhod metric by the Stoun-Skorokhod metric.   □

3. Existence of function lambda

PROPOSITION 7.2. *Let $(P^{(n)})$ be a family of probability measures on $(\mathcal{D}, \mathcal{F})$; also let sequences $(a_k)_1^\infty$ $(a_k \downarrow 0)$ and $(t_\ell)_1^\infty$ $(t_\ell \uparrow \infty)$ exist such that for any $k, \ell \in \mathbb{N}$*

$$\limsup_{\substack{c \to 0 \\ n}} P^{(n)}\left(\Delta_c^{t_\ell} > a_k\right) = 0.$$

Then for any $\varepsilon > 0$ there is a function λ_c^t with properties, enumerated in item 2, for which

$$\sup_n P^{(n)}\left(\bigcup_{c,t>0} \{\Delta_c^t > \lambda_c^t\}\right) < \varepsilon.$$

Proof. Let $c_{k,\ell} > 0$ $(k, \ell \in \mathbb{N})$, $c_{k,\ell} \downarrow 0$ with $k \to \infty$ and with $\ell \to \infty$ and

$$\sup_n P^{(n)}\left(\Delta_{c_{k,\ell}}^{t_{\ell+1}} > a_k\right) < \varepsilon\, 2^{-k-\ell}$$

Sequence $(c_{k,\ell})$ obviously exists. Assume $\lambda_{c_{k,\ell}}^{t_\ell} = a_{k-1}$; we will also continue this function in a continuous non-decreasing manner up to function λ_c^t $(c, t > 0)$. Such a function obviously exists, and $\lambda_{0+}^t = 0$ with any $t > 0$. Now for any n:

$$P^{(n)}\left(\bigcup_{c,t>0} \{\Delta_c^t > \lambda_c^t\}\right)$$

$$= P^{(n)}\left(\bigcup_{k,\ell} \bigcup_{c \in [c_{k+1,\ell}, c_{k,\ell})} \bigcup_{t \in [t_\ell, t_{\ell+1})} \{\Delta_c^t > \lambda_c^t\}\right)$$

$$\leq P^{(n)}\left(\bigcup_{k,\ell} \{\Delta_{c_{k,\ell}}^{t_{\ell+1}} > \lambda_{c_{k+1,\ell}}^{t_\ell}\}\right) \leq \sum_{k,\ell} P^{(n)}\left(\Delta_{c_{k,\ell}}^{t_{\ell+1}} > a_k\right) < \varepsilon. \qquad \square$$

4. Comparison of functionals

PROPOSITION 7.3. *For any $c, t, r > 0$, $\delta \in \mathrm{DS}(r)$ and $\xi \in \mathcal{D}$ it is fair*

(a) $h_r^t(\xi) \geq 2c \Rightarrow \Delta_c^t(\xi) \leq 2r$,

(b) $h_\delta^t(\xi) \geq 2c \Rightarrow \Delta_c^t(\xi) \leq r$.

Proof. Let $h_r^t(\xi) \geq 2c$ and $t_2 \in (0, t]$. Then there exists $k \leq N_t L_r \xi$ such that $t_2 \in [\sigma_r^k(\xi), \sigma_r^{k+1}(\xi))$, where $\sigma_r^{k+1}(\xi) - \sigma_r^k(\xi) \geq 2c$. From here either $[t_2, t_2 + c] \subset [\sigma_r^k(\xi), \sigma_r^{k+1}(\xi))$ or $[t_2 - c, t_2] \subset [\sigma_r^k(\xi), \sigma_r^{k+1}(\xi))$. Hence, $\rho(\xi(t_2), \xi(t_1)) \wedge \rho(\xi(t_2), \xi(t_3)) < 2r$, where $0 \wedge (t_2 - c) \leq t_1 \leq t_2 \leq t_3 \leq t_2 + c$. Obviously, $\rho(\xi(t_4), \xi(0)) < r$, where $0 \leq t_4 \leq c$. From here $\Delta_c^t(\xi) \leq 2r$. Property (b) is similarly proved. $\qquad \square$

5. Weak compactness

Denote

$$\nu_{r,k}^{(n)}(S) = P^{(n)}\left(X_{\sigma_r^k} \in S, \; \sigma_r^k < \infty\right) \quad (S \in \mathcal{B}(\mathbb{X})).$$

PROPOSITION 7.4. *Let* $(P^{(n)})$ *be a set of probability measures on* $(\mathcal{D}, \mathcal{F})$; *and* $(\exists (r_m), r_m \downarrow 0)$ $(\exists (t_\ell), t_\ell \uparrow \infty)$ *the following conditions be fulfilled:*

(1) $(\forall \ell, m \in \mathbb{N}) \lim_{c \to 0} \sup_n P^{(n)}(h_{r_m}^{t_\ell} < c) = 0$;

(2) $(\forall m \in \mathbb{N})$ $(\forall k \in \mathbb{N}_0) \lim_{K \uparrow \mathbb{X}} \sup_n \nu_{r_m,k}^{(n)}(\mathbb{X} \setminus K) = 0$, *where K is a compact set, i.e. a set of subprobability measures* $(\nu_{r,k}^{(n)})$ *is weakly compact.*

Then the set of measures $(P^{(n)})$ *is weakly compact.*

Proof. From propositions 7.2, 7.3 and condition (1) it follows that $(\forall \varepsilon > 0)$ there is a function λ_c^t with properties enumerated in item 2, for which $\sup_n P^{(n)}(\bigcup_{c,t>0}\{\Delta_c^t > \lambda_c^t\}) < \varepsilon$. Let $r_m \downarrow 0$. From condition (1) it follows that $(\forall \varepsilon > 0)$ $(\forall \ell, m \in \mathbb{N})$ $(\exists N_m \in \mathbb{N}_0)$

$$\sup_n P^{(n)}\left(N_{t_\ell} L_{r_m} \geq N_m\right) < \varepsilon \, 2^{-m-1},$$

(where $\mathbb{N}_0 = \mathbb{N} \cup \{0\} = \{0, 1, 2, \ldots\}$), since $\{h_r^t \geq c\} \subset \{N_t L_r \leq t/c\}$. From condition (2) it follows that $(\forall m \in \mathbb{N})$ $(\exists K_m - \text{compact set})$ $(\forall k : 0 \leq k \leq N_m)$

$$\sup_n P^{(n)}\left(X_{\sigma_{r_m}^k} \notin K_m, \; \sigma_{r_m}^k < \infty\right) < \frac{\varepsilon \, 2^{-m-1}}{N_m + 1}.$$

Let $\Delta_m = \{x : \rho(x, K_m) \leq r_m\}$. Then

$$(\forall n) P^{(n)}\left(\sigma_{\Delta_m} > t_\ell\right)$$

$$\geq P^{(n)}\left(N_{t_\ell} L_{r_m} < N_m, \; \bigcap_{k=0}^{N_m} \{X_{\sigma_{r_m}^k} \in K, \; \sigma_{r_m}^k < \infty\} \cup \{\sigma_{r_m}^k = \infty\}\right)$$

$$> 1 - \varepsilon \, 2^{-m}.$$

Let $\Delta = \bigcap_{m=1}^{\infty} \Delta_m$. It is not difficult to show that each sequence $(x_m)_1^{\infty}$, belonging to Δ, contains a sub-sequence mutually converging. Since, in addition, $\mathbb{X}$ is complete and Δ is closed, Δ is a compact set (see also [GIH 71, p. 484]). On the other hand $(\forall n \geq 1)$

$$P^{(n)}\left(\sigma_\Delta \leq t_\ell\right) = P^{(n)}\left(\inf_m \sigma_{\Delta_m} \leq t_\ell\right) \leq \sum_{\ell=1}^{\infty} P^{(n)}\left(\sigma_{\Delta_m} \leq t_\ell\right) < \varepsilon.$$

Hence, $(\forall \varepsilon > 0)\ (\forall \ell \geq 1)\ (\exists K_\ell - \text{compact set})$

$$\sup_n P^{(n)}\left(\sigma_{K_\ell} \leq \ell\right) < \varepsilon\, 2^{-\ell}, \quad K_\ell \subset K_{\ell+1}.$$

We assume that $\mathbb{X}_t = K_\ell$ with $\ell - 1 \leq t < \ell$. Then $(\mathbb{X}_t)_{t \geq 0}$ is non-decreasingly continuous from the right family of compact sets. In this case

$$P^{(n)}\left(\bigcup_{t \geq 0} \{X_t \notin \mathbb{X}_t\}\right) \leq \sum_{\ell=1}^{\infty} P^{(n)}\left(\bigcup_{\ell-1 \leq t < \ell} \{X_t \notin K_\ell\}\right)$$

$$\leq \sum_{\ell=1}^{\infty} P^{(n)}\left(\sigma_{K_\ell} < \ell\right) < \varepsilon.$$

Hence, under theorem 7.1 and sufficient condition of weak compactness (see [GIH 71, p. 429]) the set of measures $(P^{(n)})$ is weakly compact. $\qquad\square$

6. Weak compactness (the second variant)

Denote

$$\nu_{\delta,k}^{(n)} = P^{(n)}\left(X_{\sigma_\delta^k} \in S,\ \sigma_\delta^k < \infty\right) \quad \left(S \in \mathcal{B}(\mathbb{X})\right).$$

PROPOSITION 7.5. *Let* $(P^{(n)})$ *be a family of probability measures on* $(\mathcal{D}, \mathcal{F})$; *and also* $(\exists(t_\ell),\ t_\ell \uparrow \infty)\ (\exists(\delta_m) \in \mathrm{DS}(r_m),\ r_m \downarrow 0)$ *the following conditions are fulfilled:*

(1) $(\forall \ell, m \in \mathbb{N})\ \lim_{c \to 0} \sup P^{(n)}(h_{\delta_m}^{t_\ell} < c) = 0$;

(2) *at least one of two conditions is fulfilled:*

    (a) $(\forall k, m \in \mathbb{N})$ *family of subprobability measures* $(\nu_{\delta_m,k}^{(n)})$ *is weakly compact;*

    (b) $(\exists \delta \in \mathrm{DS},\ \delta$ *consists of precompact sets*$)\ (\forall \ell \in \mathbb{N})$

$$\lim_{k \to \infty} \sup_n P^{(n)}\left(\sigma_\delta^k \leq t_\ell\right) = 0.$$

Then the family of measures $(P^{(n)})$ *is weakly compact.*

Proof. (2)(a) We will show that from conditions (1) and (2)(a) it follows that $(\forall \ell, m \geq 1)$

$$\lim_{k \to \infty} \sup_n P^{(n)}\left(\sigma_{\delta_m}^k \leq t_\ell\right) = 0. \tag{7.1}$$

For this we use two obvious properties of deducing sequences:

(C_1) $(\forall \delta \in DS)$ $(\forall K - \text{compact set})$ $(\exists m \geq 1)$ $(\forall x \in K)$ $x \in \bigcup_{i=1}^{m} \Delta_i$, $\delta = (\Delta_1, \Delta_2, \ldots)$;

(C_2) $(\forall \delta \in DS, \Delta_i \neq \mathbb{X})$ $(\forall m \geq 1)$ $\delta^{(m)} \in DS$, where $\delta = (\Delta_1, \Delta_2, \ldots)$ and $\delta^{(m)} = (\Delta_m, \Delta_{m+1}, \ldots)$.

Let K_0 be a compact set. By a property C_1 $(\exists k_1 \geq 1)$

$$\{h_{\delta_m}^{t_\ell} \geq c, X_0 \in K_0\} = \{X_0 \in K_0, h_{\delta_m}^{t_\ell} \geq c, \sigma_{\delta_m}^{k_1} \geq c\}.$$

By properties C_1 and C_2, if K_1 is a compact set, then $(\exists k_2 \geq k_1)$

$$\left\{h_{\delta_m}^{t_\ell} \geq c, X_0 \in K_0, X_{\sigma_{\delta_m}^{k_1}} \in K_1, \sigma_{\delta_m}^{k_1} \leq t_\ell\right\}$$
$$= \left\{h_{\delta_m}^{t_\ell} \geq c, X_0 \in K_0, X_{\sigma_{\delta_m}^{k_1}} \in K_1, \sigma_{\delta_m}^{k_1} \leq t_\ell, \sigma_{\delta_m}^{k_2} \geq 2c\right\},$$

etc.: for any sequence of compact sets $K_0, K_1, \ldots, K_s$ there is a sequence of numbers $k_1, \ldots, k_{s+1}$ $(k_i < k_{i+1})$, for which

$$\{h_{\delta_m}^{t_\ell} \geq c, X_0 \in K_0\} \bigcap_{i=1}^{s} \left\{X_{\sigma_{\delta_m}^{k_i}} \in K_i\right\} \{\sigma_{\delta_m}^{k_s} \leq t_\ell\}$$
$$= \{h_{\delta_m}^{t_\ell} \geq c, X_0 \in K_0\} \bigcap_{i=1}^{s} \left\{X_{\sigma_{\delta_m}^{k_i}} \in K_i\right\} \{\sigma_{\delta_m}^{k_s} \leq t_\ell\} \{\sigma_{\delta_m}^{k_{s+1}} \geq (s+1)c\}.$$

Let $(\forall n)$

$$P^{(n)}\left(h_{\delta_m}^{t_\ell} < c\right) < \frac{\varepsilon}{s+2},$$

where $s = [t/c]$ (whole part of a ratio), $P^{(n)}(X_0 \notin K_0) < \varepsilon/(s+2)$ and

$$P^{(n)}\left(X_{\sigma_{\delta_m}^{k_i}} \notin K_i, \sigma_{\delta_m}^{k_i} < \infty\right) < \frac{\varepsilon}{s+2} : \quad (i = 1, \ldots, s),$$

where each k_i depends on chosen K_{i-1}. Then

$$P^{(n)}\left(\sigma_{\delta_m}^{k_{s+1}} > t\right)$$
$$\geq P^{(n)}\left(h_{\delta_m}^{t_\ell} \geq c, X_0 \in K_0, \bigcap_{i=1}^{s} \left\{X_{\sigma_{\delta_m}^{k_i}} \in K_i, \sigma_{\delta_m}^{k_i} < \infty\right\} \cup \{\sigma_{\delta_m}^{k_i} = \infty\}\right)$$
$$\geq 1 - P^{(n)}\left(h_{\delta_m}^{t_\ell} < c\right) - P^{(n)}(X_0 \notin K_0)$$
$$- \sum_{i=1}^{s} P^{(n)}\left(X_{\sigma_{\delta_m}^{k_i}} \notin K_i, \sigma_{\delta_m}^{k_i} < \infty\right) < 1 - \varepsilon.$$

Property [7.1] is proved. Further proof of weak compactness of a family of measures $(P^{(n)})$ is conducted on a sample of theorem 7.4.

(2)(b) If all $[\Delta_i]$ are compact, $(\forall s \geq 1)$ $K_s = \bigcup_{i=1}^{s}[\Delta_i]$ is a compact set and $\sigma_{K_s} \geq \sigma_{\cup_{i=1}^{s}\Delta_i} \geq \sigma_\delta^s$, where $\delta = (\Delta_1, \Delta_2, \ldots)$. From condition (2)(b) it follows that $(\forall t > 0)$

$$\lim_{s \to \infty} \sup_n P^{(n)}\left(\sigma_{K_s} \leq t\right) = 0.$$

From here in view of condition (1), as well as in proposition 7.4, weak compactness of a set of measures $(P^{(n)})$ follows.   □

7. Sufficient condition of weak convergence

Denote by $\xrightarrow{w}$ a weak convergence of (sub)probability measures in corresponding metric space.

THEOREM 7.1. *Let $(P^{(n)})_1^\infty$ and P be probability measures on $(\mathcal{D}, \mathcal{F})$; and also at least one of two conditions is fulfilled:*

(1) $(\exists (r_m)_1^\infty, \ r_m \downarrow 0)$ $(\forall k \in \mathbb{N}_0)$ $(\forall m \in \mathbb{N})$

$$P^{(n)} \circ \left(\beta_{\sigma_{r_m}^0}, \ldots, \beta_{\sigma_{r_m}^k}\right)^{-1} \xrightarrow{w} P \circ \left(\beta_{\sigma_{r_m}^0}, \ldots, \beta_{\sigma_{r_m}^k}\right)^{-1},$$

(2) $(\exists (\delta_m)_1^\infty, \ \delta_m \in DS(r_m), \ r_m \downarrow 0)$ $(\forall k \in \mathbb{N}_0)$ $(\forall m \in \mathbb{N})$

$$P^{(n)} \circ \left(\beta_{\sigma_{\delta_m}^0}, \ldots, \beta_{\sigma_{\delta_m}^k}\right)^{-1} \xrightarrow{w} P \circ \left(\beta_{\sigma_{\delta_m}^0}, \ldots, \beta_{\sigma_{\delta_m}^k}\right)^{-1}.$$

Then $P^{(n)} \xrightarrow{w} P$.

Proof. Since

$$\sigma\left(\beta_{\sigma_{r_m}^{k-1}}; \ k, m \in \mathbb{N}\right) = \sigma\left(\beta_{\sigma_{\delta_m}^{k-1}}; \ k, m \in \mathbb{N}\right) = \sigma(X_t, \ t \in \mathbb{R}_+) = \mathcal{B}(\mathcal{D})$$

(see proposition 2.13), by theorem 4 [GIH 71, p. 437], it is sufficient to prove weak compactness of a family of measures $(P^{(n)})$. From condition (1) it follows that for almost all $c > 0$ $(\forall \ell, m \in \mathbb{N})$

$$P^{(n)}\left(h_{r_m}^{t_\ell} < c\right) \longrightarrow P\left(h_{r_m}^{t_\ell} < c\right) \quad (n \longrightarrow \infty)$$

for some sequence $(t_\ell)_1^\infty$, where $t_\ell \uparrow \infty$. From here weak compactness of the family of distributions $(P^{(n)} \circ (h_{r_m}^{t_\ell})^{-1})_{n=1}^\infty$ follows and, hence,

$$\lim_{c \to 0} \sup_n P^{(n)}\left(h_{r_m}^{t_\ell} < c\right) = 0,$$

i.e. condition (1) of proposition 7.4 is fulfilled. Further from condition (1) it follows that $(\forall m \in \mathbb{N})\ (\forall k \in \mathbb{N}_0)\ \nu_{r_m,k}^{(n)} \xrightarrow{w} \nu_{r_m,k}$, where

$$\nu_{r,k}(S) = P\big(X_{\sigma_r^k} \in S,\ \sigma_r^k < \infty\big) \quad \big(S \in \mathcal{B}(\mathbb{X})\big)$$

($\nu_{r_m,k}^{(n)}$ is similarly determined). From here weak compactness of the family of measures $(\nu_{r_m,k}^{(n)})_{n=1}^\infty$ follows, i.e. condition (2) of proposition 7.4 is fulfilled. From proposition 7.4 weak compactness of the family of measures $(P^{(n)})_1^\infty$ follows. Similarly, in view of proposition 7.5, it is proved that from condition (2) weak compactness of a family of measures $(P^{(n)})_1^\infty$ follows. $\qquad\square$

8. Necessary condition of weak convergence

PROPOSITION 7.6. *Let $(P^{(n)})_1^\infty$ and P be probability measures on $(\mathcal{D}, \mathcal{F})$ and $P^{(n)} \xrightarrow{w} P$. Then there exist $(r_m)\ (r_m \downarrow 0)$ and $(\delta_m)\ (\delta_m \in \mathrm{DS}(r_m),\ r_m \downarrow 0)$ such that $(\forall k, m \in \mathbb{N})$ with $n \to \infty$*

(1) $P^{(n)} \circ (\beta_{\sigma_{r_m}^0}, \dots, \beta_{\sigma_{r_m}^k})^{-1} \xrightarrow{w} P \circ (\beta_{\sigma_{r_m}^0}, \dots, \beta_{\sigma_{r_m}^k})^{-1}$,

(2) $P^{(n)} \circ (\beta_{\sigma_{\delta_m}^0}, \dots, \beta_{\sigma_{\delta_m}^k})^{-1} \xrightarrow{w} P \circ (\beta_{\sigma_{\delta_m}^0}, \dots, \beta_{\sigma_{\delta_m}^k})^{-1}$.

Proof. By corollary 2.2 $(\exists (\delta_m)_1^\infty,\ \delta_m \in \mathrm{DS}(r_m),\ r_m \downarrow 0)\ P(\bigcap_{m=1}^\infty \Pi(\delta_m)) = 1$. Let $\varphi \in \mathcal{C}(Y^k)$ (continuous and limited function on Y^k; see item 2.9). Then by theorem 2.6 function $\varphi(\beta_{\sigma_{\delta_m}^0}, \dots, \beta_{\sigma_{\delta_m}^k})$ is continuous on $\Pi(\delta_m)$. From here, using lemma from [GIH 71, p. 437],

$$P^{(n)}\left(\varphi\left(\beta_{\sigma_{\delta_m}^0}, \dots, \beta_{\sigma_{\delta_m}^k}\right)\right) \longrightarrow P\left(\varphi\left(\beta_{\sigma_{\delta_m}^0}, \dots, \beta_{\sigma_{\delta_m}^k}\right)\right).$$

The first statement is proved similarly (see notes on items 2.33 and 2.40). $\qquad\square$

9. Weak convergence criterion

Furthermore, we assume that $e^{-\infty}(\varphi \circ X_\infty) = 0$.

THEOREM 7.2. *Let $(P^{(n)})_1^\infty$ and P be probability measures on $(\mathcal{D}, \mathcal{F})$. Then the following conditions are equivalent:*

(1) $P^{(n)} \xrightarrow{w} P$;

(2) $(\exists (\delta_m),\ \delta_m \in \mathrm{DS}(r_m),\ r_m \downarrow 0)\ (\forall k, m \in \mathbb{N})$

$$P^{(n)} \circ \left(\beta_{\sigma_{\delta_m}^0}, \dots, \beta_{\sigma_{\delta_m}^k}\right)^{-1} \xrightarrow{w} P \circ \left(\beta_{\sigma_{\delta_m}^0}, \dots, \beta_{\sigma_{\delta_m}^k}\right)^{-1};$$

(3) $(\exists (r_m)_1^\infty, \ r_m \downarrow 0) \ (\forall k, m \in \mathbb{N})$

$$P^{(n)} \circ \left(\beta_{\sigma^0_{r_m}}, \ldots, \beta_{\sigma^k_{r_m}} \right)^{-1} \xrightarrow{w} P \circ \left(\beta_{\sigma^0_{r_m}}, \ldots, \beta_{\sigma^k_{r_m}} \right)^{-1};$$

(4) $(\exists (\delta_m), \ \delta_m \in \mathrm{DS}(r_m), \ r_m \downarrow 0) \ (\forall k, m \in \mathbb{N}) \ (\forall \lambda_i > 0) \ (\forall \varphi_i \in \mathcal{C}_0)$

$$P^{(n)} \left(\prod_{i=0}^{k} e^{-\lambda_i \sigma^i_{\delta_m}} \left(\varphi_i \circ X_{\sigma^i_{\delta_m}} \right) \right) \longrightarrow P \left(\prod_{i=0}^{k} e^{-\lambda_i \sigma^i_{\delta_m}} \left(\varphi_i \circ X_{\sigma^i_{\delta_m}} \right) \right);$$

(5) $(\exists (r_m), \ r_m \downarrow 0) \ (\forall k, m \in \mathbb{N}) \ (\forall \lambda_i > 0) \ (\forall \varphi_i \in \mathcal{C}_0)$

$$P^{(n)} \left(\prod_{i=0}^{k} e^{-\lambda_i \sigma^i_{r_m}} \left(\varphi_i \circ X_{\sigma^i_{r_m}} \right) \right) \longrightarrow P \left(\prod_{i=0}^{k} e^{-\lambda_i \sigma^i_{r_m}} \left(\varphi_i \circ X_{\sigma^i_{r_m}} \right) \right).$$

Proof. Equivalence conditions (1), (2) and (3) follows from proposition 7.6. The implication (4)$\Rightarrow$(2) follows from a property of the Laplace transformation (see [FEL 67, p. 496]). The implication (1)$\Rightarrow$(4) follows from continuity on $\Pi(\delta_m)$ of the function

$$f = \prod_{i=1}^{k} e^{\lambda_i \sigma^i_{\delta_m}} \left(\varphi_i \circ X_{\sigma^i_{\delta_m}} \right) I_{\sigma^k_{\delta_m} < \infty}.$$

The equivalence (3)$\Leftrightarrow$(5) is similarly proved. $\qquad\square$

7.2. Weak convergence of semi-Markov processes

The general theorems of weak convergence of probability measures in space $\mathcal{D}$ are applied to derive sufficient conditions of weak convergence of semi-Markov processes, in particular, the conditions for a sequence of semi-Markov walks to converge weakly to a continuous semi-Markov process (see [SIL 70, SIL 74, HAR 76b, HAR 77]) are given. Alongside a time change in a Markov process, this passage to a limit in a class of stepped semi-Markov processes is a constructive method in order to obtain continuous semi-Markov processes which are not being Markovian.

10. Weak convergence of families of probability measures

Let $(P_x^{(n)})_{x \in \mathbb{X}} \ (n \in \mathbb{N})$, $(P_x)_{x \in \mathbb{X}}$ be admissible families of probability measures on $(\mathcal{D}, \mathcal{F})$. We will state that a sequence of families $\{(P_x^{(n)})\}_{n=1}^\infty$ weakly converges to a family (P_x) if $(\forall x \in \mathbb{X}) \ P_x^{(n)} \xrightarrow{w} P_x$. In this sense we understand the expression "a sequence of semi-Markov processes weakly converges to semi-Markov process". Family (P_x) is said to be weakly continuous if $(\forall x \in \mathbb{X}) \ (\forall (x_n), \ x_n \in \mathbb{X}, \ x_n \to x)$ $P_{x_n} \xrightarrow{w} P_x$.

11. Weak convergence of semi-Markov processes

THEOREM 7.3. *For the sequence of SM processes $\{(P_x^{(n)})\}_{n=1}^{\infty}$ to converge weakly to a weakly continuous SM process (P_x) it is sufficient that the following two conditions are fulfilled:*

(1) $(\forall \Delta \in \mathfrak{A}_0) (\forall \varphi \in \mathcal{C}_0) (\forall \lambda > 0) \ f_{\sigma_\Delta}(\lambda, \varphi \mid \cdot) \in \mathcal{C}_0;$

(2) $(\forall \Delta \in \mathfrak{A}_0) (\forall \varphi \in \mathcal{C}_0) (\forall \lambda > 0) (\forall K \subset \mathbb{X}, \ K - compact \ set)$

$$f_{\sigma_\Delta}^{(n)}(\lambda, \varphi \mid x) \longrightarrow f_{\sigma_\Delta}(\lambda, \varphi \mid x) \quad (n \longrightarrow \infty)$$

uniformly on $x \in K$.

Proof. By theorem 7.2 it is enough to prove that

$$(\forall k \in \mathbb{N}) \ (\forall \lambda_i > 0) \ (\forall \varphi_i \in \mathcal{C}_+) \ (\forall \Delta_i \in \mathfrak{A}_0) \ (\forall x_0 \in \mathbb{X})$$

$$P_{x_0}^{(n)} \left(\prod_{i=1}^{k} e^{-\lambda_i \tau_i \theta_{\tau_{i-1}}} \left(\varphi_i \circ X_{\tau_i} \right) \right) \longrightarrow P_{x_0} \left(\prod_{i=1}^{k} e^{-\lambda_i \tau_i \theta_{\tau_{i-1}}} \left(\varphi_i \circ X_{\tau_i} \right) \right),$$

where $\mathcal{C}_+$ is a set of all non-negative functions $\varphi \in \mathcal{C}$, $\tau_1 = \sigma_{\Delta_1}$, $\tau_i = \tau_{i-1} \dotplus \sigma_{\Delta_i}$ $(i \geq 2)$. For a semi-Markov process with $n \in \mathbb{N}_0$

$$P_{x_0}^{(n)} \left(\prod_{i=1}^{k} e^{-\lambda_i \tau_i \circ \theta_{\tau_{i-1}}} \left(\varphi_i \circ X_{\tau_i} \right) \right) = \int_{\mathbb{X}^K} \prod_{i=1}^{k} f_{\sigma_{\Delta_i}}^{(n)} \left(\lambda_i, dx_i \mid x_{i-1} \right) \varphi_i(x_i)$$

where $P_x^{(0)} = P_x$ and $f_{\sigma_\Delta}^{(0)} = f_{\sigma_\Delta}$. In order to prove convergence of the previous integral to an integral

$$\int_{\mathbb{X}^K} \prod_{i=1}^{k} f_{\sigma_{\Delta_i}} \left(\lambda_i, dx_i \mid x_{i-1} \right) \varphi_i(x_i)$$

we use the corollary from Prokhorov theorem (see [GIH 71, p. 429]): if μ_n is a sequence of finite measures, weakly converging to a measure μ_0 on $\mathcal{B}(\mathbb{X})$, then for any $\varepsilon > 0$ there is a compact set $K \subset \mathbb{X}$ such that $(\forall n \in \mathbb{N}) \ \mu_n(\mathbb{X} \setminus K) < \varepsilon$. We prove the theorem by induction. From condition (2) it follows that $(\forall \varphi \in \mathcal{C}_+)$ $(\forall x_0 \in \mathbb{X})$

$$f_{\sigma_{\Delta_1}}^{(n)} \left(\lambda_1, \varphi_1 \mid x_0 \right) \longrightarrow f_{\sigma_{\Delta_1}} \left(\lambda_1, \varphi_1 \mid x_0 \right)$$

Let $\mu_n(\varphi_{k-1}) \to \mu_0(\varphi_{k-1})$ for any $\varphi_{k-1} \in \mathcal{C}_+$, where

$$\mu_n(dx_{k-1}) = \int_{\mathbb{X}^{K-2}} \prod_{i=1}^{k-2} f_{\sigma_{\Delta_i}}^{(n)} \left(\lambda_i, dx_i \mid x_{i-1} \right) \varphi_i(x_i) f_{\sigma_{\Delta_{k-1}}}^{(n)} \left(\lambda_{k-1}, dx_{k-1} \mid x_{k-2} \right).$$

Then

$$\left| \int_{\mathbb{X}} \mu_n(dx_{k-1}) \varphi_{k-1}(x_{k-1}) f_{\sigma_{\Delta_k}}^{(n)}(\lambda_k, \varphi_k \mid x_{k-1}) \right.$$

$$\left. - \int_{\mathbb{X}} \mu_0(dx_{k-1}) \varphi_{k-1}(x_{k-1}) f_{\sigma_{\Delta_k}}(\lambda_k, \varphi_k \mid x_{k-1}) \right|$$

$$\leq \int_{\mathbb{X}} \mu_n(dx_{k-1}) \varphi_{k-1}(x_{k-1}) \left| f_{\sigma_{\Delta_k}}^{(n)}(\lambda_k, \varphi_k \mid x_{k-1}) - f_{\sigma_{\Delta_k}}(\lambda_k, \varphi_k \mid x_{k-1}) \right|$$

$$+ \left| \mu_n(\varphi'_{k-1}) - \mu_0(\varphi'_{k-1}) \right|,$$

where $\varphi'_{k-1}(x_{k-1}) = \varphi_{k-1}(x_{k-1}) f_{\sigma_{\Delta_k}}(\lambda_k, \varphi_k \mid x_{k-1})$ and, by condition (1), $\varphi'_{k-1} \in \mathcal{C}_+$. Let $K \subset \mathbb{X}$, K being compact set and $(\forall n \in \mathbb{N})$ $\mu_n(\mathbb{X} \setminus K) < \varepsilon$. Then the first summand is no more than

$$\varepsilon \, 2 \, M^2 + \sup_{x \in K} \left| f_{\sigma_{\Delta_k}}^{(n)}(\lambda_k, \varphi_k \mid x) - f_{\sigma_{\Delta_k}}(\lambda_k, \varphi_k \mid x) \right| \cdot M^{k-1},$$

where $\varphi_i \leq M$ $(i = 1, \ldots, k)$. The second summand by virtue of the inductive supposition tends to zero. In view of an arbitrary $\varepsilon > 0$ and due to uniform convergence on compact sets, the weak convergence of the family of measures is proved. Weak continuity of the limit family of measures is obvious (see theorem 7.2). □

12. Markov walk

Markov walk [KEM 70, SPI 69] with the probability kernel $Q(dx_1 \mid x)$ and step $h > 0$ is said to be a Markov chain (see item 1.5) on a countable parametrical set $\{0, h, 2h, 3h, \ldots\}$, controlled by a transition function $Q(dx_1 \mid x)$. It can also be interpreted as a semi-Markov walk, i.e. a stepped semi-Markov process (see items 1.17 and 3.5) (P_x) for which

$$P_x(\tau = nh, \, X_\tau \in S) = \left(Q(\{x\} \mid x) \right)^{n-1} Q(S \setminus \{x\} \mid x) \quad (n \in \mathbb{N}),$$

where $\tau(\xi) = \inf(t \geq 0 : \xi(0) \neq \xi(t))$. If $Q(\{x\} \mid x) = 0$, it is obvious that $P_x(\tau = h, \, X_\tau \in S) = Q(S \mid x)$.

13. Convergence of semi-Markov walks

Let (P_x) be a family of measures of semi-Markov walks with a corresponding family of transition functions (kernels) $Q(S \mid x)$ $(x \in \mathbb{X}, \, S \in \mathcal{B}(\mathbb{X}))$ of an embedded Markov chain. Then for any $x \in \mathbb{X}$ a conditional distribution of a time of the first jump $P_x(\tau \in dt \mid X_\tau)$ on set $\{\tau < \infty\}$ is determined. In the following proposition we will suppose that for any considered SM family of measures (P_x) $(\forall x \in \mathbb{X})$

$P_x(\tau < \infty) = 1$ and there exists a family of $\mathcal{B}(\mathbb{X}^2)$-measurable distribution functions $F(t \mid x, x_1)\,((x, x_1) \in \mathbb{X}^2)$, where

$$F(t \mid x, x_1) = P_x(\tau \le t \mid X_\tau = x)$$

$Q(dx_1 \mid x)$-a.s. $(t \ge 0)$. We call such a function *a two-dimensional conditional distribution function* of sojourn time at state x with passing to state x_1. Let

$$f(\lambda \mid x, x_1) = \int_0^\infty e^{-\lambda t} dF(t \mid x, x_1).$$

PROPOSITION 7.7. *Let* $\{(\widetilde{P}_x^{(n)})_{x \in \mathbb{X}}\}_{n=1}^\infty$ *be a sequence of families of measures of semi-Markov walks with the corresponding sequences of Markov kernels* $Q^{(n)}(S \mid x)$ *($Q(\{x\} \mid x) = 0$) and of two-dimensional cumulative distribution functions* $\{F^{(n)}(t \mid x_0, x_1)\}_{n=1}^\infty$ *for which* $F^{(n)}(\infty \mid x_0, x_1) = 1$. *Suppose there exists a sequence of steps* $\{h_n\}_1^\infty$ *($h_n \downarrow 0$) such that*

(1) $(\forall x \in \mathbb{X})\ P_x^{(n)} \xrightarrow{w} P_x$, *where* $(P_x^{(n)})_{x \in \mathbb{X}}$ *is a family of measures of a Markov walk, corresponding to the kernel* $Q^{(n)}(dx_1 \mid x)$ *and step* h_n, *and* (P_x) *is some admissible family of probability measures on* $(\mathcal{D}, \mathcal{F})$;

(2) for any $\lambda \ge 0$ *uniformly on* x_0 *and* x_1 *in each compact set*

$$h_n^{-1} \log f^{(n)}(\lambda \mid x_0, x_1) \longrightarrow -g(\lambda, x_0) \quad (n \longrightarrow \infty),$$

where $(\forall \lambda \ge 0)\ g(\lambda, \cdot) \in \mathcal{C}$ *and* $g(\lambda, x) \to \infty$ *as* $\lambda \to \infty$ *uniformly on all* $x \in \mathbb{X}$.

Then for any $x \in \mathbb{X}\ \widetilde{P}_x^{(n)} \xrightarrow{w} \widetilde{P}_x$, *where* $(\widetilde{P}_x)$ *is an admissible family of probability measures, obtained from the family of measures* (P_x) *under a random time change corresponding to a Laplace family of additive functionals* $(a_t(\lambda),\ t \ge 0)_{\lambda \ge 0}$, *where* $a_t(\lambda) = \int_0^t g(\lambda, X_s)\, ds$ *(see item 6.49 and theorem 6.8). If* $(P_x) \in \mathrm{SM}$, *then* $(\widetilde{P}_x) \in \mathrm{SM}$.

Proof. First of all we will prove that $(a_t(\lambda),\ t \ge 0)$ is such a family of additive functionals which determines a time change preserving a semi-Markov property. Obviously $(\forall \xi \in \mathcal{D})\ (\forall \lambda \ge 0)\ a.(\lambda \mid \xi)$ is a non-decreasing function of t continuous from the left, for which $a_0(\lambda \mid \xi) = 0$, and $(\forall \lambda > 0)\ a_t(0 \mid \xi) = 0$ (since $F^{(n)}(\infty \mid x_0, x_1) = 1$). Furthermore, $(\forall \lambda \ge 0)\ a_t(\lambda)$ is a $\mathcal{F}_t$-measurable function, since $g(\lambda, \cdot)$ is a $\mathcal{B}(\mathbb{X})$-measurable function. "Laplace property" of a functional $a_t(\lambda)$ (see item 6.61) is equivalent to complete monotonicity of $\exp(-g(\lambda, x))$ with respect to λ ($\lambda \ge 0$). However, it follows from representation

$$\exp\left(-g(\lambda, x_0)\right) = \lim_{n \to \infty} \left(f^{(n)}(\lambda \mid x_0, x_1)\right)^{1/h_n} \quad (h_n \longrightarrow 0),$$

where $(\forall n \in \mathbb{N})$ $f^{(n)}(\lambda \mid x_0, x_1)$ is a completely monotone function of λ. Actually, let us say that function $f(\lambda)$ is monotone of the order n, if $(\forall k \in \mathbb{N})$ $(\forall \lambda > 0)$ there exists derivative $\partial^k f(\lambda)/\partial \lambda^k$ and $(-1)^k (\partial^k/\partial \lambda^k) f(\lambda) \geq 0$ for all $k \leq n$. It is not difficult to show that from a complete monotonicity of function $f(\lambda)$ the monotonicity of order n of functions $(f(\lambda))^\alpha$ for all $\alpha \geq n$ follows. Hence, $\exp(-g(\lambda, x_0))$ is monotone of order n for any $n \in \mathbb{N}$, i.e. completely monotone. Furthermore, $(\forall c > 0)$ $(\forall \Delta \in \mathfrak{A})$ we have

$$a_{\sigma_\Delta}(\lambda \mid \xi) = a_c(\lambda \mid \xi) = \int_0^c g(\lambda, X_s(\xi)) \, ds \longrightarrow \infty \quad (\lambda \longrightarrow \infty)$$

uniformly on all $\xi \in \{\sigma_\Delta \geq a\}$ (since condition (2) is satisfied). At last, from boundedness of $g(\lambda, \cdot)$ it follows that $(\forall \lambda \geq 0)$ $a_t(\lambda, \xi) \to 0$ $(t \to 0)$ uniformly on all $\xi \in \mathcal{D}$. Hence, the conditions of theorem 6.8 are fulfilled, i.e. additive functional $(a_t(\lambda), t \geq 0)_{\lambda \geq 0}$ determines time change transformation, due to which a family of measures (P_x) passes to a family of measures $(\widetilde{P}_x)$, where $(\forall \delta \in \mathrm{DS})$ $(\forall k \in \mathbb{N})$ $(\forall \lambda_i > 0)$ $(\forall \varphi_i \in \mathcal{C})$

$$\widetilde{E}_x \left(\prod_{i=1}^k e^{-\lambda_i \sigma_\delta^i} \left(\varphi_i \circ X_{\sigma_\delta^i} \right) \right)$$

$$= E_x \left(\prod_{i=1}^k \exp\left(-\int_0^{\sigma_\delta^i} g(\lambda_i, X_t) \, dt \right) \left(\varphi_i \circ X_{\sigma_\delta^i} \right), \ \sigma_\delta^k < \infty \right).$$

Let $\delta_m \in \mathrm{DS}(r_m)$, $r_m \downarrow 0$ and $P_x(\bigcap_{m=1}^\infty \Pi(\delta_m)) = 1$ (such a sequence of deducing sequences exists; see corollary 2.2). Under theorem 7.2, it is enough to prove that $(\forall k, m \in \mathbb{N})$ $(\forall \lambda_i > 0)$ $(\forall \varphi_i \in \mathcal{C})$

$$\widetilde{E}_x^{(n)}\left(u_m^k \right) \longrightarrow \widetilde{E}_x\left(u_m^k \right),$$

where

$$u_m^k = \prod_{i=1}^k \exp\left(-\lambda_i \left(\sigma_{\delta_m}^i - \sigma_{\delta_m}^{i-1} \right) \right) \left(\varphi_i \circ X_{\sigma_{\delta_m}^i} \right).$$

We will prove it for $k = 1$. In a common case the proof is similar. For any $n \in \mathbb{N}$ we have

$$\widetilde{E}_x^{(n)}\left(e^{-\lambda \sigma_\Delta} \left(\varphi \circ X_{\sigma_\Delta} \right) \right) = \sum_{k=0}^\infty \widetilde{E}_x^{(n)}\left(e^{-\lambda \sigma_\Delta} \left(\varphi \circ X_{\sigma_\Delta} \right), \ \sigma_\Delta = \tau^k \right),$$

where τ^k is a time of the k-th jump. From regenerative properties of $(\widetilde{P}_x^{(n)})_{x \in \mathbb{X}}$ it follows that

$$\widetilde{E}_x^{(n)}\left(e^{-\lambda \tau^k}(\varphi \circ X_{\tau^k}), \, \sigma_\Delta = \tau^k\right)$$

$$= \widetilde{E}_x^{(n)}\left(\prod_{i=1}^{k} f^{(n)}\left(\lambda \mid X_{\tau^{i-1}}, X_{\tau^i}\right)(\varphi \circ X_{\tau^k}), \, \sigma_\Delta = \tau^k\right),$$

where $\widetilde{E}_x^{(n)}(e^{-\lambda \tau}(\varphi \circ X_\tau)) = \widetilde{E}_x^{(n)}(f(\lambda \mid X_0, X_\tau)(\varphi \circ X_\tau))$. Since $\widetilde{E}_x^{(n)}$ and $E_x^{(n)}$ coincide on sigma-algebra $\sigma(\gamma_{\tau^k}, \, k \in \mathbb{N}_0) = \sigma(X_{\tau^k}, \, k \in \mathbb{N}_0)$,

$$\widetilde{E}_x^{(n)}\left(\prod_{i=1}^{k} f^{(n)}\left(\lambda \mid X_{\tau^{i-1}}, X_{\tau^i}\right)(\varphi \circ X_{\tau^k}), \, \sigma_\Delta = \tau^k\right)$$

$$= E_x^{(n)}\left(\exp \sum_{i=1}^{k} \log f^{(n)}\left(\lambda \mid X_{\tau^{i-1}}, X_{\tau^i}\right)(\varphi \circ X_{\tau^k}), \, \sigma_\Delta = \tau^k\right)$$

$$= E_x^{(n)}\left(\exp\left(\frac{1}{h_n} \int_0^{\tau^k} \log f^{(n)}\left(\lambda \mid X_t, X_{t+h_n}\right) dt\right)(\varphi \circ X_{\tau^k}), \, \sigma_\Delta = \tau^k\right).$$

Hence,

$$\widetilde{E}_x^{(n)}\left(e^{-\lambda \sigma_\Delta}(\varphi \circ X_{\sigma_\Delta})\right)$$

$$= E_x^{(n)}\left(\exp\left(\int_0^{\sigma_\Delta} \frac{1}{h_n} \log f^{(n)}\left(\lambda \mid X_t, X_{t+h_n}\right) dt\right)(\varphi \circ X_{\sigma_\Delta}), \, \sigma_\Delta < \infty\right).$$

Since $P_x^{(n)} \xrightarrow{w} P_x$, the family of measures $(P_x^{(n)})_{n=1}^{\infty}$ is weakly compact and $(\forall \varepsilon > 0)$ $(\forall t \in \mathbb{R}_+)$ $(\forall K_t \subset \mathbb{X}, \, K_t$ – compact set$)$

$$\sup_n P_x^{(n)}\left(\bigcup_{s \leq t} \{X_s \notin K_t\}\right) < \varepsilon.$$

Besides, if $P_x(\Pi(\Delta)) = 1$, $P_x^{(n)} \circ \beta_{\sigma_\Delta}^{-1} \xrightarrow{w} P_x \circ \beta_{\sigma_\Delta}^{-1}$ and $(\forall \varepsilon > 0)$ $(\forall t_1 \in \mathbb{R}_+)$

$$\sup_n P_x^{(n)}\left(t_1 < \sigma_\Delta < \infty\right) < \varepsilon.$$

Furthermore, we find

$$\left| E_x^{(n)} \left(\exp \left(\int_0^{\sigma_\Delta} \frac{1}{h_n} \log f^{(n)} \left(\lambda \mid X_t, \, X_{t+h_n} \right) dt \right) \left(\varphi \circ X_{\sigma_\Delta} \right) \right) \right.$$

$$\left. - E_x \left(\exp \left(- \int_0^{\sigma_\Delta} g(\lambda, X_t) \, dt \right) \left(\varphi \circ X_{\sigma_\Delta} \right) \right) \right|$$

$$\leq E_x^{(n)} \left(\left| \exp \left(\int_0^{\sigma_\Delta} \frac{1}{h_n} \log f^{(n)} \left(\lambda \mid X_t, X_{t+h_n} \right) dt \right) \right. \right.$$

$$\left. \left. - \exp \left(- \int_0^{\sigma_\Delta} g(\lambda, X_t) \, dt \right) \right| \left(\varphi \circ X_{\sigma_\Delta} \right) \right)$$

$$+ \left| E_x^{(n)} \left(\exp \left(- \int_0^{\sigma_\Delta} g(\lambda, X_t) \, dt \right) \left(\varphi \circ X_{\sigma_\Delta} \right) \right) \right.$$

$$\left. - E_x \left(\exp \left(- \int_0^{\sigma_\Delta} g(\lambda, X_t) \, dt \right) \left(\varphi \circ X_{\sigma_\Delta} \right) \right) \right|.$$

The first summand of this sum is no more than a sum

$$P_x^{(n)} \left(t_1 - h_n < \sigma_\Delta < \infty \right) + P_x^{(n)} \left(\bigcup_{s \leq t_1} \{ X_s \notin K_{t_1} \} \right)$$

$$+ E_x^{(n)} \left(\left(\exp \left| \int_0^{\sigma_\Delta} \frac{1}{h_n} \log f^{(n)} \left(\lambda \mid X_t, X_{t+h_n} \right) dt + \int_0^{\sigma_\Delta} g(\lambda, X_t) \, dt \right| - 1 \right), \right.$$

$$\left. \{ \sigma_\Delta \leq t - h_n \} \cap \bigcap_{s \leq t_1} \{ X_s \in K_{t_1} \} \right),$$

which can be made as small as desired at the expense of a choice t_1 (first term), K_{t_1} (second term) and on uniform convergence on compact sets

$$\frac{1}{h_n} \log f^{(n)} \left(\lambda \mid X_t, X_{t+h_n} \right) \longrightarrow -g(\lambda, X_t),$$

making the function as small as desired in the third term. In order to prove the theorem it is enough to show that function $\exp(- \int_0^{\sigma_\Delta} g(\lambda, X_t) \, dt)$ is continuous as a function of $\xi \in \Pi(\Delta)$. Let $\xi_n \xrightarrow{PD} \xi$. Then $\sigma_\Delta(\xi_n) \to \sigma_\Delta(\xi)$. Besides, $X_t(\xi_n) \to X_t(\xi)$ in all points of continuity of ξ (see [BIL 77, p. 157]) and therefore almost everywhere with respect to the Lebesgue measure. Hence, with the same t we obtain $g(\lambda, X_t(\xi_n)) \to g(\lambda, X_t(\xi))$. From boundedness of $g(\lambda, \cdot)$, by the Lebesgue theorem (see [NAT 74, p. 120]) we obtain

$$\int_0^{\sigma_\Delta(\xi_n)} g\left(\lambda, X_t(\xi_n) \right) dt \longrightarrow \int_0^{\sigma_\Delta(\xi)} g\left(\lambda, X_t(\xi) \right) dt;$$

From here by continuity functions on $\Pi(\Delta)$ it follows that

$$\exp\left(-\int_0^{\sigma_\Delta} g(\lambda, X_t)\, dt\right).$$

$\square$

14. Example of convergence to non-Markov SM process

Let $\mathbb{X} = \mathbb{R}^1$, and $\{(\widetilde{P}_x^{(n)})\}_{n=1}^\infty$ be a sequence of SM walks. SM walk $(\widetilde{P}_x^{(n)})$ has a transition function of an embedded Markov chain

$$Q^{(n)}\left((-\infty, x) \mid x_0\right) = G\left(\sqrt{n}(x - x_0)\right),$$

where $G(x)$ is a cumulative distribution function on $\mathbb{R}^1$, and

$$\int_{-\infty}^\infty x\, dG(x) = 0, \qquad \int_{-\infty}^\infty x^2\, dG(x) = 1.$$

Consider the conditional cumulative distribution function of length of an interval between jumps of this SM walk of the form

$$F^{(n)}\left(t \mid x_0, x_1\right) = F\left(tn^{1/\alpha}\right),$$

where $0 < \alpha < 1$, and $(\forall \lambda \geq 0)\, (f(\lambda n^{-1/\alpha}))^n \to e^{-\lambda^\alpha}$, where $f(\lambda) = \int_0^\infty e^{-\lambda t} dF(t)$, i.e. F belongs to a region of attraction of the stable law with a parameter α (see [FEL 67, p. 214, 515]). Then from proposition 7.7 it follows that $(\forall x \in \mathbb{X})\, \widetilde{P}_x^{(n)} \xrightarrow{w} \widetilde{P}_x^w$, where $(\widetilde{P}_x^w)$ is a family of measures of Wiener process, transformed by a random time change with the help of an independent homogenous process with independent increments (see item 6.54) with a measure $\mathcal{U}$ on $(\Psi, \mathcal{G})$, where

$$\mathcal{U}\left(e^{-\lambda \widetilde{X}_t}\right) = e^{-\lambda^\alpha t} \quad (\lambda \geq 0).$$

In order to apply theorem 7.7 we choose $h_n = 1/n$. Then a sequence of Markov walks $\{(\widetilde{P}_x^{(n)})\}_{n=1}^\infty$, corresponding to a sequence of the kernels $(Q^{(n)}(S \mid x))$ and steps (h_n) converges weakly to Wiener process (P_x^w). Proof of weak convergence in space $(\mathcal{D}, \rho_\mathcal{D})$ follows from theorems 3 and 4 of [GIH 71, p. 486, 488]. We note that

$$n \log f\left(\lambda n^{-1/\alpha}\right) \longrightarrow -\lambda^\alpha,$$

i.e. condition (2) of proposition 7.7 is satisfied. From here $(\forall x \in \mathbb{X})\, \widetilde{P}_x^{(n)} \xrightarrow{w} \widetilde{P}_x^w$, where $(\widetilde{P}_x^w) \in$ SM, and $(\forall \Delta \in \mathfrak{A})\, (\forall \lambda > 0)\, (\forall \varphi \in \mathcal{C})$

$$\widetilde{P}_x^w\left(e^{-\lambda \sigma_\Delta}\left(\varphi \circ X_{\sigma_\Delta}\right)\right) = P_x^w\left(e^{-\lambda^\alpha \sigma_\Delta}\left(\varphi \circ X_{\sigma_\Delta}\right)\right),$$

i.e. $(\widetilde{P}_x^w)$ is a Wiener process transformed by a random time change. It represents a continuous semi-Markov process without a deterministic component in time run along the trace. Evidently it is non-Markovian. Its conditional distribution of time run along the trace is expressed as

$$e^{-\widetilde{b}(\lambda,\tau)} = \widetilde{P}_x^w\left(e^{-\lambda\tau} \mid \mathcal{F}^\circ\right) = P_x^w\left(e^{-\lambda^\alpha\tau} \mid \mathcal{F}^\circ\right) = e^{-\lambda^\alpha\tau}.$$

In order to find its Lévy measure $\widetilde{n}_\tau$ (see theorem 6.11), we consider an equation

$$\lambda^\alpha\tau = \int_{0+}^\infty \left(1 - e^{-\lambda u}\right)\widetilde{n}_\tau(du).$$

Using representation of these additive functionals in the form of curvilinear integrals by the additive functional μ_τ, we obtain

$$\lambda^\alpha\mu_\tau = \int_0^\tau \left(\int_{0+}^\infty \left(1 - e^{-\lambda u}\right)\widetilde{\nu}\left(du \mid X_{\tau_1}\right)\right)\mu\left(d\tau_1\right),$$

from here we obtain an equation with respect to $\widetilde{\nu}$:

$$\lambda^\alpha = \int_{0+}^\infty \left(1 - e^{-\lambda u}\right)\widetilde{\nu}\left(du \mid X_0\right).$$

Differentiating on λ we find a Laplace transformation of measure $u\widetilde{\nu}(du \mid X_0)$:

$$\int_{0+}^\infty e^{-\lambda u}u\widetilde{\nu}\left(du \mid X_0\right) = \alpha\lambda^{\alpha-1}.$$

For $\alpha = 1/2$ we obtain a tabulated value for this transformation [DIT 65, p. 216, formula (22.1)]; therefore

$$\widetilde{\nu}\left(du \mid X_0\right) = \frac{du}{2\sqrt{Xu^3}}.$$

Trajectory of this process represents a continuous Cantor function; the set of non-constancy of the process has a zero measure, i.e. the whole half-line without intervals of constancy of the process.

Chapter 8

Representation of a Semi-Markov Process as a Transformed Markov Process

It is natural to assume that each semi-Markov process is either itself Markovian or grows out of transformations of a time change in a Markov process. With reference to stepped semi-Markov processes this supposition makes a Lévy hypothesis, which was the area Yackel [YAC 68] was engaged in. In general the Lévy hypothesis is not justified. So, the trajectory of a homogenous stepped Markov process with two states has either no more than one jump, or an infinite number of jumps; and the trajectory of a homogenous stepped semi-Markov process with two states can have any finite number of jumps. At the same time number of jumps does not vary with a time change. There are also other specific properties of non-Markov semi-Markov invariant processes relative to time change keeping a homogenity of process in time. The problem consists of searching necessary and sufficient conditions for the semi-Markov process to be a Markov process transformed by a random time change. It is desirable to find an original Markov process and corresponding time change. In the present chapter two methods of construction of a Markov process, preserving distribution of the random trace of a given semi-Markov process, are investigated.

Firstly, for a regular semi-Markov process with a lambda-continuous family of probability measures (P_x), sufficient conditions of such construction are formulated in terms of lambda-characteristic operators of this process. With some additional conditions providing uniqueness of this representation we obtain an infinitesimal operator of a corresponding Markov process and a Laplace family of additive functionals determining the time change.

Secondly, for a semi-Markov process with a regular additive functional of time run along the trace we formulate sufficient conditions for the corresponding Markov process to be constructed in terms of parameters of Lévy expansion for a conditional process with independent increment (time run process). In order to prove Markovness of the constructed semi-Markov family of measures we use the theorem on regeneration times which is applied to a sequence of regeneration times of the semi-Markov process with trajectories without intervals of constancy, converging from above to a non-random time instant.

Constructed by the second method Markov process, which is determined by its trace distribution and by its conditional distribution of time run along the trace, we use this to derive formulae of stationary distributions of the original semi-Markov process.

8.1. Construction of a Markov process by the operator of a semi-Markov process

With the help of the Hille-Iosida theorem, a Markov process with an infinitesimal operator determined with the help of lambda-characteristic operator of the given semi-Markov process is constructed [VEN 75, GIH 73, DYN 63, ITO 63].

1. Definitions

Let

$$\mathcal{C}_K = \left\{ f \in \mathcal{C}_0(\mathbb{X}) : (\forall \varepsilon > 0) \left\{ |f| \geq \varepsilon \right\} \in \mathfrak{K} \right\},$$

where $\mathfrak{K}$ is a set of all compact subsets of space $\mathbb{X}$;

$$\mathcal{C}_R = \left\{ f \in \mathcal{C}_0(\mathbb{X}) : (\exists f_1 \in \mathcal{C}_K) (\exists c \in \mathbb{R}) \, f = f_1 + c \right\}.$$

A Markov process is said to be
(a) stochastically continuous if $(\forall f \in \mathcal{C}_0) (\forall x \in \mathbb{X})$

$$T_t(f \mid x) \longrightarrow f(x) \quad (t \downarrow 0),$$

where $T_t(f \mid x) = E_x(f \circ X_t)$;
(b) Feller process if $(\forall t \geq 0) (\forall f \in \mathcal{C}_0)$

$$T_t f \in \mathcal{C}_0;$$

(c) regular if $(\forall t \geq 0) (\forall f \in \mathcal{C}_K)$

$$T_t f \in \mathcal{C}_K$$

(see [GIH 73, p. 156, 160, 170]). We call semi-Markov process (P_x) lambda-regular if $(\forall \lambda > 0) (\forall f \in \mathcal{C}_K)$

$$R_\lambda f \in \mathcal{C}_K$$

(see item 3.22). Let $\mathcal{C}(x)$ be a set of all measurable functions $f : \mathbb{X} \mapsto \mathbb{R}$ continuous at the point x $(x \in \mathbb{X})$.

2. The Hille-Iosida theorem

THEOREM 8.1. *For the linear operator A to be an infinitesimal operator of a stochastically continuous, Feller-type, regular Markov process, it is sufficient to fulfill the following conditions:*

(1) A is determined on a set of functions $V \subset \mathcal{C}_R$ everywhere dense in topology of uniform convergence in space $\mathcal{C}_R$;

(2) $(\forall f \in V)$ if $f(x_0) \geq 0$ and $(\forall x \in \mathbb{X}) f(x) \leq f(x_0)$, then $A(f \mid x_0) \leq 0$ (maximum principle);

(3) $(\forall \lambda > 0) (\forall \varphi \in \mathcal{C}_R) \forall \psi \in V (\lambda \mathcal{E} - A)\psi = \varphi$, where $\mathcal{E}$ is an operator of identical transformation.

Proof. See [ITO 63, p. 27]; [GIH 73, p. 172, 180].   □

3. Construction of a Markov process

We call a domain (region of determination) of an infinitesimal operator A the set of all $\varphi \in \mathbb{B}$ for which

$$A^t(\varphi \mid x) \equiv \frac{1}{t}\left(E_x(\varphi \circ X_t) - \varphi(x)\right) \longrightarrow A(\varphi \mid x)$$

uniformly on $x \in \mathbb{X}$. We will designate all values, functions and sets, derived from $(\overline{P}_x)$, by a letter with an overline. For example,

$$\overline{R}_{\lambda_0}(\varphi \mid x) = \overline{E}_x\left(\int_0^\infty e^{-\lambda t}(\varphi \circ X_t)\, dt\right)$$

$$= \int_D \left(\int_0^\infty e^{-\lambda t}(\varphi \circ X_t)\, dt\right) d\overline{P}_x.$$

THEOREM 8.2. *Let $(P_x) \in$ SM be a λ-regular and λ-continuous family; let the following conditions be fulfilled:*

(1) $(\forall \varphi \in \mathcal{C}_K) \lambda R_\lambda \varphi \to \varphi$ with $\lambda \to \infty$ uniformly on $\mathbb{X}$;

(2) $(\forall \lambda > 0)\ \mathbb{I} \in V_\lambda(\mathbb{X})$, $(\forall x \in \mathbb{X})\ A_\lambda(\mathbb{I} \mid x) < 0$, $A_\lambda(\mathbb{I} \mid x) \to 0$ with $\lambda \to 0$;
(3) $(\forall x \in \mathbb{X})\ (\forall \lambda > 0)\ A_\lambda$ is a pseudo-local operator at point x (see item 3.27).

Then $(\forall \lambda_0 > 0)$ there exists a non-breaking Markov process $(\overline{P}_x)$ with an infinitesimal operator $\overline{A} = -\lambda_0 A_0/A_{\lambda_0}\mathbb{I}$, with is defined on a domain $\overline{V} = V_0(\mathbb{X})$, and $(\forall \varphi \in C_0)$

$$\overline{R}_{\lambda_0}\varphi = R_{\lambda_0}\varphi.$$

Proof. By theorems 3.3 and 3.4, $(\forall \varphi \in C_0)\ (\forall \lambda > 0)\ R_\lambda \varphi \in V_\lambda(\mathbb{X})$, and

$$A_\lambda R_\lambda \varphi = \left(A_0 + A_\lambda \mathbb{I}E\right) R_\lambda \varphi = \varphi R_\lambda \mathbb{I}/\lambda.$$

Since $\mathbb{I} \in V_\lambda(\mathbb{X})$ and $A_{\lambda_0}\mathbb{I} < 0$, $R_\lambda \varphi \in V_0(\mathbb{X}) = \overline{V}$. In this case $(\forall \varphi \in C_K)\ (\forall c \in \mathbb{R})$ $\lambda R_\lambda(\varphi + c) = \lambda R_\lambda \varphi + c \to \varphi + c$ uniformly on $x \in \mathbb{X}$. From here and by lambda-regularity of family (P_x) it follows that set $\overline{V}$ is dense in C_R in topology of uniform convergence. Furthermore, $(\forall \varphi \in V(\mathbb{X}))$ if $\varphi(x_0) \geq 0$ and $(\forall x \in \mathbb{X})\ \varphi(x_0) \geq \varphi(x)$,

$$\frac{-\lambda_0}{A_{\lambda_0}(\mathbb{I} \mid x_0)} A_0\left(\varphi \mid x_0\right)$$

$$= \frac{-\lambda_0}{A_{\lambda_0}(\mathbb{I} \mid x_0)} \lim_{r \to 0} \frac{E_{x_0}\left(\varphi \circ X_{\sigma_r},\ \sigma_r < \infty\right) - \varphi(x_0)}{m_r(x_0)} \leq 0.$$

For any $(\forall \varphi \in C_R)$, equation $(\lambda_0 E - \overline{A})\psi = \varphi$ has an evident solution: $\psi = R_{\lambda_0}\varphi$. Hence, by theorem 8.1, there exists a unique Markov process $(\overline{P}_x)$ (which is a stochastically continuous regular Feller process) such that $\overline{A}$ is its infinitesimal operator and $(\forall \varphi \in C)\ \overline{R}_{\lambda_0}\varphi = R_{\lambda_0}\varphi$. The process $(\overline{P}_x)$ is not a breaking process, since $\overline{R}_{\lambda_0}\mathbb{I} = 1/\lambda_0$.   □

8.2. Comparison of an original process with a transformed Markov process

Laplace family of additive functionals is constructed. The Markov family, constructed in the previous section, is transformed by a random time change. The identity of distributions for both original and constructed semi-Markov processes is proved.

4. Additive functional

Consider a function

$$a_t(\lambda) = \lambda_0 \int_0^t \left(\frac{A_\lambda \mathbb{I}}{A_{\lambda_0}\mathbb{I}} \circ X_s\right) ds \quad (\lambda, \lambda_0 > 0),$$

PROPOSITION 8.1. *Let the conditions of theorem 8.2 be fulfilled, and* $(\forall x \in \mathbb{X})$ $A_\lambda(\mathbb{I} \mid x) \to \infty$ *as* $\lambda \to \infty$. *Then* $(a_t(\lambda))$ $(\lambda > 0)$ *is the Laplace family of additive functionals (see items 6.48 and 6.49).*

Proof. From all properties of Laplace family of additive functionals the only complete monotonicity is non-obvious. In order to prove the latter we can use the criterion of complete monotonicity from lemma 5.6, item 5.13, which expresses the condition of complete monotonicity in terms of finite differences. Furthermore, function $e^{-\lambda t}$ is completely monotone, and linear operations of integration do not change a sign of differences, and a limit of a sequence of values of constant signs is a value of the same sign. □

5. Family of transformed measures

By theorem 6.8, the family $(a.(\lambda \mid \cdot))_{\lambda>0}$ determines a random time change, i.e. a family of measures (Q_x) on $\mathfrak{B}$, where $Q_x(\mathcal{H}) = 1$ and $(\forall x \in \mathbb{X})$ $\overline{P}_x = Q_x \circ q^{-1}$. Since $(\overline{P}_x)$ is a strictly Markov family, then $(\overline{P}_x) \in SM$, hence, by theorem 6.9 $(\widetilde{P}_x) \in SM$, where $\widetilde{P}_x = Q_x \circ u^{-1}$. It remains to be proved that $(\widetilde{P}_x) = (P_x)$. In order to prove this, it is sufficient to establish that $(\forall \Delta \in \mathfrak{A}, [\Delta] \in \mathfrak{K})$ $(\forall n \in N)$ $(\forall \lambda_i > 0)$ $(\forall \varphi_i \in C)$

$$\widetilde{R}_{\sigma\Delta}(\lambda_1, \ldots, \lambda_n; \varphi_1, \ldots, \varphi_n) = R_{\sigma\Delta}(\lambda_1, \ldots, \lambda_n; \varphi_1, \ldots, \varphi_n),$$

where

$$\widetilde{R}_{\sigma\Delta}(\lambda_1, \ldots, \lambda_n; \varphi_1, \ldots, \varphi_n \mid x)$$

$$\equiv \widetilde{E}_x \left(\int_{s_n < \sigma\Delta} \prod_{i=1}^{n} \left(e^{-\lambda_i t_i} \left(\varphi_i \circ X_{s_i} \right) dt_i \right) \right),$$

$s_k = \sum_{i=1}^{k} t_i$ (see item 3.21). In what follows all values, functions and sets derived from $(\widetilde{P}_x)$ are designated with a "wave" above the letter.

6. Domain of identity of operators

Consider a lambda-characteristic operator of the second kind

$$\mathcal{A}_\lambda(\varphi \mid x) = \lim_{r \to 0} \mathcal{A}_\lambda^r(\varphi \mid x)$$

(see item 3.26) where $(\forall x \in \mathbb{X})$ $(\forall \lambda > 0)$ $(\forall \varphi \in \mathbb{B})$

$$\mathcal{A}_\lambda^r(\varphi \mid x) = \lambda \frac{E_x \left(e^{-\lambda \sigma_r} \left(\varphi \circ X_{\sigma_r} \right) \right) - \varphi(x)}{E_x \left(1 - e^{-\lambda \sigma_r} \right)}$$

and $\mathcal{V}_\lambda(x)$ is a set of all $\varphi \in \mathbb{B}$, for which the preceding limit exists.

PROPOSITION 8.2. *Let $\varphi \in \overline{V} \cap \mathcal{C}$, $\overline{A}\varphi \in \mathcal{C}$; and conditions of theorem 8.2 be fulfilled; and $(\forall \lambda > 0)\ \mathcal{A}_\lambda \mathbb{I} \in \mathcal{C}$. Then*

(1) $\varphi \in \mathcal{V}_\lambda \cap \widetilde{\mathcal{V}}_\lambda$ and $\mathcal{A}_\lambda(\varphi) = \widetilde{\mathcal{A}}_\lambda(\varphi)$;

(2) the following conditions are fulfilled:

   (a) $(\forall K \in \mathfrak{K})\ (\exists M > 0)\ (\exists r_0 > 0)\ (\forall x \in K)\ (\forall r,\ 0 < r < r_0)$

$$\widetilde{\mathcal{A}}_\lambda^r(\varphi \mid x) \le M;$$

   (b) $(\forall x \in \mathbb{X})\ \widetilde{\mathcal{A}}_\lambda^r(\varphi \mid x) \to \widetilde{\mathcal{A}}_\lambda(\varphi \mid x)\ (r \to 0)$;

   (c) $\widetilde{\mathcal{A}}_\lambda(\varphi) \in \mathcal{C}$.

Proof. (1) We have $(\forall x \in \mathbb{X})\ (\forall \varphi \in \overline{V} \cap \mathcal{C})\ \varphi \in V_0 \cap \mathcal{C} = \bigcap_{\lambda > 0} \mathcal{V}_\lambda \cap \mathcal{C}$,

$$\mathcal{A}_\lambda(\varphi \mid x) = -\lambda \frac{A_0(\varphi \mid x)}{A_\lambda(\mathbb{I} \mid x)} - \lambda \varphi(x)$$

$$= \lambda \frac{A_{\lambda_0}(\mathbb{I} \mid x)\overline{A}(\varphi \mid x)}{\lambda_0 A_\lambda(\mathbb{I} \mid x)} + \lambda \varphi(x).$$

(2) On the other hand, according to definition of a time change with the help of Laplace family of additive functionals (item 6.52), we have $(\forall x \in \mathbb{X})\ (\forall \varphi \in \overline{V} \cap \mathcal{C},$ $\overline{A}\varphi \in \mathcal{C})$

$$\widetilde{\mathcal{A}}_\lambda^r(\varphi \mid x) = \frac{\lambda}{\widetilde{E}_x(1 - e^{-\lambda \sigma_r})} \widetilde{E}_x\big(e^{-\lambda \sigma_r}(\varphi \circ X_{\sigma_r}) - \varphi \circ X_0\big)$$

$$= \frac{\lambda}{\overline{E}_x(1 - e^{-a_{\sigma_r}(\lambda)})} \overline{E}_x\big(e^{-a_{\sigma_r}(\lambda)}(\varphi \circ X_{\sigma_r}) - (\varphi \circ X_0)\big)$$

$$= \frac{\lambda}{\overline{E}_x(1 - e^{-a_{\sigma_r}(\lambda)})} \overline{E}_x\big(e^{-g_\lambda(x)\sigma_r}(\varphi \circ X_{\sigma_r}) - (\varphi \circ X_0)\big)$$

$$+ \frac{\lambda}{\overline{E}_x(1 - e^{-a_{\sigma_r}(\lambda)})} \overline{E}_x\big((e^{-a_{\sigma_r}(\lambda)} - e^{-g_\lambda(x)\sigma_r})(\varphi \circ X_{\sigma_r})\big)$$

$$= \frac{\lambda}{g_\lambda(x)} \overline{\mathcal{A}}_{g_\lambda(x)}^r(\varphi \mid x)\alpha_r(x) + \varepsilon_r(x),$$

where $g_\lambda = \lambda_0 A_\lambda \mathbb{I} / A_{\lambda_0} \mathbb{I}$,

$$\alpha_r(x) = \overline{E}_x\big(1 - e^{-g_\lambda(x)\sigma_r}\big) / \overline{E}_x\big(1 - e^{-a_{\sigma_r}(\lambda)}\big)$$

and $\varepsilon_r(x)$ is a remainder. By Dynkin's lemma [ITO 63, p. 47],

$$\overline{A}^r_{g_\lambda(x)}(\varphi \mid x) \equiv \frac{g_\lambda(x)}{\overline{E}_x\left(1 - e^{-g_\lambda(x)\sigma_r}\right)}\overline{E}_x\left(e^{-g_\lambda(x)\sigma_r}\left(\varphi \circ X_{\sigma_r}\right) - \left(\varphi \circ X_0\right)\right)$$

$$= \frac{g_\lambda(x)}{\overline{E}_x\left(1 - e^{-g_\lambda(x)\sigma_r}\right)}\overline{E}_x\left(\int_0^{\sigma_r} e^{-g_\lambda(x)t}\left(\overline{A}\varphi - g_\lambda(x)\varphi\right) \circ X_t \, dt\right)$$

$$= \overline{A}(\varphi \mid x) - g_\lambda(x)\varphi(x) + \varepsilon'_r;$$

$$\varepsilon'_r = \frac{g_\lambda(x)}{\overline{E}_x\left(1 - e^{-g_\lambda(x)\sigma_r}\right)}\overline{E}_x\left(\int_0^{\sigma_r} e^{-g_\lambda(x)t}\left(\left(\overline{A}\varphi - g_\lambda(x)\varphi\right) \circ X_t\right.\right.$$

$$\left.\left. - \left(\overline{A}\varphi - g_\lambda(x)\varphi\right) \circ X_0\right) dt\right),$$

$$\left|\varepsilon'_r\right| \leq \sup_{x_1 \in B(x,r)} \left(\left|\overline{A}(\varphi \mid x_1) - \overline{A}(\varphi \mid x)\right| + g_\lambda(x)\left|\varphi(x_1) - \varphi(x)\right|\right) \longrightarrow 0$$

uniformly on any compact set. It is also evident that

$$\overline{g}_{\lambda,r}(x)/\widehat{g}_{\lambda,r}(x) \leq \alpha_r(x) \leq \widehat{g}_{\lambda,r}(x)/\overline{g}_{\lambda,r}(x),$$

where

$$\overline{g}_{\lambda,r}(x) = \inf_{x_1 \in B(x,r)} g_\lambda(x), \qquad \widehat{g}_{\lambda,r}(x) = \sup_{x_1 \in B(x,r)} g_\lambda(x).$$

Since $g_\lambda \in \mathcal{C}$ and $\min\{\lambda, \lambda_0\} \leq g_\lambda \leq \max\{\lambda, \lambda_0\}$, $\alpha_r(x) \to 1$ as $r \to 0$ uniformly on x. Furthermore

$$\left|\varepsilon_r\right| \leq \sup|\varphi|\lambda\frac{\overline{E}_x\left(e^{-\overline{g}_{\lambda,r}(x)\sigma_r} - e^{-\widehat{g}_{\lambda,r}(x)\sigma_r}\right)}{\overline{E}_x\left(1 - e^{-\overline{g}_{\lambda,r}(x)\sigma_r}\right)}$$

$$\leq \sup|\varphi|\lambda\left(\frac{\widehat{g}_{\lambda,r}(x)}{\overline{g}_{\lambda,r}(x)} - 1\right)$$

and, hence, $\varepsilon_r \to 0$ $(r \to 0)$ uniformly on x. From here

$$\widetilde{A}^r_\lambda(\varphi \mid x) \longrightarrow \frac{\lambda}{g_\lambda(x)}\left(g_\lambda(x)\varphi(x) - \overline{A}(\varphi \mid x)\right)$$

uniformly on x on each compact set and

$$\widetilde{A}_\lambda(\varphi \mid x) = \lambda\varphi(x) - \frac{\lambda A_{\lambda_0}(\mathbb{I} \mid x)\overline{A}(\varphi \mid x)}{\lambda_0 A_\lambda(\mathbb{I} \mid x)}. \qquad \square$$

7. Original and transformed processes

THEOREM 8.3. *Let* $(P_x) \in SM$; $(\exists (\Delta_m)_{m=1}^{\infty}, \Delta_m \in \mathfrak{A}, [\Delta_m] \in \mathfrak{K}, \mathbb{X} = \bigcup \Delta_n)$ $(\forall m \in \mathbb{N})$ (P_x) *be a lambda-continuous on interval* $[0, \sigma_{\Delta_m})$ *and lambda-regular family; and besides, let the following conditions be fulfilled:*

(1) $(\forall \varphi \in C_K)$ $\lambda R_\lambda \varphi \to \varphi$ *as* $\lambda \to \infty$ *uniformly on* $\mathbb{X}$;

(2) $(\forall \lambda > 0)$ $\mathbb{I} \in V_\lambda(\mathbb{X})$, $(\forall x \in \mathbb{X})$ $A_\lambda(\mathbb{I} \mid x) < 0$, $A_\lambda(\mathbb{I} \mid x) \to 0$ *as* $\lambda \to 0$, $A_\lambda(\mathbb{I} \mid x) \to -\infty$ *as* $\lambda \to \infty$;

(3) $(\forall \lambda > 0)$ $A_\lambda \mathbb{I} \in C$;

(4) $(\forall x \in \mathbb{X})$ $(\forall \lambda > 0)$ A_λ *is a pseudo-local operator at the point* x.

Then for any $(\forall \lambda_0 > 0)$ *there exists and only one Markov family* $(\overline{P}_x)$, *from which* (P_x) *is being obtained as a result of a random time change, and such that* $(\forall \varphi \in C_0)$ $\overline{R}_{\lambda_0} \varphi = R_{\lambda_0} \varphi$. *An infinitesimal operator of this process is* $\overline{A} = -\lambda_0 A_0 / A_{\lambda_0} \mathbb{I}$, *and time change corresponds to Laplace family of additive functionals* $(a_t(\lambda), t \geq 0)$ $(\lambda > 0)$, *where* $(\forall \xi \in \mathcal{D})$

$$a_t(\lambda \mid \xi) = \lambda_0 \int_0^t \frac{A_\lambda(\mathbb{I} \mid X_s \xi)}{A_{\lambda_0}(\mathbb{I} \mid X_s \xi)} ds.$$

Proof. Let $\psi \in C_0$. Check that for $\varphi = R_{\sigma_\Delta}(\lambda; \psi)$ the conditions of proposition 8.2 are fulfilled. Obviously, $\varphi \in \overline{V} \cap C_0$, and since $A_\lambda \varphi = -\psi$,

$$\overline{A}\varphi = -\lambda_0 A_0 \varphi / A_{\lambda_0} \mathbb{I} = -\lambda_0 (A_\lambda \varphi - \varphi A_\lambda \mathbb{I}) / A_{\lambda_0} \mathbb{I}$$

$$= \lambda_0 A_\lambda \mathbb{I}(-\psi + \varphi \lambda) / (\lambda A_{\lambda_0} \mathbb{I}).$$

Therefore, $\overline{A}\varphi \in C_0$. Furthermore, from pseudo-locality of operators A_λ and $\widetilde{A}_\lambda$ it follows

$$A_\lambda R_{\sigma_\Delta}(\lambda; \psi) = \widetilde{A}_\lambda \widetilde{R}_{\sigma_\Delta}(\lambda; \psi) = -\psi,$$

and also, by proposition 8.2,

$$A_\lambda R_{\sigma_\Delta}(\lambda; \psi) = \widetilde{A}_\lambda R_{\sigma_\Delta}(\lambda; \psi),$$

from here

$$\widetilde{A}_\lambda R_{\sigma_\Delta}(\lambda; \psi) = \widetilde{A}_\lambda \widetilde{R}_{\sigma_\Delta}(\lambda; \psi).$$

Substitute these expressions as arguments in operator $\widetilde{R}_{\sigma_\Delta}(\lambda; \cdot)$. The left part of the obtained equality by propositions 3.9 and 8.2 is equal to $-R_{\sigma_\Delta}(\lambda; \psi)$. The right part is equal to $-\widetilde{R}_{\sigma_\Delta}(\lambda;\psi)$. Furthermore, we prove by induction. We will use denotations of item 3.24:

$$R(1,n) = R_{\sigma_\Delta}\big(\lambda_1,\ldots,\lambda_n;\ \varphi_1,\ldots,\varphi_n\big)$$

and so on. Let for all $k \leq n-1$ it be proved that $R(1,k) = \widetilde{R}(1,k)$. Then, by theorem 3.4, we have

$$\mathcal{A}_{\lambda_1} R(1,n) = -\Phi_{1,n}, \quad \widetilde{\mathcal{A}}_{\lambda_1}\widetilde{R}(1,n) = -\widetilde{\Phi}_{1,n}, \quad \Phi_{1,n} = \widetilde{\Phi}_{1,n}.$$

Besides, as well as in for $n = 1$ we convince ourselves that function $\varphi = R(1,n)$ satisfies the conditions of proposition 8.2; from here we obtain $\widetilde{\mathcal{A}}_{\lambda_1} R(1,n) = \mathcal{A}_{\lambda_1} R(1,n)$ and consequently $\widetilde{\mathcal{A}}_{\lambda_1} R(1,n) = \widetilde{\mathcal{A}}_{\lambda_1}\widetilde{R}(1,n)$. Again substituting these values into operator $\widetilde{R}_{\sigma_\Delta}(\lambda,\cdot)$, we obtain $-R(1,n)$ on the left side of the equality. Further, the boundedness of $\widetilde{\mathcal{A}}^r_{\lambda_1}\widetilde{R}(1,n)$ and convergence to a limit follow from the proof of theorem 3.4. From here, by proposition 3.9, we obtain $-\widetilde{R}(1,n)$ on the right side of the equality. $\qquad\square$

8. Note about strict semi-Markov property

From theorem 6.9 it follows that conditions of theorem 8.3 are sufficient for family (P_x) to be strictly semi-Markovian (item 6.55). Really, (P_x) turns out from a strictly Markov family with the help of time change keeping all regeneration times which belong to IMT. Since $(\forall \tau \in \mathrm{IMT})\ \tau \in \mathrm{RT}(\overline{P}_x)$, we obtain $\tau \in \mathrm{RT}(P_x)$, i.e. (P_x) is a strictly semi-Markov process.

8.3. Construction of a Markov process by parameters of the Lévy formula

An analytical form of a conditional transition generating function of a distribution of a time run along the trace from theorem 6.11 (Lévy formula) gives us another method of representation of a semi-Markov process in the form of a transformed Markov process. An idea consists of looking for a transformation providing absence of intervals of constancy in trajectories of the process, but preserving the distribution of a trace. Such a transformed process appears to be Markovian. In order to prove Markovness of the constructed family of measures the theorem on regeneration times should be used, which is applied to a sequence of Markov times converging to a determinate instant of time. In this case a time change transforming the constructed Markov process into the original semi-Markov process is evident.

9. Process without intervals of constancy

Our next task consists of constructing a Markov process with the same random trace as in the original semi-Markov process. In other words, measures of the original semi-Markov process and the constructed Markov process have to coincide on the sigma-algebra $\mathcal{F}^\circ$. For this aim it is sufficient in the original process to change the distribution of a time run without varying the trace. Consider a family of kernels depending on a parameter $\lambda > 0$ of the form

$$\overline{f}_G(\lambda, S \mid x) \equiv E_x\big(\exp\big(-\overline{b}(\lambda, \tau)\big); \, X_\tau \in S\big)$$

$(G \in \mathfrak{A}, \, S \in \mathcal{B}(\mathbb{X}), \, x \in \mathbb{X})$, where $\tau = \sigma_G$, $\overline{b}(\lambda, \tau)$ is equal to λA_τ, and A_τ is some non-negative $\mathcal{F}^\circ$-measurable additive functional. A semi-Markov process with such a parameter in a Lévy expansion (if it exists) does not contain any interval of constancy (its Lévy measure is equal to zero). An example of such an additive functional is

$$\mu_\tau = a_\tau + \int_{0+}^{\infty} u n_\tau(du).$$

We will obtain a semi-Markov process with this additive functional as a limit of a sequence of semi-Markov processes with additive functionals $b(\lambda_k, \tau)/\lambda_k$ ($\lambda_k > 0$, $\lambda_k \to 0$). Thus, we denote

$$b^{(k)}(\lambda, \tau) = \lambda b(\lambda_k, \tau)/\lambda_k. \qquad [8.1]$$

PROPOSITION 8.3. *A family of kernels $f_G^{(k)}(\lambda, S \mid x)$ with parameter [8.1] is the family of transition generating functions of a semi-Markov process.*

Proof. Obviously, kernel $f_G^{(k)}(\lambda, S \mid x)$ with parameter [8.1] is an image of Laplace transformation of some semi-Markov kernel $F_G^{(k)}(dt \times S \mid x)$. In order to prove the proposition it is enough to check that the family of kernels $(F_G^{(k)})$ satisfies the conditions of proposition 4.2. Show the system of kernels $(f_G^{(k)})$ satisfies the Markov equation for transition generating functions. Actually, for $G \subset G_1$ and $\tau_1 = \sigma_{G_1}$ we have

$$f_{G_1}^{(k)}(\lambda, \varphi \mid x)$$
$$= E_x\big(\exp\big(-\lambda b(\lambda_k, \tau \dot{+} \tau_1)/\lambda_k\big)\big(\varphi \circ X_{\tau \dot{+} \tau_1}\big)\big)$$
$$= E_x\big(\exp\big(-\lambda\big(b(\lambda_k, \tau) + b(\lambda_k, \tau_1) \circ \theta_\tau\big)/\lambda_k\big)\big(\varphi \circ X_{\tau \dot{+} \tau_1}\big)\big)$$
$$= E_x\big(E_{X_\tau}\big(\exp\big(-\lambda b(\lambda_k, \tau_1)/\lambda_k\big)\big(\varphi \circ X_{\tau_1}\big)\big)\exp\big(-\lambda b(\lambda_k, \tau)/\lambda_k\big)\big)$$
$$= \int_{\mathbb{X}} f_{G_1}^{(k)}(\lambda, \varphi \mid x_1) f_G^{(k)}(\lambda, dx_1 \mid x).$$

From here admissibility of the family of transition function follows (see item 4.13). Hence, for any $x \in \mathbb{X}$, a projective family of measures is determined on sub-sigma-algebras of the sigma-algebra $\mathcal{F}$ (see Chapter 4). The condition of proposition 4.2(e) means convergence $f_{\tau_n}^{(k)}(\lambda, \mathbb{X} \mid x) \to 0$ as $n \to \infty$, where $\tau_n = \sigma_\delta^n$, δ is a deducing sequence of open sets, $\lambda > 0$, $x \in \mathbb{X}$. From the property of a deducing sequence it follows that $(\forall x \in \mathbb{X})\,(\forall \lambda > 0)\; P_x$-a.s. $\exp(-b(\lambda, \tau_n)) \to 0\;(n \to \infty)$. Thus, at least one of three components: a_{τ_n}, $\int_{0+}^{1} un_{\tau_n}\,du$, $\int_{1}^{\infty} n_{\tau_n}\,du$, tends to infinity P_x-a.s. From here it follows that function $b^{(k)}(\lambda, \tau_n)\; P_x$-a.s. tends to infinity,

$$b^{(k)}(\lambda, \tau_n) \geq \lambda a_{\tau_n} + \frac{\lambda}{\lambda_k}\left(1 - e^{-\lambda_k}\right)\left(\int_{0+}^{1} un_{\tau_n}(du) + \int_{1}^{\infty} n_{\tau_n}(du)\right),$$

i.e. condition (e) is fulfilled. The condition of proposition 4.2(d) means convergence $(\forall G \in \mathfrak{A}_1)\,(\forall \lambda > 0)\,(\forall x \in G)$

$$\int_{\mathbb{X}} f_G^{(k)}(\lambda, \mathbb{X} \mid x_1)\, f_{G-r}(0, dx_1 \mid x) \longrightarrow 1 \quad (r \longrightarrow 0),$$

where $\mathfrak{A}_1 \subset \mathfrak{A}$ is a rich enough family of open sets. We have

$$1 - \int_{\mathbb{X}} f_G^{(k)}(\lambda, \mathbb{X} \mid x_1)\, f_{G-r}(0, dx_1 \mid x)$$

$$= \int_{\mathbb{X}} \left(1 - f_G^{(k)}(\lambda, \mathbb{X} \mid x_1)\right) f_{G-r}(0, dx_1 \mid x)$$

$$= \int_{\mathbb{X}} E_{x_1}\left(1 - \exp\left(-b^{(k)}(\lambda, \sigma_G)\right)\right) f_{G-r}(0, dx_1 \mid x).$$

For $\lambda < \lambda_k$ using inequality

$$\left(1 - \exp(-\lambda u)\right)/\lambda > \left(1 - \exp\left(-\lambda_k u\right)\right)/\lambda_k,$$

we obtain $b^{(k)}(\lambda, \sigma_G) \leq b(\lambda, \sigma_G)$, from here follows the required property. For $\lambda > \lambda_k$; we use the inequality

$$1 - \exp\left(-\lambda u/\lambda_k\right) \leq \lambda\left(1 - \exp(-u)\right)/\lambda_k$$

and we also obtain the required property, since in the case of the original process it is fulfilled for a rich enough class of open sets. Hence, the constructed projective family of measures for any $x \in \mathbb{X}$ can be extended up to a probability measure $P_x^{(k)}$ on $(\mathcal{D}, \mathcal{F})$ and, by theorem 4.6, the family of this measures is semi-Markovian.   □

Consider a family of semi-Markov transition generating functions $(\overline{f}_G)\,(G \in \mathfrak{A})$ where

$$\overline{f}_G(\lambda, S \mid x) \equiv E_x\left(\exp\left(-\lambda \mu_\tau\right); X_\tau \in S\right).$$

PROPOSITION 8.4. *The family of kernels $\overline{f}_G(\lambda, S \mid x)$ is a family of transition generating functions of some semi-Markov process.*

Proof. This family of generating functions for any $x \in \mathbb{X}$ determines a projective family of probability measures

$$\left(\overline{P}_x \circ \left(\beta_{\sigma_r^0}, \ldots, \beta_{\sigma_r^n}\right)^{-1}\right) \quad (r > 0, \, n \geq 1),$$

where

$$\beta_\tau(\xi) = \begin{cases} \infty, & \tau(\xi) = \infty, \\ \left(\tau(\xi), X_\tau(\xi)\right), & \tau(\xi) < \infty. \end{cases}$$

Since P_x-a.s. $b^{(k)}(\lambda, \tau) \to \lambda\mu_\tau$ as $k \to \infty$ the distribution $f_G^{(k)}(\lambda, \cdot \mid x)$ converges weakly to $\overline{f}_G(\lambda, \cdot \mid x)$ and hence, by theorem 7.3, the sequence of distributions $(P_x^{(k)} \circ (\beta_{\sigma_r^0}, \ldots, \beta_{\sigma_r^n})^{-1})$ converges weakly to $(\overline{P}_x \circ (\beta_{\sigma_r^0}, \ldots, \beta_{\sigma_r^n})^{-1})$. Under theorem 7.2, from this convergency, firstly, weak compactness of the family of distributions $(P_x^{(k)})$ follows and, secondly, weak convergence of this sequence to the limit $\overline{P}_x$ follows (it is a projective limit of the family of distributions $(\overline{P}_x \circ (\beta_{\sigma_r^0}, \ldots, \beta_{\sigma_r^n})^{-1})$), which is a probability measure on $\mathcal{D}$, and the family of these measures is consistent as a family distributions of a semi-Markov process. $\square$

10. Markovness of transformed family of measures

By our construction, the obtained semi-Markov family of measures determines a process without intervals of constancy. Our hypothesis is that without any additional conditions, such a process is Markovian. At the present time we do not know how to prove this hypothesis. In order to prove Markovness of the constructed process we use some assumption on regularity of the original process.

THEOREM 8.4. *Let for any $\lambda > 0$, $G \in \mathfrak{A}$, and continuous bounded function φ on $\mathbb{X}$ a transition generating function $f_G(\lambda, \varphi \mid \cdot)$ be continuous. Then: 1) the semi-Markov process with the transition generating function $f_G^{(k)}(\lambda, S \mid x)$ is a Markov process (see proposition 8.3); 2) the semi-Markov process with the transition generating function $\overline{f}_G(\lambda, S \mid x)$ is a Markov process (see proposition 8.4).*

Proof. On the basis of theorem 7.2 we conclude that $(P_x^{(k)})$ is a weakly continuous family. Prove that the semi-Markov family of measures $(\overline{P}_x)$ (and also $(P_{(k)})$ for any k) is also weakly continuous. By Lévy representation for a first exit time of the transformed process from set G, it follows that $\overline{P}_x$-a.s. this time is equal to μ_{σ_G}. On the other hand, by Lévy formula $\sigma_G = a_{\sigma_G} + \int_{0+}^{\infty} u P(du, \sigma_G)$. In this expression two last summands depend monotonically on G. In particular, by replacing G with G^{-r} (G^{+r}) they both increase (decrease) as $r \downarrow 0$. The same conclusion is fair with respect to the intensity of the Poisson measure, i.e. Lévy measures $n(du, \sigma_G)$, and also that of additive functional $b(\lambda, \sigma_G)$. Hence $\overline{P}_x(\Pi(G)) = 1$ for those G, for which $P_x(\Pi(G)) = 1$, where $\Pi(G)$ is a set of trajectories ξ, exiting correctly from G. The same relates to any finite sequence of open sets and a set of functions, exiting correctly from these sets. We obtain the family of functions of the form

$$\prod_{i=1}^{n} \exp\left(-\lambda\mu_{\tau_i}\right)\varphi_i\left(X_{\tau_i}\right) \quad \left(\tau_i = \sigma_{G_1,\dots,G_i}, \varphi_i \in C_0(\mathbb{X})\right)$$

is continuous on a full measure set and, consequently, an integral of this function on measure P_x is continuous according to x. It implies lambda continuity of the family $(\overline{P}_x)$ (and also its weak continuity). Furthermore, for a given deducing sequence δ and $t > 0$ we define a Markov time τ_δ^t:

$$\tau_\delta^t(\xi) = \sigma_\delta^k(\xi) \Longleftrightarrow \sigma_\delta^{k-1}(\xi) < t \le \sigma_\delta^k(\xi),$$

which is a regeneration time for the semi-Markov family $(\overline{P}_x)$. Besides, the sequence of such Markov times constructed for δ with decreasing ranks as $\operatorname{rank}\delta \to 0$, $\overline{P}_x$-a.s. converges to t. From here it follows that the fixed time $t \in \mathbb{R}_+$ is a regeneration time of the family $(\overline{P}_x)$, i.e. this family is Markov.   □

Possibility of constructing such *an associated Markov process* for which any intrinsic Markov time τ coincides with μ_τ – a conditional expectation of τ with respect to a trace in the original semi-Markov process – implies interesting consequences for a distribution of value μ_τ. It is well-known that for a Markov process of diffusion type, ratio $E_x((\sigma_G)^2)/E_x(\sigma_G)$ tends to zero as $G \downarrow x$. For a semi-Markov process of diffusion type, which is not Markovian, this fact generally is not true. Only boundedness of this ratio can be proved. However, for semi-Markov process convergence to zero of the ratio $P_x((\mu_{\sigma_G})^2)/P_x(\mu_{\sigma_G})$ is true since for its associated Markov process $\mu_{\sigma_G} = \sigma_G$. From here, in particular, fairness of representation [8.5] follows (see below) as well as that of a similar assumption while deriving the Ito formula.

8.4. Stationary distribution of semi-Markov process

Representation of a semi-Markov process as a Markov process transformed by a time change makes it possible to solve a problem of existence of a stationary distribution for a SM process and to express it in terms of the associated Markov process. In the present section a stationary distribution for a continuous SM process is investigated. A proof of existence of such a distribution is based on well-known theorems on a stationary distribution for stepped SM processes (see Koroljuk and Turbin [KOR 76], Shurenkov [SHU 89]). For this aim we use uniform convergence of a sequence semi-Markov walks to a continuous semi-Markov process.

11. Three-dimensional distribution

In what follows we will be interested in intervals of constancy of trajectory ξ. Remember that $R_t^-(\xi)$ and $R_t^+(\xi)$ are lengths of the left and the right parts of an interval of constancy of trajectory ξ, covering a point t; if such an interval is absent, we assume $R_t^-(\xi) = R_t^+(\xi) = 0$ (see 1.18).

Consider distribution $P_x^t(B) = P_x(\theta_t^{-1}B)$ $(B \in \mathcal{F})$. A semi-Markov process is said to be *ergodic* if for any $x \in \mathbb{X}$ and $\xi \in \mathcal{D}$ P_x-a.s. there exists a limit $\lim_{t\to\infty}(1/t)\int_0^t f(\theta_t\xi)\,dt$, which is not dependent on a trajectory where f is a bounded measurable function on $\mathcal{D}$. We will consider a rather simpler variant of ergodicity when distribution P_x^t tends weakly to some distribution P^∞ which is the same for different x.

For a Markov process we have $P_x^t(B) = E_x(P_{X_t}(B))$, and thus convergence of a measure to a limit reduces to convergence to a limit of its one-dimensional distribution $H_t(S \mid x) \equiv P_x(X_t \in S)$ $(S \in \mathcal{B}(\mathbb{X}))$. Evidently, for a weakly continuous Markov family of measures (P_x) from weak convergence $H_t(\cdot \mid x) \xrightarrow{w} H_\infty$ it follows that

$$P_x^t(f) \longrightarrow \int_{\mathbb{X}} P_y(f)H_\infty(dy),$$

where f is a continuous bounded functional on $\mathcal{D}$.

For a semi-Markov process a fixed instant of time is not (in general) its regeneration time and the preceding arguments are not applicable. However, for any $t \in \mathbb{R}_+$ a sequence of regeneration times of this process can be constructed which converges from above to a function $t + R_t^+$. It follows that if the family of measures (P_x) of a semi-Markov process is weakly continuous, the time $t + R_t^+$ is a regeneration time for this process. On the other hand, for some classes of semi-Markov processes a pair (X_t, R_t^-) is a Markov process (see, e.g., Gihman and Skorokhod [GIH 73]). Taking into account a special role of processes R_t^- and R_t^+, we will investigate a three-dimensional distribution

$$G_t(S \times A_1 \times A_2 \mid x) = P_x(X_t \in S, \ R_t^- \in A_1, \ R_t^+ \in A_2)$$

$(A_1, A_2 \in \mathcal{B}(\mathbb{R}_+))$. Evidently, a measure P_x^t for semi-Markov process with a weakly continuous family of measures (P_x) tends weakly to a limit, if for any $x \in \mathbb{X}$ as $t \to \infty$ the three-dimensional distribution $G_t(\cdot \mid x)$ tends weakly to some distribution G_∞, which is the same for any x.

12. Shurenkov theorem

Consider a transition function $F(dt \times dx_1 \mid x)$ of a stepped semi-Markov process. Let $(\forall x \in \mathbb{X})$ $F(\mathbb{R}_+ \times \mathbb{X} \mid x) = 1$; F^n be the n-th iteration of the kernel $F \equiv F^1$. A marginal kernel $H(dy \mid x) = F(\mathbb{R}_+ \times dy \mid x)$ is a transition function of the embedded Markov chain. A Markov chain is said to be ergodic if for any $x \in \mathbb{X}$

$$\frac{1}{n}\sum_{k=1}^n H^k(\cdot \mid x) \xrightarrow{w} H^\infty,$$

where H^k is the k-th iteration of the kernel $H \equiv H^1$; H^∞ is a probability measure on $\mathbb{X}$, for which a stationarity condition is fulfilled:

$$H^\infty(A) = \int_{\mathbb{X}} H(A \mid x) H^\infty(dx_1) \quad (A \in \mathcal{B}(\mathbb{X})).$$

Let

$$U(A \times S \mid x) = \sum_{n=0}^{\infty} F^n(A \times S \mid x),$$

where $F^0(A \times S \mid x) = I(A \times S \mid x)$, an indicator function. Let $Q^n(A \mid x) = F^n(A \times \mathbb{X} \mid x)$ $(n \geq 0, A \in \mathcal{B}(\mathbb{R}_+))$ and $Q \equiv Q^1$. Let $q_n(t \mid x)$ $(t \geq 0)$ be a density of an absolutely continuous component of measure $Q^n(\cdot \mid x)$. The family of distributions $(Q^n(\cdot \mid x))$ $(x \in \mathbb{X})$ is said to be *non-singular* with respect to measure ℓ on $\mathbb{X}$, if

$$\int_{\mathbb{X}} \int_0^\infty q_n(t \mid x) \, dt \, \ell(dx) > 0.$$

Denote

$$M = \int_{\mathbb{X}} \int_0^\infty t Q(dt \mid x) H^\infty(dx).$$

For a stepped semi-Markov process the following theorem is fair.

THEOREM 8.5. *Let the embedded Markov chain of a given semi-Markov process be ergodic with a stationary probability distribution H^∞. Let $0 < M < \infty$ and besides there exists $n \geq 1$ such that the family of distributions $(Q^n(\cdot \mid x))$ is non-singular with respect to the measure H^∞. Let for a measurable real function ψ, determined on $\mathbb{R}_+ \times \mathbb{X}$, the following conditions be fulfilled:*

(1) $H_*^\infty\{x \in \mathbb{X} : \sup_{t \geq 0} \int_{\mathbb{X}} \int_0^t \psi(t - s, y) U(ds \times dy \mid x) < \infty\} > 0$, where H_*^∞ is an internal measure of the measure H^∞;

(2) $\lim_{t \to \infty} \psi(t, x) = 0$ for H^∞-almost all x;

(3) $\int_{\mathbb{X}} \int_0^\infty \psi(t, x) \, dt \, H^\infty(dx) < \infty$;

(4) the family of $\mathcal{B}(\mathbb{X})$-measurable functions $(\psi(t, \cdot))$ $(t \geq 0)$ has a H^∞-integrable majorant.

Then it is fair

$$\lim_{t \to \infty} \int_{\mathbb{X}} \int_0^t \psi(t - s, y) U(ds \times dy \mid x)$$

$$= \frac{1}{M} \int_{\mathbb{X}} \int_0^\infty \psi(t, y) \, dt \, H^\infty(dy).$$

[8.2]

Proof. See Shurenkov [SHU 89, p. 127]. $\qquad\qquad\square$

13. Ergodicity and stationary distribution

Now we return to general semi-Markov processes. According to the results of Shurenkov [SHU 89], in order to prove ergodicity of a process it is sufficient to find embedded in it an appropriate Markov renewal process (a so called Markov inter- ference of chance). Let $\tau \equiv \tau_1$ be a regeneration time of a semi-Markov family of probability measures; (τ_n) be a sequence of iterated Markov times. In order to apply the Shurenkov theorem we have to propose that for any x P_x-a.s. $\tau_n < \infty$ and $\tau_n \to \infty$. This would be the case, for example, if $\tau = \sigma_A \dotplus \sigma_B$, where A and B are two open subsets of $\mathbb{X}$ such that $\mathbb{X} \setminus A \subset B$. Note that an alternate sequence of sets $(A, B, A, B, \ldots)$ is a partial case of a deducing sequence. Thus, if all τ_n are finite the sequence of pairs (τ_n, X_{τ_n}) $(n \geq 0, \tau_0 = 0)$ composes a Markov renewal process. Suppose that the semi-Markov transition function $F_\tau(A \times S \mid x) \equiv P_x(\tau \in A, \; X_\tau \in S)$ and its derived kernels $F_{\tau_n} \equiv F_\tau^n$, $H_{\tau_n} \equiv H_\tau^n$, Q_{τ_n} satisfy conditions of theorem 8.5 with a limit distribution H_τ^∞ and an average meaning of a length of the interval between neighboring regeneration times

$$M_\tau = \int_{\mathbb{X}} \int_0^\infty t Q_\tau(dt \mid x) H_\tau^\infty(dx).$$

Consider the three-dimensional distribution. We have

$$G_t\big(S \times A_1 \times A_2 \mid x\big)$$

$$= \sum_{k=0}^\infty P_x\big(X_t \in S, \; R_t^- \in A_1, \; R_t^+ \in A_2, \; \tau_k \leq t < \tau_{k+1}\big)$$

$$= \sum_{k=0}^\infty \int_0^t P_x\big(\tau_k \in ds, \; X_t \in S, \; R_t^- \in A_1, \; R_t^+ \in A_2, \; t < s + \tau \circ \theta_{\tau_k}\big)$$

$$= \sum_{k=0}^\infty \int_0^t P_x\big(\tau_k \in ds, \; X_{t-s} \circ \theta_{\tau_k} \in S, \; R_{t-s}^- \circ \theta_{\tau_k} \in A_1,$$

$$R_{t-s}^+ \circ \theta_{\tau_k} \in A_2, \; t - s < \tau \circ \theta_{\tau_k}\big)$$

$$= \sum_{k=0}^\infty \int_0^t E_x\big(P_{X(\tau_k)}\big(X_{t-s} \in S, \; R_t^- \in A_1, \; R_t^+ \in A_2, \; t - s < \tau\big); \; \tau_k \in ds\big)$$

$$= \int_{\mathbb{X}} \int_0^t P_y\big(X_{t-s} \in S, \; R_{t-s}^- \in A_1, \; R_{t-s}^+ \in A_2, \; t - s < \tau\big) U(ds \times dy \mid x).$$

Let $\psi_\tau(t, y) = P_y(X_t \in S, \; R_t^- \in A_1, \; R_t^+ \in A_2, \; t < \tau)$. Check for this function fulfillment of conditions of theorem 8.5. It is a bounded function converging to zero as $t \to \infty$. Since $\psi_\tau(t, y) \leq P_y(t < \tau)$, the integral included in condition (1) is not more than 1; an integral of ψ on t is not more than $m_\tau(y) \equiv P_y(\tau)$ and therefore it is H_τ^∞-a.s. finite. Hence, there exists a limit of probability $G_t(S \times A_1 \times A_2 \mid x)$ as $t \to \infty$ and this limit is equal to

$$G_\infty(S \times A_1 \times A_2) = \frac{1}{M_\tau} \int_{\mathbb{X}} \int_0^\infty \psi_\tau(t, x)\, dt\, H_\tau^\infty(dx).$$

14. Representation in terms of Lévy expansion

Our next aim is to express the integral $\int_0^\infty \psi_\tau(t, x)\, dt$ in terms of Lévy expansion for a transition generating function of a semi-Markov process. For $\lambda > 0$ and a bounded function φ we consider Lévy expansion for transition generating function f_τ:

$$f_\tau(\lambda, \varphi \mid x) \equiv E_x\big(\exp(-\lambda\tau)\varphi(X_\tau)\big) = E_x\big(\exp\big(-b(\lambda, \tau)\big)\varphi(X_\tau)\big),$$

where $b(\lambda, \tau)$ is a $\mathcal{F}^\circ$-measurable additive functional

$$b(\lambda, \tau) = \lambda a_\tau + \int_{0+}^\infty \big(1 - e^{-\lambda u}\big) n_\tau(du) + B(\lambda, \tau) \qquad [8.3]$$

(see theorem 6.11), where a_τ is a $\mathcal{F}^\circ$-measurable additive functional; $n_\tau(S)$ ($S \in \mathcal{B}(\mathbb{R}_+)$) is a $\mathcal{F}^\circ$-measurable additive functional being a measure on argument S (Lévy measure). Measure n_τ is an intensity of Poisson field $\mathcal{P}_\tau$ on a sequence of states (trace) determining position and length of "Poisson" intervals of constancy. An additive functional $B(\lambda, \tau)$ is determined by "conditionally fixed" intervals of constancy, and for a given measure P_x it is of the form

$$B(\lambda, \tau) = -\sum_{\tau_i < \tau} \log P_x\big(e^{-\lambda\sigma_0} \circ \theta_{\tau_i} \mid \mathcal{F}^\circ\big),$$

where $(\tau_i + \sigma_0)$ is either finite, or an infinite countable collection of $\mathcal{F}^\circ$-measurable Markov times with respect to a widened stream of sigma-algebras; σ_0 is a first exit time from an initial point. In a simple case which we will consider below, times τ_i themselves are Markov times, namely, first hitting times of some points of the space of states. Sojourn times at these points are independent random values with distributions depending on $i \in \mathbb{N}$.

From the previous expansion of the generating function it follows Lévy representation for a Markov time τ follows (see item 6.62, formula [6.4]):

$$\tau = a_\tau + \int_{0+}^\infty u \mathcal{P}_\tau(du) + \sum_{\tau_i < \tau} \sigma_0 \circ \theta_{\tau_i}. \qquad [8.4]$$

Furthermore, everywhere, except a concluding example, we will suppose that conditionally fixed intervals of constancy are absent.

Consider an integral

$$G_\infty(\varphi; A_1 \times A_2) \equiv \int_{\mathbb{X}} \varphi(x) G_\infty(dx \times A_1 \times A_2),$$

where φ is a continuous bounded real function on $\mathbb{X}$. For this integral the following representation is fair

$$G_\infty(\varphi; A_1 \times A_2) = \frac{1}{M_\tau} \int_{\mathbb{X}} \int_0^\infty \widetilde{\psi}_\tau(t, x) \, dt \, H_\tau^\infty(dx),$$

where $\widetilde{\psi}_\tau(t, x) = P_x(\varphi(X_t); R_t^- \in A_1, R_t^+ \in A_2, t < \tau)$. We have

$$\int_0^\infty \widetilde{\psi}_\tau(t, x) \, dt = E_x\left(\int_0^\tau \varphi(X_t) I_{A_1}(R_t^-) I_{A_2}(R_t^+) \, dt\right).$$

We split the interval $[0, \tau)$, depending on a trajectory, on a finite number N_d of parts by Markov times $\bar{\tau}_k$ in such a manner that on each of these intervals the trajectory does not exit from some open set of a small diameter d. Then function $\varphi(X_t)$ has a small variation on this interval. Thus, the previous value can be found as a limit of sums

$$E_x\left(\sum_{k=1}^{N_d} \varphi(X_{\bar{\tau}_{k-1}}) \int_{\bar{\tau}_{k-1}}^{\bar{\tau}_k} I_{A_1}(R_t^-) I_{A_2}(R_t^+) \, dt\right).$$

Using Lévy representation [8.4] for times $\bar{\tau}_k$, we can write the integrals in another view, at least for two partial cases. If $A_1 = A_2 = \{0\}$,

$$\int_{\bar{\tau}_k}^{\bar{\tau}_{k+1}} I_{A_1}(R_t^-) I_{A_2}(R_t^+) \, dt = a_{\bar{\tau}_{k+1}} - a_{\bar{\tau}_k}.$$

If $A_1 = (r, \infty)$ and $A_2 = (s, \infty)$,

$$\int_{\bar{\tau}_k}^{\bar{\tau}_{k+1}} I_{A_1}(R_t^-) I_{A_2}(R_t^+) \, dt = \int_{s+r}^\infty (u - s - r) \mathcal{P}_{[\bar{\tau}_k, \bar{\tau}_{k+1})}(du),$$

where $\mathcal{P}_{[\bar{\tau}_k, \bar{\tau}_{k+1})}(du) = \mathcal{P}_{\bar{\tau}_{k+1}}(du) - \mathcal{P}_{\bar{\tau}_k}(du)$. Using semi-Markov properties of the process the Poisson measure can be replaced by a measure of its intensity. As a result we can write the sum

$$E_x\left(\sum_{k=1}^{N_d} \varphi(X_{\bar{\tau}_{k-1}}) c_{[\bar{\tau}_k, \bar{\tau}_{k+1})}\right),$$

where $c_{[\bar{\tau}_k, \bar{\tau}_{k+1})} = c_{\bar{\tau}_{k+1}} - c_{\bar{\tau}_k}$ and

$$c_\tau = \begin{cases} a_\tau, & A_1 = A_2 = \{0\}, \\ \displaystyle\int_{r+s}^\infty (u - r - s) n_\tau(du), & A_1 = (r, \infty), \ A_2 = (s, \infty). \end{cases}$$

Passing to a limit as $d \to 0$, we obtain a representation of an integrand in form of a curvilinear integral by $\mathcal{F}^{\circ}$-measurable additive functional c_τ (see item 6.65):

$$\int_0^\infty \tilde{\psi}_\tau(t,x)\,dt = E_x\left(\int_0^\tau \varphi(X_{\tau_1})\,dc_{\tau_1}\right).$$

This representation accepts a more simple form if additive functionals a_τ and $n_\tau(B)$ themselves are representable in view of curvilinear integrals by the same additive functional. It is not difficult to prove that function

$$A_\lambda \mathbb{I}(x) \equiv \lim_{\Delta \downarrow x} P_x\left(\exp\left(-\lambda\sigma_\Delta\right) - 1\right)/P_x\left(\sigma_\Delta\right)$$

can be represented in Lévy form

$$-A_\lambda \mathbb{I}(x) = \lambda\alpha(x) + \int_{0+}^\infty \left(1 - e^{-\lambda u}\right)\nu(du \mid x),$$

where $\alpha(x)$ is some non-negative function, and $\nu(du \mid x)$ is some family of measures depending on $x \in \mathbb{X}$. It is not difficult to give sufficient conditions for the function $A_\lambda \mathbb{I}$ continuous in some neighborhood of the point x to have the following representation

$$b(\lambda,\tau) = \int_0^\tau \left(-A_\lambda \mathbb{I} \circ X_{\tau_1}\right)d\mu_{\tau_1}, \qquad [8.5]$$

where μ is an additive functional: $\mu_\tau = a_\tau + \int_0^\infty u n_\tau(du)$ (see item 10, note on theorem 8.4). In what follows we suppose that this representation is fair; and thus

$$b(\lambda,\tau) = \int_0^\tau \left(\lambda\alpha(X_{\tau_1}) + \int_{0+}^\infty \left(1 - e^{-\lambda u}\right)\nu(du \mid X_{\tau_1})\right)d\mu_{\tau_1}. \qquad [8.6]$$

In this case

$$G_\infty\left(\varphi; A_1 \times A_2\right) = \frac{1}{M_\tau}\int_{\mathbb{X}} E_x\left(\int_0^\tau \varphi(X_{\tau_1})\gamma(X_{\tau_1})\,d\mu_{\tau_1}\right)H_\tau^\infty(dx), \qquad [8.7]$$

where

$$\gamma(x) = \begin{cases} \alpha(x), & A_1 = A_2 = \{0\}, \\ \int_{r+s}^\infty (u - r - s)\nu(du \mid x), & A_1 = (r,\infty),\ A_2 = (s,\infty). \end{cases}$$

Thus, the three-dimensional stationary distribution of semi-Markov process is expressed through a one-dimensional process:

$$G_\infty\left(\varphi; A_1 \times A_2\right) = G_\infty\left(\varphi\gamma; \mathbb{R}_+ \times \mathbb{R}_+\right).$$

We have obtained a stationary distribution characterizing (with a family of probability measures (P_x)) the stationary measure P^∞ of a semi-Markov process. In order
to give this distribution more closed form and get rid of formal dependence on the
choice of τ, we use an associated Markov process and express the three-dimensional
stationary distribution of a semi-Markov process through stationary distribution of a
Markov process.

15. Markov representation of stationary distribution

Consider the original semi-Markov family of measures (P_x) and the Markov fami-
lies of measures $(P_x^{(k)})$ and $(\overline{P}_x)$, constructed in the previous section. An expectation
of a Markov time can be found by differentiating the corresponding Laplace transformation image on parameter λ for $\lambda = 0$. From here

$$m_\tau(x) = E_x\left(a_\tau + \int_{0+}^\infty u n_\tau(du)\right),$$

$$m_\tau^{(k)}(x) = E_x\left(a_\tau + \frac{1}{\lambda_k}\int_{0+}^\infty (1 - e^{\lambda_k u})n_\tau(du)\right),$$

$$\overline{m}_\tau(x) = E_x(\mu_\tau) = m_\tau(x).$$

Let $\mu_\tau^{(k)} = E_x^{(k)}(\tau \mid \mathcal{F}^\circ)$ and $\overline{\mu}_\tau = \overline{E}_x(\tau \mid \mathcal{F}^\circ)$. From here

$$\mu_\tau^{(k)} = a_\tau + \frac{1}{\lambda_k}\int_{0+}^\infty (1 - e^{\lambda_k u})n_\tau(du), \quad \overline{\mu}_\tau = \mu_\tau.$$

According to the assumption, additive functional $\mu_\tau^{(k)}$ can be written in the form of a
curvilinear integral:

$$\mu_\tau^{(k)} = \int_0^\tau \left(\alpha(X_{\tau_1}) + \frac{1}{\lambda_k}\int_{0+}^\infty (1 - e^{-\lambda_k u})\nu(du \mid X_{\tau_1})\right)d\mu_{\tau_1}.$$

From the arbitrariness of τ we conclude that the following rule of replacing an additive
functional is fair:

$$d\mu_{\tau_1}^{(k)} = \left(\alpha(X_{\tau_1}) + \frac{1}{\lambda_k}\int_{0+}^\infty (1 - e^{-\lambda_k u})\nu(du \mid X_{\tau_1})\right)d\mu_{\tau_1},$$

$$d\mu_{\tau_1} = \left(\alpha(X_{\tau_1}) + \frac{1}{\lambda_k}\int_{0+}^\infty (1 - e^{-\lambda_k u})\nu(du \mid X_{\tau_1})\right)^{-1} d\mu_{\tau_1}^{(k)}.$$

Hence the stationary three-dimensional distribution of semi-Markov process [8.7]
obtained above can be represented in two forms. Firstly, it is

$$G_\infty(\varphi; A_1 \times A_2)$$

$$= \frac{1}{M_\tau}\int_X E_x\left(\int_0^\tau \varphi(X_{\tau_1})\gamma(X_{\tau_1})\beta^{(k)}(X_{\tau_1})\,d\mu_{\tau_1}^{(k)}\right)H_\tau^\infty(dx),$$

where

$$\beta^{(k)}(x) = \left(\alpha(x) + \frac{1}{\lambda_k} \int_{0+}^{\infty} \left(1 - e^{-\lambda_k u}\right) \nu(du \mid x) \right)^{-1}.$$

Secondly,

$$G_{\infty}\left(\varphi; A_1 \times A_2\right) = \frac{1}{M_{\tau}} \int_{\mathbb{X}} E_x \left(\int_0^{\tau} \varphi(X_{\tau_1}) \gamma(X_{\tau_1}) \, d\overline{\mu}_{\tau_1} \right) H_{\tau}^{\infty}(dx).$$

Since the curvilinear integral is $\mathcal{F}^{\circ}$-measurable and P_x, $P_x^{(k)}$ and $\overline{P}_x$ coincide on $\mathcal{F}^{\circ}$, then passing from a curvilinear integral to a common one, internal integrals in these representations can be represented as

$$E_x^{(k)} \left(\int_0^{\tau} \varphi(X_t) \gamma(X_t) \beta^{(k)}(X_t) \, dt \right), \qquad \overline{E}_x \left(\int_0^{\tau} \varphi(X_t) \gamma(X_t) \, dt \right)$$

correspondingly. Assume that families of measures $(P_x^{(k)})$ and $(\overline{P}_x)$, and functions $\psi^{(k)}(t, x)^{(k)} \equiv E_x^{(k)}(\varphi(X_t) \gamma(X_t) \beta^{(k)}(X_t))$ and $\overline{\psi}(t, x) \equiv \overline{E}_x(\varphi(X_t) \gamma(X_t))$ satisfy conditions of theorem 8.5 for the same Markov time τ. Obviously, these assumptions touch upon only a time run along the trace, but the trace itself remains fixed. In particular, these Markov processes correspond to the same embedded Markov chain with a stationary distribution H_{τ}^{∞}. In this case there exist one-dimensional stationary distributions $G_{\infty}^{(k)}$ and $\overline{G}_{\infty}$ of the constructed Markov processes; and integrals with respect to these distributions are

$$G_{\infty}^{(k)}\left(\varphi \gamma \beta^{(k)}\right) = \frac{1}{M_{\tau}^{(k)}} \int_{\mathbb{X}} E_x^{(k)} \left(\int_0^{\tau} \varphi(X_t) \gamma(X_t) \beta^{(k)}(X_t) \, dt \right) H_{\tau}^{\infty}(dx),$$

$$\overline{G}_{\infty}(\varphi \gamma) = \frac{1}{M_{\tau}} \int_{\mathbb{X}} \overline{E}_x \left(\int_0^{\tau} \varphi(X_t) \gamma(X_t) \, dt \right) H_{\tau}^{\infty}(dx)$$

correspondingly, where

$$M_{\tau}^{(k)} = \int_{\mathbb{X}} m_{\tau}^{(k)}(x) \, H_{\tau}^{\infty}(dx),$$

$$\overline{M}_{\tau} = \int_{\mathbb{X}} \overline{m}_{\tau}(x) \, H_{\tau}^{\infty}(dx) = M_{\tau}.$$

From here it follows the interesting Markov representations of stationary distribution of semi-Markov process

$$G_{\infty}\left(\varphi; A_1 \times A_2\right) = \frac{M_{\tau}^{(k)}}{M_{\tau}} G_{\infty}^{(k)}\left(\varphi \gamma \beta^{(k)}\right),$$

$$G_{\infty}\left(\varphi; A_1 \times A_2\right) = \overline{G}_{\infty}(\varphi \gamma).$$

In this case we need not make too strong suppositions on Markov families of measures as in theorem 8.5. We will prove that distributions $G_\infty^{(k)}$ and $\overline{G}_\infty$ are already stationary distributions of constructed Markov processes.

LEMMA 8.1. *For any semi-Markov family (P_x) with a stationary distribution H_τ^∞ of an embedded Markov chain, corresponding to a regeneration time τ, for any $n \geq 1$ and a measurable function φ it is fair*

$$\int_X \int_0^\infty E_x\big(\varphi(X_t);\ t < \tau_n\big)\, dt\, H_\tau^\infty(dx)$$

$$= n \int_X \int_0^\infty E_x\big(\varphi(X_t);\ t < \tau\big)\, dt\, H_\tau^\infty(dx),$$

in particular

$$M_{\tau_n} \equiv \int_X E_x\big(\tau_n\big)\, H_\tau^\infty(dx) = n M_\tau$$

Proof. We have

$$\int_0^\infty E_x\big(\varphi(X_t);\ t < \tau_{n+1}\big)\, dt = \int_0^\infty E_x\big(\varphi(X_t);\ t < \tau_n\big)\, dt$$

$$+ \int_0^\infty E_x\big(\varphi(X_t);\ \tau_n \leq t < \tau_{n+1}\big)\, dt.$$

The second integral is equal to

$$\int_0^\infty \int_0^t \overline{E}_x\big(\varphi(X_t);\ \tau_n \in dt_1,\ t < t_1 + \tau \circ \theta_{\tau_n}\big)\, dt.$$

Substituting into integral $X_t = X_{t-t_1} \circ \theta_{\tau_n}$ and using a regeneration condition for Markov time τ_k, we obtain

$$\int_0^\infty \int_0^t E_x\big(E_{X(\tau_n)}\big(\varphi(X_{t-t_1});\ t - t_1 < \tau\big);\ \tau_n \in dt_1\big)\, dt$$

$$= \int_0^\infty \int_0^\infty E_x\big(E_{X(\tau_n)}\big(\varphi(X_t);\ t < \tau\big);\ \tau_n \in dt_1\big)\, dt$$

$$= \int_0^\infty E_x\big(E_{X(\tau_n)}\big(\varphi(X_t);\ t < \tau\big)\big)\, dt.$$

Integrating both integrals on x and using stationary property of measure H_τ^∞, we obtain

$$\int_X \int_0^\infty E_x\big(\varphi(X_t);\ t < \tau_{n+1}\big)\, dt\, H_\tau^\infty(dx)$$

$$= \int_X \int_0^\infty E_x\big(\varphi(X_t);\ t < \tau_n\big)\, dt\, H_\tau^\infty(dx)$$

$$+ \int_X \int_0^\infty E_x\big(\varphi(X_t);\ t < \tau\big)\, dt\, H_\tau^\infty(dx).$$

$\qquad\qquad\qquad\qquad\qquad\qquad\qquad\qquad\qquad\qquad\qquad\qquad\qquad\square$

The following proposition, formulated for $(\overline{P}_x)$, is fair for any Markov process, obtained from a given semi-Markov process with the help of time change, in particular, for $(P_x^{(k)})$.

PROPOSITION 8.5. *One-dimensional distribution $\overline{G}_\infty$ is a stationary distribution for the constructed Markov process.*

Proof. For any measurable bounded function φ and $h > 0$ it is fair

$$\int_X \overline{E}_x\big(\varphi(X_h)\big)\overline{G}_\infty(dx)$$

$$= \frac{1}{\overline{M}_\tau} \int_X \int_0^\infty \overline{E}_x\big(\overline{E}_{X_t}\big(\varphi(X_h)\big);\ t < \tau\big)\, dt\, H_\tau^\infty(dx).$$

According to lemma 8.1, for any $n \geq 1$ the previous expression can be written as

$$\frac{1}{n\overline{M}_\tau} \int_X \int_0^\infty \overline{E}_x\big(\overline{E}_{X_t}\big(\varphi(X_h)\big);\ t < \tau_n\big)\, dt\, H_\tau^\infty(dx).$$

From Markov property it follows that the internal integral is equal to

$$\int_0^\infty \overline{E}_x\big(\varphi(X_{t+h});\ t < \tau_n\big)\, dt$$

$$= \int_0^\infty \overline{E}_x\big(\varphi(X_{t+h});\ t + h < \tau_n\big)\, dt + \varepsilon_1$$

$$= \int_h^\infty \overline{E}_x\big(\varphi(X_t);\ t < \tau_n\big)\, dt + \varepsilon_1$$

$$= \int_0^\infty \overline{E}_x\big(\varphi(X_t);\ t < \tau_n\big)\, dt + \varepsilon_1 - \varepsilon_2,$$

where

$$\varepsilon_1 = \int_0^\infty \overline{E}_x\big(\varphi(X_{t+h}); \, t < \tau_n \leq t+h\big)\, dt,$$

$$\varepsilon_2 = \int_0^h \overline{E}_x\big(\varphi(X_t); \, t < \tau_n\big)\, dt.$$

Obviously, $\varepsilon_2 \leq h \sup |\varphi|$. Furthermore,

$$\varepsilon_1 \leq \sup |\varphi| \int_0^\infty \overline{P}_x\big(t < \tau_n \leq t+h\big)\, dt$$

$$= \sup |\varphi| \left(\int_0^\infty \overline{P}_x\big(t < \tau_n\big)\, dt - \int_h^\infty \overline{P}_x\big(t < \tau_n\big)\, dt \right) \leq h \sup |\varphi|.$$

From here

$$\int_{\mathbb{X}} \overline{E}_x\big(\varphi(X_h)\big)\overline{G}_\infty(dx)$$

$$= \frac{1}{nM_\tau} \int_{\mathbb{X}} \left(\int_0^\infty \overline{E}_x\big(\varphi(X_t); \, t < \tau_n\big)\, dt + \varepsilon_1 - \varepsilon_2 \right) H_\tau^\infty$$

$$= \overline{G}_\infty(\varphi) + \big(\varepsilon_1 - \varepsilon_2\big)/\big(nM_\tau\big).$$

Since n can be as large as desired, the following equality is proved

$$\int_{\mathbb{X}} \overline{P}_x\big(\varphi(X_h)\big)\overline{G}_\infty(dx) = \overline{G}_\infty(\varphi).$$

Hence, by arbitrariness of h and φ, the stationarity of measure $\overline{G}_\infty$ is proved.   □

EXAMPLE 8.1. Consider a monotone continuous semi-Markov process for $\mathbb{X} = \mathbb{R}_+$. Let ζ_x be the first hitting time of level x. The process $\xi(t)$ $(t \in \mathbb{R}_+)$ is said to be an inverse process with independent positive increments if $\zeta_x(\xi)$ $(x \in \mathbb{R}_+)$ is a process with independent positive increments. The process is determined on the positive half-line by a family of measures (P_x) determined on a space of non-decreasing positive Cantor functions. For this process it is interesting to determine a stationary distribution of the pair (R_t^-, R_t^+). We will also consider a fraction part of the value of the process at the instant t. Factually we pass from a monotone process to a non-monotone ergodic process $\widetilde{\xi}(t) \equiv \xi(t) \mod 1$ with a space of states $[0, 1)$ and a family of measures $(\widetilde{P}_x)$. A natural embedded Markov renewal process for such a process is generated by a sequence of jumps from unit to zero: $(\tau_1, \tau_2, \ldots)$ where $\tau \equiv \tau_1 = \sigma_0 \dotplus \sigma_{(0,1)}$ is the first hitting zero, except the initial position, i.e. the first return to zero if the initial position is zero (here σ_0 is the first exit time from the initial position. The embedded Markov chain consists of repeating the same meaning, zero: $H_\tau^\infty(\{0\}) = 1$. Using the Lévy formula for a homogenous process with independent positive increments we obtain for $0 < y < x < 1$

$$E_y\left(\exp\left(-\lambda\sigma_{[0,x)}\right)\right) = \exp\left(-(x-y)\left(\lambda\alpha + \int_{0+}^{\infty}\left(1 - e^{-\lambda u}\right)\nu(du)\right)\right),$$

where $\alpha \geq 0$ and $\nu(du)$ is a measure on $(0,\infty)$ (Lévy measure). Let

$$G_t\left(S \times A_1 \times A_2 \mid y\right) = \widetilde{P}_y\left(X_t \in S,\ R_t^- \in A_1,\ R_t^+ \in A_2\right).$$

According to formula [8.7] we obtain

$$G_\infty\left((0, x) \times A_1 \times A_2\right) = x\gamma/M,$$

where

$$M \equiv \widetilde{P}_0\left(\tau_1\right) = \alpha + \int_{0+}^{\infty} u\,\nu(du),$$

$$\gamma = \begin{cases} \alpha, & A_1 = A_2 = \{0\}, \\ \displaystyle\int_{r+s}^{\infty}(u - r - s)\nu(du), & A_1 = (r, \infty),\ A_2 = (s, \infty). \end{cases}$$

EXAMPLE 8.2. Consider a monotone continuous semi-Markov process representing an inverse process with independent positive increments $\xi(t)$ $(t \in \mathbb{R}_+)$, in which all intervals of constancy are conditionally fixed. For such a process

$$\zeta_x = a_x + \sum_{x_i < x} t_i,$$

where a_x is a non-decreasing function of $x \in \mathbb{R}_+$, $\mathcal{X} \equiv (x_i)$ is a sequence of points from set $\mathbb{R}_+$, and (t_i) is a sequence of positive random values. Distribution of value t_i depends only on x_i: $P_0(t_i < t) = F_{x_i}(t)$ $(t > 0)$. Suppose that $a_x = \alpha x$ $(\alpha > 0)$; the set $\mathcal{X}$ is periodic with a period 1 (for any $t \geq 0$ we assume $\mathcal{X} \cap [t + 1, \infty) = (\mathcal{X} \cap [t, \infty)) + 1$); and also a family of distribution functions is periodic with a period 1 ($F_{x+1} = F_x$ $(x \geq 0)$). We will consider the fraction part of a value of the process at instant t, i.e. the ergodic process $\widetilde{\xi}(t) \equiv \xi(t) \mod 1$ with the space of states $[0, 1)$. For $0 < x < 1$ a stationary probability $G_\infty((0, x) \times A_1 \times A_2)$ of the process is equal to $\omega(x)/M$ where

$$M = \alpha + \sum_{x_i < 1} \int_0^{\infty} t\,dF_{x_i}(t),$$

$$\omega(x) = \begin{cases} \alpha x, & A_1 = A_2 = \{0\}, \\ \displaystyle\sum_{x_i < x}\int_{r+s}^{\infty}(t - r - s)\,dF_{x_i}(t), & A_1 = (r, \infty),\ A_2 = (s, \infty). \end{cases}$$

Chapter 9

Semi-Markov Model of Chromatography

In this chapter we consider applications of the theory of continuous semi-Markov processes. These processes happen to be suitable mathematical models for various natural phenomena. We consider application of this model to the study of transfer of matter through a porous medium. This transfer and the separation of mixtures of substances connected with it is the important constituent of such natural phenomena as metasomatism [KOR 82], filtration of oil and gas [LEI 47], percolation of liquid and gas through constructional materials, etc. In the most pure aspect this process is exhibited in chromatography [GID 65, GUT 81, HOF 72]. The theoretical treatments of transfer processes are conducted within the framework of the appropriate applied disciplines and frequently lead to overlapped outcomes. For example, there is an explicit parallelism in outcomes of the theory of metasomatic zonality [KOR 82] and chromatography [HOF 72]. There are common features in different variants of the process of transfer, which can be separate from concrete engineering applications. The main peculiarity of such a process is the semi-Markov character of movement of particles composing the substance which is transferred by a solvent. We will explain this character in this chapter and will provide some formulae to evaluate parameters of such a movement.

The stochastic model we deal with begins with the supposition that the observable measure of distribution of substance is an expectation of some random measure which is a random point field in a given region of space. Each point of this field can be interpreted as a molecule or particle keeping a wholeness when being transferred by a solvent. The exact interpretation is not important for a model, since only the averaged magnitudes are observed. The point field and its average, generally speaking, depend on time. This dependence is related to the movement of particles. The stochastic model assumes such a movement under the rule of some random process.

The essential supposition of a model is the assumption of independence of random processes corresponding to various particles. With these suppositions two laws determine a measure of the matter in which we are interested:

(1) expectations of initial and boundary point field;

(2) individual rules of driving of a separate particle.

In the present chapter the basic attention is given to a substantiation and application of the semi-Markov law of driving of separate particles and corollaries from this supposition.

The continuous semi-Markov process is an appropriate model for a process of transfer of matter with a solvent through a porous medium due to the practical absence of inertia of driven particles making it possible to apply the premise of the theory of semi-Markov processes, and also due to a character of trajectories of particles of transferable substance. The experimental fact is that during the transfer of matter through a porous medium in each instant particles can be found both in a solution and in the substance of a filter in absorbed condition. Only by driving with time stoppings on the immovable phase can such distribution of particles between phases be explained. In other words, the continuous trajectories of driving have to contain intervals of constancy. How we know this is one of the peculiarities of trajectories of continuous (non-Markov) semi-Markov processes.

9.1. Chromatography

A brief overview of the basic concepts of chromatography is given: its method, instrument realization and registered signal.

1. Method

Chromatography is a physico-chemical method of separation based on distribution of divided components between two phases: motionless (filter) and mobile (eluent). A mobile phase flows continuously through a motionless phase. The purpose of separation is to cause the preparative selection of substances as pure components of a mixture. Another aim is physico-chemical measurements. Separation of complicated mixtures when moving along the surface of an absorbent occurs due to distinctions of intermolecular interactions (various absorbableness) and the consequent determination of divided components on an exit of a device with the help of special detectors. Chromatography was developed by the Russian botanist M.S. Tswett in 1901–1903 during a study of the structure of chlorophyll and the mechanism of photosynthesis (see [GID 65, YAS 76, GOL 74], etc.).

2. Device

A central part of a traditional chromatograph is a column, i.e. a narrow cylindrical vessel, completed by an absorbent. It solves the main problem: separation of components of a mixture. All remaining devices in the chromatograph are intended either for registration of divided components on the exit from the column, or for creation of stable conditions for the column. In analytical chromatography in the majority of cases an exhibiting variant of chromatography is used, in which an inert liquid or gas-carrier continuously passes through the column with a constant (liquid) or increasing (gas) velocity. A device, which injects an analyzable test in the column, is situated at the beginning of the column. The test injected in an eluent stream begins to move to the exit from a column. The components of a mixture, which are weakly sorbable, passes through a column with a greater velocity and exits a column earlier. As a rule, for chromatographic separation they use convertable physical adsorption, where adsorbing substances can be desorbed by a stream of liquid or the gas-carrier.

Column chromatography is possible to classify as one-dimensional. There exists a planar chromatography in which the role of a column is played by a sheet of a paper or a layer of another porous material on which the liquid moves in two-dimensional space and the registration is carried out on tracks of components on a surface of a material.

3. Chromatogram

The stream of eluent including desorbing components passes through a feeler of the detector, the signal of which is registered. The signal curve of the detector depending on time is called a *chromatogram*. In Figure 9.1, a chromatogram $x(t)$ of mixture of substances borrowed from [GOL 74, p. 85] is represented.

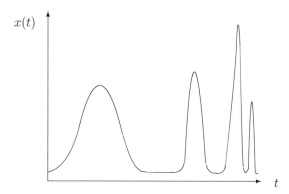

Figure 9.1. *Chromatogram of a mixture of substances with different parameters of delay*

A typical chromatogram of one absorbing substance is represented by a curve of the bell-shaped asymmetric form whose maximum is shifted with respect to the origin of coordinates (which corresponds to the instant the test is injected). With small asymmetry, which is usually connected to a large displacement, this curve is well approximated by a Gaussian curve. Thus, basic characteristics of this curve are shift, height of curve peak, width of a bell-shaped part and asymmetry. The area limited by this curve is directly proportional to a quantity of the analyzable substance. Therefore, it is natural to consider parameters of a normalized curve, i.e. a curve which is equal to the registered one, divided by a size of this area. The normalized curve can be interpreted as a density function of some random variable.

In statistical terms the shift is the first initial moment of this distribution, the width is mean-square deviation (radical of a variance), and the asymmetry is the third central moment. The height of peak in this case is a value derived from the previous parameters and actual form of a curve. An example of such a derived value is a height equivalent to a theoretical plate (HETP), connecting a variance and shift. Below we will give the exact definition of HETP in terms of a mathematical model and a formula of its approximate value used in evaluation.

A large difference in shifts with small variances of distributions corresponding to two various adsorbing substances is desirable for separation of these substances. Therefore, efforts of investigators were directed on studying the parameters of a chromatogram, on physico-chemical properties of a filter and eluent, chromatograph design and techniques of operation and, in particular, a velocity of eluent running through a filter. One such dependence, the van Deemter formula, is derived below as a corollary from the offered semi-Markov model of chromatography.

9.2. Model of liquid column chromatography

Continuous monotone semi-Markov processes with trajectories without terminal absorption in a filter are considered. Such a process is an inverse process with independent positive increments. The Levy formula for these processes and its corollaries are analyzed. One of corollaries is the presence of intervals of constancy, which are interpreted as a delay of a particle with a temporal adsorption. The probability performances of this delay are given. The "height equivalent to a theoretical plate" and its approached value used for evaluation of quality of chromatographic separation is determined.

4. Monotone process

To begin with, we consider the most simple but practically important case of process $\xi(t)$, when the adsorbing particle moves in one-dimensional space $\mathbb{X} = \mathbb{R}^1$ with monotone movement without terminal absorption in a filter. This case is realized, for example, in long and narrow chromatography columns with a liquid mobile phase without chemical interaction between the analyzable substance and a substance of the filter. In this case the semi-Markov property turns into the requirement for a particle to have independent sojourn times in intervals of the column which are mutually disjoint: if $\tau_x(\xi)$ is a random time to reach section x of the column and $0 < a < b < c$, then $\tau_b - \tau_a$, $\tau_c - \tau_b$ are independent random values. From here it follows that the random process $(\tau(x))_{x \geq 0}$ is a process with independent increments (here and further $\tau(x) \equiv \tau_x$), and process $\xi(t)$ is a so called inverse process with independent positive increments.

5. Lévy formula

Let (P_x) be a family of distributions of this process. For anyone $\lambda > 0$ the Lévy formula is fair (see [ITO 65] and also Chapter 5, item 73)

$$E_0\big(\exp\big(-\lambda\tau(x)\big)\big) = \exp\big(-b(\lambda, x)\big),$$

where

$$b(\lambda, x) = \lambda a\big([0, x)\big) + \int_{0+}^{\infty} \big(1 - e^{-\lambda u}\big) n\big(du \times [0, x)\big)$$

$$- \sum_{0 \leq y_i < x} \log P_{y_i}\big(e^{-\lambda \sigma_0}\big),$$

where $a(dx)$ is a locally finite absolutely continuous measure on axis $\mathbb{X}$, and $n(du \times dx)$ is a measure on a half-plane $(0, \infty) \times (-\infty, \infty)$, called a Lévy measure, (y_i) is a countable set of points on axis $\mathbb{X}$, and σ_0 is the first exit time from an initial point, where $E_{y_i}(\sigma_0 > 0) > 0$.

6. The Lévy representation and its interpretation

The most interesting corollary from the Lévy formula is the Lévy representation: P_x-a.s.

$$\tau(x) = a\big([0, x)\big) + \int_{0+}^{\infty} u P\big(du \times [0, x)\big) + \sum_{y_i < x} c_i,$$

where $P(du \times dx)$ is a Poisson measure (a random point field of a special view) on a half-plane $(0, \infty) \times (-\infty, \infty)$ with an intensity measure $n(du \times dx)$, $c_i \equiv \sigma_0 \circ \theta_{\tau(y_i)}$ is a random variable, the random variables in set (c_i) are mutually independent and independent of Poisson field P.

The interpretation of Lévy representation is interesting and is the most attractive feature of semi-Markov model chromatography. The determinate part

$$a\big([x_1, x_2)\big) = a\big([0, x_2)\big) - a\big([0, x_1)\big)$$

can be interpreted as a result of driving of particles with a liquid and can be connected to a velocity $v(x)$ of moving of a liquid-carrier:

$$a\big([x_1, x_2)\big) = \int_{x_1}^{x_2} \frac{1}{v(x)}\, dx,$$

where $1/v(x)$ is a density of the measure $a(dx)$ with respect to the Lebesgue measure on axis $\mathbb{X}$. In a spatially-homogenous case $v(x) \equiv v$ and $a([0, x)) = x/v$. The random part consists of intervals of constancy of two sorts. At first, it is a sum of random delays in random points, the duration and position of which on a segment $[x_1, x_2)$ is determined by a Poisson measure $\mathcal{P}$. Secondly, it is a sum of independent random delays in fixed points from a countable set (y_i). For homogenous in space process $\xi(t)$ the delay of the second sort of intervals of constancy is absent. Furthermore, we will be limited to processes without these sort of delays. As we see, the emerging of a random delay follows from the sole supposition that sojourn times in mutually disjoint intervals are independent random values.

We note that the magnitudes of intervals of constancy can be infinitely small; however, the summarized length of these small intervals can be comparable with the summary length of large intervals. This will be the case when for any $\varepsilon > 0$ $n((0, \varepsilon] \times [x_1, x_2)) = \infty$. Nowadays experimental data about magnitudes of intervals of constancy are not known and consequently there are no reasons to neglect a case of infinite intensity.

7. Density of Lévy measure

The Lévy measure can be represented as an integral of some conditional one-dimensional measure $\nu(du \mid x)$:

$$n\big(B \times [x_1, x_2)\big) = \int_{x_1}^{x_2} \nu(B \mid x)\, dx,$$

which in a homogenous case turns into equality $n(B \times [0, x)) = \nu(B)x$, where $\nu(B) = \nu(B \mid 0)$. From here in a homogenous case we have

$$b(\lambda, 0, x) = x\left(\frac{\lambda}{v} + \int_{0+}^{\infty} \big(1 - e^{-\lambda u}\big)\nu(du)\right).$$

8. Moments of distribution of delay

A residual $\tau(x) - a([0, x))$ in a monotone case is referred to as a delay, where $a([0, x))$ is a passage time of eluent through chromatographic column. Thus, it is a random part of Lévy representation (in this case, Poisson).

$$\tau(x) - a([0, x)) = \int_{0+}^{\infty} u\mathcal{P}(du \times [0, x)).$$

Moments of distribution of a random variable $\tau(x)$ can be obtained from the Lévy formula by differentiation on λ with $\lambda = 0$:

$$E_0\tau(x) \equiv \int_0^{\infty} tP_0(\tau(x) \in dt) = -\frac{\partial}{\partial\lambda} E_0 e^{-\lambda\tau(x)}\Big|_{\lambda=0}$$

$$= a([0, x)) + \int_{0+}^{\infty} un(du \times [0, x)),$$

$$\sigma^2(\tau(x)) \equiv E_0(\tau(x) - E_0\tau(x))^2 = \int_{0+}^{\infty} u^2 n(du \times [0, x)).$$

We also obtain the 3rd and 4th central moment:

$$E_0(\tau(x) - E_0\tau(x))^3 = \int_{0+}^{\infty} u^3 n(du \times [0, x)),$$

$$E_0(\tau(x) - E_0\tau(x))^4 = \int_{0+}^{\infty} u^4 n(du \times [0, x)) + 3\sigma^4(\tau(x)).$$

It is not difficult to count up other moments of distribution of random variable $\tau(x)$. From the fourth order all central moments depend on x in a non-linear way.

Note that in the considered case of monotone moving the important characteristic of a filter, the magnitude $H_\tau \equiv \sigma^2(\tau(x))/x$, which in a homogenous case is equal to $\int_{0+}^{\infty} u^2 \nu(du)$, depends on a velocity v of a carrying liquid only in the case when the conditional Lévy measure $\nu(du)$ depends on v. The form of this dependence does not follow from the semi-Markov model itself and serves an additional supposition, which can be justified only with the help of experiments. It seems to be reasonable to suppose for measure ν to be deformed in inverse proportion to a velocity eluent:

$$\nu(B) = \frac{\nu_1(B)}{v}. \tag{9.1}$$

In this case the relationship between determinate and random parts of a delay does not depend on a velocity eluent: the longer a particle moves, the longer it remains motionless on sorbent. The velocity of the eluent plays a role of a scale parameter for a process.

9. Height equivalent to a theoretical plate

Alongside a variance $\sigma^2(\tau(x))$, the variance of a random variable $\xi(t)$ may be of interest. Its exact value in terms of a Lévy measure is expressed with the awkward formula following from an inversion of a Laplace transformation. To estimate this variance the approximate equality: $\sigma(\xi(t_x)) \cong \sigma(\tau(x))v_c$ can be used, where t_x is obtained from the condition $E_0\xi(t_x) = x$ and $v_c = x/E_0\tau(x)$ [GID 65]. In a homogenous case we have

$$v_c = \left(\frac{1}{v} + \int_{0+}^{\infty} u\nu(du)\right)^{-1} = vR,$$

where R is a factor of a diminution of a velocity of transferable substance comparatively with respect to a velocity of the carrier. In particular, if condition [9.1] is satisfied, R does not depend on a velocity.

The magnitude $\sigma^2(\xi(t_x))/x$ in [GID 65] is called a height equivalent to a theoretical plate (HETP). This title is borrowed from the practice of a distillation, in which the separation of substances happens as a result of multiple transfusions of solutions from each plate into the following one, where plates compose an infinite sequence. In an outcome of these transfusions the test located in the first plate of a series is spread on all plates and the maxima of concentration also moves along a series. We will not describe this procedure explicitly, and will not justify the definition of HETP for chromatography. For practical purposes, in particular, for an evaluation of quality chromatograph, an approximation of HETP is used, namely

$$H \equiv \left(\sigma(\tau(x))v_c\right)^2/x. \tag{9.2}$$

The higher the chromatograph quality, the smaller H will be. In a one-dimensional monotone case we have $H = H_\tau R^2 v^2$ and, if condition [9.1] is satisfied

$$H = R^2 v \int_{0+}^{\infty} u^2\nu_1(du),$$

i.e. it is directly proportional to an eluent velocity.

Note that formula [9.2] is deduced without the supposition of a monotonicity of process $\xi(t)$. We use it below when analyzing continuous semi-Markov processes of diffusion type.

9.3. Some monotone Semi-Markov processes

The family of measures $(\nu(du \mid x))_{x \in X}$ of monotone processes determines a so-called field of delay. This field in a common case is infinite-dimensional because measure $\nu(\cdot \mid x)$ is characterized by an infinite set of parameters, for example, moments. For practical applications it is preferable to use families of measures with a finite number of parameters.

10. A Gut and Ahlberg model

The example of a field with finite intensity is considered in [GUT 81]. In this work, devoted to a problem of sums of series with a random number of summands, the stochastic model of chromatography was offered. In this model a particle moves with stops in such a way that uniform moving intervals with a velocity v alternate with intervals of constancy. The supposition about an exponential distribution of length of a uniform moving interval (with a parameter αv), which is accepted in this work, refers these processes to a class of homogenous monotone continuous semi-Markov processes. The considered process has a finite intensity of a random point field of jumps of function $\tau(x)$ and an exponential distribution of magnitudes of jumps, which are intervals of constancy of function $\xi(t)$ (with a parameter β). The parameter of an exponential function in Lévy expansion for this process is of the form

$$b(\lambda, 0, x) = \frac{x\lambda}{v} + x\alpha \int_0^\infty \left(1 - e^{-\lambda u}\right)e^{-\beta u}du.$$

Parameters v, α and β of this process are functions of the first three moments of distribution of random variable $\tau(x)$:

$$E_0\tau(x) = x\left(\frac{1}{v} + \frac{\alpha}{\beta}\right), \quad \sigma^2(\tau(x)) = \frac{2\alpha x}{\beta^2},$$

$$E_0\left(\tau(x) - E_0\tau(x)\right)^3 = \frac{6\alpha x}{\beta^3},$$

which can be estimated using a chromatogram.

11. Inverse gamma-process

An inverse gamma-process (without drift) is a monotone semi-Markov process with independent positive increments of random function $\tau(x)$, distributed according to gamma-distribution:

$$\frac{1}{\Gamma(x\alpha)}h^{x\alpha}t^{x\alpha-1}e^{-ht} \quad (x > 0),$$

where $\Gamma(x)$ is a gamma-function, $\alpha > 0$ is a form parameter and $h > 0$ is a scale parameter. The parameter of an exponential Lévy function of this process has the form

$$x\alpha \ln \frac{h + \lambda}{h} = x\alpha \int_0^\infty \left(1 - e^{-\lambda u}\right)\frac{e^{-hu}}{u}du$$

(see [PRU 81, formula 2.3.19.28]). For practical purposes it is reasonable to supplement this parameter with a term $x\lambda/v$, where v is velocity of eluent. This supplemented parameter

$$b(\lambda, 0, x) = \frac{x\lambda}{v} + x\alpha \int_0^\infty \left(1 - e^{-\lambda u}\right) \frac{e^{-hu}}{u} \, du$$

corresponds to an inverse gamma-process with drift, for which

$$E_0\tau(x) = \frac{x}{v} + \frac{x\alpha}{h}, \qquad \sigma^2\big(\tau(x)\big) = \frac{x\alpha}{h^2}, \qquad E_0\big(\tau(x) - E_0\tau(x)\big)^3 = \frac{2x\alpha}{h^3}.$$

The principal advantage of an inverse gamma-process with drift in comparison with other monotone semi-Markov models is its simplicity, sufficient flexibility due to its three parameters, and also its well-known properties. Tables of its values can be found, for example, in [PAG 63].

12. Process of maximum values

The large subclass of inverse processes with independent positive increments is made with processes of record values connected with semi-Markov non-monotone processes. For continuous one-dimensional process $\xi(t)$ with an initial point $x = \xi(0)$ the process $\eta(t) = \max(\xi(s) : 0 \le s \le t)$ is called a process of maximum values. In addition, if $\xi(t)$ is a semi-Markov process, $\eta(t)$ is semi-Markovian too. Besides it is a monotone process and therefore an inverse process with independent positive increments.

Consider, for example, a continuous semi-Markov process such as a homogenous Brownian motion with drift. The density of its one-dimensional distribution has a view

$$p_t(x) = \frac{1}{\sqrt{2\pi dt}} \exp\left(-\frac{(x - \mu t)^2}{2dt}\right).$$

The semi-Markov transition generating function of process $h_G(\lambda, x)$ (where $G = (a, b)$ and $a < x < b$) is expressed by the formula

$$h_G(\lambda, x) = e^{A(b-x)} \frac{\sinh(x - a)\sqrt{A^2 - 2B(\lambda)}}{\sinh(b - a)\sqrt{A^2 - 2B(\lambda)}},$$

where $A = \mu/d$ and $B(\lambda) = -\lambda/d$, (see items 5.22 and 5.23). The generating time function of the first exit of the corresponding process of maximum values from an interval $(-\infty, b)$ can be received as a limit of the above function as a to $-\infty$. Taking into account the values of factors, we obtain

$$E_0\big(e^{-\lambda\tau(x)}\big) = \exp\left(-x\sqrt{\frac{2}{d}}\left(\sqrt{\frac{\mu^2}{2d} + \lambda} - \frac{\mu}{\sqrt{2d}}\right)\right). \qquad [9.3]$$

Consider a case $\mu \geq 0$. In this case the process $\eta(t)$ does not have an infinite interval of constancy: $P_0(\tau(x) = \infty) = 0$. We need to find parameters of the Lévy formula, i.e. to find unknown measures $a(\cdot)$ and ν in the equation

$$\lambda a\big([0, x)\big) + x \int_{0+}^{\infty} \left(1 - e^{-\lambda u}\right)\nu(du) = x\sqrt{\frac{2}{d}}\left(\sqrt{\frac{\mu^2}{2d} + \lambda} - \frac{\mu}{\sqrt{2d}}\right). \qquad [9.4]$$

By dividing both parts of the equation by λ and tending λ to infinity, we obtain $a\big([0, x)\big) = 0$. It is easy to check that the function

$$\frac{1}{\sqrt{2\pi d\, u^3}} \exp\left(-u\frac{\mu^2}{2d}\right). \qquad [9.5]$$

is a density of a Lévy measure in equation [9.4]. The P_0-density of the distribution of the first hitting time of level x is of the form

$$p_{\tau(x)}(t) = \frac{x}{\sqrt{2\pi t^3 d}} \exp\left(-\frac{(x - \mu t)^2}{2td}\right)$$

(see [BOR 96, formula 1.1.4]). Three first moments of distribution $\tau(x)$ for this process look like

$$E_0\tau(x) = \frac{x}{\mu}, \qquad \sigma^2\big(\tau(x)\big) = \frac{xd}{\mu^3}, \qquad E_0\big(\tau(x) - E_0\tau(x)\big)^3 = \frac{3x\,d^2}{\mu^5}. \qquad [9.6]$$

With $\mu = 0$, i.e. for a process without drift, we obtain a monotone, slowly growing process: the average first passage time of any level is equal to infinity, though the probability of its reaching is equal to 1.

13. Process of maximum values with negative drift

From a semi-Markov property it follows that for $0 < x_1 < x$

$$E_0 e^{-\lambda\tau(x)} = E_0 e^{-\lambda\tau(x_1)} E_{x_1} e^{-\lambda\tau(x)},$$

from here $E_0(\tau(x) < \infty) = E_0(\tau(x_1) < \infty)E_{x_1}(\tau(x) < \infty)$ and hence, firstly,

$$E_0\big(e^{-\lambda\tau(x)} \mid \tau(x) < \infty\big) = E_0\big(e^{-\lambda\tau(x_1)} \mid \tau(x_1) < \infty\big)E_{x_1}\big(e^{-\lambda\tau(x)} \mid \tau(x) < \infty\big)$$
$$= \exp\big(-b(\lambda, 0, x)\big),$$

(see item 5) and, secondly,

$$E_0\big(\tau(x) < \infty\big) = \exp\big(-c\big([0, x)\big)\big),$$

where $c(dx)$ is a measure on axis X. We suppose that this measure (as well as measure $a(dx)$) is absolutely continuous with respect to a Lebesgue measure. Hence, in this case, the Lévy formula is completed by one term (not depending on λ)

$$
E_{x_1}\left(e^{-\lambda\tau(x_2)}\right) = \exp\left(- c\big([x_1, x_2]\big) - \lambda a\big([x_1, x_2]\big) \right. \tag{9.7}
$$
$$
\left. - \int_{0+}^{\infty} \left(1 - e^{-\lambda u}\right) n\big(du \times [x_1, x_2]\big) \right);
$$

yet a distribution of $\tau(x)$ is subprobability; and the moments received in item 8 are moments of the measure $P_0(\cdot \mid \tau(x) < \infty)$, i.e. conditional moments. However, distribution of the $\xi(t)$ itself is probability for any $t \geq 0$:

$$
P_0\big(\xi(t) < x\big) = P_0\big(\tau(x) = \infty\big) + P_0\big(\tau(x) < \infty\big) P_0\big(\xi(t) < x \mid \tau(x) < \infty\big).
$$

In a limit as $t \to \infty$ they obtain a distribution of absorbed particles which for a spatial-homogenous case is exponential. This limit distribution is appropriate for moving the substance along a one-dimensional path with absorbtion when the substance from a point-wise source is brought by an eluent with a constant velocity.

Consider expression [9.3] under condition $\mu < 0$. This process finally stops at a random instant and on a random level. Since $E_0(e^{-\lambda\tau(x)}) = E_0(e^{-\lambda\tau(x)}; \tau(x) < \infty)$, from formula [9.3] for $\lambda = 0$ we obtain an exit probability on level x:

$$
P_0\big(\tau(x) < \infty\big) = \exp\big(- 2x|\mu|/d\big) < 1.
$$

According to item 12, formula [9.7] for this process passes in the expression

$$
E_0\big(\exp\big(- \lambda\tau(x)\big)\big)
$$
$$
= \exp\left(- 2x|\mu| - x\int_{0+}^{\infty} \left(1 - e^{\lambda u}\right)\frac{1}{\sqrt{2\pi d\, u^3}}\exp\left(- u\frac{\mu^2}{2d}\right) du\right).
$$

From equality

$$
E_0\big(\exp\big(- \lambda\tau(x - \varepsilon)\big),\ \tau(x) = \infty\big)
$$
$$
= E_0\big(\exp\big(- \lambda\tau(x - \varepsilon)\big)\big) P_{x-\varepsilon}\big(\tau(\varepsilon) = \infty\big)
$$

($\varepsilon > 0$) we obtain a conditional generating function of a time when the infinite interval of constancy begins:

$$
P_0\big(\exp\big(- \lambda\tau(x - 0)\big) \mid \tau(x) = \infty\big) = \exp\left(- x\left(\sqrt{2\frac{\lambda}{d} + \mu^2} - |\mu|\right)\right),
$$

although this time is not a Markov time (see item 2.2).

9.4. Transfer with diffusion

We have to take into account diffusion when investigating gaseous chromatography. In this case they use a column chromatograph connected to a vessel with a compressed gas-eluent. Let $b > 0$ is the length of the column. Near the open end of the column a detector is placed in order to record the first appearance of the substance of the test. It makes it possible to consider the recorded function as the distribution density of the first hitting time of level b for a process of diffusion type.

It is natural to consider a Markov diffusion process as a possible model for movement of a gas particle. In this case the movement is controlled by a backward differential Kolmogorov equation:

$$\frac{\partial f}{\partial t} = \frac{1}{2} d(x) \frac{\partial^2 f}{\partial x^2} + \mu(x) \frac{\partial f}{\partial x},$$

where $f = f(t, x) = p_t(S \mid x)$ is a transition function of process $(S \in \mathcal{B}(\mathbb{X}))$. However, as it follows from formulae [9.6] under the conditions of the Markov model, it is impossible to explain separation of different substances in the gaseous chromatograph. Actually, in a gas mixture all the particles have the same drift μ. Thus, they may differ only due to their local variance d. However, the first moment does not depend on d, i.e. different kinds of particles cannot be separated at the end point of the column. In order to be exact we have to note that formulae [9.6] correspond to spatial-homogenous cases, but the same outcome is fair in common cases as well.

14. Semi-Markov model

The semi-Markov model seems to be a well adapted model in order to learn the gaseous chromatography phenomenon. We know that the transition generating function of a semi-Markov process of diffusion type is controlled by equation [5.6] (Chapter 5). Consider a coefficient $B(\lambda, x)$ of this equation. By theorem 5.6 its derivative $-\partial B(\lambda, x)/\partial \lambda$ is a completely monotone function on $\lambda \geq 0$. Consider a limit

$$\gamma(x) = \lim_{\lambda \to \infty} \left(-\partial B(\lambda, x)/\partial \lambda \right).$$

An important element of our model of gaseous chromatography is that $\gamma(x) > 0$ for any $x \leq b$. This means coefficient $B(\lambda, x)$ contains a non-degenerated linear part such that

$$c(\lambda, x) \equiv -B(\lambda, x) - \gamma(x)\lambda$$

is a completely monotone function on $\lambda \geq 0$. A Markov diffusion process with the transition generating function controlled by equation

$$\frac{1}{2} f'' + A(x)f' - \gamma(x)\lambda f = 0, \tag{9.8}$$

is said to be *the support Markov process* for the given semi-Markov process. Both these processes have the same distribution for their random trace. Trajectories of the original process differ from those of the support Markov process due to presence of Poisson intervals of constancy, corresponding to Lévy measure $\nu(du \mid x)$, where

$$c(\lambda, x) = \int_{0+}^{\infty} \left(1 - e^{\lambda u}\right)\nu(du \mid x).$$

The Markov process corresponding to equation [9.8] has Kolmogorov coefficients as follows:

$$\mu(x) = A(x)/\gamma(x), \qquad d(x) = 1/\gamma(x).$$

We know that in the one-dimensional case, coefficient $\mu(x)$ determines a group velocity $V(x)$ of the system of independent homogenous particles controlled by this equation. Our assumption is that for different kinds of particles their coefficients μ of support Markov processes are identical, and coincide with those of the gas-eluent. Thus, we suppose that $\mu(x) = V(x)$. Parameter $V(x)$ is a macroscopic velocity of the eluent flow through a porous adsorbent medium along the column. This parameter essentially depends on x, because a decrease of pressure from the entry to the exit of the column is a necessary condition for gas to flow through the column.

Evaluation of the gas velocity in cylindrical vessel (column) is based on the Boyle-Mariott law

$$pv = Cm, \tag{9.9}$$

where p is the pressure of gas, v is volume of gas, C is a coefficient depending on a chemical composition and temperature of gas, and m is mass of gas. Representing the volume as the product, VSt, of the gas velocity, V, in a given cross-section of the vessel, by area, S, of the cross-section, and by a time, t, we obtain the equality

$$pV = Cm_0/S, \tag{9.10}$$

where $m_0 = m/t$ is a normed expenditure of gas mass, which is constant at every cross-section of the vessel under usual stationary condition of work. In order to derive dependence of the velocity on distance, x, of a cross-section from the entry to the column we use the Gagen-Poiseil equation [GOL 74, p. 71]:

$$V = -kp', \tag{9.11}$$

where k is a coefficient depending on the porous medium inside the column and chemical composition of the gas. Differentiating the constant product pV with respect to x and substituting p' from [9.11] we obtain equation $-V^2/k + pV' = 0$. Taking into account [9.10] we derive equation $V'/V^3 = c$, where $c = S/(kCm_0)$. A solution to this equation is $V(x) = (C_1 - 2c_1 x)^{-1/2}$. In terms of pressure, by [9.10], this means that $p(x) = (Cm_0/S)(C_1 - 2c_1x)^{1/2}$. Hence $C_1 = p^2(b)/(Cm_0/S)^2 + 2c_1b$, where under usual conditions of work $p(b)$ is equal to atmospheric pressure. Thus,

$$V(x) = \frac{Cm_0/S}{\sqrt{p^2(b) + 2(b - x)Cm_0/(kS)}}, \qquad [9.12]$$

The denominator of this expression represents a pressure at cross-section x of the column. In particular, $p^2(0) = p^2(b) + 2bCm_0/(kS)$. Under condition $m_0 > 0$ the velocity is positive. It follows that $P_x(\tau_b < \infty) = 1$ for any $x < b$. Equation [5.6] (Chapter 5) can now be rewritten as

$$\frac{1}{2}f'' + V(x)\gamma f' - (\gamma\lambda + c(\lambda))f = 0, \qquad [9.13]$$

where $V(x)$ is given by formula [9.12]. We will suppose that $\gamma = \gamma(0), c(\lambda) = c(\lambda, 0)$ (not dependent on x). This supposition is natural for a thermostated column – the usual condition of work.

15. About boundary conditions

The random value we are interested in is the first exit time of a particle from the interval determined by the length, b, of the column. How should this time be interpreted? In the case of a semi-Markov process of diffusion type, the particle, starting at point $x = 0$, theoretically returns to the area of negative values many times in spite of the positive group velocity of gas-eluent at point $x = 0$. This velocity is made by a device creating pressure at the start of the column. We take into consideration that the particle movement in the column does not vary if the column is virtually prolonged in the area of negative values in such a way that the pressure and group velocity of gas at point $x = 0$ are invariable. Mathematically this means considering equation [5.6] on interval $(-\infty, b)$, replacing velocity $V(x)$ determined by [9.12] on its analytical prolongation on the set of negative values, i.e. considering formula [9.12] for negative x. In this case the time we are interested in gains a precise sense. It is $\tau_b = \sigma_{(-\infty, b)}$ with respect to probability measure P_0. For any positive γ and $x < b$ $P_x(\tau_b < \infty, X_{\tau_b} = b) = 1$. This makes it possible to reduce the problem to analysis of function $h_{(-\infty, b)}(\lambda; x)$ (see item 5.1). Unfortunately, we do not know an analytical form of a solution to equation [9.13]. However, in order to find moments of random value τ_b we do not need this analytical form, since from equation [9.13] a differential equation can be derived for any $M_{k,a}(x)$ ($k \geq 1$), where $M_{k,a}(x) = E_x((\sigma_{(a,b)})^k)$. It follows from representation

$$M_{k,a}(x) = (-1)^k \frac{\partial^k E_x\big(\exp\big(-\lambda\sigma_{(a,b)}\big)\big)}{(\partial\lambda)^k}\bigg|_{\lambda=0}$$

and from differential equation [9.13] differentiated by parameter λ.

16. Approximate solutions

Before analyzing equation [9.13] with variable velocity $V = V(x)$ we consider its rough approximation, which is a solution of a similar equation but with a constant coefficient V. According to the theory of linear differential equations with constant coefficients, a general solution to the equation

$$\frac{1}{2} f'' + V\gamma f' - (\gamma\lambda + c(\lambda)) f = 0$$

is $f = C_1 \exp(\alpha_1 x) + C_2 \exp(\alpha_2 x)$, where

$$\alpha_1 = -V\gamma - \sqrt{(V\gamma)^2 + 2(\gamma\lambda + c(\lambda))},$$

$$\alpha_2 = -V\gamma + \sqrt{(V\gamma)^2 + 2(\gamma\lambda + c(\lambda))}.$$

The constants C_1, C_2 can being obtained from boundary conditions for the desired partial solution. We are interested in a partial solution which is the generating function

$$f(\lambda \mid x) \equiv E_x\big(\exp\big(-\lambda\tau_b\big)\big).$$

From the boundedness of this solution as $x \to -\infty$ it follows that $C_1 = 0$. The second constant can be found from condition $1 = f(\lambda \mid b) = C_2 \exp(\alpha_2 b)$. Hence

$$f(\lambda \mid x) = \exp\Big((b-x)\Big(V\gamma - \sqrt{(V\gamma)^2 + 2(\gamma\lambda + c(\lambda))}\Big)\Big).$$

With the help of the inverse Laplace transformation the distribution density of τ_b can be found, which is represented by a unimodal function with positive asymmetry. Moments of this distribution can be found by differentiating function $f(\lambda \mid x)$ with respect to λ at point $\lambda = 0$. We obtain

$$M_1(x) \equiv E_x(\tau_b) = -\frac{\partial}{\partial\lambda} f(\lambda \mid x)\Big|_{\lambda=0} = (b-x)\frac{\gamma + \delta_1}{V\gamma},$$

where $\delta_1 \equiv (\partial c(\lambda)/\partial\lambda)|_{\lambda=0} = \int_0^\infty u\nu(du)$,

$$M_2(x) \equiv E_x\big((\tau_b)^2\big) = \frac{\partial^2}{(\partial\lambda)^2} f(\lambda \mid x)\Big|_{\lambda=0}$$

$$= \left(\frac{(b-x)(\gamma+\delta_1)}{V\gamma}\right)^2 + (b-x)\frac{\delta_2}{V\gamma} + (b-x)\frac{(\gamma+\delta_1)^2}{(V\gamma)^3},$$

where $\delta_2 = -(\partial^2 c(\lambda)/(\partial\lambda)^2)|_{\lambda=0} = \int_0^\infty u^2\nu(du)$. From here

$$\widetilde{M}_2(x) \equiv E_x\big(\tau_b - M_1(x)\big)^2 = M_2(x) - \big(M_1(x)\big)^2$$

$$= (b-x)\frac{\delta_2}{V\gamma} + (b-x)\frac{(\gamma+\delta_1)^2}{(V\gamma)^3}.$$

We are interested in moments of distribution of τ_b:

$$m_1(b) = M_1(0) = b\frac{\gamma + \delta_1}{V\gamma},$$

$$\mu_2(b) = \widetilde{M_2}(0) = b\frac{\delta_2}{V\gamma} + b\frac{(\gamma + \delta_1)^2}{(V\gamma)^3}.$$

Note that formula [9.2] for the liquid chromatograph HETP can be rewritten as

$$H \equiv H(0, b) = \frac{b\mu_2(b)}{m_1^2(b)}. \qquad [9.14]$$

Thus, in this case,

$$H(0, b) = \frac{\delta_2}{(\gamma + \delta_1)^2}V\gamma + \frac{1}{V\gamma},$$

which precisely corresponds to the well-known van Deemter formula for gaseous chromatography [GOL 74].

17. Precise solutions

Continue investigation of equation [9.13] with variable coefficient $V(x)$.

First moment

Note that function

$$f_{\sigma_{(a,b)}}(\lambda, \{a, b\} \mid x) = g_{(a,b)}(\lambda; x) + h_{(a,b)}(\lambda; x) \equiv E_x\big(\exp\big(-\lambda\sigma_{(a,b)}\big)\big)$$

satisfies equation [9.13]. Substituting this function in the equation and differentiating members of the equation with respect to λ at point $\lambda = 0$, we obtain the differential equation

$$\frac{1}{2}M''_{1,a} + V(x)\gamma M'_{1,a} + (\gamma + \delta_1) = 0, \qquad [9.15]$$

since $c(0) = 0$ and $f_{\sigma_{(a,b)}}(0, \{a, b\} \mid x) = 1$. In addition $M_{1,a}(a) = M_{1,a}(b) = 0$.

Let us denote $\varphi(x) = M'_{1,a}(x)$ and find a partial solution of the equation

$$\frac{1}{2}\varphi' + V(x)\gamma\varphi + (\gamma + \delta_1) = 0,$$

where according to formula [9.12] $V(x) = A/\sqrt{B - x}$, $A = \sqrt{Cm_0k/2S}$, $B = p^2(0)kS/2Cm_0$. Look for a partial solution of the form $\overline{\varphi}(x) = A_1 + A_2\sqrt{B - x}$.

Substituting this expression in the equation and equating coefficients of every power of the root to zero we obtain two equations with respect to A_1 and A_2, from where $A_1 = -(\gamma+\delta_1)/(4A^2\gamma^2)$, $A_2 = -(\gamma+\delta_1)/(A\gamma)$. A general solution of the homogenous equation is of the form

$$\varphi_0(x) = C_1 \exp\left(-2\gamma \int_a^x V(s)\, ds\right)$$
$$= C_1 \exp\left(-4A\gamma(\sqrt{B-a} - \sqrt{B-x})\right),$$

where C_1 is an arbitrary constant. Consequently

$$M_{1,a}(x) = C_2 + \int_a^x \left(\overline{\varphi}(s) + \varphi_0(s)\right) ds$$
$$= C_2 + \int_a^x \left(\overline{\varphi}(s) + C_1 \exp\left(-4A\gamma(\sqrt{B-a} - \sqrt{B-s})\right)\right) ds.$$

The arbitrary constant C_2 of the general solution is determined from condition $0 = M_{1,a}(a) = C_2$. The second boundary condition determines C_1:

$$0 = M_{1,a}(b) = \int_a^b \left(\overline{\varphi}(s) + C_1 \exp\left(-4A\gamma(\sqrt{B-a} - \sqrt{B-s})\right)\right) ds,$$

from where

$$C_1 = -\int_a^b \overline{\varphi}(s)\, ds \left(\int_a^b \exp\left(-4A\gamma(\sqrt{B-a} - \sqrt{B-s})\right) ds\right)^{-1}.$$

We obtain

$$M_{1,a}(x) = \int_a^x \overline{\varphi}(s)\, ds - \int_a^b \overline{\varphi}(s)\, ds \frac{\int_a^x \exp\left(-4A\gamma(\sqrt{B-a} - \sqrt{B-s})\right) ds}{\int_a^b \exp\left(-4A\gamma(\sqrt{B-a} - \sqrt{B-s})\right) ds}$$
$$= -\int_x^b \overline{\varphi}(s)\, ds + \int_a^b \overline{\varphi}(s)\, ds \frac{\int_x^b \exp\left(4A\gamma\sqrt{B-s}\right) ds}{\int_a^b \exp\left(4A\gamma\sqrt{B-s}\right) ds}.$$

In this expression

$$\int_a^b \overline{\varphi}(s)\, ds = A_1(b-a) + \frac{2}{3}A_2\left((B-a)^{3/2} - (B-b)^{3/2}\right),$$

$$\int_a^b \exp\left(4A\gamma\sqrt{B-s}\right) ds = \frac{2}{4A\gamma}\left(\left(\sqrt{B-a} - \frac{1}{4A\gamma}\right) \exp\left(4A\gamma\sqrt{B-a}\right)\right.$$
$$\left. - \left(\sqrt{B-b} - \frac{1}{4A\gamma}\right) \exp\left(4A\gamma\sqrt{B-b}\right)\right).$$

The ratio of the two latter expressions tends to zero as $a \to -\infty$. Thus, the expectation in which we are interested is equal to

$$M_1(x) = \lim_{a \to -\infty} M_{1,a}(x) = -A_1(b-x) - \frac{2}{3}A_2\big((B-x)^{3/2} - (B-b)^{3/2}\big)$$

$$= (b-x)(\gamma+\delta_1)/(4A^2\gamma^2) + \frac{2}{3}\big((B-x)^{3/2} - (B-b)^{3/2}\big)(\gamma+\delta_1)/(A\gamma)$$

$$= \frac{\gamma+\delta_1}{A\gamma}\left(\frac{b-x}{4A\gamma} + \frac{2}{3}\big((B-x)^{3/2} - (B-b)^{3/2}\big)\right).$$

[9.16]

Second moment

Double differentiation of equation [9.13] with respect to λ at point $\lambda = 0$ brings us to equation

$$\frac{1}{2}M_{2,a}'' + V(x)\gamma M_{2,a}' + \delta_2 + 2(\gamma+\delta_1)M_{1,a}(x) = 0.$$

Denoting $\varphi_a = M_{2,a}'$, we obtain equation

$$\frac{1}{2}\varphi_a' + V(x)\gamma\varphi_a + \delta_2 + 2(\gamma+\delta_1)M_{1,a}(x) = 0,$$

which can be solved in a standard way. Let us assume that we know a partial solution $\overline{\varphi}_a$ of this equation which increases with a power rate as $a \to -\infty$. Then, according to previous evaluation, we obtain $M_2(x) = \lim_{a \to -\infty}(-1)\int_x^b \overline{\varphi}_a(s)\,ds$. In particular, if the integrand is uniformly limited on the region of integration, then $M_2(x) = -\int_x^b \overline{\varphi}(s)\,ds$, where $\overline{\varphi}(x) = \lim_{a \to -\infty}\overline{\varphi}_a(x)$. On the other hand, $\overline{\varphi}$ is a partial solution of the equation

$$\frac{1}{2}\varphi_a' + V(x)\gamma\varphi_a + \delta_2 + 2(\gamma+\delta_1)M_1(x) = 0,$$

which it is easier to construct by a standard way. In this case a free term has a view $A_1 + A_2(B-x) + A_3(B-x)^{3/2}$, where

$$A_1 = -\delta_2 - \frac{(\gamma+\delta_1)^2(B-b)}{2A^2\gamma^2} - \frac{4(\gamma+\delta_1)^2(B-b)^{3/2}}{3A\gamma},$$

$$A_2 = \frac{(\gamma+\delta_1)^2}{2A^2\gamma^2}, \qquad A_3 = \frac{4(\gamma+\delta_1)^2}{3A\gamma}.$$

Let us look for $\widetilde{\varphi}$ of the form $\sum_{k=0}^{4} F_k(B-x)^{k/2}$. The standard procedure of uncertain coefficients gives meanings

$$F_0 = \frac{1}{4A^2\gamma^2}\left(-A_1 - \frac{3}{8A^2\gamma^2}\left(A_2 + \frac{A_3}{A\gamma}\right)\right),$$

$$F_1 = \frac{1}{A\gamma}\left(-A_1 - \frac{3}{8A^2\gamma^2}\left(A_2 + \frac{A_3}{A\gamma}\right)\right),$$

$$F_2 = \frac{3}{4A^2\gamma^2}\left(-A_2 - \frac{A_3}{A\gamma}\right),$$

$$F_3 = \frac{1}{A\gamma}\left(-A_2 - \frac{A_3}{A\gamma}\right),$$

$$F_4 = -\frac{A_3}{A\gamma}.$$

Integrating the partial solution we obtain

$$M_2(x) = -F_0(b-x) - \sum_{k=1}^{4}\frac{F_k}{k/2+1}\left((B-x)^{k/2+1} - (B-b)^{k/2+1}\right).$$

This expression gives the variance of τ_b as a function of the initial point $x \in [0,b)$: $\widetilde{M}_2(x) = M_2(x) - M_1^2(x)$. After elementary but awkward transformations we obtain

$$\widetilde{M}_2(x) = \frac{\delta_2}{A\gamma}\left(\frac{b-x}{4A\gamma} + \frac{2}{3}\left((B-x)^{3/2} - (B-b)^{3/2}\right)\right)$$

$$+ \frac{(\gamma + \delta_1)^2}{A^2\gamma^2}\left(\frac{11}{64}\frac{b-x}{A^4\gamma^4} + \frac{11}{24}\frac{(B-x)^{3/2} - (B-b)^{3/2}}{A^3\gamma^3}\right.$$

$$\left. + \frac{5}{8}\frac{(B-x)^2 - (B-b)^2}{A^2\gamma^2} + \frac{2}{5}\frac{(B-x)^{5/2} - (B-b)^{5/2}}{A\gamma}\right).$$

$$[9.17]$$

Letting $x = 0$ in formulae [9.16] and [9.17] and using previous denotations we obtain expressions for expectation and variance of the first exit time of a particle from a column of length b:

$$m_1(b) = M_1(0) = E_0(\tau_b);$$

$$\mu_2(b) = \widetilde{M}_2(0) = E_0\left(\tau_b - m_1(b)\right)^2.$$

Thus,

$$m_1(b) = \frac{\gamma + \delta_1}{A\gamma}\left(\frac{b}{4A\gamma} + \frac{2}{3}\left(B^{3/2} - (B-b)^{3/2}\right)\right),$$

$$\mu_2(b) = \frac{\delta_2}{A\gamma}\left(\frac{b}{4A\gamma} + \frac{2}{3}\left(B^{3/2} - (B-b)^{3/2}\right)\right)$$

$$+ \frac{(\gamma + \delta_1)^2}{A^2\gamma^2}\left(\frac{11b}{64A^4\gamma^4} + \frac{11\left(B^{3/2} - (B-b)^{3/2}\right)}{16A^3\gamma^3}\right.$$

$$+ \left.\frac{5\left(B^2 - (B-b)^2\right)}{8A^2\gamma^2} + \frac{2\left(B^{5/2} - (B-b)^{5/2}\right)}{5A\gamma}\right).$$

Analogously, initial and central moments of the distribution of τ_b of any order $k > 2$ can be evaluated.

Derived characteristics

Using inequality

$$bB^{k-1} < B^k - (B-b)^k < bkB^{k-1} \quad (0 < b < B,\ k > 1),$$

we see that $m_1(b)$ and $\mu_2(b)$ increase asymptotically linearly as functions of b. It implies that, as well as in the case of liquid chromatography, the resolution

$$\rho(b) \equiv m_1(b)/\sqrt{\mu_2(b)},$$

of gaseous chromatograph has an order $\sqrt{b}$ as $b \to \infty$.

In order to analyze how HETP depends on velocity we define a local HETP as an integral characteristic of a chromatograph column of a small length h, beginning at point x. According to formula [9.14] its HETP is

$$H(x, x+h) = h\frac{E_x\left(\tau_{x+h} - E_x\left(\tau_{x+h}\right)\right)^2}{\left(E_x\left(\tau_{x+h}\right)\right)^2}.$$

As h tends to zero, we obtain a desired local characteristic

$$H(x) = \frac{\psi_2(x)}{\left(\psi_1(x)\right)^2},$$

where

$$\psi_1(x) = -\frac{\partial}{\partial x}M_1(x), \qquad \psi_2(x) = -\frac{\partial}{\partial x}\widetilde{M}_2(x).$$

For a model with a constant velocity the local HETP coincides with the integral HETP, [9.14]. In the case of variable velocity we have

$$\psi_1(x) = \frac{\gamma + \delta_1}{A\gamma}\left(\frac{1}{4A\gamma} + (B-x)^{1/2}\right),$$

$$\psi_2(x) = \frac{\delta_2}{A\gamma}\left(\frac{1}{4A\gamma} + (B-x)^{1/2}\right)$$

$$+ \left(\frac{\gamma + \delta_1}{A\gamma}\right)^2\left(\frac{11}{64}\frac{1}{A^4\gamma^4} + \frac{11}{16}\frac{(B-x)^{1/2}}{A^3\gamma^3} + \frac{5}{4}\frac{B-x}{A^2\gamma^2} + \frac{(B-x)^{3/2}}{A\gamma}\right).$$

According to our denotation we have $(B-x)^{1/2} = A/V(x)$. Expressing powers of this difference in terms of velocity we obtain

$$H(x) = \frac{\delta_2 A\gamma}{(\gamma + \delta_1)^2}\left(\frac{1}{4A\gamma} + \frac{A}{V}\right)^{-1}$$

$$+ \left(\frac{11}{64}\frac{1}{A^4\gamma^4} + \frac{11}{16}\frac{1}{A^2\gamma^3 V} + \frac{5}{4}\frac{1}{\gamma^2 V^2} + \frac{A^2}{\gamma V^3}\right)\left(\frac{1}{4A\gamma} + \frac{A}{V}\right)^{-2}.$$

The first term of this sum increases as a function of $V(x)$. The second term decreases. Hence, for any x, there exists a minimum of this expression. This formula can be considered as a refined variant of the van Deemter formula for gaseous chromatography, taking into account different velocities in different cross-sections of the column. In this case the functional NTP (number of theoretical plates) can be defined as

$$Z(b) = \int_0^b \frac{dx}{H(x)}.$$

9.5. Transfer with final absorption

We will consider a continuous monotone semi-Markov process for which the trajectory has an infinite interval of constancy, i.e. in some instants it stops forever (we distinguish cases of a final stopping and breakage of a trajectory). The final (terminal) stopping can be interpreted as a sedimentation of a particle in a filter, for example, in an outcome of chemical transformation (chem-sorption). Such processes are interesting from the point of view of geological applications (see [GOL 81, KOR 82]).

18. Semi-Markov process with final stopping.

We are interested in continuous processes stopping forever at some random time $\zeta = \zeta(\xi)$. We also state that at time ζ an infinite interval of constancy with a value

$\xi(t) = \xi(\zeta)$ begins for all $t \geq \zeta(\xi)$. This time, in general, depends on the future: the event $\{\zeta \leq t\}$ is not an element of $\mathcal{F}_t$. In other words, this time is not a Markov time. However, for a semi-Markov process, a distribution of this time and of the value of a trajectory on the infinite interval of constancy can be expressed in terms of semi-Markov transition functions as a limit of an outcome of stepped approximation. Obviously, $\zeta(L_r\xi) \leq \zeta(\xi)$ and $\zeta(L_r\xi) \to \zeta(\xi)$ as $r \to 0$. Besides, $X_\zeta(L_r\xi) \to X_\zeta\xi$. The latter is trivial if ζ is a point of continuity of the function ξ. However, it is true even if ζ is a point of discontinuity of function ξ, because in this case there exists r_0 such that for all $r < r_0$ $\zeta(\xi) = \zeta(L_r\xi)$. Let $F_\zeta(A \times S \mid x) = P_x(\zeta \in A, \ X_\zeta \in S)$, $H_\zeta(S \mid x) = F_\zeta(\mathbb{R}_+ \times S \mid x)$. Thus, measure $H_\zeta(S \mid x)$ is a weak limit of a sequence of measures $H_\zeta^r(S \mid x) \equiv P_x(\zeta \circ L_r < \infty, \ X_\zeta \circ L_r \in S)$ as $r \to 0$. Find a distribution of a pair $(\zeta \circ L_r, \ X_\zeta \circ L_r)$:

$$P_x\big(\zeta \circ L_r < t, \ X_\zeta \circ L_r \in S\big)$$

$$= \sum_{n=0}^{\infty} P_x\big(\sigma_r^n < t, \ \sigma_r \circ \theta_{\sigma_r^n} = \infty, \ X_{\sigma_r^n} \in S\big)$$

$$= \sum_{n=0}^{\infty} \int_S F_{\sigma_r^n}\big([0,t) \times dx_1 \mid x\big)\big(1 - H_r\big(\mathbb{X} \mid x_1\big)\big)$$

$$= \int_S U_r\big([0,t) \times dx_1 \mid x\big)\big(1 - H_r\big(\mathbb{X} \mid x_1\big)\big),$$

where

$$H_r(S \mid x) = F_{\sigma_r}\big(\mathbb{R}_+ \times S \mid x\big),$$

$$U_r\big([0,t) \times S \mid x\big) = \sum_{n=0}^{\infty} F_{\sigma_r^n}\big([0,t) \times S \mid x\big).$$

The latter measure on $\mathbb{R}_+ \times \mathbb{X}$ represents a measure of intensity of a random locally-finite integer-valued measure $N([0,t) \times S)$, counting a number of hittings of two-dimensional points of the marked point process $(\sigma_r^n, X_{\sigma_r^n})_{n=0}^{\infty}$ into set $[0,t) \times S$. In particular,

$$H_\zeta^r(S \mid x) = \int_S Z_r\big(dx_1 \mid x\big)\big(1 - H_r\big(\mathbb{X} \mid x_1\big)\big), \qquad [9.18]$$

where $Z_r(S \mid x) = U_r(\mathbb{R}_+ \times S \mid x)$. In general, the latter measure tends to infinity, and the integrand in [9.18] tends to zero, if $r \to 0$. In some cases, it is possible to find asymptotics of these functions.

Note that the right part of formula [9.18] determines distribution of the limit point of the trajectory for $\zeta < \infty$ as well as for $\zeta = \infty$; in the latter case if a limit $\lim \xi(t)$ as $t \to \infty$ exists P_x-a.s. In the following example we assume that $P_x(\zeta < \infty) = 1$.

In this case the limit point is a stopping point, and for any x measure $H_\zeta(S \mid x)$ is a probability measure. It is not difficult to give and justify sufficient conditions for this property to be fulfilled.

19. Diffusion and non-diffusion types of movement

We will consider two principal subclasses of the class of SM processes: diffusion and non-diffusion types. A distinction between these subclasses is exposed when analyzing distribution of the first exit from a small neighborhood of the initial point. A one-dimensional continuous process has the simplest description (see Chapter 5). In this section we consider functions g and h defined in item 5.1 with $\lambda = 0$, changing the denotation of coefficients of expansion a little. Let $a < x < b$ and

$$g_{(a,b)}(x) = P_x\big(\sigma_{(a,b)} < \infty,\ X_{\sigma_{(a,b)}} = a\big),$$

$$h_{(a,b)}(x) = P_x\big(\sigma_{(a,b)} < \infty,\ X_{\sigma_{(a,b)}} = b\big).$$

For diffusion process the following expansion holds (see item 5.12)

$$g_{(x-r,x+r)}(x) = \frac{1}{2} - \frac{1}{2}b(x)r - \frac{1}{2}c(x)r^2 + o\big(r^2\big),$$

$$h_{(x-r,x+r)}(x) = \frac{1}{2} + \frac{1}{2}b(x)r - \frac{1}{2}c(x)r^2 + o\big(r^2\big),$$

where $b(x)$ is a shift parameter determining tendency of movement to the right (if $b(x) > 0$) or to the left (if $b(x) < 0$); $c(x) > 0$ is a parameter of stopping the movement. In this case

$$P_x\big(\zeta \geq \sigma_{(a,b)}\big) \leq P_x\big(\sigma_{(a,b)} < \infty\big) = g_{(a,b)}(x) + h_{(a,b)}(x) = 1 - c(x)r^2 + o\big(r^2\big).$$

For a non-diffusion ("smooth") process

$$g_{(x-r,x+r)}(x) = 1 - c(x)r + o(r), \qquad h_{(x-r,x+r)}(x) = o(r),$$

or

$$h_{(x-r,x+r)}(x) = 1 - c(x)r + o(r), \qquad g_{(x-r,x+r)}(x) = o(r).$$

Parameter $c(x)$ determines stopping of the movement as well although in another manner than in the case of diffusion. For a d dimensional space $d \geq 2$ the distinction between a diffusion and non-diffusion character of movement is in principle the same. For diffusion process for the given point $x \in \mathbb{R}^d$ there exists a non-degenerate linear map of space $\mathbb{R}^d$ on itself preserving the point x immovable and such that in a new space distribution of the point of the first exit from a small ball neighborhood with the center in x is uniform on the surface of the ball (in first approximation); in second approximation to this distribution some terms relating to a shift or stopping inside the ball are added (see items 5.27 and 5.34). For a non-diffusion process at almost all points of space distribution of the point of the first exit from a small ball neighborhood in first approximation is concentrated on intersection of the surface of the ball with a space of less dimension. If the dimension of this space is equal to 1 the distribution in first approximation is concentrated at one point, determining a unique sequence of states (trace) passing through this point; in second approximation only the term determining stopping inside the ball is added. We call the latter case the smooth character of the process.

20. Accumulation equations for diffusion type process

Consider a continuous semi-Markov process of diffusion type and its "kernel of accumulation" $H_\zeta(S \mid x)$, i.e. a probability measure depending on the initial point of the process. We will derive for this kernel forward and backward differential equations with respect to output (S) and input (x) arguments correspondingly. We begin with deriving the backward equation. According to item 5.36 for any twice differentiable function φ there exists a limit

$$A_0^o(\varphi \mid x) \equiv \lim_{r \to 0} \frac{1}{r^2}\big(H_r(\varphi \mid x) - \varphi(x)\big)$$

$$= \frac{1}{2}\sum_{ij} a^{ij}(x)\varphi_{ij}''(x) + \sum_{i=1}^{d} b^i(x)\varphi_i'(x) - c(x)\varphi(x),$$

where $d \geq 1$ is a dimension of Euclidean space $\mathbb{X}$; $(a^{ij}(x))$ is a symmetric positive definite matrix of coefficients which can depend on x; in addition, the sum of diagonal terms of this matrix is equal to 1; $(b^i(x))$ is a vector field of coefficients; $c(x)$ is a positive function; φ_i', φ_{ij}'' are partial derivatives of φ on coordinates with numbers i and j correspondingly. For an arbitrary function $\xi \in \mathcal{D}$ and $R > 0$ we consider a function $L^R\xi$ such that $L^R\xi(t) = \xi(0)$, if $t < \sigma_R(\xi)$, and $L^R\xi(t) = \xi(t)$, if $t \geq \sigma_R(\xi)$. Let

$$H_\zeta^R(S \mid x) = P_x\big(X_\zeta \circ L^R \in S; \; \zeta \circ L^R < \infty\big).$$

Then

$$H_\zeta^R(\varphi \mid x) = \varphi(x)\big(1 - H_R(\mathbb{X} \mid x)\big) + \int_{\mathbb{X}} H_R(dy \mid x)H_\zeta(\varphi \mid y). \qquad [9.19]$$

Subtract from both parts t $H_\zeta(\varphi \mid x)$ and divide by R^2. Then, assuming function $H_\zeta(\varphi \mid x)$ to be continuous and twice differentiable (this assumption can be justified as well) we obtain for $R \to 0$ on the right part of equality

$$\frac{1}{2d}\varphi(x)c(x) + \frac{1}{2d}A_0^o\big(H_\zeta(\varphi) \mid x\big),$$

and that on the left part a limit of the expression

$$(H_\zeta^R(\varphi \mid x) - H_\zeta(\varphi \mid x))/R^2 = \frac{1}{R^2} P_x((\varphi(x) - \varphi(X_\zeta)); \; \zeta < \sigma_R)$$

$$\leq \max\left\{|\varphi(x) - \varphi(x_1)| : |x - x_1| \leq R\right\} \frac{1}{R^2}(1 - H_R(\mathbb{X} \mid x)),$$

which is equal to zero. From here we discover that function $H_\zeta(\varphi) \equiv H_\zeta(\varphi \mid \cdot)$ satisfies the equation

$$\frac{1}{2}\sum_{ij} a^{ij} \left(H_\zeta(\varphi)\right)''_{ij} + \sum_{i=1}^{d} b^i \left(H_\zeta(\varphi)\right)'_i - cH_\zeta(\varphi) + c\varphi = 0. \qquad [9.20]$$

This equation is called a *backward accumulation equation*. An interesting solution tends to zero on infinity. From the maximum principle it follows that such a solution is unique.

Derive the forward accumulation equation. In this case instead of kernel $H_\zeta(S \mid x)$ we investigate its average with respect to a particular measure. Let $\mu(x)$ be the density of probability distribution on $\mathbb{X}$ and

$$H_\zeta(S \mid \mu) = \int_{\mathbb{X}} H_\zeta(S \mid x)\mu(x)\,dx.$$

Averaged kernels $H_\zeta^r(S \mid \mu)$ and $Z_r(S \mid \mu)$ and measure P_μ have a similar sense. According to equation [9.18] we have

$$H_\zeta^r(\varphi \mid \mu) = \int_{\mathbb{X}} Z_r(dx \mid \mu)\varphi(x)\left(1 - H_r(\mathbb{X} \mid x)\right).$$

For any continuous function φ the left part of this equality tends to a limit $H_\zeta(\varphi \mid \mu)$. On the right part we have $(1 - H_r(\mathbb{X} \mid x))/r^2 \to c(x)$. Assume that function $c(x)$ is continuous. Then as $r \to 0$ there exists a weak limit $W(S \mid \mu)$ of measure $r^2 Z_r(S \mid \mu)$, and, consequently,

$$H_\zeta(\varphi \mid \mu) = \int_{\mathbb{X}} W(dx \mid \mu)\varphi(x)c(x).$$

Assume that there exists a density $h_\zeta(x \mid \mu)$ of measure $H_\zeta(S \mid \mu)$. Then a density $w(x \mid \mu)$ of the measure $W(S \mid \mu)$ exists, and in addition $h_\zeta(x \mid \mu) = w(x \mid \mu)\,c(x)$.

According to definition of function Z_r (see [9.18]) we have

$$Z_r(\varphi \mid \mu) = \sum_{k=0}^{\infty} E_\mu\big(\varphi(X_{\sigma_r^k}); \ \sigma_r^k < \infty\big)$$

$$= \varphi(\mu) + \sum_{k=1}^{\infty} E_\mu\big(\varphi(X_{\sigma_r}) \circ \theta_{\sigma_r^{k-1}}; \ \sigma_r^{k-1} < \infty, \ \sigma_r \circ \theta_{\sigma_r^{k-1}} < \infty\big)$$

$$= \varphi(\mu) + \sum_{k=0}^{\infty} E_\mu\big(E_{X(\sigma_r^k)}(\varphi(X_{\sigma_r}); \ \sigma_r^k < \infty)\big)$$

$$= \varphi(\mu) + \int_{\mathbb{X}} Z_r(dx \mid \mu) E_x\big(\varphi(X_{\sigma_r}); \ \sigma_r < \infty\big),$$

where $\varphi(\mu) = \int_{\mathbb{X}} \mu(x)\varphi(x)\,dx$. From here we obtain an equation

$$\varphi(\mu) + \int_{\mathbb{X}} Z_r(dx \mid \mu)\big(E_x(\varphi(X_{\sigma_r}); \ \sigma_r < \infty) - \varphi(x)\big) = 0. \qquad [9.21]$$

Assume the function $A_0^o(\varphi \mid x)$ is continuous on $\mathbb{X}$. Then, by multiplying measure Z_r by r^2, and dividing the integrated difference on r^2, and passing to a limit as $r \to 0$, we obtain equation

$$\varphi(\mu) + \int_{\mathbb{X}} W(dx \mid \mu)\, A_0^o(\varphi \mid x) = 0.$$

The integral in this expression can be represented in the form

$$\int_{\mathbb{X}} w(x \mid \mu)\left(\frac{1}{2}\sum_{ij} a^{ij}(x)\varphi_{ij}''(x) + \sum_{i=1}^{d} b^i(x)\varphi_i'(x) - c(x)\varphi(x)\right) dx.$$

Assume function $\varphi(x)$ and all its partial derivatives of the first and second order tend to zero on infinity. Then, by applying the rule of integrating by parts, we obtain the other representation of the integral

$$\int_{\mathbb{X}} \varphi(x)\left(\frac{1}{2}\sum_{ij} (a^{ij}(x)w(x \mid \mu))_{ij}'' - \sum_{i=1}^{d} (b^i(x)w(x \mid \mu))_i' - c(x)w(x \mid \mu)\right) dx.$$

Using the possibility of an arbitrary choice of function φ we conclude that

$$\frac{1}{2}\sum_{ij} (a^{ij}w(\mu))_{ij}'' - \sum_{i=1}^{d} (b^i w(\mu))_i' - cw(\mu) + \mu = 0,$$

where $w(\mu) = w(\cdot \mid \mu)$ and $w(x \mid \mu) = \int_{\mathbb{X}} w(x \mid x_1)\mu(x_1)\,dx_1$. We will obtain from this equation *a forward equation of accumulation* if we substitute into it

$w(x \mid \mu) = h_\varsigma(x \mid \mu)/c(x)$. As a result we obtain

$$\frac{1}{2} \sum_{ij} \left(h_\varsigma(\mu)a^{ij}/c\right)''_{ij} - \sum_{i=1}^{d} \left(h_\varsigma(\mu)b^i/c\right)'_i - h_\varsigma(\mu) + \mu = 0. \qquad [9.22]$$

A limitation for a solution of this equation is that for any initial probability density μ a solution $h_\varsigma(\mu)$ is also a density of a probability distribution, i.e. it is non-negative and integrable, and the integral of it on $\mathbb{X}$ is equal to 1. Such a solution is unique. In differential equation theory equation [9.22] is said to be conjugate to equation [9.20].

Note that in theory of Markov processes with breakage when a time of break is identified with hitting time to the point of infinity, the kernel $H_\varsigma(S \mid x)$ is interpreted as a distribution of a point of the process immediately before break (depending on the initial point of the process). Therefore equations [9.20] and [9.22] can be derived by methods of the theory of Markov processes. So we have demonstrated "semi-Markov method" of proof on the base of semi-Markov interpretation.

21. Accumulation equations for smooth type processes

Semi-Markov processes of smooth type can be defined by the following property of distribution of the first exit point from a small ball neighborhood of the initial point of the process:

$$H_r(\varphi \mid x) = \varphi(x + r\bar{b})\big(1 - c(x)r + o(r)\big) \quad (r \longrightarrow 0),$$

where φ is a continuous function; $\bar{b} = \bar{b}(x) = (b^i(x))$ is a point on the surface of the unit sphere. The family $(\bar{b}(x))$ $(x \in \mathbb{X})$ determines a field of directions in the space $\mathbb{X}$. We suppose the vector function $\bar{b}(x)$ to be continuous and the function $c(x)$ to be continuous and positive. Under such assumptions for all φ there exists the limit

$$\alpha_0(\varphi \mid x) \equiv \lim_{r \to 0} \frac{1}{r}\big(H_r(\varphi \mid x) - \varphi(x)\big) = \sum_{i=1}^{d} b^i(x)\varphi'_i(x) - c(x)\varphi(x).$$

To derive the backward accumulation equation we use equation [9.19] with another normalizing factor. From this follows the identity $\alpha_0(H_\varsigma(\varphi) \mid x) + c(x)\varphi(x) = 0$, which implies equation

$$\sum_{i=1}^{d} b^i \big(H_\varsigma(\varphi)\big)'_i - cH_\varsigma(\varphi) + c\varphi = 0. \qquad [9.23]$$

This is *the backward accumulation equation* for processes of smooth type. We are interested in bounded non-negative solutions tending to zero on infinity if the parametric function is the same. In this connection $\max H_\varsigma(\varphi) \leq \max \varphi$.

To derive the forward accumulation equation we use equation [9.21]. In this case measure $rZ_r(S \mid \mu)$ tends weakly to some finite measure $Y(S \mid \mu)$. Its density $y(x \mid \mu)$ (if it exists) is connected with the density of the measure $H_\varsigma(S \mid \mu)$ by equality: $h_\varsigma(x \mid \mu) = y(x \mid \mu)c(x)$. From identity [9.21] we obtain equation

$$\varphi(\mu) + \int_{\mathbb{X}} y(x \mid \mu)\alpha_0(\varphi \mid x)\, dx = 0.$$

Substituting the value of operator α_0 in the left part and integrating by parts we take identity

$$\int_{\mathbb{X}} \varphi(x) \left(-\sum_{i=1}^{d} \left(y(x \mid \mu)b^i(x)\right)'_i - c(x)y(x \mid \mu) + \mu(x) \right) dx = 0,$$

from which we obtain equation

$$-\sum_{i=1}^{d} \left(y(\mu)b^i\right)'_i - c\, y(\mu) + \mu = 0.$$

Replacing function y with $h_\varsigma(\mu)$ we get *the forward accumulation equation* for processes of smooth type:

$$\sum_{i=1}^{d} \left(h_\varsigma(\mu)b^i/c\right)'_i + h_\varsigma(\mu) - \mu = 0. \qquad [9.24]$$

The required solution is a probability density if function μ is the same.

22. Equations with regard to central symmetry

Backward accumulation equations relate to a problem of reconstruction of the initial distribution of matter on the basis of the given result of its transportation. In this book we are not going to solve this problem. In what follows we investigate forward accumulation equations with rather simple fields of coefficients. They answer the matter of how the matter around a source is distributed.

Assume the set of coefficients of the differential equations to satisfy the principle of central symmetry with respect to the origin of coordinates. We consider a particular case of such a set of coefficients, namely

$$a^{ij} = \frac{1}{d}\delta^{ij}, \qquad b^i(x) = b(r)\frac{x^i}{r}, \qquad c = c(r),$$

where x^i is the i-th coordinate of vector x; $r = \sqrt{\sum_{i=1}^{d}(x^i)^2}$ is the length of x; δ^{ij} is the Kronecker symbol ($\delta^{ij} = 0$ if $i \neq j$, and $\delta^{ii} = 1$); $a(r)$, $b(r)$ and $c(r)$

are continuous positive functions of $r \geq 0$ (for smooth type processes $b(r) = 1$). Hence the vector field at any point but the origin is directed along the beam going from the origin through this point (*radial vector field*). Such a field represents an idealized picture of velocity directions inside a laminar liquid stream flowing from a point source in d-dimensional space ($d \geq 1$).

Consider the forward accumulation equation for a process of diffusion type with function μ of degenerate form. Let μ be the unit loading at the origin of coordinates which we denote as $\bar{0}$. Hence, $\int_{\mathbb{X}} \varphi(x)\mu(x)\, dx = \varphi(\bar{0})$, where φ is any continuous bounded function. Evidently in this case the density $h = h_\zeta(\cdot \mid \mu)$ represents a centrally symmetric function $h = h(r)$. We have

$$\left(ha^{ii}/c\right)'_i = \frac{1}{d}(h/c)'\frac{\partial r}{\partial x^i} = \frac{1}{d}(h/c)'\frac{x^i}{r},$$

$$\left(ha^{ii}/c\right)''_{ii} = \frac{1}{d}\frac{(h/c)'}{r} + \frac{1}{d}\left(\frac{(h/c)''}{r} - \frac{(h/c)'}{r^2}\right)\frac{\left(x^i\right)^2}{r},$$

$$\left(hx^i b/(rc)\right)'_i = \frac{hb/c}{r} + \left(\frac{(hb/c)'}{r} - \frac{(hb/c)}{r^2}\right)\frac{\left(x^i\right)^2}{r}.$$

Here and in what follows denotations f', f'' (without fixing arguments the functions are differentiated by) relate to derivatives of f by r. Therefore we obtain

$$\sum_{ij}\left(ha^{ij}/c\right)''_{ij} = \sum_{i=1}^{d}\left(ha^{ii}/c\right)''_{ii} = \frac{(h/c)'}{r} + \frac{1}{d}\left(\frac{(h/c)''}{r} - \frac{(h/c)'}{r^2}\right)r$$

$$= \frac{1}{d}(h/c)'' + \frac{d-1}{d}\frac{(h/c)'}{r},$$

$$\sum_{i=1}^{d}\left(hx^i b/(rc)\right)'_i = dhb/(rc) + \left(\frac{(hb/c)'}{r} - \frac{(hb/c)}{r^2}\right)r$$

$$= (hb/c)' + (d-1)\frac{hb/c}{r}.$$

Equation [9.22] is transformed into equation

$$\frac{1}{2d}(h/c)'' + \frac{d-1}{2d}\frac{(h/c)'}{r} - (hb/c)' - (d-1)\frac{hb/c}{r} - h = 0 \quad (r > 0). \quad [9.25]$$

Correspondingly, equation [9.24] for a smooth process is transformed into equation

$$(h/c)' + (d-1)\frac{h/c}{r} + h = 0 \quad (r > 0). \quad [9.26]$$

The unit loading at the origin of coordinates affects properties of solutions of equations [9.25] and [9.26]. Under our suppositions equation [9.22] can be represented in the form

$$\nabla^2(ha/c) - \text{div}(h\bar{b}/c) - h + \mu = 0,$$

where $\nabla^2 u$ is the Laplacian applied to function u; $r\,div\,\bar{v}$ is the divergence of vector field $\bar{v}$; $\bar{b}$ is the vector field with coordinates b^i. Integrate all the members of this equation over a small ball neighborhood (of radius R) of the origin of coordinates. Under our supposition the integral of the last term is equal to 1 for any $R > 0$. The integral of the third term tends to zero (if h is bounded or tends to infinity not very quickly as its argument tends to $\bar{0}$). The first and second terms remain. From the theory of differential equations it follows that the integral of the divergence of the given vector field over a ball is equal to the integral of function $h\,b/c$ over the surface of the ball, i.e. it is equal to $R^{d-1}\omega_d\,h(R)\,b(R)/c(R)$, where ω_d is the area of the unit ball surface in d-dimensional space. The integral of $\nabla^2(h/c)$ over the ball is equal to the integral of function $-(h/c)'$ over the surface of the ball, i.e. it is equal to $-R^{d-1}\omega_d(h/c)'(R)$. Therefore, the following condition must hold

$$R^{d-1}\omega_d\,(h/c)'(R) + R^{d-1}\omega_d\,h(R)\,b(R)/c(R) \longrightarrow 1. \qquad [9.27]$$

This is true if function b is bounded in a neighborhood of the origin and function h/c has a pole in point $\bar{0}$ of the corresponding order. Namely

$$h(r)/c(r) \sim \begin{cases} (-\log r)/(2\pi), & d = 2 \\ 1/\big(r^{d-2}(d-2)\omega_d\big), & d \geq 3, \end{cases} \qquad [9.28]$$

Note that near such a pole the function does not tend to infinity very quickly. In this case the integral of the divergence tends to zero as $R \to 0$. Thus, the desired centrally symmetric solution of the forward accumulation equation has to be of order [9.28] and consequently it depends on the rate of $c(r)$ at the zero point. If function b is not bounded in a neighborhood of zero it can be the case that the second term of the equation determines the order of the solution. This term is determining for equation [9.26] when the second derivatives are absent. To assign the order of solutions at the origin means in fact to assign initial conditions for equations [9.25] and [9.26]. However, it is not convenient to use these conditions for drawing graphs of solutions because of their instability. We search for positive and integrable solutions on the positive semi-axis: $\int_0^\infty h(r)\,r^{d-1}\,dr < \infty$, and therefore, $h(r) \to 0$ ($r \to \infty$). Besides under "time inversion" solutions of the equations become stable. Hence solutions of equations [9.25] and [9.26] can be found with the help of a computer on any finite interval (ε, T) $(0 < \varepsilon < T)$ replacing $r \mapsto T - r$. Below we give some examples of how to choose parameters of equations [9.25] and [9.26], which have a physical interpretation, and also their solutions in analytical (if possible) or graphical form (the result of a computer's work).

23. Choice of coefficients

In spite of this, solutions of accumulation equations depend on the ratios a^{ij}/c and b^i/c, the interpretation of these ratios is more natural if one considers the fields (a^{ij}), (b^i) and c separately because each of them has a specific physical sense. It is possible that some of them to depend on another.

Generally, liquids and gases play a role of carrier for a particle of a matter. As a rule liquids determine the smooth type of a process, and gases the diffusion types. We take the simplest supposition with respect to diffusion. We consider it to be constant on all the space. The reason for such a choice follows from the interpretation of diffusion as a result of chaotic movement of molecules in the stable thermal field. It is possible to give other interpretations of diffusion, for example, turbulence. However, taking into account turbulence would change our model and the form of relating differential equations. In this book we consider only laminar streams of liquids and gases.

In the diffusion case the velocity $\bar{v}$ of the carrier matter determines a tendency of movement of a particle but not its actual shift. In this case we take the field of coefficients $\bar{b}$ to be equal to the field of velocities. In the smooth case the velocity field plays the main role in transportation of a particle. In this case we suppose $\bar{b} = \bar{v}/|\bar{v}|$. This means that these fields coincide in directions but differ in values of vectors. In both types of processes values of velocities can affect the coefficient of absorption c. In the centrally symmetric case we assume the velocity field to be $\bar{v}(x) = v(r)\, x/r$, where $v(r)$ depends on the dimension of the space and on the loss of carrier matter during transportation of the particle. The choice is based on the principle of balance of matter under stationary activity of the point source.

For incompressible liquid and without loss of liquid from the system (e.g., it could be transformed to another aggregate state) the equation of balance has the form:

$$\operatorname{div} \bar{v} = 0, \qquad\qquad [9.29]$$

hence $v' = -(d-1)\, v/r$ and, consequently, $v(r) = v(1)/r^{d-1}$. If the incompressible liquid flows on the plane and goes out from the system as vapor at a rate of v_1 per unit square, the velocity on the surface decreases faster. The balance of carrier matter has the form: $v_0 = 2\pi r\, v(r) + v_1 \pi r^2$, where v_0 is activity of the source in point $\bar{0}$, hence $v(r) = (v_0 - v_1 \pi r^2)/(2\pi r)$.

In this case (in contrast to the previous one when the liquid covers the plane wholly) the circle determined by condition $v(r) > 0$ is the domain of the carrier liquid. Hence $r_{max} = \sqrt{v_0/(\pi\, v_1)}$ is the radius of this circle.

In three-dimensional space the velocity of the particle transported by incompressible liquid varies inversely proportionally to the squared distance from the source:

$v(r) = v(1)/r^2$. In this case it is difficult to give a reasonable interpretation to loss of the carrier matter, and we do not consider it.

For a gaseous carrier the velocity of the particle depends on pressure $p = p(r)$ which depends on the resistance of the medium and it decreases according to the distance from the source. When it passes through the homogenous porous medium, the balance equation for the matter has the form:

$$\mathrm{div}(p\bar{v}) = 0, \qquad [9.30]$$

which follows from the Boyle-Mariott law. Hence, in a centrally symmetric case, we obtain the equation: $(p\,v)' + (d-1)\,p\,v/r = 0$ and consequently, $pv = p(1)v(1)/r^{d-1}$. On the other hand from the Gaghen-Poiseil law for laminar flowing of gas the velocity of the stream is proportional to the gradient of pressure (see [GOL 74]):

$$\bar{v} = -k\nabla p, \qquad [9.31]$$

where $k > 0$ is a coefficient depending on the properties of gas and the porous medium; ∇p is a vector with coordinates p_i'. Using the value of product pv, we obtain the differential equation with respect to p: $pp' = -k_1/r^{d-1}$, where $k_1 = p(1)\,v(1)/k$. Hence $p^2 = p_\infty^2 + 2\,k_1/((d-2)\,r^{d-2})\,(d \geq 3)$, where p_∞ is the pressure of gas in an infinitely removed point, for example, atmospheric pressure. Therefore, for space $(d = 3)$, we obtain

$$v = \frac{m}{r^{3/2}\sqrt{r+n}},$$

where $m = p(1)v(1)/p_\infty$, $n = 2p(1)v(1)/(p_\infty k)$.

We consider the field of absorption $c(r)$ to be of two forms. Firstly, this is a constant field $c(r) \equiv c_1 = $ const. We call such a field *strong*. In this case, probability of absorption only depends on the distance passed (possibly on the number of collisions of the transported particle with molecules of immovable phase). Secondly, this is a field varying inversely proportionally to the velocity of a carrier: $c(r) = c_1/b(r)$. We call such a field *weak*. In a weak field probability of absorption in some interval of immovable phase depends not only on the length of this interval but on the time it takes for the particle to interact with the immovable phase. The second dependence seems to be more plausible. It is verified indirectly in the theory of chromatography (see item 15). It is possible to have any intermediate laws of interaction, but these will not be considered here.

24. Transportation with liquid carrier

The liquid carrier can flow out of the source and flood over the horizontal surface. In this case we deal with a two-dimensional accumulation problem [HAR 78].

If the carrier is not lost as a result of evaporation, its radial velocity decreases due to the geometry of the plane. If the carrier is being lost by evaporation its radial velocity decreases faster and reaches a value of zero at a finite distance from the source. The liquid carrier can penetrate through a three-dimensional region filled with porous matter. In this case the carrier is not lost; the radial velocity decreases only due to geometry of space. Besides, we consider two forms of absorption fields for each dimension: strong and weak. We do not take into account diffusion for the liquid carrier. Therefore in this case we deal with first order differential equations.

For dimension $d = 2$ three types of coefficients are investigated:

(1) $c(r) = c_1$ – flood over without evaporation and strong absorption field. In this case the solution of equation [9.26] has the form

$$h = C\frac{1}{r}\exp\left(-c_1 r\right).$$

The distribution density of the accumulated matter has an acute maximum at the point of the source.

(2) $c = c_1 r$ – flood over without evaporation and weak absorption field.

$$h = C\exp\left(-\frac{c_1 r^2}{2}\right).$$

This is the unique case when the accumulated matter has the normal distribution as a result of transporting matter by a liquid carrier.

(3) $c = 2c_1\pi r/(v_0 - v_1\pi r^2)$ $(0 < r < \sqrt{v_0/\pi v_1})$ – flood over with evaporation and strong absorption field. In this case the solution of equation [9.26] has the form

$$h = C\left(v_0 - v_1\pi r^2\right)^{c_1/v_1 - 1}.$$

There is clearly expressed dependence of the distribution form on the ratio of the absorption and evaporation coefficients. If $c_1/v_1 > 1$ the matter is being accumulated in the form of a cupola above the source. If $c_1/v_1 < 1$ the matter is concentrated near the boundary of a circular domain forming a ring with a sharp border.

For dimension $d = 3$ two types of coefficients are investigated:

(1) $c = c_1$ – penetrating of liquid into three-dimensional volume and strong absorption field. The solution of equation [9.26] has the form:

$$h = C\frac{1}{r^2}\exp\left(-c_1 r\right).$$

The distribution density of such a form has an acute maximum above the source.

(2) $c = c_1 r^2$ – penetrating of liquid into three-dimensional volume and weak absorption field. The solution of equation [9.26] has the form:

$$h = C \exp\left(-\frac{c_1 r^3}{3}\right).$$

This distribution is similar to the normal distribution but it has a more clearly expressed boundary between large and small values of the density (see item 15).

25. Transportation with gaseous carrier

We consider an intrusion of a gaseous carrier into a porous medium, where the gas moves with deceleration less than that of a liquid. Accumulation of the matter corresponds to two types of fields of coefficients:

(1) $b = m/r^{3/2}\sqrt{r+n}$, $c = c_1$ – strong absorption field. In this case equation [9.25] has the form

$$h'' + h'\left(\frac{2}{r} - 6b\right) - h\left(\frac{3nb}{r(r+n)} + 6c_1\right) = 0.$$

Its analytical solution is not known. For graphs of solutions, see Figures 9.2 and 9.3. There is an acute maximum at the place of the source; the rate of decreasing of the density depends on the value c_1.

(2) $b = m/r^{3/2}\sqrt{r+n}$, $c = c_1/b$ – weak absorption field. In this case equation [9.25] has the form $h'' - Fh' + Gh = 0$, where

$$F = \frac{6r + 4n}{r(r+n)},$$

$$G = -\frac{5r^2 + 6rn + (9/4)n^2}{r^2(r+n)^2} + \frac{b(36r + 24n)}{r(r+n)} - \frac{6c_1}{b}.$$

Its analytical solution is not known. For graphs of solutions, see Figures 9.4 and 9.5. The distribution has a "crater" at the place of the source. The radius of the ring of maximum values depend on c_1 (the left pictures correspond to smaller values of c_1).

To obtain graphics we use a standard algorithm for approximate solving systems of first order differential equations, solved with respect to derivatives. This algorithm is realized with the computer ("Stend" program by Prof. V.A. Proursin). In all the pictures profiles of distribution densities are shown as functions of the distance from the source at zero (on the left).

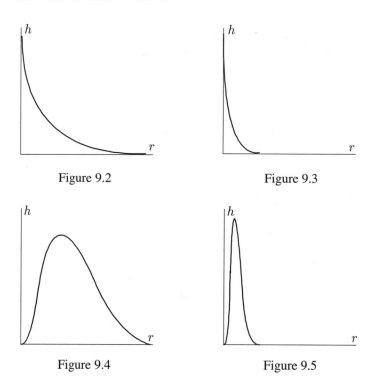

Figure 9.2   Figure 9.3

Figure 9.4   Figure 9.5

26. Conclusion

Using methods of the theory of semi-Markov processes we obtained the forward and backward accumulation equations. The density of the measure of accumulated matter is proportional to the distribution density of the process at the time of stopping. When analyzing the process of accumulation two problems arise: forward and backward. The backward problem is to reconstruct a source on the basis of an observable distribution. For this aim the backward equation can be used. The main content of this section is to derive forward accumulation equations for diffusion and smooth types of processes and to solve them in the case of circular symmetry. The solutions answer the question of how the matter around a source is distributed. We consider a point source in two- and three-dimensional homogenous and isotropic medium. It is shown that under some combinations of the model parameters zones of accumulated matter can have either the maximum or a local minimum of concentration at the place of the source. In the latter case increased concentration zones form concentric rings or spheres around the source. The radii of these rings and spheres depend on the rate and character of absorption of moving particles by the substance of the filter. Thus, this property can imply separation of components of a mixture which differ in their absorption rates. Such a conclusion may have a geological sense [GOL 81].

Bibliography

[ALD 78a] ALDOUS D., *Weak convergence of stochastic processes for processes viewed in the Strasbourg manner*, Cambridge, England, 1978.

[ALD 78b] ALDOUS D., "Stopping times and tightness", *Ann. Probab.*, vol. 6, no. 2, p. 1–6, 1978.

[ARK 79] ARKIN V.I., EVSTIGNEEV I.V., *Probability Models of Control and Economy Dynamics*, Nauka, Moscow, 1979 (in Russian).

[BIL 77] BILLINGSLY P., *Convergence of Probability Measures*, Nauka, Moscow, 1977 (in Russian).

[BIR 84] BIRKHOFF G., *Theory of Lattices*, Nauka, Moscow, 1984 (in Russian).

[BLU 62] BLUMENTHAL R.M., GETOOR R.K., MCKEAN H.P., JR., "Markov processes with identical hitting distributions", *Illinois J. Math.*, vol. 6, no. 3, p. 402–421, 1962.

[BLU 68] BLUMENTHAL R.M., GETOOR R.K., *Markov Processes and Potential Theory*, Academic Press, New York, 1968.

[BOR 96] BORODIN A.N., SALMINEN P., *Handbook of Brownian Motion – Facts and Formulae*, Birkhäuser Verlag, Basel, Boston, Berlin, 1996.

[CHA 79] CHACON R.V., JAMISON B., "Processes with state-dependent hitting probabilities and their equivalence under time changes", *Adv. Math.*, vol. 32, no. 1, p. 1–36, 1979.

[CHA 81] CHACON R.V., LE JAN Y., WALSH J.B., "Spatial trajectories", *Lect. Not. Math.*, vol. 850, 1981.

[CHE 73] CHEONG C.K., DE SMIT J.H.A., TEUGELS J.L., *Bibliography on Semi-Markov Processes*, Core discussion papers, no. 7310, Louvain, Brussels, 1973.

[CIN 79] ÇINLAR E., "On increasing continuous processes", *Stoch. Proc. Appl.*, no. 9, p. 147–154, 1979.

[COX 67] COX D., SMITH V., *Renewal Theory*, Sov. radio, Moscow, 1967 (in Russian).

[DAR 53] DARLING D.A., SIEGERT A.J.F., "The first passage problem for a continuous Markov process" *Ann. Math. Statist.*, vol. 24, p. 624–639, 1953.

[DEL 72] DELLACHERIE C., *Capacités et Processus Stochastiques*, Berlin, Heidelberg, New York, 1972.

[DIT 65] DITKIN V.A., PRUDNIKOV A.P., *Handbook on Operation Calculus*, High school, Moscow, 1965 (in Russian).

[DIT 66] DITKIN V.A., PRUDNIKOV A.P., *Operation Calculus*, High school, Moscow, 1966 (in Russian).

[DYN 59] DYNKIN E.B., *Foundation of Theory of Markov Processes*, Fizmatgiz, Moscow, 1959 (in Russian).

[DYN 63] DYNKIN E.B., *Markov Processes*, Fizmatgiz, Moscow, 1963 (in Russian).

[DYN 67] DYNKIN E.B., YUSHKEVICH A.A., *Theorems and Tasks on Markov Processes*, Nauka, Moscow, 1967 (in Russian).

[DYN 75] DYNKIN E.B., YUSHKEVICH A.A., *Controlable Markov Processes and Their Applications*, Nauka, Moscow, 1975 (in Russian).

[FEL 67] FELLER W., *Introduction to Probability Theory and Its Applications*, Mir, Moscow, 1967 (in Russian).

[GID 65] GIDDINGS J.C., *Dynamics of Chromatography*, vol. 1, Marcel Decker, Inc., New York, 1965.

[GIH 71] GIHMAN I.I., SKOROKHOD A.V., *Theory of Random Processes*, vol. 1, Nauka, Moscow, 1971 (in Russian).

[GIH 73] GIHMAN I.I., SKOROKHOD A.V., *Theory of Random Processes*, vol. 2, Nauka, Moscow, 1973 (in Russian).

[GIH 75] GIHMAN I.I., SKOROKHOD A.V., *Theory of Random Processes*, vol. 3, Nauka, Moscow, 1975 (in Russian).

[GIL 89] GILBARG D., TRUDINGER N., *Elliptic Differential Equations with Partial Derivatives of Second Order*, Nauka, Moscow, 1989 (in Russian).

[GOL 74] GOLBERT C.A., VIGDERGAUZ M.S., *Course of Gas Chromatography*, Chemistry, Moscow, 1974 (in Russian).

[GOL 81] GOLUBEV V.S., *Dynamics of Geo-Chemical Processes*, Nedra, Moscow, 1981 (in Russian).

[GUB 72] GUBENKO L.G., STATLANG E.S., "On controlable semi-Markov processes", *Cybernetics*, no. 2, p. 26–29, 1972 (in Russian).

[GUT 75] GUT A., "Weak convergence and first passage times", *J. Appl. Probab.*, vol. 12, no 2, p. 324–334, 1975.

[GUT 81] GUT A., AHLBERG P., "On the theory of chromatography based upon renewal theory and a central limit theorem for randomly indexed partial sums of random variables", *Chemica Scripta*, vol. 18, no 5. p. 248–255, 1981.

[HAL 53] HALMOSH P., *Theory of Measure*, Mir, Moscow, 1953 (in Russian).

[HAN 62] HANT G.A., *Markov Processes and Potentials*, Inostr. Literatury, Moscow, 1962 (in Russian).

[HAR 69] HARLAMOV B.P., "Characterization of random functions by random pre-images", *Zapiski nauch. semin. LOMI*, vol. 12, p. 165–196, 1969 (in Russian).

[HAR 71a] HARLAMOV B.P., "On the first time of exit of a random walk on line from an interval from interval", *Math. notes.*, vol. 9, no. 6, p. 713–722, 1971 (in Russian).

[HAR 71b] HARLAMOV B.P., "Determining of random processes by streams of first entries", *Reports to AS USSR*, vol. 196, no.2, p. 312–315, 1971 (in Russian).

[HAR 72] HARLAMOV B.P., "Random time change and continuous semi-Markov processes", *Zapiski nauch. semin. LOMI*, vol. 29, p. 30–37, 1972 (in Russian).

[HAR 74] HARLAMOV B.P., "Random processes with semi-Markov streams of first entries", *Zapiski nauch. semin. LOMI*, vol. 41, p. 139–164, 1974 (in Russian).

[HAR 76a] HARLAMOV B.P., "On connection between random curves, time changes, and regeneration times of random processes", *Zapiski nauch. semin. LOMI*, vol. 55, p. 128–194, 1976 (in Russian).

[HAR 76b] HARLAMOV B.P., "On convergence of semi-Markov walks to continuous semi-Markov process", *Probability theory and its applications*, vol. 21, no. 3, 497–511, 1976 (in Russian).

[HAR 77] HARLAMOV B.P., "Property of correct exit and a limit theorem for semi-Markov processes", *Zapiski nauch. semin. LOMI*, vol. 72, p. 186–201, 1977 (in Russian).

[HAR 78] HARLAMOV B.P., "On a mathematical model of accumulation of accessory minerals in sedimental deposits", in *Investigation on Mathematical Geology*, p. 80–89, Nauka, Leningrad, 1978 (in Russian).

[HAR 79] HARLAMOV B.P., YANIMYAGI V.E., "Construction of homogeneous Markov process with given distributions of the first exit points", *Zapiski nauch. semin. LOMI*, vol. 85, p. 207–224, 1979 (in Russian).

[HAR 80a] HARLAMOV B.P., "Additive functionals and time change preserving semi-Markov property of process", *Zapiski nauch. semin. LOMI*, vol. 97, p. 203–216, 1980 (in Russian).

[HAR 80b] HARLAMOV B.P., "Deducing sequences and continuous semi-Markov processes on the line", *Zapiski nauch. semin. LOMI*, vol. 119, p. 230–236, 1980 (in Russian).

[HAR 83] HARLAMOV B.P., "Transition functions of continuous semi-Markov process", *Zapiski nauch. semin. LOMI*, vol. 130, p. 190–205, 1983 (in Russian).

[HAR 89] HARLAMOV B.P., "Characteristic operator and curvilinear integral for semi-Markov process", *Zapiski nauch. semin. LOMI*, vol. 177, p. 170–180, 1989 (in Russian).

[HAR 90] HARLAMOV B.P., "On statistics of continuous Markov processes: semi-Markov approach" In *Probability Theory and Mathematical Statistics. Proc. 5th Vilnius Conf.*, p. 504–511, 1990.

[HAR 99] HARLAMOV B.P., "Optimal prophylaxis policy for systems with partly observable parameters", in *Statistical and Probabilistic Models in Reliability*, D.C. Ionescu and N. Limnios (eds.), Birkhauser, Boston, p. 265–278, 1999.

[HAR 02] HARLAMOV B.P., "Absolute continuity of measures in class of semi-Markov processes of diffusion type", *Zapiski nauch. semin. POMI*, vol. 294, p. 216–244, 2002 (in Russian).

[HOF 72] HOFMANN A., "Chromatographic theory of infiltration metasomatism and its application to feldspars", *Amer. J. Sci.*, vol. 272, p. 69–90, 1972.

[ITO 60] ITÔ K., *Probability Processes*, vol. 1, Inostr. Literatury, Moscow, 1960.

[ITO 63] ITÔ K., *Probability Processes*, vol. 2, Inostr. Literatury, Moscow, 1963.

[ITO 65] ITÔ K., McKEAN H.P., *Diffusion Processes and Their Sample Paths*, Berlin, Heidelberg, New York, 1965.

[KEI 79] KEILSON J., *Markov Chain Models – Rarity and Exponentiality*, Springer Verlag, New York, Heidelberg, Berlin, 1979.

[KEL 68] KELLY G., *General Topology*, Nauka, Moscow, 1968 (in Russian).

[KEM 70] KEMENY G., SNELL G., *Finite Markov Chains*, Nauka, Moscow, 1970 (in Russian).

[KNI 64] KNIGHT F., OREY S., "Construction of a Markov process from hitting probabilities", *J. Math. Mech.*, vol. 13, no. 15, p. 857–874, 1964.

[KOL 36] KOLMOGOROV A.N., *Foundation of Probability Theory*, ONTI, Moscow, 1936 (in Russian).

[KOL 38] KOLMOGOROV A.N., "On analytical methods in probability theory", *Progress of math. sci.*, no. 5, p. 5–41, 1938 (in Russian).

[KOL 72] KOLMOGOROV A.N., FOMIN S.V., *Elements of Theory of Functions and Functional Analysis*, Nauka, Moscow, 1972 (in Russian).

[KOR 74] KOROLYUK V.S., BRODI S.M., TURBIN A.F., "Semi-Markov processes and their application", in *Probability Theory. Mathematical Statistics. Theoretical Cybernetics*, VINITI, Moscow, vol. 2, p. 47–98, 1974 (in Russian).

[KOR 76] KOROLYUK V.S., TURBIN A.F., *Semi-Markov Processes and Their Application*, Naukova dumka, Kiev, 1976 (in Russian).

[KOR 82] KORZHINSKI D.S., *Theory of Metasomatic Zonality*, Nauka, Moscow, 1982 (in Russian).

[KOR 91] KORLAT A.N., KUZNETSOV V.N., NOVIKOV M.M., TURBIN A.F., *Semi-Markov Models of Restorable Systems and Queueing Systems*, Steenca, Kisheneu, 1991.

[KOV 65] KOVALENKO I.N., "Queueing theory" In *Results of Science. Probability Theory*, VINITI, Moscow, p. 73–125, 1965 (in Russian).

[KRY 77] KRYLOV N.V., *Controlable Processes of Diffusion Type*, Nauka, Moscow, 1977 (in Russian).

[KUZ 80] KUZNETSOV S.E. "Any Markov process in Borel space has transition function", *Probab. theory and its application*, vol. 25, no. 2, p. 389–393, 1980.

[LAM 67] LAMPERTI J., "On random time substitutions and Feller property", in *Markov Processes and Potential Theory*, John Wiley & Sons, New York, London, Sydney, p. 87–103, 1967.

[JAN 81] LE JAN Y., "Arc length associated with a Markov process", *Adv. Math.*, vol. 42, no. 2, p. 136–142, 1981.

[LEI 47] LEIBENZON L.S., *Motion of Natural Liquids and Gases in Porous Medium*, Nedra, Moscow, 1947.

[LEV 54] LÉVY P., "Processus semi-Markoviens", in *Proc. Int. Congr. Math. (Amsterdam, 1954)*, vol. 3, p. 416–426, 1956.

[LIP 74] LIPTSER R.SH., SHIRYAEV A.N., *Statistics of Random Processes*, Nauka, Moscow, 1974 (in Russian).

[LIP 86] LIPTSER R.SH., SHIRYAEV A.N. *Theory of Martingales*, Nauka, Moscow, 1986 (in Russian).

[LOE 62] LOEV M., *Probability Theory*, Inostr. Literatury, Moscow, 1962.

[MAI 71] MAINSONNEUVE B., "Ensembles régénératifs, temps locaux et subordinateurs", *Lect. Not. Math.*, vol. 191, p. 147–170, 1971.

[MAI 74] MAINSONNEUVE B., "Systémes régénératifs", *Astérisque*, vol. 15, 1974.

[MAI 77] MAIN KH., OSAKI S., *Markov Decision Processes*, Nauka, Moscow, 1977 (in Russian).

[MEY 67] MEYER P.-A., "Processus de Markov", *Lect. Not. Math.*, vol. 26, 1967.

[MEY 73] MEYER P.-A., *Probability and Potentials*, Mir, Moscow, 1973 (in Russian).

[MIK 68] MIKHLIN S.G., *Course of Mathematical Physics*, Nauka, Moscow, 1968 (in Russian).

[MUR 73] MÜRMANN M.G., "A semi-Markovian model for the Brownian motion", *Lect. Not. Math.*, vol. 321, p. 248–272, 1973.

[NAT 74] NATANSON I.P., *Theory of Functions of Real Variables*, Nauka, Moscow, 1974 (in Russian).

[NEV 69] NEVEU G., *Mathematical Foundation of Probability Theory*, Mir, Moscow, 1969 (in Russian).

[NOL 80] NOLLAU V., *Semi-Markovsche Prozesse*, Akad. Verlag, Berlin, 1980.

[PAG 63] PAGUROVA V.I., *Tables of Non-Complete Gamma-Function*, Nauka, Moscow, 1963 (in Russian).

[PRO 77] PROTTER PH., "Stability of the classification of stopping times", *Z. Wahrsch. verw. Geb.*, vol. 37, p. 201–209, 1977.

[PRU 81] PRUDNIKOV A.P., BRYCHKOV YU.A., MARICHEV O.I., *Integrals and Series*, Nauka, Moscow, 1981 (in Russian).

[PYK 61] PYKE R. "Markov renewal processes: difinitions and preliminary properties", *Ann. Math. Statist.*, vol. 32, no. 4, p. 1231–1242, 1961.

[ROB 77] ROBBINS G., SIGMUND D., CHOU I., *Theory of Optimal of Stopping Rules*, Nauka, Moscow, 1977 (in Russian).

[ROD 71] RODRIGUEZ D.M., "Processes obtainable from Brownian motion by means of random time change", *Ann. Math. Statist.*, vol. 42, no. 1, p. 176–189, 1971.

[RUS 75] RUSAS G., *Contiguality of Probability Measures*, Mir, Moscow, 1975 (in Russian).

[SAN 54] SANSONE G., *Common Differential Equations*, vol. 2, Inostr. Literatury, Moscow, 1954 (in Russian).

[SER 71] SERFOZO R.F., "Random time transformations of semi-Markov processes", *Ann. Math. Statist.*, vol. 42, no. 1, p. 176–188, 1971.

[SHI 71a] SHIH C.T., "Construction of Markov processes from hitting distributions", *Z. Wahrsch. verw. Geb.*, vol. 18, p. 47–72, 1971.

[SHI 71b] SHIH C.T., "Construction of Markov processes from hitting distributions. II", *Ann. Math. Statist.*, vol. 42, no. 1, p. 97–114, 1971.

[SHI 76] SHIRYAEV A.N., *Statistical Sequential Analysis*, Nauka, Moscow, 1976 (in Russian).

[SHI 80] SHIRYAEV A.N., *Probability*, Nauka, Moscow, 1980 (in Russian).

[SHU 77] SHURENKOV V.M., "Transformations of random processes preserving Markov property", *Probability theory and its applications*, vol. 22, no. 1, p. 122–126, 1977 (in Russian).

[SHU 89] SHURENKOV V.M., *Ergodic Markov Processes*, Nauka, Moscow, 1989 (in Russian).

[SIL 70] SIL'VESTROV D.S., "Limit theorems for semi-Markov processes and their applications", *Probability theory and math. statistics*, vol. 3, p. 155–194, 1970.

[SIL 74] SIL'VESTROV D.S., *Limit theorems for complex random values*, High school, Kiev, 1974.

[SKO 56] SKOROKHOD A.V., "Limit theory for random processes", *Probability theory and its applications*, vol. 1, no. 3, p. 289–319, 1956 (in Russian).

[SKO 58] SKOROKHOD A.V., "Limit theory for Markov processes", *Probability theory and its applications*, vol. 3, no. 3, p. 217–264, 1958 (in Russian).

[SKO 61] SKOROKHOD A.V., *Investigation on Theory of Random Processes*, Kiev. Univ., Kiev, 1961 (in Russian).

[SKO 70] SKORNYAKOV L.A., *Elements of lattice theory*, Nauka, Moscow, 1970 (in Russian).

[SMI 55] SMITH W.L., "Regenerative stochastic processes", *Proc. Roy. Soc. Ser. A*, vol. 232. p. 6–31, 1955.

[SMI 66] SMITH W.L., "Some peculiar semi-Markov process", in *Proc. 5th Berkeley Sympos. Math. Statist. and Probab.* vol. 2, Part 2, p. 255–263, 1965–1966.

[SOB 54] SOBOLEV S.L., *Equations of Mathematical Physics*, Gostekhizdat, Moscow, 1954 (in Russian).

[SPI 69] SPITSER F., *Principles of Random Walks*, Mir, Moscow, 1969 (in Russian).

[STO 63] STONE C., "Weak convergence of stochastic processes defined on a semifinite time interval", *Proc. Amer. Math. Soc.*, vol. 14, no. 5, p. 694–696, 1963.

[TAK 80] TAKSAR M.I., "Regenerative sets on real line", *Lect. Not. Math.*, vol. 784, p. 437–474, 1980.

[TEU 76] TEUGELS J.L., "A bibliography on semi-Markov processes", *J. Comput. Appl. Math.*, no. 2, p. 125–144, 1976.

[VEN 75] VENTSEL A.D., *Course of Random Process Theory*, Nauka, Moscow, 1975 (in Russian).

[VIN 90] VINOGRADOV V.N., SOROKIN G.M., KOLOKOL'NIKOV M.G., *Abrasive Wear*, Mashinostroenie, Moscow, 1990 (in Russian).

[VOL 58] VOLKONSKI V.A., "Random time change in strictly Markov processes", *Probab. theory and its appl.*, vol. 3, no. 3, p. 332–350, 1958 (in Russian).

[VOL 61] VOLKONSKI V.A., "Construction of non-homogeneous Markov process with the help of random time change", *Probab. theory and its appl.*, vol. 6, no. 1, p. 47–56, 1961 (in Russian).

[YAC 68] YACKEL J., "A random time change relating semi-Markov and Markov processes", *Ann. Math. Statist.*, vol. 39, no. 2, p. 358–364, 1968.

[YAS 76] YASHIN YA.I., *Physico-Chemical Foundation of Chromatographical Separation*, Chemistry, Moscow, 1976 (in Russian).

Index